PRESTRESSED CONCRETE ANALYSIS AND DESIGN

Ohio River Assymetrical Stayed Girder Bridge with a Main Span of 900 ft (275 m). (*Courtesy Arvid Grant and Associates, Inc.*)

PRESTRESSED CONCRETE ANALYSIS AND DESIGN
Fundamentals

Antoine E. Naaman

Professor of Structural Design
University of Illinois at Chicago Circle

McGraw-Hill Book Company

New York St. Louis San Francisco Auckland Bogotá Hamburg
Johannesburg London Madrid Mexico Montreal New Delhi
Panama Paris São Paulo Singapore Sydney Tokyo Toronto

This book was set in Times Roman.
The editors were Diane D. Heiberg and Julienne V. Brown;
the production supervisor was Diane Renda.
The cover was designed by Carla Bauer.
Halliday Lithograph Corporation was printer and binder.

PRESTRESSED CONCRETE ANALYSIS AND DESIGN

Fundamentals

1234567890 HDHD 898765432

ISBN 0-07-045761-1

Library of Congress Cataloging in Publication Data

Naaman, Antoine E.
    Prestressed concrete analysis and design.

    Includes bibliographical references and index.
    1. Prestressed concrete construction.
I. Title.
TA683.9.N3   624.1'8341   81-19300
ISBN 0-07-045761-1   AACR2

To
the memory of my parents
and
Ingrid, Patrice, and Charles
for their love

CONTENTS

Appendixes

Index

PREFACE

Although historically prestressed concrete has experienced a slower start than reinforced concrete and its development has followed a different path, it has now evolved into a reliable technology and has established itself as a major structural material. Prestressed concrete has made significant contributions to the precast manufacturing industry, the cement industry, and the construction industry as a whole. This has led to an enormous array of structural applications from bridges to nuclear power vessels, from buildings serving every use and occupancy to ships, and from lowly products such as ties and piles to monumental TV towers, and offshore drilling platforms. Seldom is a major project planned today without prestressed concrete being considered as one of the viable alternative solutions. A careful analysis of future trends indicates a substantial increase in the use of prestressed concrete. This is also supported by recent developments in partially prestressed concrete, which integrates both reinforced and prestressed concrete and treats them as the extreme boundaries of the same system. It has become almost inevitable that each material be considered separately while consideration of their combination, which may represent a better solution, is neglected.

A similar trend is expected at the educational level. Design courses in prestressed concrete will be more widely offered at universities and may be moved from the list of technical electives to the list of required courses in structural engineering curricula. It is also likely that reinforced and prestressed concrete will be offered together as part of the same general course or course sequence on concrete structures, hence essentially covering partially prestressed (of partially reinforced) concrete.

This book is written for students and professionals. It is meant as a strong teaching text, as well as a reference tool for practicing engineers. It emphasizes the fundamental concepts of analysis and design and provides the student with a sufficiently strong basis to allow the handling of everyday design problems and the tackling of the more complex ones with confidence. A particular effort is made throughout to synthesize and condense the essential information and to give an overview of the directions in which the design is proceeding. Self-sufficient logical design flow charts summarizing the step-by-step design procedure and containing all necessary design equations are often presented. They reduce the burden of guesswork and iterative search encountered in the design process, and are essential when programmable calculators and computers are used. Important formulas and equations are also condensed in tables for ready use. To provide a correlation with reinforced concrete design and to help engineers already familiar with reinforced concrete, the case of partially prestressed (or partially reinforced) concrete is often addressed. Selected references are given at the end of each chapter. An attempt was made to include not only necessary readings but also most recent research work in the United States for up-to-date information. Current specifications of the ACI building code and many requirements of the AASHTO and AREA specifications are integrated in the text. When appropriate guidance is missing from the code suggestions are made to accommodate at best the intent of the code.

Whenever possible, widely accepted symbols such as those used in the ACI code are adopted and all symbols used in the text are defined and summarized for easy reference in App. A. A consistent sign convention is followed throughout, allowing rigorous treatments when needed. This is essential, for instance, in the case of continuous beams where the sign of secondary moments cannot be visualized *a priori* and must be derived from the analysis.

Because of the inevitable future conversion from U.S. customary units to the International System of Units (SI), all important tables, figures, and design information, as well as dimensionally inconsistent equations, are given in dual units. However, because the prestressed concrete industry is not on the verge of change from U.S. customary units to SI units, all examples are treated in U.S. units to allow students and professionals to keep in touch with current practice. In addition, SI conversion factors and SI equivalents for some dimensionally inconsistent equations used in various flow charts are given in App. B.

The text is organized into 14 chapters, grouped into three parts according to their intended function:

1. The first three chapters contain essential design information and reference data. They provide a general background on materials properties, design philosophy, and codes.
2. Chapters 4 to 8 develop the fundamental basis and underlying principles for the analysis and design of prestressed concrete. They include analysis and design for flexure by the working stress design method with an introduction to optimum design (Chap. 4), analysis and design for flexure by the ultimate strength design method with full coverage of partial prestressing (Chap. 5), design for shear and torsion and their combined effects with flexure (Chap. 6), design for deflection control with treatment of partially prestressed cracked sections and the incremental time-step method to predict long-term effects (Chap. 7), and prediction of prestress losses either by lump sum estimates or by the incremental time-step procedure (Chap. 8).

3. Chapters 9 to 14 address the particular analysis and design aspects of structural elements or systems in various applications of prestressed concrete. They cover composite beams (Chap. 9), continuous beams (Chap. 10), slabs (Chap. 11), tensile members and cylindrical tanks (Chap. 12), short and slender columns (Chap. 13), and bridges (Chap. 14).

A set of appendixes is also given at the end of the book, including a list of symbols (A), SI conversion factors (B), technical information on some post-tensioning systems (C), bearing pads and flat jacks (D), excerpts from the AASHTO specifications (E), and answers to selected problems (F).

This book can cover up to two courses in prestressed concrete at the senior undergraduate level and at the graduate level. If the material is to be covered in one course only, the following organization is suggested:

Chapters 1 to 3.   Cover very fast (up to four lecture hours). The student will have the opportunity to review the material in conjunction with other chapters.
Chapter 4.   Omit end-zone design (4.16, 4.17).
Chapter 5.   Omit Sec. 5.7.
Chapter 6.   Omit Sec. 6.17 on combined torsion and shear.
Chapter 7.   Omit the incremental time-step method and the case of cracked sections.
Chapter 8.   Cover lump sum estimates only.
Chapters 9 and 14; Chaps. 10 and 11.   Should be covered at least partly, as they address the most common applications of prestressed concrete.

The sections omitted in the above outline can be treated in a first course, if time allows, or in a second course with Chaps. 12 and 13 on tensile and compressive members. More advanced topics, such as partial prestressing, nonlinear analysis, analysis of beams with unbonded tendons, optimum design, design for fire, cracking, fatigue, and earthquake resistant design can also be addressed in an advanced graduate course or in special research projects and will be covered in a second volume to follow.

In writing this text, I have attempted to organize and to convey what I have learned and practically experienced in prestressed concrete to date. It is my sincere hope that those who will seek knowledge in it will not be disappointed.

Antoine E. Naaman

ACKNOWLEDGMENTS

Many people deserve my gratitude for their contribution to this volume and I would like to thank them for their help. I am particularly indebted to Daniel P. Jenny of the Prestressed Concrete Institute for his careful review of and constructive comments on the entire manuscript. Invaluable suggestions were also made by Professors E. J. Nawy, T. T. C. Hsu, M. K. Tadros, A. H. Mattock, and P. Balaguru who all reviewed portions of the manuscript.

Special thanks are due to my former student, K. Visalvanitch, who reviewed the arithmetic of the entire manuscript with devotion and competence; to many students who, with or without their knowledge, have assisted throughout the years in checking examples and problems; to George D. Nasser of the Prestressed Concrete Institute for his generous, unconditional help in suggesting and providing photographs and figures from PCI publications; to Professor S. P. Shah, who was among the few to encourage me to undertake this project; and, finally, to my wife Ingrid for her careful typing and editing of the initial manuscript and for her support during this demanding endeavor.

Material taken and/or adapted from the American Concrete Institute (ACI) building code and other ACI Recommended Practices appears frequently throughout the text. The courteous cooperation of the American Concrete Institute is gratefully acknowledged.

Many other individuals and organizations have given me their permission to adapt, use, or reproduce material from their publications in this book; I thank them for their courtesy. They are:

Abam Engineers, Inc.
American Association of State Highway and Transportation Officials
Arvid Grant and Associates
Bureau BBR, Ltd. (Switzerland)
California Department of Transportation
Canadian Prestressed Concrete Institute

CCL Systems, Ltd. (U.K.)
CN Tower Limited (Canada)
Concrete Technology Corp.
C.V.G.—EDELCA (Venezuela)
Dywidag Systems International, Inc.
Engineering News-Record
Figg and Muller Engineers, Inc.
FIP, Fédération Internationale de la Précontrainte (U.K.)
Freyssinet International, Inc.
Ingenieurburo H. Rigendinger (Switzerland)
Inryco Inc., Post-Tensioning Division
Kurt Orban Company, Inc.
Mr. R. L'Hermite (France)
Material Service Corporation
Norwegian Contractors Group (Norway)
Portland Cement Association
Post-Tensioning Institute
Sika Chemical Corporation
Somerset Wire Company, Ltd. (U.K.)
Sumitomo Electric Industries, Ltd. (Japan)
Superior Concrete Accessories, Inc.
Supreme Products Division, The Measuregraph Company
TNO Metal Research Institute (The Netherlands)
U.S. Army Corps of Engineers, (Sacramento District)
University of Illinois at Chicago Circle
VSL Corporation
Mr. R. J. Wheen

Antoine E. Naaman

PRESTRESSED CONCRETE
ANALYSIS AND DESIGN

Gantre Valley Bridge. Box girder construction with partially prestressed concrete stays and a main span of 571 ft (174 m). *(Courtesy Ingenieurburo H. Rigendinger, Switzerland.)*

CHAPTER 1

PRINCIPLE AND METHODS
OF PRESTRESSING

1.1 INTRODUCTION

Prestressing is the deliberate creation of permanent internal stresses in a structure or system in order to improve its performance. Such stresses are designed to counteract those induced by external loadings. Prestressing generally involves at least two materials, the stressor and the stressee which, when acting together, perform better than either one taken separately. Prestressing is a principle. The French mathematician Henri Poincaré once said: "A principle is neither true nor false, it is convenient." The principle of prestressing is indeed very convenient and has been widely applied. Its application to steel and concrete is relatively recent but has taken by far the widest proportions.

The application of prestressing to concrete is in a way a natural result. Concrete is strong in compression and weak in tension. For design purposes its tensile resistance is discounted. Prestressing the concrete would produce compressive stresses, either uniform or nonuniform, which will counteract tensile stresses induced by external loadings. The original concept attempted to counteract tensile stresses entirely, thus producing a crack-free material during service. However, it has since evolved to counteract only in part externally induced tensile stresses, thus allowing tension and controlled cracking in a way similar to reinforced concrete. This has led to what is called partial prestressing.

Partially prestressed concrete occupies the whole spectrum of the reinforcing range between fully reinforced and fully prestressed concrete. In a way it is a combination of both. Today it has become difficult to talk about either material

1

separately, without talking about their combination. As essentially they use the same basic components, steel and concrete, their historical development will eventually be addressed simultaneously. However, herein we will mostly focus on the development of prestressed concrete.

1.2 EXAMPLES OF PRESTRESSING

Examples of prestressing are numerous among manufactured tools and products. Indeed, some are very old and illustrate the principle of prestressing. The hunter's bow is prestressed by the string to achieve a sharp recoil action during ejection of the arrow. The dried wooden staves forming a wooden barrel are prestressed by tightening metal bands around them. When the barrel is filled with liquid, the wooden staves expand, the prestress is increased and leakage is prevented. To prevent the relative movement between the iron tire and the wooden rim of a cartwheel, the tire is fitted around the rim while in a heated state. Upon cooling, contraction of the tire produces a permanent prestress in the form of radial compression on the rim. The blade of a frame saw is prestressed (in tension) by twisting a rope at the opposite end of the frame (Fig. 1.1). During the cutting operation the blade is pushed or pulled through the wood being cut. Part of the blade is subjected

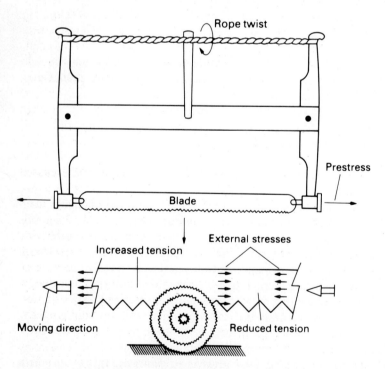

Figure 1.1 Prestressing in a frame saw.

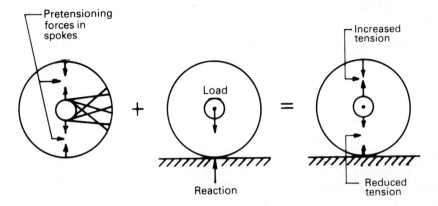

Figure 1.2 Prestressing in a bicycle wheel.

to external compression while the other is subjected to tension. The compression is counteracted by the internal pretension, thus no buckling occurs while tension adds up to the existing tension and the blade remains rigid. The spokes of the wheel of a bicycle are put in tension between the outer and inner rims (Fig. 1.2). When the load is applied, the lower spokes decompress but remain in tension while the upper spokes undergo increased tension. As all spokes remain in tension, the wheel keeps its rigidity. Tempered glass offers another example of prestressing whereas upon forming, the glass sheet is rapidly cooled. The skin, which cools first, gains rigidity and is slowly compressed by the core during its own hardening. Putting the skin in compression reduces the occurrence of surface cracks and decreases the risk of fracture under loading.

1.3 HISTORY OF PRESTRESSED CONCRETE

The first application of prestressing to concrete appears to be due to P. H. Jackson, an engineer from California. In 1886 (Ref. 1.1) he obtained a U.S. patent for tightening steel tie rods in artificial stones (concrete blocks) and concrete arches used for slabs and roofs (Fig. 1.3). At about the same time, in 1888, C. E. W. Doehring from Germany also obtained a patent (Ref. 1.2) for prestressing concrete slabs with metal wires. However, the performance of the first prestressed concrete structural elements was hindered by the low steel strengths available at the time: because of the relatively low steel stresses used and the relatively high prestress losses due to creep and shrinkage of the concrete, the prestress would soon vanish. Retensioning was suggested by G. R. Steiner (United States, 1908) to overcome this problem, while other researchers such as J. Mandl and M. Koenen (Ref. 1.3) of Germany attempted to identify and quantify prestress losses.

However, it was the French engineer Eugene Freyssinet (Ref. 1.4) who first grasped the importance of prestress losses and proposed ways to overcome them.

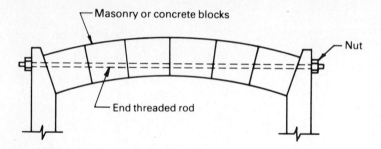

Figure 1.3 Jackson's first patent on prestressed concrete.

Based on his experience in building arch bridges (at Vendre in 1907 and Plougastel in 1927) and prestressing them by external jacking at the crown to facilitate form-work removal, he suggested that very high strength steels and high elongations must be used in prestressed concrete (Ref. 1.5). High steel elongations would not be entirely counteracted by the shortening of the concrete due to creep and shrinkage. Later, in 1940, he introduced his first prestressing system, a wedge-anchored cable with 12 wires which is still in use today. Thereafter he designed and built many bridges in prestressed concrete, starting with the bridge at Luzancy, France, in 1941. At that point the applications of prestressed concrete took off with much vigor in France and Europe and leaped into full competition with structural steel.

Although prestressed concrete planks and fence posts were produced by R. E. Dill in the United States since 1925 (Ref. 1.6), it was only in 1949/1950 that the first prestressed concrete bridge, the 155-foot-span (47-meter) Walnut Lane Bridge in Philadelphia, was built.

Simultaneously to and in continuation of the developments brought by E. Freyssinet many researchers contributed greatly to the full expansion of pre-stressed concrete. They include G. Magnel (Ref. 1.7) of Belgium, Y. Guyon of France (Ref. 1.8), P. Abeles of England (Ref. 1.9) who developed the concept of partial prestressing, F. Leonhardt of Germany (Ref. 1.10), V. V. Mikhailov of Russia, and T. Y. Lin of the United States (Ref. 1.11) to whom we owe the design method of load balancing which is so convenient for indeterminate structures.

Many prestressing systems and techniques were also developed and today prestressed concrete is widely accepted and used. Numerous books and textbooks on the design and construction of prestressed concrete structures were written and many associations and institutes contribute to the advancement of the state of the art on prestressed concrete (Refs. 1.12 to 1.24).

More than four decades of experience have given prestressed concrete a proven record of reliable performance. At present, applications of prestressed con-crete essentially occur in every structural element or building system: bridges, building components such as beams, slabs, and columns, pipes and piles, pave-ments, ties, tanks, tunnels, stadia, nuclear power vessels, TV towers, floating storage and offshore structures. Some examples are shown in Figs. 1.4 and 1.5.

(*a*) The Pine Valley Creek Bridge was built by the cast-in-place segmental cantilever construction with a main span of 450 ft (138 m). (*Courtesy California Department of Transporation.*)

(*b*) The Parrotts Ferry Bridge, California, has a main span of 640 ft (195 m), the longest lightweight concrete span in the world. (*Courtesy U.S. Army Corps of Engineers.*)

Figure 1.4 Examples of prestressed concrete bridges.

Figure 1.5 Examples of precast prestressed building structures. *(Courtesy Prestressed Concrete Institute.)*

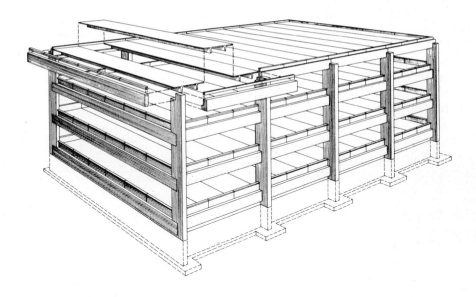

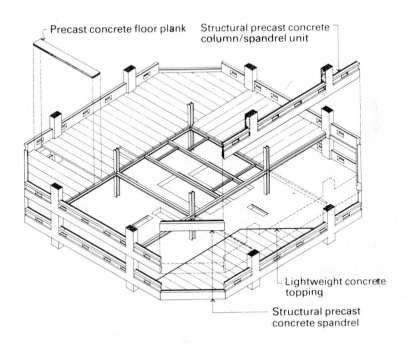

Precast concrete floor plank   Structural precast concrete column/spandrel unit

Lightweight concrete topping

Structural precast concrete spandrel

Figure 1.6 Typical building systems using precast prestressed concrete. *(Courtesy Prestressed Concrete Institute.)*

Prestressed concrete bridges have reached span lengths previously considered exotic and higher limits are expected. The Parrotts Ferry Bridge in California has a main span of 640 ft (195 m) while that of the Pasco-Kennewick cable stayed bridge is 981 ft (299 m). Bulk bridge applications in the United States were in the span range of 50 to 150 ft (15 to 46 m) with the precast prestressed I girder and box girder bridges extensively used in America's Interstate highway system, as well as in most highways and secondary roads. Similar extensive usage of precast prestressed hollow-cored slabs, T and double T beams of spans of up to 100 ft (31 m) is observed in the U.S. building market where numerous building systems are used (Fig. 1.6) making prestressed concrete fully competitive in many sectors of the construction industry.

1.4 PRESTRESSING METHODS

Several methods and techniques of prestressing are available. However, except for chemical prestressing, most can be classified within two major groups: pretensioning and posttensioning. Some methods are specifically identified with a particular application but nevertheless belong to one of the above groups.

1. Pretensioning

In pretensioning the prestressing tendons (wires, strands) are stretched to a predetermined tension and anchored to fixed bulkheads or molds. The concrete is poured around the tendons, cured, and upon hardening the tendons are released. As the bond between the tendons and the concrete resists the shortening of the tendons, the concrete gets compressed. The prefix "pre" in pretensioning refers to the fact that the tendons are put in tension *prior* to hardening of the concrete.

In order to stretch the tendons hydraulic jacks are generally used. Once the predetermined elongation is reached, the tendons are anchored to the bulkhead using anchors similar to those described for posttensioning. Anchors for individual strands are also called chucks. A typical chuck anchor is shown in Fig. 1.7.

Another technique for stretching the tendons is the electrothermal prestressing. Here high-strength deformed bars are heated by means of an electric current to between 250 and 450°C, then placed and anchored at the ends of the molds or a pretensioning bed. Electrothermal prestressing is not used in the United States but has been used in the Soviet Union and Eastern Europe. Its applications are limited.

Depending on the pretensioned structural elements produced, the profile of the tendons is either straight (Fig. 1.8) such as in hollow-cored slabs or allows for one or two deflection points (also called draping or hold-down points) such as in bridge girders (Fig. 1.9). Draping is generally achieved by pulling or pushing down part of the tendons to the desired position. A sketch of the "hold-down" procedure and typical draping devices is shown in Fig. 1.10.

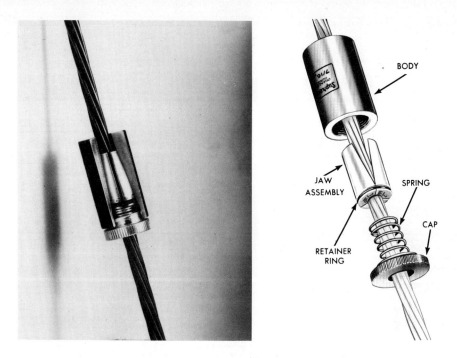

Figure 1.7 Typical chuck anchor for a single tendon. *(Courtesy Supreme Products Corporation.)*

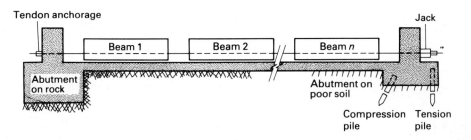

Figure 1.8 Typical pretensioning bed and abutments showing beams with straight tendons.

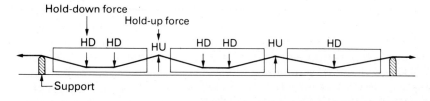

Figure 1.9 Typical pretensioning tendons profile with one or two draping points.

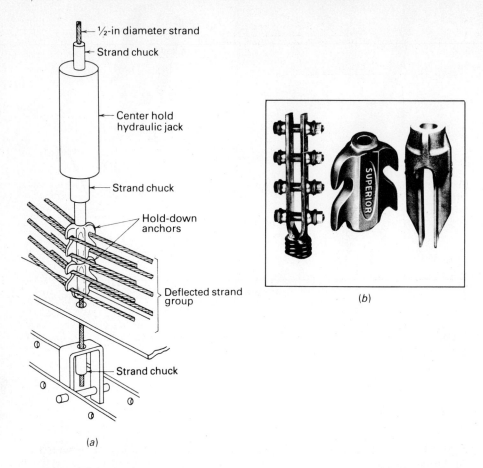

½-in diameter strand

Strand chuck

Center hold
hydraulic jack

Strand chuck

Hold-down
anchors

Deflected strand
group

Strand chuck

(b)

(a)

Figure 1.10 Typical "hold-down" procedure and typical hold-down anchors. *(Courtesy Superior Concrete Accessories Inc.)*

Small-diameter tendons are generally used in pretensioning to allow for the bond between steel and concrete to develop over a short distance. Most popular sizes in the United States are the $\frac{3}{8}$-in- (9.5-mm-) and the $\frac{1}{2}$-in- (12.5-mm-) diameter strands.

Pretensioning is the method mostly used for the production of precast prestressed concrete elements in the United States because it offers great potential for mechanization. Efficient long-line production techniques with casting bed lengths of up to 600 ft (182 m) where individual elements are cast end to end are preferred, as they require a single tensioning operation. Elements of standardized cross sections are mass produced yet customized by varying the length of each element and by placing inserts, holes, or blockouts for the mechanical or electrical distribution systems. Accelerated curing often permits early removal of the elements and daily reuse of the forms (24-hour production cycle). Excellent quality control and optimum use of labor and materials are achieved.

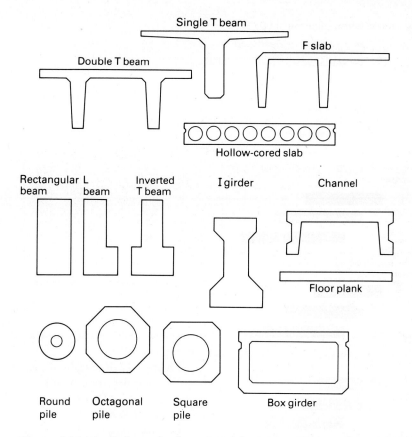

Figure 1.11 Typical standard sections of precast prestressed concrete products in the United States.

Typical elements and member cross sections aimed at particular applications were developed and standardized. The most common standard shapes in the United States are shown in Fig. 1.11. Spans of up to 150 ft (46 m) are not uncommon and are mostly limited by transportation and erection constraints. Load tables and charts were developed by the industry where for a given standard shape, external load, and span, the most appropriate section and tendon arrangement can be readily selected (Ref. 1.25). A typical example is shown in Fig. 1.12 for hollow-cored slabs.

2. Posttensioning

In posttensioning the tendons are stressed and anchored at the ends of the concrete member *after* the member has been cast and has attained sufficient strength.

Commonly, a mortar-tight metal tube or duct (also called sheath) is placed along the member before concrete casting. The tendons may have been preplaced loose inside the sheath prior to casting or could be placed after hardening of the

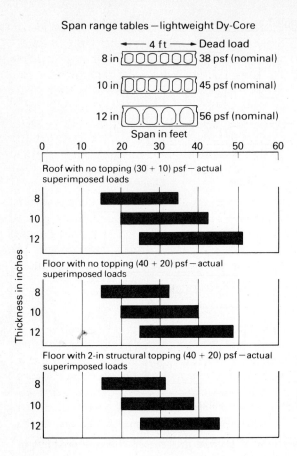

Figure 1.12 Example of load chart for prestressed hollow-cored slabs. *(Courtesy Material Service Corporation.)*

concrete. After stressing and anchoring, the void between each tendon and its duct is filled with a mortar grout which subsequently hardens. Grouting ensures bonding of the tendon to the surrounding concrete, improves the resistance of the member to cracking, and reduces the risks of corrosion for the steel tendons. Figure 1.13 gives a perspective of some typical posttensioning operations.

The above posttensioning technique implies using what are commonly called " bonded tendons." If the duct is filled with grease instead of grout, the bond would be prevented throughout the length of the tendon and the tendon force would apply to the concrete member only at the anchorages. This leads to " unbonded tendons." Unbonded tendons are generally coated with grease or bituminous material, wrapped with waterproof paper or placed inside a flexible plastic hose, and placed in the forms prior to concrete casting. When the concrete reaches sufficient strength, the tendons are stressed and anchored. They remain unbonded throughout their length and during service of the structure. This technique is being increasingly used in slab systems of residential and parking structures with several bays because of its extreme efficiency and economy. The tendons are put in tension at the periphery of the slab and may span up to 10 consecutive bays.

The tendons generally used in posttensioning are made of wires, strands, or bars. Bars are tensioned one at a time, wires and strands can be tensioned singly or in groups. In one of the Freyssinet systems 12 wires or strands forming a tendon can be pulled simultaneously. Up to 170 wires with 0.25-in (6.35-mm) diameter, can form a single tendon in the BBRV system and up to 31 strands with 0.6-in (15.2-mm) diameter can form a single tendon in the VSL system. These tendons carry very large forces. Tendons with a capacity of up to 1000 tons are commonly used in nuclear vessels. They often need specialized jacking and anchoring equipment. Hydraulic jacks are normally used and, with the tendons and anchorages, they all are often an integral part of the posttensioning system selected. These systems are reviewed in Sec. 1.5.

Although posttensioning can be used in precast prestressed operations, it is most useful in cast-in-place construction where large building and bridge girders cannot be transported, and for customized structures which need tensioning on the job site. Its application in unusual projects, such as nuclear power vessels, TV towers, and offshore structures has become a must and will certainly continue to expand.

3. Chemical Prestressing

Pretensioning and posttensioning represent two groups to which most prestressing techniques belong. It was mentioned above that electrothermal prestressing is in effect a pretensioning method. Chemical prestressing, however, does not belong to any of the above groups. In chemical prestressing the tendons are preplaced untensioned in the forms and the concrete is poured. Due to the special expansive cement used, the concrete, instead of shrinking, expands after curing and during hardening. As the steel is bonded to the concrete, it stretches with it, thus undergoing tension and inducing compression in the concrete.

The first modern development of expansive cement and its potential application to prestressing appeared in the work of the French engineer H. Lossier (Ref. 1.26) in 1946. The idea was further developed in the Soviet Union (Ref. 1.27) and in the United States, mainly at the University of California, Berkeley (Ref. 1.28). It was believed that chemical prestressing can be applied to elements with low levels of prestress, such as pavements and slabs. However, due to the difficulty of controlling the expansion of the concrete, which occurs in all directions, chemical prestressing did not develop in the United States beyond the research field. Nevertheless, expansive cement is being applied to shrinkage-compensating concretes used in long slabs to reduce or eliminate shrinkage joints.

1.5 PRESTRESSING SYSTEMS

It was mentioned above that tensioning the tendons may be achieved in several ways. The most common tensioning systems are mechanical. They are generally protected by patents. The knowledge of the system used is very helpful when detailing the steel and the location of end anchorages.

Figure 1.13 Sequential steps in posttensioning operations. (*Courtesy Inryco Inc., Post Tensioning Division.*)

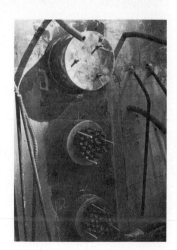

The basic principles used in these systems are few and essentially similar. The details vary. Patents have been taken on the method of applying the prestress, the type of jack used, the method or device used to anchor the tendons, the number and diameter of wires or strands forming a tendon, and the like.

Typical anchorage systems are shown in Fig. 1.14. Some are based on the principle of direct bearing. These include threaded bars anchored with nut and plate such as for the Dywidag system or wires with preformed end buttons bearing on a plate through an anchor head such as the BBRV system. In the buttoned wires systems sufficient accuracy is needed in estimating the exact length of the tendons before and after tensioning. Such difficulty is overcome if an anchor system based on wedge action or wedge and grip action is used. The wedge may accommodate simultaneously several wires or strands on its outer periphery such as in the Freyssinet system, or it may grip in sandwich a single bar or strand such as in the Cona system. Several wedges holding one strand each can have the same anchor head, as in the VSL system. Dead anchors are also available. They are encased directly in the concrete and are generally used in short members when tensioning is needed at one end only. Some of these prestressing systems are described in more detail in App. C. Additional information, technical data, and even design aids are frequently available from the various manufacturers of these systems.

1.6 PARTICULAR PRESTRESSING TECHNIQUES

Most prestressing methods have been classified in Sec. 1.5 as belonging to the pretensioning or the posttensioning group. Other classifications can be made according to particular attributes. Such an attribute may be, for instance, whether the tendons are bonded or unbonded. Some particular tenchiques are described below.

1. External Prestressing

Contrary to internal pretressing which implies that the tendons are in contact with the concrete, external prestressing refers to prestressing in which the force is externally applied. This can be done using hydraulic or flat jacks placed between the abutments and the ends of a concrete member (Fig. 1.15). Flat jacks are thin steel or neoprene bags which when inflated exert a very high force over a very small distance (Fig. 1.16). Several of these can be superimposed to obtain a greater movement. Inflation of the jack is achieved using water or grout under pressure. When the required movement is reached, the grout is allowed to harden. Flat jacks are very efficient and economical. However, with time, creep losses in the concrete and movements due to temperature differentials hinder their effectiveness. Thus, they cannot be reliably used as a permanent solution for prestressing. Some design information on a commercially available flat jack system is given in App. D.

External prestressing also implies that tendons are placed outside the concrete member. In improving the load-carrying capacity of some old bridges, external tendons were placed outside the bridge girders, tensioned, and anchored at their

First Freyssinet wedge cone
for 12 wires.

Freyssinet wedge cone for
12 strands.

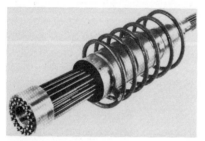

BBRV anchorage for
buttonhead wires.

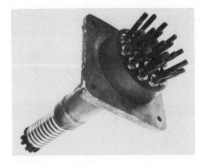

Freyssinet multistrand
K range anchorage.

Dywidag threaded bar anchorage.

VSL multistrand type E anchorage.

Inrico Cona monostrand
anchorage.

CCL systems multistrand
anchorage.

Figure 1.14 Typical anchorages used in various prestressing systems.

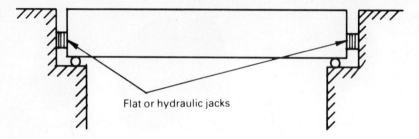

Flat or hydraulic jacks

Figure 1.15 External prestressing: jacking against abutments.

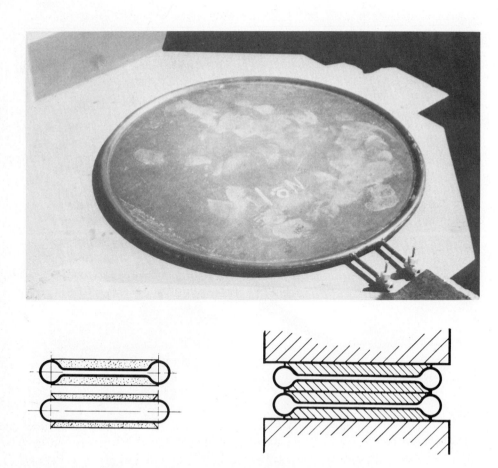

Figure 1.16 Typical flat jacks. *(Courtesy Freyssinet International Inc.)*

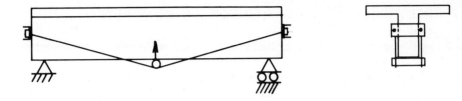

Figure 1.17 Example of external prestressing to strengthen an old bridge.

ends (Fig. 1.17). External prestressing is also used on concrete tanks or pipes as described next.

2. Circular Prestressing

Compared to linear prestressing, which generally refers to elongated elements such as beams, bridges, and piles, the term circular prestressing applies to pipes, pressure vessels, and tanks. Although no analytical difficulty exists in designing circular structures, some particular techniques were developed to prestress them efficiently. The most common one is the wire-wound technique in which the unreinforced concrete core of a pipe or tank is wrapped with a wire under tension, thus creating the uniform radial compression needed for prestressing the core. Upon completion the steel wire is protected with a layer of mortar usually applied by "shotcreting" or "guniting." To rapidly and economically place the wire under tension, it is drawn through a die (the preload method) or a special differential winding system is used.

3. Stage Stressing

Stage stressing refers to the application of the prestressing force in stages. This is often done to avoid overstressing the concrete in its early age or when dead loads are applied in stages and their effect must be counteracted in steps. Stage stressing is generally achieved by fully stressing part of the tendons at each stage.

4. Partial Prestressing

Partial prestressing generally implies a combination of prestressed and non-prestressed reinforcement, both contributing to the resistance of the member. The aim is to allow tension and cracking under full service loads while ensuring adequate ultimate strength. According to one definition, partial prestressing can also be achieved without nonprestressed reinforcement by using a relatively low effective prestress in the prestressing tendons. Better control of camber and deflection, increased ductility, and cost savings are some of its advantages. Partial prestressing is covered in detail in Vol 2.

1.7 PRESTRESSED VERSUS REINFORCED CONCRETE

In its earliest development prestressed concrete used to be mostly compared to reinforced concrete. The currently increasing awareness and use of partially prestressed concrete (or partially reinforced concrete) renders such comparison somewhat inappropriate. Prestressed concrete and reinforced concrete are in effect the two extreme boundaries of the same system. They cannot be considered competitors because they complement each other in function and in application. Since prestressed concrete has become in part a manufactured product (the precast prestressed industry) and since it permits increasingly longer spans, it is mostly competing with steel construction rather than with reinforced concrete. Note that the use of concrete offers inherent advantages such as fire resistance, high insulation qualities, low maintenance, low energy requirement, versatility, etc., which are common to both reinforced and prestressed concrete.

Following are some of the often cited advantages of prestressed concrete compared to reinforced concrete.

1. Prestressed concrete uses high-strength steel and concrete. Thus, it takes advantage of materials with superior qualities and, everything else being equal, needs smaller quantities of materials than reinforced concrete. Steel strengths of up to 300 ksi (2100 MPa) and concrete strengths of up to 12 ksi (84 MPa) are being used at present in the United States. Note that doubling the concrete strength, say from 5 to 10 ksi (35 to 70 MPa) will increase the cost of concrete by about 30 percent only.
2. In prestressed concrete, the entire concrete section is active in resisting the load, while in reinforced concrete only the uncracked part of the section is active.
3. Because of 1 and 2 above, prestressed concrete members are lighter, more slender, and aesthetically more appealing than their reinforced concrete counterparts. Their lighter weight is particularly important in long-span girders and bridges where the dead load is a dominant design factor.
4. As prestressed concrete is crackless, it provides better protection than reinforced concrete against corrosion of the steel in aggressive environments and is more suitable for fluid-retaining structures such as tanks and nuclear vessels.
5. Prestressed concrete (fully or partially) provides means for effective deflection control, especially under long-term sustained loading.
6. Prestressed concrete has better shear resistance than reinforced concrete due to the slope of the tendons near the supports and to the precompression which reduces diagonal tension. Thus it will require lesser stirrups.
7. It is often claimed that prestressed concrete structures have an inherent safety as they undergo the most severe loading during initial tensioning of the steel. In effect they are pretested. If they pass this first test, they are likely to perform well under service loads.

One cannot make a priori a general statement on whether prestressed concrete is more or less economical than reinforced concrete. Prestressed concrete requires a higher level of technology and often such technology is simply not available where the structure is to be built. Prestressed concrete uses less, but higher quality, materials than reinforced concrete and, at least in posttensioning, needs relatively expensive anchorages. Assuming the technology is available, its comparative cost depends very much on the type of structure and the design criteria. There are applications such as long-span bridges in which reinforced concrete cannot compete with prestressed concrete. But there are also areas where their range of applicability overlaps and the choice for either one should be based not only on initial cost but also on other costs and performance criteria. The initial cost of the structure is no longer an all-important factor. It is the author's experience that for structural applications which fall within the capabilities of both reinforced and prestressed concrete, the initial costs using either material are often less than 10 percent apart. For instance, total construction time might render a precast prestressed structure substantially more economical than its reinforced counterpart. Many tradeoffs must be considered before making a decision as some of the advantages of prestressed concrete may not always be needed.

1.8 EXAMPLE

The following example, in which reinforced and prestressed concrete solutions are compared, illustrates some of the features of prestressed concrete. Let us assume that a simply supported slab is to be built. Both reinforced and prestressed concrete are potential alternatives. Consider the following dimensional and materials properties (Fig. 1.18):

Slab depth $h = 12$ in (30.5 cm); unit slab width considered in the computations $b = 12$ in (30.5 cm); distance from extreme compressive fiber to centroid of tensile reinforcement $d_s = d_p = 10$ in (25.4 cm); concrete compressive strength $f'_c = 5000$ psi (34.5 MPa); maximum allowable concrete compressive stress in service $\bar{\sigma}_{cs} = 2000$ psi (13.8 MPa); yield strength of reinforcing steel $f_y = 60$ ksi (414 MPa); maximum allowable working stress of reinforcing steel $f_s = 24$ ksi (165.6 MPa); effective stress in prestressing steel $f_{pe} = 160$ ksi (1104 MPa); span length $l = 30$ ft (9.14 m).

The concrete is a lightweight concrete with a unit weight of 106.67 lb/ft³ (16.76 kN/m³) and the live load is assumed uniform with magnitude of 106.67 lb/ft² (5.11 kN/m²). These values have been selected so as to lead to simple round numbers for the computed stresses.

The maximum moment at midspan due to the deadweight of the slab is given by:

$$M_D = \frac{106.67}{12} \times \frac{(30 \times 12)^2}{8} = 144,000 \text{ lb-in}$$

and the corresponding maximum stresses on the top and bottom fibers of the concrete section, assumed uncracked, are:

$$\sigma = \pm \frac{6M_D}{bh^2} = \pm \frac{6 \times 144,000}{12 \times 12^2} = \pm 500 \text{ psi (3.45 MPa)}$$

As the live load moment M_L has the same magnitude as the dead load moment, it will also generate stresses of the same magnitude in the uncracked section. Several cases can be considered:

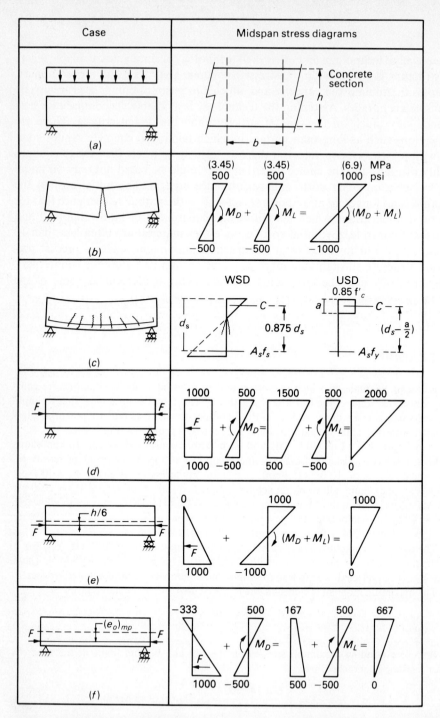

Figure 1.18 Typical schemes to increase the tensile resistance of a concrete element. (a) Example slab. (b) Plain concrete. (c) Reinforced concrete. (d) Prestressed concrete: axial prestressing. (e) Eccentric prestressing. (f) Prestressing at maximum practical eccentricity.

1. **Nonreinforced slab**
   The magnitude of the extreme fiber stress due to the application of combined dead and live loads (Fig. 1.18*b*) is 1000 psi (6.9 MPa). This stress is acceptable in compression but is higher than the tensile strength of concrete in flexure (modulus of rupture). Hence cracking in the tensile zone and subsequent collapse of the slab will occur.

2. **Reinforced concrete slab: WSD approach**
   Let us determine the required amount of steel reinforcement using the Working Stress Design approach, assuming cracked section and linear elastic analysis. If we estimate the lever arm between the center of compression and the center of tension (Fig. 1.18*c*) at $0.875 \, d_s$, the required area of steel reinforcement can be obtained from

$$A_s = \frac{M_D + M_L}{0.875 \, d_s \, \bar{f}_s} = \frac{288,000}{0.875 \times 10 \times 24,000} = 1.371 \text{ in}^2$$

3. **Reinforced concrete slab: USD approach**
   If the Ultimate Strength Design approach according to the ACI code procedure is used, the area of steel reinforcement can be obtained from solving the two equations of equilibrium of the section at ultimate capacity. The reinforcement is designed to generate an internal moment resistance equal to the specified strength design moment given by:

$$M_u = 1.4 \, M_D + 1.7 \, M_L = 446,400 \text{ lb-in}$$

The equations of force and moment equilibrium of the section at ultimate, assuming yielding of the steel, are given (Fig. 1.18*c*) by:

$$A_s \, f_y = 0.85 \, f'_c ba$$

and

$$\phi A_s \, f_y \left(d_s - \frac{a}{2} \right) = M_u = 446,400$$

The above two equations have two unknowns a and A_s. Assuming $\phi = 0.9$ and solving for A_s leads to:

$$A_s = 0.871 \text{ in}^2$$

Note that the steel area using USD is substantially smaller than that using WSD.

4. **Prestressed concrete slabs: uniform prestress**
   Let us assume that no tension is allowed in the slab. If the prestressing steel passes by the centroid of the section, the stress induced by the prestressing force is uniform (Fig. 1.18*d*). Its magnitude should be at least equal to 1000 psi (6.9 MPa) to counteract the maximum tensile stress σ_b on the bottom fiber due to dead and live loads. Thus:

$$\sigma_g = -\sigma_b = (\sigma_b)_F = 1000 \text{ psi}$$

$$F = \sigma_g \times A_c = 1000 \times 144 = 144,000 \text{ lb}$$

$$A_{ps} = \frac{F}{f_{pe}} = \frac{144,000}{160,000} = 0.9 \text{ in}^2$$

where σ_g and $(\sigma_b)_F$ are the stresses at the centroid and at the bottom fiber of the concrete section, due to the prestressing force F, and A_{ps} is the area of prestressed reinforcement. Note, from Fig. 1.18*d*, that the effect of prestressing leads to an increase in compression on the top fiber of concrete. Such increase is not necessary as it adds to the compressive stresses induced by external loads. A more efficient distribution may be achieved by placing the prestressing force at an appropriate eccentricity as described next.

5. **Prestressed concrete slab: eccentric prestress (central kern)**
Let us assume that the prestressing force is placed at the lower limit of the middle third of the section. Its eccentricity e_o is equal to $h/6$. For a rectangular section the middle third of the section also bounds the central kern. In this case a triangular stress distribution with zero stress (no tension) on the top fiber is induced. If we want a stress of 1000 psi (6.9 MPa) at the bottom fiber, the average stress at the centroid of the section will be 500 psi (3.45 MPa) and the required prestressing force will be:

$$F = \sigma_g A_c = 500 \times 144 = 72,000 \text{ lb}$$

The corresponding area of prestressing steel is:

$$A_{ps} = \frac{F}{f_{pe}} = \frac{72,000}{160,000} = 0.45 \text{ in}^2$$

Its value is half that obtained for case 4. This result shows the importance of appropriately placing the prestressing force to achieve higher efficiency where needed.

6. **Prestressed concrete slab: eccentric prestress (maximum eccentricity)**
A prestressing force F placed at a constant eccentricity e_o from the centroid of the section is equivalent to an axial force of magnitude F, passing by the centroid, and a moment of magnitude Fe_o. The corresponding stress on the bottom fiber of the section is given by:

$$(\sigma_b)_F = \frac{F}{A_c} + \frac{Fe_o}{(bh^2/6)}$$

which for a rectangular section can be written as:

$$(\sigma_b)_F = \frac{F}{A_c}\left(1 + \frac{6e_o}{h}\right)$$

It is clear that for a given force F the higher the eccentricity e_o, the higher the stress $(\sigma_b)_F$. Inversely given a required value of $(\sigma_b)_F$, the higher the eccentricity e_o, the smaller the prestressing force F and the higher the savings on the prestressing steel.

Let us assume that the maximum practical value of e_o to ensure a sufficient concrete cover of the steel is 4 in (10.16 cm) and that no tension is allowed in the section. The minimum prestressing force required to incur a stress of 1000 psi (6.9 MPa) on the bottom fiber is obtained by solving the following equation:

$$1000 = \frac{F}{144}\left(1 + \frac{6 \times 4}{12}\right)$$

for which $F = 48,000$ lb.
The corresponding area of prestressing steel is

$$A_{ps} = \frac{48,000}{160,000} = 0.30 \text{ in}^2$$

Note that, although the prestressing force acting alone produces tension on the top fiber

(Fig. 1.18*f*), because the dead load acts immediately upon application of the prestress, the resulting stress (at midspan) is still a compression. In order to avoid tension at the supports, the steel profile is generally draped upward instead of being left straight as assumed in this example.

7. **Prestressed concrete slab allowing tension**
   If a tension of 212 psi (1.46 MPa) is allowed on the bottom fiber under full service load, the equation given in the previous case would be written as:

$$(1000 - 212) = \frac{F}{144}\left(1 + \frac{6 \times 4}{12}\right)$$

for which $F = 37,824$ lb and

$$A_{ps} = \frac{37,824}{160,000} = 0.236 \text{ in}^2$$

Remarks: It is often said that one of the advantages of using prestressed concrete is that the dead load can be carried, entirely or in part, free. This can be illustrated by the above example assuming no tension is allowed.

Let us compute the prestressing force which, if placed at the lower limit of the central kern, would produce a compressive stress on the bottom fiber equal and opposite to that induced by the live load only. Thus:

$$500 = \frac{F}{144}\left(1 + \frac{6 \times 2}{12}\right)$$

for which $F = 36,000$ lb and $A_{ps} = 36,000/160,000 = 0.225$ in^2.

Using the above force, we can increase the eccentricity e_o to its maximum practical value of 4 in (10.16 cm). The corresponding stress on the bottom fiber due to F becomes equal to 750 psi instead of 500 psi. This increment of 250 psi will, in our case, balance 50 percent of the tensile stress produced by the dead load. We would have essentially carried 50 percent of the dead load at no charge. The other 50 percent will be carried by increasing the prestressing force to its value given in case 5. In some instances where the ratio of live load moment to dead load moment is large, the entire dead load can be carried free. This often happens for small span lengths which are then described as "less than critical" spans (Ref. 1.8).

The reader may want to check that if the live load in the above example is doubled, the span becomes "less than critical," that is, the entire dead load may be carried free.

1.9 A LOOK AT THE FUTURE

Prestressed concrete has passed way beyond the developmental stage and has established itself as a major structural material. Similarly, prestressing techniques have evolved into a reliable technology. Prestressed concrete has made major contributions to the construction and the cement industries. It has lead to an incredible array of structural applications from bridges to nuclear power vessels and offshore structures. Seldom is a major project planned today without considering prestressing as one of the viable alternative solutions. Moreover, some unusual structures cannot be conceived without prestressed concrete (Fig. 1.19).

(*a*) **ARCO floating terminal for liquid petroleum gas.** (*Courtesy Concrete Technology Corporation, Tacoma.*)

(*b*) **Prestressing the spillway gate anchor at Guri Dam, first stage.** (*Courtesy C. V. G.-Edelca, Venezuela.*)

26

(*c*) **Condeep offshore oil platform—Statfjord Platform A.** (*Courtesy Norwegian Contractors, Oslo.*)

(*d*) **Canadian National Tower, Toronto.** (*Courtesy CN Tower Limited, Toronto.*)

Figure 1.19 Unusual structures using prestressed concrete.

A careful analysis of what is to come by the end of this century indicates that the future of prestressed concrete is very bright. Constructed facilities will keep expanding not only in volume but also in reach. To achieve daring limits, prestressed concrete will be needed.

Current market penetration for prestressed concrete in the United States is estimated at about 10 percent and it is believed that a penetration of 30 percent will be achieved eventually. (Ref. 1.29). At the educational level the same trend is expected. Design courses in prestressed concrete will very likely move from the list of elective optional courses to the list of required courses in civil engineering curricula.

As pointed out earlier, prestressed concrete structures contain all the beneficial attributes of concrete as a material plus those inherent to prestressing. The current widespread use of prestressed concrete in developed countries is startling. It is expected to remain one of the strongest construction systems on the market.

REFERENCES

1.1. P. H. Jackson: U.S. Patent No. 357999 (1886/88).

1.2. C. E. W. Doehring: German Patent No. 53548 (1888/89).

1.3. M. Koenen: German Patent No. 249007 (1912).

1.4. E. Freyssinet and J. Seailles: French Patent No. 680 547 (1928).

1.5. E. Freyssinet: *Une Revolution dans les Techniques du Beton*, Librairie de L'Enseignement Technique, Leon Eyrolles, Editor, Paris, 1936.

1.6. R. E. Dill: U.S. Patent No. 1684663 (1925/28).

1.7. G. Magnel: *Prestressed Concrete*, 3d ed., revised and enlarged, Concrete Publications Ltd., London, 1954.

1.8. Y. Guyon: *Prestressed Concrete*, vols. 1 and 2, John Wiley & Sons, New York, 1953, 1960. Also *Limit State Design of Prestressed Concrete*, translated from French, Halstead and Wiley, 1972.

1.9. P. Abeles: *An Introduction to Prestressed Concrete*, vols. 1 and 2, Concrete Publications Ltd., London, 1964.

1.10. F. Leonhardt: *Prestressed Concrete Design and Construction*, English translation, Wilhelm Ernst and Sohn, Berlin, 1964, (1st ed., 1955, 2d ed., 1962 in German).

1.11. T. Y. Lin: *Design of Prestressed Concrete Structures*, 2d ed., John Wiley & Sons, New York, 1963. Also revised and enlarged 3d ed. by T. Y. Lin and N. Burns, 1981.

1.12. K. Preston and N. Sollenberger: *Modern Prestressed Concrete*, McGraw-Hill Book Company, New York, 1967.

1.13. V. V. Mikhailov: *Prestressed Concrete Structures : Theory and Design*, translated from Russian by J. H. Dixon, Cement and Concrete Association, London, 1969.

1.14. N. Khachaturian and G. Gurfinkel: *Prestressed Concrete*, McGraw-Hill Book Company, New York, 1969.

1.15. Ben C. Gerwick, Jr.: *Construction of Prestressed Concrete Structures*, Wiley-Interscience, New York, 1971.

1.16. G. S. Ramaswamy: *Modern Prestressed Concrete*, Pitman Publishing, London, 1976.

1.17. J. R. Libby: *Modern Prestressed Concrete: Design Principles and Construction Methods*, 2d ed., Van Nostrand-Reinhold, New York, 1977.

1.18. A. H. Nilson: *Design of Prestressed Concrete*, John Wiley & Sons, New York, 1978.

1.19. Federation Internationale de la Precontrainte (FIP), London, England.
1.20. Prestressed Concrete Institute, Chicago, Illinois 60606, U.S.
1.21. Post-Tensioning Institute, Phoenix, Arizona 85013, U.S.
1.22. American Concrete Institute, Detroit, Michigan 48219, U.S.
1.23. Portland Cement Association, Skokie, Illinois 60076, U.S.
1.24. Cement and Concrete Association, London, England.
1.25. *Design Handbook Precast Prestressed Concrete*, 2d ed., Prestressed Concrete Institute, Chicago, 1978.
1.26. H. Lossier: "Cements with controlled expansion and their application to prestressed concrete," *The Structural Engineer*, vol. 24, no. 10, October 1946, pp. 505–534.
1.27. V. V. Mikhailov: "New developments in self-stressed concrete," *Proceedings of the World Conference on Prestressed Concrete*, San Francisco, 1957.
1.28. T. Y. Lin and A. Klein: "Chemical prestressing of concrete elements using expanding cements," *ACI Journal Proceedings*, vol. 60, no. 9, September 1963, pp. 1187–1218.
1.29. W. Burr Bennet, Jr.: "Prestressing: The technique that makes concrete fully competitive," *Concrete Construction*, vol. 25, no. 2, February 1980, pp. 125–131.

PROBLEMS

1.1 Give an example of a system, device, instrument, or procedure in which the principle of prestressing is used. Describe it schematically. Explain how the prestress is applied and how it counteracts stresses induced during service. Draw schematically the corresponding forces and/or stresses.

1.2 Using the results obtained for the example described in Sec. 1.8, estimate the cost of the slab (say per unit area or unit strip) for each of the cases 2 to 5 studied. For this find out and use unit costs prevailing in the geographic area where you live. Two approaches should be considered:

1. Using global unit costs such as dollars per cubic yard of finished concrete in place, dollars per pound of reinforcing steel in place, and dollars per pound of prestressing steel in place.
2. Using unit costs broken down into two categories: material and labor.
   Draw conclusions from your findings.

Production of precast prestressed concrete hollow-cored slabs. (*Courtesy Material Service Corporation.*)

CHAPTER 2

PRESTRESSING MATERIALS: STEEL AND CONCRETE

Prestressed concrete utilizes high-quality materials, namely high-strength steel and concrete. In addition, ordinary reinforcing steel is being increasingly used in partially prestressed concrete. The full coverage of the characteristics of each component, steel or concrete, is a treatise by itself. Their composition, properties, and manufacturing processes have been extensively described in the technical literature. This chapter gives only a brief overview of their most important characteristics. Information relevant to the understanding of prestressed concrete's structural behavior is provided. A direct approach is used to give the designer enough technical information to allow him to proceed in most common design problems. More on the modeling of the materials and the structure can be found in Chap. 8 and Vol. 2. Although other materials (such as fiberglass tendons) can be used for prestressing, only steel and concrete are considered next.

2.1 REINFORCING STEELS

Nonprestressed reinforcing steel is commonly used in prestressed concrete structures as shear reinforcement or supplementary reinforcement in regions of high local stresses and deformations. In partial prestressing it contributes to the main structural resistance to an extent equal to that of prestressing steel. Detailed information on reinforcing steels can be found in many books and textbooks on reinforced concrete, as well as in many standards and code specifications (Refs. 2.1 to 2.3). The Concrete Reinforcing Steel Institute keeps up-to-date information on the state of the art of reinforcing steels.

Table 2.1 ASTM standard reinforcing bars

	Nominal diameter		Nominal area		Nominal weight lb/ft	Nominal weight kg/m
Bar†	in	mm	in^2	mm^2		
3	$\frac{3}{8}$ = 0.375	9.52	0.11	71	0.376	0.56
4	$\frac{1}{2}$ = 0.500	12.70	0.20	129	0.668	0.994
5	$\frac{5}{8}$ = 0.625	15.88	0.31	200	1.043	1.552
6	$\frac{3}{4}$ = 0.750	19.05	0.44	284	1.502	2.235
7	$\frac{7}{8}$ = 0.875	22.22	0.60	387	2.044	3.042
8	1.0 = 1.00	25.40	0.79	510	2.670	3.973
9	$1\frac{1}{8}$ = 1.128	28.65	1.00	645	3.400	5.060
10	$1\frac{1}{4}$ = 1.270	32.26	1.27	819	4.303	6.404
11	$1\frac{3}{8}$ = 1.410	35.81	1.56	1006	5.313	7.907
14	$1\frac{3}{4}$ = 1.693	43.00	2.25	1452	7.650	11.38
18	$2\frac{1}{4}$ = 2.257	57.33	4.00	2581	13.600	20.24

† Based on the number of eighths of an inch that add up to the nominal diameter.

Some of the most important design characteristics of reinforcing steels are summarized in Tables 2.1 to 2.3. Table 2.1 provides information on the type and sizes of standard ASTM reinforcing bars used in the United States. Although smooth bars can be obtained, at present they are mostly deformed to improve their bonding properties. Welded wire fabrics and meshes are being increasingly used as main reinforcement in reinforced concrete slabs or supplemental reinforcement in

Table 2.2 Characteristics of a selected set of reinforcing wires used in welded wire meshes

	Nominal diameter		Nominal area‡	
Wire designation†	in	mm	in^2	mm^2
W2	0.159	4.039	0.020	12.9
W4, D4	0.225	5.715	0.040	25.8
W6, D6	0.276	7.010	0.060	38.7
W8, D8	0.319	8.103	0.080	51.6
W10, D10	0.356	9.042	0.100	64.5
W12, D12	0.390	9.906	0.120	77.4
W14, D14	0.422	10.719	0.140	90.3
W20, D20	0.504	12.802	0.200	129
W30, D30	0.618	15.697	0.300	193.6

† W for smooth, D for deformed.
‡ Wire density = 490 lb/ft^3 or 7850 kg/m^3.

Table 2.3 ASTM standardized properties of reinforcing steels

Reinforcing steel type	ASTM standard	Grade	Minimum yield strength		Minimum tensile strength	
			ksi	MPa	ksi	MPa
Reinforcing bars	A615 & A617	40	40	276	70	483
	A615 & A617	60	60	414	90	620
	A616	50	50	345	80	552
	A616	60	60	414	90	620
	A706	60	60	414	80	552
Wire						
Smooth	A82	....	70	483	80	552
Deformed	A496	....	75	517	85	586
Welded wire fabric						
Smooth	A185	....	65	448	75	517
Deformed	A497	....	70	483	80	552

prestressed slabs. They improve substantially the cracking performance of slabs prestressed with unbonded tendons. Typical sizes of wires, smooth or deformed, used separately or in welded meshes, are given in Table 2.2. Many sizes in between those shown are also available. In practice, welded meshes having any reasonable size opening of square or rectangular shape are manufactured as standard product or, on request, for a particular design. Common wire spacings in one of the two principal directions of the mesh are 2, 3, 4, 6, 8, 10, and 12 inches. More details can be obtained from local suppliers or the Wire Reinforcement Institute.

The minimum tensile properties of reinforcing bars and wires are summarized in Table 2.3. Also shown is the corresponding ASTM standard designation. Although several grades are available, the most common grade in the United States is the A615 grade 60 deformed steel which gives a minimum yield strength of 60 ksi (414 MPa).

Typical stress-strain curves of reinforcing steels are shown in Fig. 2.1. Two different shapes can be identified: one, typical of most reinforcing bars, shows a well-defined yielding behavior and the other, typical of wires and a smaller category of bars, does not indicate a well-defined yielding. For this latter type the yield stress is defined as the stress corresponding to a total strain of 0.0035 (ACI) or 0.005 (ASTM). In all cases the steel shows a significant strain-hardening behavior and sufficient ductility before failure.

Reinforcing steels exhibit good resistance to fatigue. Typical stress range versus number of cycles to failure curves are shown in Fig. 2.2. It can be generally observed that an endurance limit to fatigue seems to exist. On the basis of an

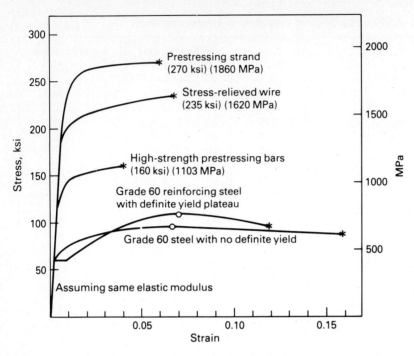

Figure 2.1 Typical stress-strain curves of reinforcing and prestressing steels.

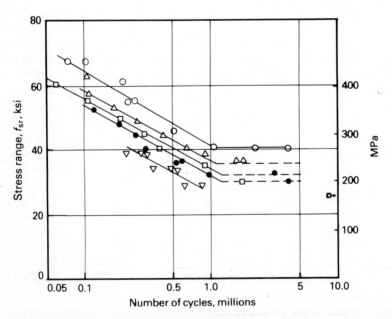

Figure 2.2 Representative fatigue test results for North American Bars. (*Ref. 2.4, Courtesy Portland Cement Association.*)

extensive series of tests on North American bars (Ref. 2.4) the following conclusions were drawn: the fatigue strength that is the acceptable stress range for more than two million loading cycles is practically independent of the grade of steel but depends on the minimum applied stress and the degree of stress concentration induced in deformed bars during the deformation process. Other factors were also found to influence the fatigue strength such as the nominal diameter and the yield strength but to a lesser extent. The following relationship was proposed for use in design:

$$f_{sr} = 21 - \frac{f_{min}}{3} + 8\,\frac{r}{h} \qquad \text{in ksi}$$

or
$$f_{sr} = 145 - \frac{f_{min}}{3} + 55\,\frac{r}{h} \qquad \text{in MPa} \qquad (2.1)$$

where f_{sr} = safe stress range for more than two million loading cycles
   f_{min} = minimum stress
   r/h = ratio of base radius r to depth of rolled-on deformation (a value of 0.3 may be used in absence of specific information)

For values of acceptable stress ranges at a number of cycles smaller than two million, a number of best fit relationships of a semilogarithmic form were also proposed in Ref. 2.4 and can be used for design.

2.2 PRESTRESSING STEELS

The importance of using, in prestressed concrete, high-strength steels that permit high elongations was mentioned in Chap. 1. Otherwise, due to prestress losses, the steel stress will decrease substantially or vanish with time. As prestress losses in the steel (not including friction) can often approach 60 ksi (414 MPa), it is clear that minimum tensile strengths substantially higher than this value are needed to achieve effective prestressing. Other desirable properties are needed. Ideally a tendon material should not only have high strength but also (1) remain elastic up to relatively high stresses, (2) show sufficient ductility before failure, (3) have good bonding properties, low relaxation, good resistance to fatigue and corrosion, and (4) be economical and easy to handle. Clearly a tradeoff should be made, and high-strength steels offer a good compromise.

1. Types of Prestressing Tendons

Three types of steel tendons are used in prestressed concrete: wires, strands (or cables) made with several wires, and bars. Typical shapes and commonly available diameters are shown in Fig. 2.3; other particular shapes are given in Fig. 2.4. Prestressing tendons used in the United States are manufactured to meet minimum ASTM specifications described in the following standards: A421 for uncoated stress-relieved wires, A416 for uncoated seven-wire stress-relieved strands, and

Kind	Size (Diameter)		Shape
	mm	in.	
Plain round wir	2.0~9.0	0.08~0.360	
Indented wire	5.0~7.0	0.200~0.276	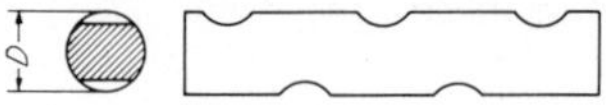
Sumi-Twist	7.3~13.0	0.276~0.512	
Two-ply wire	2.9×2	0.114×2	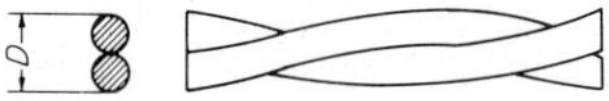
Seven-wire strand	6.2~15.2	0.250~0.600	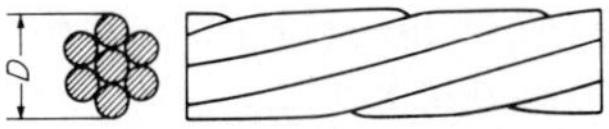
Nineteen-wire strand	17.8~21.8	0.700~0.860	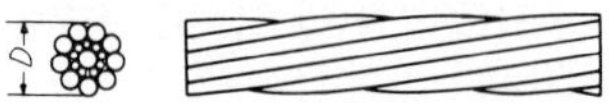
Round bar	9.2~32.0	0.362~1.260	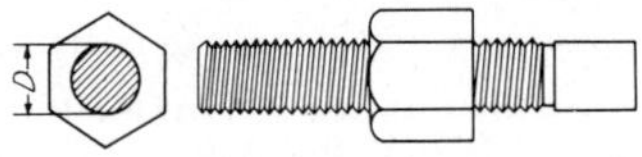
Threaded bar	23.0~32.0	0.906~1.260	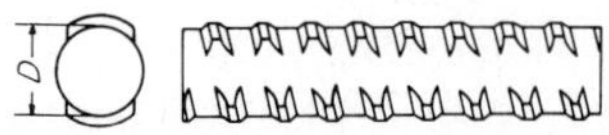

Figure 2.3 Common shapes and diameters of prestressing tendons. *(Courtesy Sumitomo's Tensioning Materials, Japan.)*

A722-75 for uncoated high-strength steel bars. Although these standards would guarantee minimum properties, actual properties depend greatly on the manufacturing process. Local manufacturers and suppliers should be consulted for particular details.

The high tensile strength and adequate ductility of prestressing steels are generally obtained by using (1) high carbon hot rolled alloy steel (2) cold drawn or

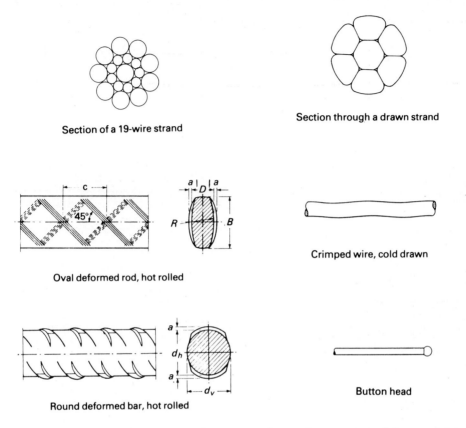

Section of a 19-wire strand

Section through a drawn strand

Oval deformed rod, hot rolled

Crimped wire, cold drawn

Round deformed bar, hot rolled

Button head

Figure 2.4 Shapes of some particular prestressing tendons. *(Adapted from Ref. 2.5.)*

deformed carbon steel, preferably tempered, and (3) hot rolled and heat treated carbon steel. Some of these treatments are clarified in the following section and in Fig. 2.5. When alloys are used, they comprise manganese, silicon, chromium, and, to a lesser extent, nickel.

Most prestressing wires are produced by the cold working (drawing or rolling) process. Wires are manufactured with different cross-sectional shapes and surface conditions: round or oval, smooth or indented, ribbed, twisted, or crimped. When cut to size, round wires used in some posttensioning systems can have button heads formed at their ends (Fig. 2.4). Typical characteristics of prestressing wires most frequently used in the United States are shown in Table 2.4.

Prestressing strands are produced from several wires. In the seven-wire strand, six peripheral wires are wound helically over a central wire which has a slightly higher diameter than the others. Because strands are made with relatively small-diameter wires, they are much easier to handle (more flexible) than a single tendon of the same nominal diameter and they achieve superior properties due to better quality control. Typical characteristics of prestressing strands used in the United

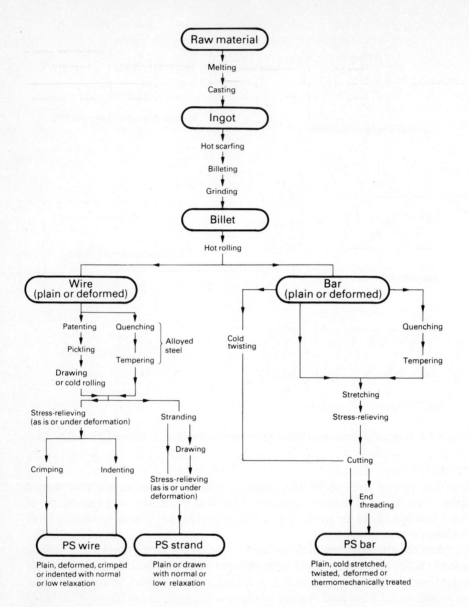

Figure 2.5 Flow chart illustrating typical production processes for prestressing steels.

States are given in Table 2.4. Most popular is the $\frac{1}{2}$-inch diameter strand with a minimum tensile strength of 270 ksi (1860 MPa). One variety of seven-wire strand is the die-formed or drawn strand (Fig. 2.4). Drawing tightens the wires against each other and leads to a smoother outside surface. Mechanical properties such as strength and relaxation are enhanced. Common upper and lower tolerance limits

Table 2.4 Typical characteristics of stress-relieved prestressing wires and strands

Prestressing steel	ASTM type or grade	Nominal diameter		Nominal area†		Minimum tensile strength, f_{pu}	
		in	mm	in²	mm²	ksi	MPa
Stress-relieved wires (ASTM A421)	WA,†BA‡	0.192	4.88	0.0289	18.7	250	1725
	WA	0.196	4.98	0.0302	19.4	250	1725
	BA	0.196	4.98	0.0302	19.4	240	1655
	WA, BA	0.25	6.35	0.0491	31.6	240	1655
	WA	0.276	7.01	0.0598	38.7	235	1622
Stress-relieved 7-wire strands (ASTM A416)	Grade 250	0.25	6.35	0.036	23.22	↑	↑
		0.313	7.94	0.058	37.42		
		0.375	9.53	0.080	51.61		
		0.438	11.11	0.108	69.68	250	1725
		0.500	12.54	0.144	92.90		
		0.600	15.24	0.216	139.35	↓	↓
	Grade 270	0.375	9.53	0.085	54.84	↑	↑
		0.438	11.11	0.115	74.19		
		0.500	12.54	0.153	98.71	270	1860
		0.563	14.29	0.192	123.87		
		0.600	15.24	0.216	139.35	↓	↓

† Tendon density = 490 lb/ft³ or 7850 kg/m³.

‡ Type WA wire for wedge or grip anchoring and type BA for button-head anchoring.

on strand area and mass are +4 and −2 percent. The equivalent area is generally obtained from weighing assuming a density of 490 lb/ft³ (7850 kg/m³).

Prestressing bars are manufactured with a smooth or ribbed surface. Smooth bars can be end-threaded mechanically to be used with anchoring systems based on nut and plate. The ribs in a ribbed bar can act as a thread such as in the Dywidag system. Thus the bar can be anchored anywhere along its length. Prestressing bars are generally made with alloy steel heat treated to achieve desirable properties. Common diameters and grades used in the United States are summarized in Table 2.5.

2. Production Process

A typical flow chart of the manufacturing process of prestressing steels is shown in Fig. 2.5. It is self-explanatory. Some particular procedures need to be explained and will clarify some terms used in the preceding section:

1. *Patenting.* A heat treatment process which gives the hot rolled steel a suitable metallurgical structure leading to improved homogeneity, strength, and toughness.

Table 2.5 Typical characteristics of prestressing bars

Type	ASTM grade	Nominal diameter		Nominal area†		Minimum tensile strength f_{pu}	
		in	mm	in²	mm²	ksi	MPa
Smooth alloyed steel bars (ASTM A722-75)	145	0.750	19.05	0.442	283.9	↑ 145 ↓	↑ 1000 ↓
		0.875	22.22	0.601	387.1		
		1.0	25.40	0.785	503.2		
		1.125	28.57	0.994	638.7		
		1.250	31.75	1.227	793.5		
		1.375	34.92	1.485	954.8		
	160	0.750	19.05	0.442	283.9	↑ 160 ↓	↑ 1104 ↓
		0.875	22.22	0.601	387.1		
		1.0	25.40	0.785	503.2		
		1.125	28.57	0.994	638.7		
		1.250	31.75	1.227	793.5		
		1.375	34.92	1.485	954.8		
Deformed bars‡	......	0.625	15.87	0.280	180.6	157	1083
		1.0	25.4	0.852	548.4	150	1035
		1.0	25.4	0.852	548.4	160	1104
		1.25	31.75	1.295	835.5	150	1035
		1.25	31.75	1.295	835.5	160	1104
		1.50	34.92	1.630	1051.6	150	1035

† Bar density = 490 lb/ft³ or 7850 kg/m³.

‡ Adapted from the PCI Handbook.

2. *Pickling.* Chemical or mechanical removal of the hard scale of oxidized iron formed on the surface of the rod.

3. *Quenching and tempering.* A heat treatment process in which the steel is heated to about 800°C followed by rapid cooling (such as oil quenching) and subsequent tempering at about 450°C.

4. *Cold drawing.* A process in which the tendon is pulled through a die to reduce its cross-sectional area and improve its strength.

5. *Stress-relieving.* Heating to an appropriate temperature, generally less than 500°C, to improve the mechanical properties, specifically ductility. Stress-relieving under prescribed conditions of deformation leads to improved relaxation characteristics (low relaxation tendons).

6. *Stabilizing.* Heating the tendon under a high tension sufficient to produce a permanent elongation. It improves most mechanical properties and leads to low relaxation.

7. *Indenting.* Introduction of regular dents at the periphery of a wire to improve its bonding characteristics.

8. *Crimping.* Consists of a series of regular bends of the wire in one plane to improve its bonding characteristics.

3. Mechanical and Stress-Strain Properties

The mechanical properties of prestressing steels of direct use in design must generally satisfy a number of minimum requirements. Such properties include the tensile strength and corresponding failure strain, the yield point, the proportional limit, and the modulus of elasticity. Steel tendons are manufactured to achieve actual properties generally superior to the minimum properties required by various standards. For instance, the minimum required (or specified) strength, f_{pu}, also called characteristic strength, corresponds to a value below which no more than five percent of all test data will lie. It is associated with a standard deviation of 40 MPa (5.8 ksi) (Ref. 2.5).

A summary of ASTM minimum requirements related to the relevant mechanical properties of prestressing steels is given in Table 2.6. It can be seen that for a given characteristic strength f_{pu} the minimum yield strength should not be less than $0.85 f_{pu}$ for all tendons except deformed bars and that their ultimate strain should not be less than four percent. More details on actual properties and their variability can be found in Vol. 2, where the nonlinear analysis of prestressed concrete members is addressed.

Typical stress-strain curves of prestressing steels in uniaxial tension are plotted in Fig. 2.1 and compared with those of reinforcing steels. It can be observed that prestressing steels (1) offer substantially higher strengths than reinforcing steels accompanied by lower failure strains, (2) do not show a well-defined yielding behavior, and (3) have a relatively high proportional limit. Their stress-strain curve can be represented by three successive portions: an initial linear elastic portion up to the proportional limit, a nonlinear portion with gradually decreasing slope, and a final almost linear strain-hardening portion with a small positive slope leading to failure.

Because yielding of prestressing steels is not well defined, their yield strength is determined according to a strain criterion. Current ASTM standards specify that the yield stress of prestressing wires and strands should correspond to a total strain of one percent, while a total strain of 0.7 percent is recommended for prestressing bars (Table 2.5). Another often used method to determine the yield stress is called the offset-strain method. In this method the yield stress corresponds to a permanent plastic strain of 0.2 percent (or 0.1 percent depending on the standard used). An example illustrating the determination of yield stress is shown in Fig. 2.6. Also shown is the determination of the proportional limit.

Theoretically the proportional limit corresponds to the point at which the initial portion of the stress-strain curve deviates from linearity. In practice, however, the porportional limit is generally determined by the stress corresponding to a permanent strain of 0.01 percent (or 0.02 percent).

The modulus of elasticity of steel is essentially a constant independent of

Table 2.6 Minimum properties of prestressing tendons

Prestressing steel	ASTM standard	ASTM minimum specified		Other expected properties		
		Yield strength	Elongation at failure	Elastic modulus		Prop. limit f_{pp} at 0.01% offset strain
				ksi	MPa	
Wires	A421	$f_{py} = 0.85 f_{pu}$ at $\varepsilon_{py} = 0.010$	$\varepsilon_{pu} = 0.040$ gauge length 10 in (254 mm)	29,000	200,000	75 to 85% of minimum specified strength
Strands	A416	$f_{py} = 0.85 f_{pu}$ at $\varepsilon_{py} = 0.010$	$\varepsilon_{pu} = 0.040$ gauge length 24 in (610 mm)	27,000	186,000	70 to 75% of minimum specified strength
Bars	A722–75	Smooth bars $f_{py} = 0.85 f_{pu}$ at $\varepsilon_{py} = 0.0070$ Deformed bars $f_{py} = 0.80 f_{pu}$ at $\varepsilon_{py} = 0.070$	$\varepsilon_{pu} = 0.040$ gauge length 20 bar diameter	28,000	193,000	60 to 65% of minimum specified strength

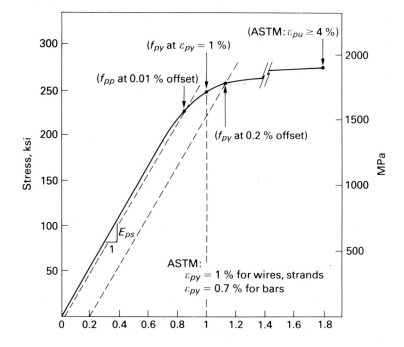

Figure 2.6 Typical determination of yield strength for a prestressing steel.

strength. However, slight differences exist between the elastic moduli of pre-stressing steels. Straight wires (as reinforcing bars) generally show the highest modulus (Table 2.6). Because of alloying the modulus of prestressing bars is smaller. As strands are manufactured by spiral winding, their equivalent modulus is smaller than that of the wire component. Typical values are given in Table 2.6 but the manufacturing process may influence the order shown and actual data must be used when available.

4. Relaxation

Relaxation is the loss of tension with time in a stressed tendon maintained at constant length and temperature. Similarly to creep, which describes the change in strain with time at a constant stress, relaxation results from the adaptation of the material to an externally applied constraint. The relaxation losses of prestressing steels are generally negligible for stresses smaller than 50 percent of ultimate strength. However, they increase rapidly with an increase in stress and temperature.

Typical variations of relaxation loss with time at various levels of initial stress and at various temperatures are shown in Figs. 2.7 and 2.8. It can be generally observed that relaxation continues with time but at a decreasing rate. However,

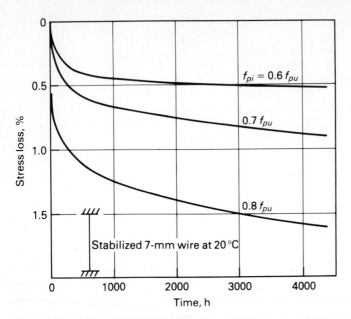

Figure 2.7 Typical effect of initial stress on relaxation of prestressing steel. *(Adapted from data by GKN, Somerset, Wire & Strand, Cardiff, Great Britain.)*

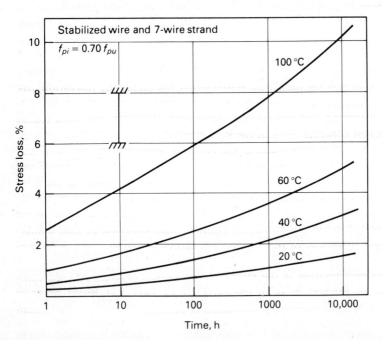

Figure 2.8 Typical effect of temperature on relaxation of prestressing steels. *(Adapted from data by GKN, Somerset, Wire & Strand, Cardiff, Great Britain.)*

Table 2.7 Typical relaxation of prestressing steels at 1000 hours

	f_{pi}/f_{pu}		
	0.6	0.7	0.8
FIP Commission Normal relaxation (level 1)	4.5%	8%	12%
Low relaxation (level 2)	1%	2%	4.5%
ASTM A416 and A421 Low relaxation	......	$\leq 2.5\%$	$\leq 3.5\%$

there is evidence (Ref. 2.6) that the relative influence of temperature within the ambient range diminishes with time while the influence of initial strsss remains the same. In practice the long-term relaxation of prestressing steels is determined from short-term tests. Most commonly the relaxation loss is experimentally determined at 1000 hours and a multiplier is used to estimate the relaxation at the end of the life of the structure. Relaxation depends on the type and grade of steel. However, for the purpose of design, prestressing steels are divided into two groups having either normal or low relaxation properties. Low relaxation is about 25 percent of normal relaxation. Typical relaxation losses at 1000 hours recommended by the FIP Commission on Prestressing Steels (Ref. 2.5) are given in Table 2.7 in percent of initial stress. Note that the higher the initial stress, the higher the relaxation loss. A multiplier of three is suggested to estimate normal life relaxation of 50 years. The corresponding relaxation loss can be very significant. However, one must realize that in a prestressed concrete element the steel stress decreases with time not only because of relaxation but also because the concrete shortens due to creep and shrinkage. The tendon will shorten an equal amount and its initial stress level will decrease. As a result, the apparent relaxation loss is smaller than the pure relaxation otherwise obtained on a reference tendon maintained at constant length. The effect of creep or relaxation with time on the state of stress and strain in the material is illustrated in Fig. 2.9 and could apply to steel, concrete, or their combination. A detailed treatment of this interaction can be found in Chap. 8 on prestress losses.

An extensive investigation of the relaxation properties of stress-relieved wires was carried out by Magura et al. in the early sixties. (Ref. 2.7). Based on their results, they proposed the following relationship to predict, in function of the initial stress, the stress in the prestressing steel at any time t

$$f_{ps}(t) = f_{pi}\left[1 - \frac{\log t}{10}\left(\frac{f_{pi}}{f_{py}} - 0.55\right)\right] \tag{2.2}$$

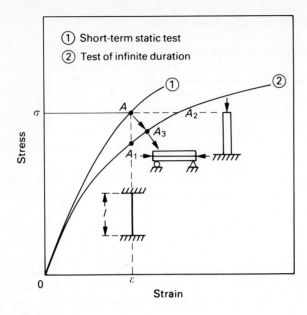

Figure 2.9 Effect of relaxation and creep on the state of stresses and strains.

where t is in hours and is not less than one hour, log t is to the base 10, and the ratio f_{pi}/f_{py} is not less than 0.55. For low relaxation steel a denominator of 45 is suggested instead of 10 under the log t term. As a first approximation Eq. (2.2) can also be used for prestressing strands and bars. For a given time t it is a quadratic function of the initial stress.

The stress loss due to pure relaxation can be computed from

$$\Delta f_{pR}(t) = f_{pi} - f_{ps}(t) = f_{pi} \frac{\log (t)}{10} \left(\frac{f_{pi}}{f_{py}} - 0.55 \right) \tag{2.3}$$

Here too a denominator of 45 is suggested for low relaxation steel. Note that the loss in percent of the initial stress can be obtained by dividing the two sides of the equation by f_{pi}.

It was pointed out above that because of the effects of creep and shrinkage in a prestressed concrete member, the apparent relaxation of the steel will be smaller than the pure relaxation otherwise obtained on a reference strand and predicted by Eqs. (2.2) and (2.3). In lieu of a more exact analysis (see Chap. 8) the following relationship suggested by the FIP Commission on Prestressing Steels (Refs. 2.5) can also be used to estimate the apparent life relaxation:

$$(\Delta f_{pR})_{\text{app}} = (\Delta f_{pR})_{\text{pure}} \left[1 - 2 \frac{\Delta f_{pS} + \Delta f_{pC}}{f_{pi}} \right] \tag{2.4}$$

where Δf_{pC} and Δf_{pS} are the life losses of stress in the prestressing steel due to creep and shrinkage of the concrete. Their values must be estimated and possibly revised after a first iteration, as in effect creep losses depend on the stress in the steel which itself depends on relaxation.

5. Effects of Temperature

The mechanical properties of prestressing steels do not change appreciably when the ambient temperature varies over a reasonable range. This is generally the case for the strength, yield point, proportional limit, and elastic modulus. Even though a change in temperature from 20 to 40°C (Fig. 2.8) seems to increase relaxation at 1000 hours by 50 percent, it may be assumed (Refs. 2.5, 2.6) when extrapolating over a long period of time that the relaxation loss at 40°C is about equal to that at 20°C. The only difference is that at higher temperature the relaxation loss will be reached at an earlier time.

Extreme temperature conditions can seriously affect the mechanical properties of prestressing steels (Ref. 2.8). A sharp decrease in temperature will lead to an improvement in strength and modulus (Fig. 2.10), but a deterioration in ductility and impact resistance takes place. It is estimated that decreasing the temperature from 20 to −200°C will increase tensile and yield strengths by about 20 percent (Ref. 2.9). Similarly, relaxation properties are substantially improved at low temperatures (Fig. 2.11). However, increasing the temperature produces opposite effects. Above 200°C a rapid deterioration can be expected (Fig. 2.10). This means that in the presence of fire a critical temperature may be reached whereas external stresses will overcome the strength of the structure and collapse will follow. In designing for possible service at high temperature, it may be more appropriate to use alloy steel which undergoes a reduction in strength at temperatures above 350°C but, provided the heating period is not too long, regains its strength when cooled. Under similar conditions, cold worked and heat treated steels will suffer serious permanent reduction in strength. Note that if alloy steel is kept at temperatures between 350 and 550°C for sufficient periods of time (of the order of a few hours), it tends to become extremely brittle. More detailed information on the structural resistance of prestressed concrete to fire is given in Vol. 2.

lecture upto Here

6. Fatigue

Fatigue is the process of deterioration of the mechanical properties of materials under fluctuating stresses. Such stresses are generally induced by the repetitive application of live loads on the structure. The resistance of a material to fatigue is often described by an *S-N* curve, where *S* represents the stress range in the material and *N* the number of cycles to failure, also called fatigue life. Below a certain stress range, termed endurance limit, some materials do not fail up to a quasi infinite number of cycles. Prestressing steels do not seem to have an endurance limit (Refs. 2.10 to 2.13). At present a fatigue life of two million cycles is considered a minimum and a life of up to ten million cycles is designed for in some particular applications. Typical *S-N* data for prestressing tendons are plotted in Fig. 2.12 (Ref. 2.14). In prestressed concrete, the minimum stress in the steel is relatively high (50 to 60 percent of ultimate). As the *S-N* curve does not take into account the effect of minimum stress, a Goodman diagram can be used for the particular fatigue life considered in the design. A typical Goodman diagram for prestressing

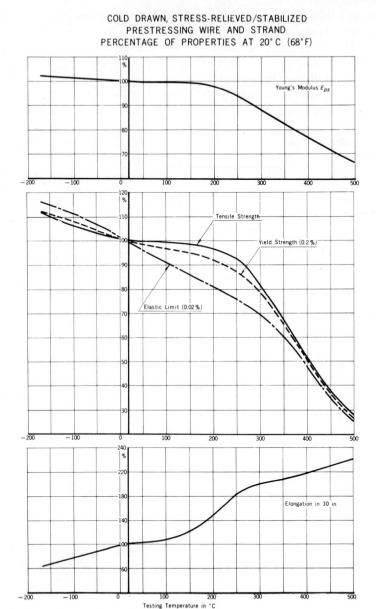

Figure 2.10 Effect of temperature on various properties of prestressing steels.
(Courtesy Shinko Wire Company Ltd., Japan.)

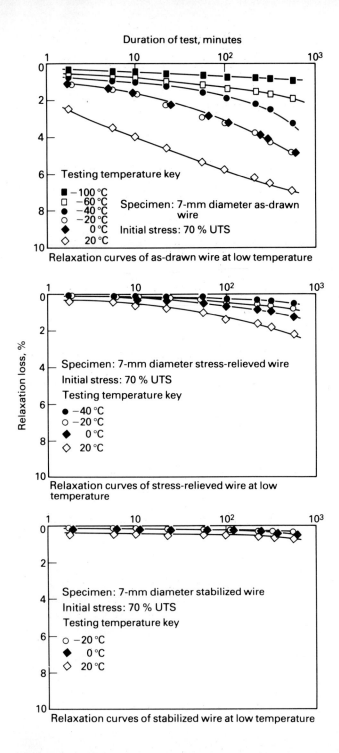

Figure 2.11 Effect of low temperature on relaxation of prestressing steels. *(From Ref. 2.9, courtesy Federation Internationale de la Précontrainte.)*

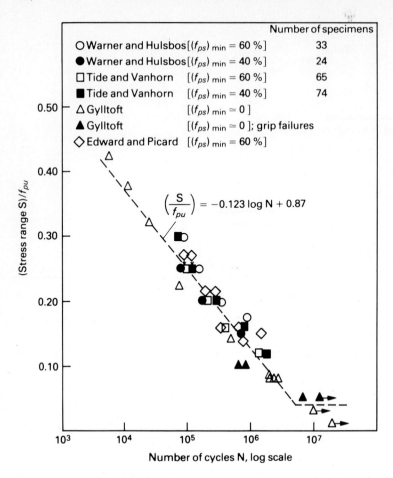

Figure 2.12 Typical *S–N* curve for prestressing tendons *(Ref. 2.14).*

wires and strands proposed by Ekberg et al. (Ref. 2.14) is shown in Fig. 2.13 assuming a fatigue life of two million cycles. It can be seen that at a minimum stress of about 55 percent of ultimate, a fluctuating stress range of about 13 percent is acceptable. The FIP Commission on Prestressing Steels suggests the use of a Smith diagram which is a modification of the Goodman diagram plotted with the mean stress instead of the minimum stress (Fig. 2.14).

ACI Committee 215, charged to study fatigue in concrete structures, (Ref. 2.16) recommends a maximum stress range of $0.10 f_{pu}$ for prestressing wires and $0.12 f_{pu}$ for strands. Such limits are substantially higher than stress range values observed in fully prestressed uncracked members. However, if cracking is allowed or if partial prestressing is used, the above limitations may become critical in the design. In European practice a stress range of up to 200 MPa (29 ksi) is considered acceptable for prestressing tendons while a safe design value below 150 MPa (22 ksi) is

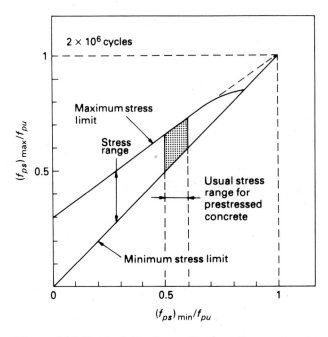

Figure 2.13 Typical Goodman diagram for prestressing wires and strands *(from Ref. 2.15).*

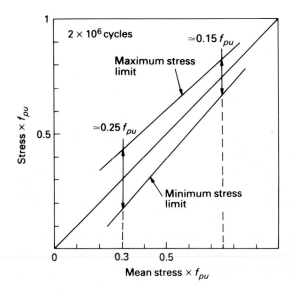

Figure 2.14 Typical Smith diagram for prestressing steels *(Ref. 2.5).*

preferred. A stress range limit of 80 MPa (11.6 ksi) is recommended by the FIP Commission on Prestressing Steels (Ref. 2.5) for the anchorage-tendon system.

The above recommendations are derived from relatively high frequency cyclic tests. If cyclic loading is not continuously applied, the fatigue resistance tends to improve with the extent of the rest periods.

7. Corrosion

The magnitude and consequences of corrosion in prestressing steels are much more severe than in reinforcing steels. This is not only because high-grade steel is more susceptible to corrosion but also because the diameter of prestressing tendons is relatively small. Thus, even a small uniform corrosive layer or a corroded spot can substantially reduce the cross-sectional area of the steel, induce stress concentrations, and eventually lead to premature failure. Exposure of unprotected prestressing steels to normal environments, even for short periods of time (a few months) can lead to an appreciable decrease in their tensile properties and a greater reduction in their fatigue resistance. Prestressing steels are generally susceptible to two major types of corrosion: electrochemical corrosion and stress corrosion. In the electrochemical corrosion an aqueous solution must be present, even in the form of a thin film, and air (oxygen) is needed. Stress corrosion is a type of corrosion causing brittleness in the steel under certain conditions of stresses and environments. Sensitivity to corrosion is different for different steels.

Corrosion is very difficult to quantify, thus leaving the designer with his engineering judgment in providing against corrosion. Qualitatively, however, the effect

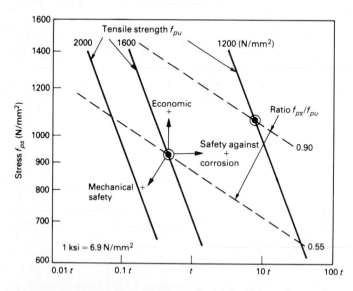

Figure 2.15 Relative lifetime of prestressing steels in a corrosive medium. *(Courtesy TNO Metal Research Institute, Netherlands.)*

of important parameters on the rate of corrosion is best illustrated in a graph. Figure 2.15 (Ref. 2.5) shows the relative lifetime up to fracture of a cold-drawn stress-relieved wire placed in a particular medium causing corrosive cracking as a function of strength and stress-to-strength ratio. It can be seen that the higher the strength and the higher the applied stress, the smaller the lifetime of the wire. Figure 2.15 also illustrates the tradeoffs that must be made in prestressed concrete between safety against corrosion, mechanical safety, and economy. The higher the strength and the higher the prestress, the more economical the structure, but also the more susceptible to corrosion it will be.

To guard against corrosion prestressing steels must be transported and stored in closed waterproof containers and protected from damage and contamination during handling. When bonded tendons are used, as soon as the tensioning operation is completed good quality grout should be injected in order to guard against water pockets and contaminants. A sufficient concrete cover will add to the protection of the tendons from corrosion. The concrete and admixtures used should contain the least amounts of calcium chloride. Unbonded tendons should be protected by an anticorrosive material such as asphalt, grease, oil, or a combination of grease and plastic tubing. The consequences of corrosion can be so serious that all appropriate precautions must be taken to prevent it or to reduce its effects.

2.3 CONCRETE

Extensive technical information on concrete and its use exists (Refs. 2.17 to 2.20). Concrete is a versatile composite material of a very complex nature, yet it can be approached at any desired level of sophistication. The simplest is when only its compressive strength is specified for design. The technical level at which concrete must be approached is higher in prestressed concrete than in reinforced concrete. In designing a prestressed concrete structure, the designer does not only consider the strength of concrete but also its time-dependent properties such as creep and shrinkage. He has to assess their effects on the loss of prestress in the steel and the long-term deformation or deflection of the structure. In practice the corresponding procedure must be reduced to a manageable level.

1. Composition

The main components of concrete are Portland cement, aggregates (fine and coarse), and water. Five types of standard Portland cements are recognized in the United States. They are manufactured in accordance with ASTM Standard C150-74. Type I is a general purpose cement commonly used when the special properties of the other four types are not needed. Type II is for general concrete construction exposed to moderate sulfate attack or where moderate heat of hydration is required. Type III is for use where high early strength is required. It leads to a seven-day strength about equal to that achieved at 28 days by a Type I cement. Type IV cement is for use when low heat of hydration is desired, and type V when high sulfate resistance is needed. The first three types of cement come with an A variety, such as type IIIA, when air entrainment is desired.

To achieve higher strength properties high cement contents, low water-to-cement ratios, and good quality aggregates are used. Normal weight coarse aggregates include granite, basalt, limestone quartzite, and slate. They lead to a normal weight concrete with a density about 150 lb/ft^3 (23.6 kN/m^3). Structural lightweight aggregates include expanded shale clay and slate and lead to lightweight concretes of density between 100 lb/ft^3 and 120 lb/ft^3 (15.7 kN/m^3 to 18.9 kN/m^3). Sands, either normal weight or lightweight, are used as fine aggregates.

Admixtures are often added to fresh concrete to achieve a particular effect. Water-reducing admixtures and superplasticizers allow a substantial reduction in the water-cement ratio for the same workability of the mix, thus leading to increased strengths. They are also used to achieve a desired strength at a lower cement content. Air entraining agents are commonly used in cold climates to improve the freeze-thaw resistance of concrete. In posttensioned construction expanding agents are added to the grout to ensure proper filling of the ducts. Many

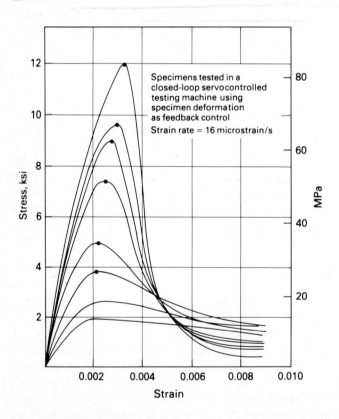

Specimens tested in a closed-loop servocontrolled testing machine using specimen deformation as feedback control

Strain rate = 16 microstrain/s

Figure 2.16 Typical stress-strain curves of normal weight concrete in compression. *(Adapted from Ref. 2.21.)*

other admixtures are available for use in concrete. Care should be taken to avoid admixtures that contain calcium chlorides as they cause corrosion of the reinforcement.

2. Stress-Strain Curve

Typical stress-strain curves of normal weight concretes in uniaxial compression (Ref. 2.21) are shown in Fig. 2.16. They comprise two main portions: an ascending portion up to the peak point (maximum stress) and a descending portion. The strength corresponds to the maximum stress. The ascending portion can itself be separated for the purpose of design into two parts: a quasi-linear portion up to about 40 percent of the strength followed by a nonlinear portion reaching gradually to the peak point. Note that the higher the strength of concrete, the higher the slope at the origin of the ascending portion. Thus, the higher the strength, the higher the elastic modulus. The tangent at the origin of the stress-strain curve gives the initial modulus (Fig. 2.17). A line joining the origin to any point of the curve gives the secant modulus. The value of the secant modulus at 45 percent of the maximum stress is recommended by ACI for use in design. In the nonlinear portion of the stress-strain curve, the total strain is made of an elastic strain and a permanent irrecoverable strain mostly attributed to microcracking. It is generally

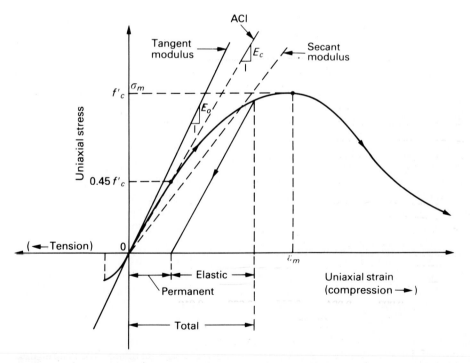

Figure 2.17 Determination of modulus from the stress-strain curve of concrete.

observed that unloading from any point of the curve is achieved as a first approximation along a line parallel to that giving the modulus.

Going back to Fig. 2.16 it can be observed that the higher the concrete strength, the steeper is the descending portion of the stress-strain curve. A steep descending portion is often associated with brittleness or lack of ductility of the material. Concretes of lower strengths show higher material ductilities. However, structural ductility can be obtained even with high-strength concretes as pointed out in Vol. 2 where the nonlinear analysis of prestressed concrete members is addressed. Another interesting characteristic of the descending portion is the presence of an inflection point. After the inflection point the curve tends asymptotically toward the strain axis, suggesting that very large plastic strains can be sustained by concrete.

Structural lightweight concretes have stress-strain curves similar in shape to those of normal weight concretes. However, for identical strengths their elastic modulus is smaller and the strain at the peak point is larger. It is important to realize that the shape of the stress-strain curve of concrete depends not only on the type of aggregate but also on many parameters including the testing technique, the strain rate, and the amount of confinement. More details are given in Vol. 2.

Compressive strengths of up to 10 ksi (69 MPa) have been achieved using structural lightweight concretes and up to 26 ksi (180 MPa) using normal weight concretes. Most common strength ranges are given next.

3. Mechanical Properties

The compressive strength is the most important design property of concrete. Many other properties can be related to the compressive strength. For some applications only the desired compressive strength is specified in the design. In the United States practice the compressive strength is obtained from testing 6 × 12 cylinders at the age of 28 days. When high-early-strength cement (type III) is used, an age of seven days is commonly considered. The compressive strength corresponds to the maximum stress at the peak point of the stress-strain curve. The strength obtained from a cylinder test falls within a range of 70 to 90 percent of the strength obtained from a cube test. The higher the strength, the higher the ratio. It is generally assumed in design that the actual compressive strength is equal to the design specified strength f'_c. In effect, the definition of f'_c is such that a random compression test on the material should lead with a probability of 90 percent to a strength higher than or equal to f'_c.

Most common values of f'_c for prestressed concrete structures in the United States are between 5 and 7 ksi (35 and 48 MPa) whether normal weight or lightweight concrete is used. Higher compressive strengths of 9 and 11 ksi (62 and 76 MPa) are used with normal weight concretes but to a much lesser extent. These strengths are on the average substantially higher than those specified for ordinary reinforced concrete construction.

Other useful mechanical properties of concrete can be related to its compressive strength. They include its tensile strength determined from a uniform tensile

Table 2.8 Common mechanical properties of structural concretes

Property	Concrete type	Observed range†	Common design value U.S. Units†	Common design value SI units‡
Direct tensile strength f'_{tc}	Normal weight	$-3\sqrt{f'_c}$ to $-5\sqrt{f'_c}$	$-3\sqrt{f'_c}$ or $-\frac{1}{3}\sqrt{\gamma_c f'_c}$	$-0.25\sqrt{f'_c}$ or $-0.0069\sqrt{\gamma_c f'_c}$
	Lightweight	$-2\sqrt{f'_c}$ to $-3.5\sqrt{f'_c}$	$-2\sqrt{f'_c}$ or $-\frac{1}{3}\sqrt{\gamma_c f'_c}$	$-0.17\sqrt{f'_c}$ or $-0.0069\sqrt{\gamma_c f'_c}$
Split cylinder tensile strength	Normal weight	$-6\sqrt{f'_c}$ to $-7\sqrt{f'_c}$	$-6\sqrt{f'_c}$ or $-0.6\sqrt{\gamma_c f'_c}$	$-0.50\sqrt{f'_c}$ or $-0.00138\sqrt{\gamma_c f'_c}$
	Lightweight	$-4\sqrt{f'_c}$ to $-5\sqrt{f'_c}$	$-4\sqrt{f'_c}$ or $-0.5\sqrt{\gamma_c f'_c}$	$-0.33\sqrt{f'_c}$ or $-0.00115\sqrt{\gamma_c f'_c}$
Modulus of rupture f_r	Normal weight	$-7.5\sqrt{f'_c}$ to $-12\sqrt{f'_c}$	$-7.5\sqrt{f'_c}$	$-0.62\sqrt{f'_c}$
	Lightweight	$-5\sqrt{f'_c}$ to $-9\sqrt{f'_c}$	$-0.75(7.5\sqrt{f'_c})$, lightweight sand $-0.85(7.5\sqrt{f'_c})$, normal sand	$-0.47\sqrt{f'_c}$, lightweight sand $-0.53\sqrt{f'_c}$, normal sand
Modulus of elasticity E_c	Normal weight	$E_c = (27 \text{ to } 35)\gamma_c^{3/2}\sqrt{f'_c}$ *[33 → normal]*	$33\gamma_c^{3/2}\sqrt{f'_c}$ or $57{,}000\sqrt{f'_c}$	$0.043\,\gamma_c^{1.5}\sqrt{f'_c}$ or $4730\sqrt{f'_c}$
	Lightweight	$(25 \text{ to } 30)\gamma_c^{3/2}\sqrt{f'_c}$ *[$E_{ci} = 33\times\gamma^{3/2}\sqrt{f'_{ci}}$]*	$33\gamma_c^{3/2}\sqrt{f'_c}$	$0.043\,\gamma_c^{1.5}\sqrt{f'_c}$
Poisson's ratio	Normal weight	0.15 to 0.20	0.20	0.20
	Lightweight	0.15 to 0.20	0.20	0.20

† Valid only for f'_c in pounds per square inch and γ_c in pounds per cubic foot. Tension negative, compression positive.

‡ f'_c in MPa and γ_c in kilograms per cubic meter.

test, its tensile strength determined from the split cylinder test, its modulus of rupture or tensile strength determined from a flexural test, and its elastic modulus. The usual ranges of experimentally observed values of these properties in function of the compressive strength are summarized in Table 2.8 for normal weight and structural lightweight concrete. The lower limit is generally adopted in design.

Although the compressive strength of concrete at 28 days is used as a reference, the actual compressive strength varies with time and depends on many other variables, including the type of cement and the curing history; the same is true of the other mechanical properties which are related to the compressive strength. Typical variations with time of the compressive strengths of identical concretes made with different types of cements are shown in Fig. 2.18 (Ref. 2.22). It is estimated that the strength at the end of a life span can be 20 to 30 percent higher than that at 28 days. The variation of early age strength is important in prestressed concrete structure as the prestress is often applied while the concrete is still young.

Several properties of concrete other than the compressive strength are also time dependent. They include shrinkage and creep. The knowledge of these properties will allow the prediction with time of prestress losses in the steel, stresses and strains in the section, and long-term deformations or deflections of the structure. Shrinkage and creep are covered in the next sections. Their variation with time is treated much in the same way as the compressive strength.

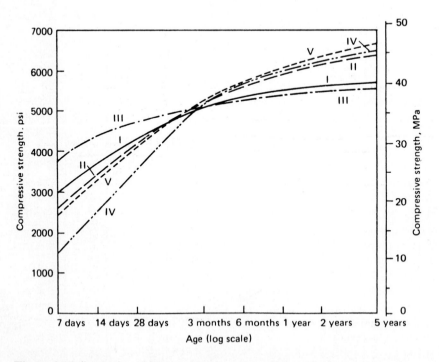

Figure 2.18 Typical variations of concrete strength with time *(Ref. 2.22).*

In general, the property of interest (strength, shrinkage, or creep) is related to a reference value by a time function. For instance, the strength of concrete at any time t can be put in the following form:

$$f'_c(t) = g(t)f'_c(28) \tag{2.5}$$

where $g(t)$ is a time function and $f'_c(28)$ is the reference strength. Different types of time functions such as logarithmic or exponential functions have been used. Some are given in Chap. 8 on prestress losses. Only one type of function, a fractional function initially proposed by Ross (Ref. 2.23) and recommended by ACI Committee 209 (Ref. 2.24) will be described in this chapter. It was further extended by Branson and Kripanarayanan (Ref. 2.25) and leads itself to sufficient flexibility in modeling actual behavior. The general form of this time function as used in the prediction of strength, shrinkage, and creep is:

$$g(t) = \frac{t^a}{b + ct^a} \tag{2.6}$$

where t is the time in days and a, b, and c are constants; a and c can take on the value of one. Recommended numerical values of these constants for the strength of concrete, shrinkage strain, and creep coefficient are given in Table 2.9 for moist-cured and steam-cured concrete. They can be used in absence of more specific data.

4. Shrinkage

Concrete contains more water than is strictly required by the chemical hydration reaction of the cement. The excess water is called free water. Its loss through evaporation leads to gradual shortening of the member with time, described as shrinkage. As the member shortens, the prestressing steel loses part of its prestress. This is called shrinkage loss. The evaluation of shrinkage loss as part of the total prestress losses is an important step in design.

Shrinkage depends on many variables. Most important are the amount of free water, the relative humidity of the environment, the ambient temperature, the type of aggregates used, and the size and shape of the structural member. Shrinkage is assumed independent of loading and would not take place if the concrete is kept at 100 percent relative humidity. Shrinkage in concrete results primarily from the shrinkage of the cement paste; the aggregates shrink very little. Thus, differential internal stresses are created in the structure of concrete leading to compression on the aggregates and tension in the paste accompanied by microcracking.

Mechanically, shrinkage is described by a shortening strain called shrinkage strain. The shrinkage strain of concrete $\varepsilon_s(t)$ under constant environmental conditions increases with time and tends asymptotically toward a final maximum value called ultimate shrinkage strain ε_{SU} (Fig. 2.19). For all practical purposes the ultimate shrinkage strain is assumed to occur at the estimated life of the structure. The rate of increase of $\varepsilon_s(t)$ is highest at early ages. Practically, for moist-cured concrete, about 50 percent of total shrinkage occurs within a month and about 90

Table 2.9 Recommended relationships for some time-dependent properties of concrete

Property	Relationship†	Values of constants		
		Moist-cured concrete	Steam-cured concrete	
Compressive strength	For $t \geq 1$ day $$f'_c(t) = \frac{t}{b + ct} f'_c(28)$$ Same for normal and lightweight concrete	Type I cement $b = 4$ $c = 0.85$ Type III cement $b = 2.30$ $c = 0.92$	Type I cement $b = 1$ $c = 0.95$ Type III cement $b = 0.70$ $c = 0.98$	
Shrinkage strain	$$\varepsilon_S(t) = \frac{t}{b + t} \varepsilon_{SU} K_{SH} K_{SS}$$ Same for normal and lightweight concretes using type I or type III cements K_{SH} humidity correction factor K_{SS} shape and size factor	$40\% \leq H \leq 80\%$ $b = 35$ $t \geq 7$ days $K_{SH} = 1.40 - 0.01H$ $K_{SS} =$ see Table 2.10‡ $80\% \leq H \leq 100\%$ $b = 35$ $t \geq 7$ days $K_{SH} = 3 - 0.03H$ $K_{SS} =$ see Table 2.10‡	$40\% \leq H \leq 80\%$ $b = 55$ $t \geq 1$ to 3 days $K_{SH} = 1.40 - 0.01H$ $K_{SS} =$ see Table 2.10‡ $80\% \leq H \leq 100\%$ $b = 55$ $t \geq 1$ to 3 days $K_{SH} = 3 - 0.03H$ $K_{SS} =$ see Table 2.10‡	
Creep coefficient	$$C_c(t) = \frac{t^{0.60}}{10 + t^{0.60}} C_{CU} K_{CH} K_{CA} K_{CS}$$ K_{CH} humidity correction factor K_{CA} age at loading factor K_{CS} shape and size factor t_A age at loading	$t, t_A \geq 7$ days and $H \geq 40\%$ $K_{CA} = 1.25\, t_A^{-0.118}$ $K_{CH} = 1.27 - 0.0067H$ $K_{CS} =$ see Table 2.10‡	$t, t_A \geq 1$ to 3 days and $H \geq 40\%$ $K_{CA} = 1.13\, t_A^{-0.095}$ $K_{CH} = 1.27 - 0.0067H$ $K_{CS} =$ see Table 2.10‡	

† Valid in all systems of units; H given in percent.

‡ In lieu of Table 2.10, the author recommends the use of $K_{SS} = K_{CS} = 1.14 - 0.09(V/S)$ where (V/S) is the volume-to-surface ratio of the member.

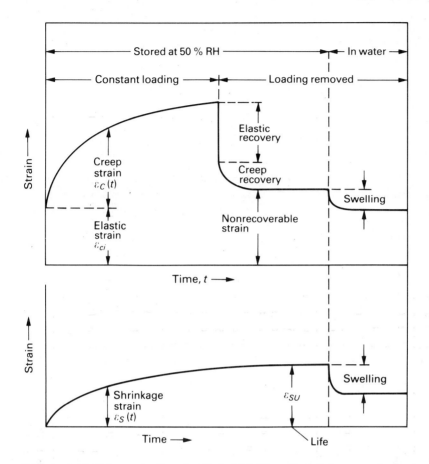

Figure 2.19 Schematical variation with time of shrinkage and creep strain of concrete.

percent within a year of exposure. Note (Fig. 2.19) that once shrinkage has occurred complete recovery will not take place, even if the member is placed again in water.

The shrinkage strain $\varepsilon_S(t)$ at any time t is generally related to the ultimate shrinkage strain ε_{SU} by a time function. The time function recommended by ACI Committee 209 (Ref. 2.24) is of the same form as that described earlier for predicting the strengths. It is given in Table 2.9 where conditions of application are clarified and where other correction factors have been included in the overall relationship between $\varepsilon_S(t)$ and ε_{SU}. The factor K_{SH} is a correction factor which accounts for the influence of the relative humidity on shrinkage strain. K_{SS} is a correction factor which depends on the size and shape of the member, often described by the ratio of surface to volume. K_{SS} can be omitted in the calculations if the least dimension of the member is less than 6 in (15 cm). However, the PCI

Table 2.10 Size and shape factors for creep and shrinkage (*Ref. 2.26*)

Volume-to-surface ratio		Size and shape factor	
in	cm	Creep	Shrinkage
1	2.54	1.05	1.04
2	5.1	0.96	0.96
3	7.6	0.87	0.86
4	10.2	0.77	0.77
5	12.7	0.68	0.69
6	15.2	0.68	0.60

committee charged to study prestress losses (Ref. 2.26) recommends the use of the shape and size factors and suggests the numerical values summarized in Table 2.10. Note that expressions for K_{SS} have also been recommended in Ref. 2.24 for various sizes of members. It is felt, however, that the PCI committee approach is more direct and simpler to apply in all cases. The author believes that further simplification can be achieved by using the same size and shape factor for creep and shrinkage. He suggests the expression given in the footnote of Table 2.9. As great uncertainty is associated with the predictions of shrinkage and creep, such approximation is certainly acceptable.

The ultimate shrinkage strain ε_{SU} of concretes used for prestressing varies generally between 0 and 10×10^{-4} in/in. (Strain is nondimensional and is the same in all systems.) It is zero for shrinkage compensating concrete and approaches 10×10^{-4} for lightweight concrete. Average values of 8×10^{-4} and 7.3×10^{-4} are suggested in Ref. 2.25 for moist-cured and steam-cured concretes, respectively, assuming standard conditions at 40 percent relative humidity. These are about twice those derived from the recommendations of the British code CP-110. It is the author's opinion that average values of ε_{SU} of 6×10^{-4} and 4×10^{-4}, respectively, can be used under standard conditions for moist-cured and steam-cured normal weight concretes commonly used for prestressing. Alternatively, ε_{SU} can be evaluated from several prediction equations found in the technical literature. These equations are generally expressed as a simple function of one variable such as relative humidity, total amount of mixing water, or modulus of elasticity (see Sec. 8.7).

5. Creep

Creep is the time-dependent strain in excess of elastic strain induced in a material subjected to a sustained stress. Contrary to shrinkage, creep is caused by loading.

Typical increase under constant load of creep strain $\varepsilon_C(t)$ with time is shown in Fig. 2.19. If the load is removed after a period of time, the initial elastic strain is recovered immediately. A part of the creep strain is gradually recovered with time and the remaining part is irrecoverable. If the load is maintained, the creep strain tends asymptotically toward a maximum value ε_{CU} called ultimate creep strain. It is considered attained at the end of the life of the structure. A simple rheological model to simulate creep behavior is shown in Fig. 2.20. A large body of information exists on the creep behavior of concrete and its modeling (Refs. 2.27–2.30). In practice, manageable procedures have been devised to account for creep. This is particularly relevant to prestressed concrete members, as they "creep" under the effect of prestress, inducing significant stress losses in the steel and long-term deflections of the member.

The creep of concrete depends on many other factors than time. They include the relative humidity, the type of aggregate, the mix proportions, the age of the concrete at the time of loading, and the size and shape of the member.

The creep strain of a given concrete is considered proportional to the initially applied stress. At any time t, the creep strain $\varepsilon_C(t)$ can be related to the initial elastic strain by a factor called creep coefficient, thus:

$$C_C(t) = \frac{\varepsilon_C(t)}{\varepsilon_{ci}} \qquad (2.7)$$

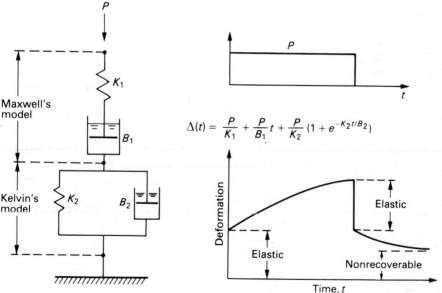

$$\Delta(t) = \frac{P}{K_1} + \frac{P}{B_1}t + \frac{P}{K_2}(1 + e^{-K_2 t/B_2})$$

Figure 2.20 Burger's rheological model of concrete.

As with the life span of the structure $\varepsilon_C(t)$ tends toward ε_{CU} and, as ε_{ci} is a constant, $C_C(t)$ tends toward a maximum value C_{CU} called ultimate creep coefficient and given by:

$$C_{CU} = \frac{\varepsilon_{CU}}{\varepsilon_{ci}} \tag{2.8}$$

Eliminating ε_{ci} from Eqs. (2.7) and (2.8) leads to:

$$\frac{C_C(t)}{C_{CU}} = \frac{\varepsilon_C(t)}{\varepsilon_{CU}} \tag{2.9}$$

Thus, the same time function can represent the variation of the creep strain or the creep coefficient. Similarly to the time function used for strength and shrinkage, the time function recommended for creep in Ref. 2.24 is reproduced in Table 2.9 where conditions of application are clarified. The relationship shown relates the creep coefficient at any time t to the ultimate creep coefficient. Several correction factors have been added to facilitate its direct use in design. The factor K_{CA}, called creep maturity coefficient, takes into account the age of concrete t_A at the time of loading. K_{CH} accounts for the relative humidity and K_{CS} for the shape and size of the member. The values of K_{CS} suggested in Ref. 2.26 are given in Table 2.10, while the effect of age at loading is illustrated in Fig. 2.21 (Ref. 2.27). Note that the equations initially proposed by Branson and Kripanarayanan were developed assuming a

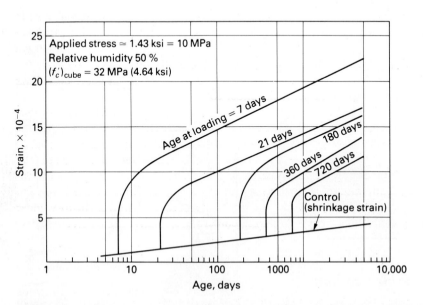

Figure 2.21 Effect of age at loading on creep of concrete. (*Adapted from Ref. 2.27.*)

Table 2.11 Typical values of ultimate creep coefficients (*adapted from Ref. 2.1*)

Compressive strength		Ultimate creep coefficient†
psi	MPa	C_{CU}
3000	20.7	3.1
4000	27.6	2.9
5000	34.5	2.65
6000	41.4	2.4
7000	48.3	2.2
8000	55.2	2.0

† Interpolation acceptable for inbetween values.

concrete slump less than 4 in (10 cm) and a minimum member thickness of 6 in (15 cm) or less. An average value of the ultimate creep coefficient C_{CU} of 2.35 was also proposed for standard conditions and when specific data are not available. Other values of C_{CU} suggested in Ref. 2.1 are given in Table 2.11 in function of the compressive strength.

As pointed out earlier, the creep of concrete leads not only to losses of stress in the prestressing steel but also to long-term deformations or deflection of the structure. This is illustrated in Fig. 2.22 (Ref. 2.28) where the deflection of prestressed and partially prestressed concrete beams are plotted versus time. Note that the deflection curve for a constant load is very similar in shape to the curve describing the variation of creep strain with time under constant conditions.

In describing the concrete strain variations with time due to creep and shrinkage, it was essentially assumed that constant environmental conditions such as in a laboratory environment existed. However, actual conditions where continuous variations in temperature and humidity prevail will lead to fluctuating curves with much less defined limiting values. This is particularly the case, for instance, for segmentally built bridges where each segment has a different age, is loaded at different times and at different loadings, and undergoes its own strain history under the combined effects of creep, shrinkage, temperature, relative humidity, and external loads.

6. Fatigue

It is generally observed that concrete in direct compression can sustain for about ten million cycles, a fluctuating stress between 0 and 50 percent of its static compressive strength. The fatigue strength of concrete for other types of stresses is

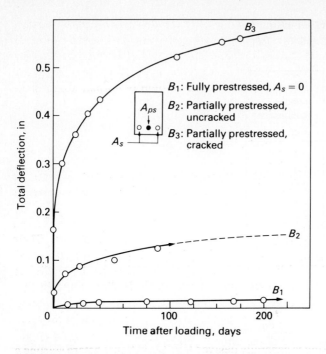

Figure 2.22 Typical variation of deflection with time under sustained loading. *(Adapted from Ref. 2.31.)*

estimated at 55 percent of the corresponding static strength. Such stress fluctuations are not often encountered in practice. Thus, failure of concrete by fatigue is not very common. For the purpose of design, the stress range limit suggested by ACI Committee 215 on Fatigue (Ref. 2.16) and in Ref. 2.29 can be used:

$$f_{cr} = 0.4\, f'_c - \frac{f_{min}}{2} \tag{2.10}$$

where f'_c is the compressive strength and f_{min} the minimum stress.

7. Effects of Temperature

Concrete heated to up to 200°C does not undergo significant physical or chemical changes. It generally loses all its free water at 100°C (212°F) and, except for some microcracking, maintains most of its mechanical characteristics. The strength properties of concrete generally increase with a substantial drop in temperature. This improvement depends, however, on the water content of the concrete and is much higher for moist concrete than for dry concrete. Compressive strength increases of more than 200 percent have been reported for moist-cured concretes tested at −150°C.

The coefficient of thermal expansion of concrete depends on many factors, including the mix proportions and the type of aggregate. An average value of 5.5×10^{-6} per degree Fahrenheit (12×10^{-6} per degree Celsius) can be used for normal weight concrete. For lightweight concrete a range of 4 to 6×10^{-6} per degree Fahrenheit is observed and depends on the amount of normal weight sand used.

8. Steam Curing

Curing involves the retention of sufficient free water in the concrete to facilitate the process of hydration. At ambient temperature a moist environment is generally maintained to ensure curing.

Mass produced precast prestressed concrete members must be removed from their molds as soon as possible to ensure an efficient production process. Their strength must be sufficient (0.7 to $0.8 f'_c$) at that time to allow the transfer of the prestress. This can be achieved by steam curing. Steam curing can be applied at low (atmospheric), high, or intermediate low-high pressures. It allows a production cycle of 24 hours. Low-pressure steam curing is most popular in the United States. A typical cycle of temperature within the steam enclosure versus time is shown in Fig. 2.23. The temperature of the concrete member follows with a time delay and smoother transitions. Curing is generally delayed four to six hours before starting a

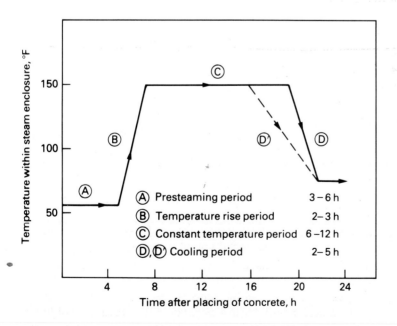

Figure 2.23 Typical low-pressure steam–curing cycle.

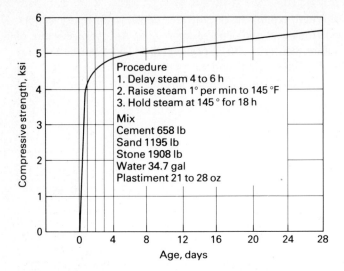

Figure 2.24 Typical time-strength curve for steam-cured concrete. *(Courtesy Sika Chemical Corporation.)*

slow temperature rise (40 to 60°F per hour). The length of the maximum temperature period and the subsequent rates of temperature drops may be selected in many ways. Commonly, the total time between concrete casting and steam shutoff is about 18 hours. Typical variation of strength of steam-cured concrete versus concrete age is shown in Fig. 2.24 and illustrates the values achievable at one day. Steam curing leads not only to improved strength but also to better shrinkage and creep characteristics of the concrete as noted in Table 2.9.

2.4 CONCLUDING REMARKS

The materials properties summarized in this chapter are essential elements to the full understanding of prestressed concrete behavior. They help clarify the basis for many of the design procedures treated in the coming chapters. Not all of these properties are of immediate use in design. However, the more final the design and the more complex the structure, the bigger is the need to account for actual behavior in the design.

Two additional items could have been treated in this chapter: prestress losses and allowable stresses. Prestress losses depend not only on material properties but also on structural behavior. Their immediate and thorough treatment would confuse the reader newly acquainted with prestress concrete. It is left to Chap. 8. Allowable stresses are related to the strength properties of the materials. They are not properties by themselves: they are arbitrary design limits and are given in the next chapter where different design philosophies are explained.

REFERENCES

2.1. G. Winter and A. H. Nilson: *Design of Concrete Structures*, 9th ed., McGraw-Hill Book Company, New York, 1979.

2.2 Steel Feinforcement Properties and Availability, Report of ACI Committee 439, *ACI Journal*, vol. 74, no. 10, October 1977, p. 481.

2.3. Concrete Reinforcing Steel Institute: *Manual of Standard Practice*, 20th ed., Chicago, Illinois, 1980.

2.4. T. Helgason, J. M. Hanson, N. J. Somes, W. G. Corley, and E. Hognestad: "Fatigue Strength of High Yield Reinforcing Bars," PCA Report to the Highway Research Board, NCHRP, March 1975, 371 pp.

2.5. "Report on Prestressing Steels: Types and Properties," FIP Commission on Prestressing Steels and Systems, FIP, Wexham Spring, England, August 1976, 18 pp.

2.6. W. Podolny Jr. and T. Melville: "Understanding the Relaxation in Prestressing," *PCI Journal*, vol. 14, no. 4, August 1969, pp. 43–54.

2.7. D. D. Magura, M. A. Sozen, and C. P. Siess: "A Study of Stress Relaxation in Prestressing Reinforcement," *PCI Journal*, vol. 9, no. 2, April 1964, pp. 13–57.

2.8. A. Doi, H. Tomoika, Y. Tanaka, and M. Duraudi: "Physical Properties of Prestressing Steel at Low and Elevated Temperatures," Report No. DRG75-04, Shinko Wire Co. Ltd., Amagasaki, Japan.

2.9. A State of the Art Report on Materials of Prestressing and Reinforcement for Use Under Cryogenic Conditions, FIP Commission on Prestressing Steels and Systems, FIP Notes, May/June 1979, pp. 4–11.

2.10. R. F. Warner and C. L. Hulsbos: "Fatigue Properties of Prestressing Strands," *PCI Journal*, vol. 11, no. 1, February 1966, pp. 32–52.

2.11. R. H. R. Tide and D. A. VanHorn: "A Statistical Study of the Static and Fatigue Properties of High Strength Prestressing Strand," Fritz Engineering Laboratory Report no. 309.2, Lehigh University, June 1966.

2.12. K. M. Price and A. D. Edwards: "Fatigue Strength of Prestressed Concrete Flexural Members," *Proceedings, Institute of Civil Engineers*, London, vol. 47, October 1970, pp. 205–226.

2.13. A. D. Edwards and A. Picard: "Fatigue Characteristics of Prestressing Strands," *Proceedings, Institute of Civil Engineers*, London, vol. 53, September 1972, pp. 323–336.

2.14. A. E. Naaman: "Fatigue of Partially Prestressed Beams" preprint from a symposium on "Fatigue in Concrete Structures," American Concrete Institute Annual Convention, Puerto Rico, Sept. 1980, 23 pp. To be published, American Concrete Institute, Detroit, 1982.

2.15. C. G. Ekberg, R. E. Walther, and R. G. Slutter: "Fatigue Resistance of Prestressed Concrete Beams in Bending," *Journal of the Structural Division, ASCE*, July 1957, Paper no. 1304, pp. 1–17.

2.16. "Consideration for Design of Concrete Structures Subjected to Fatigue Loading," Report by ACI Committee 215, *ACI Journal*, vol. 71, no. 3, March 1974, pp. 97–121.

2.17. A. M. Neville: *Properties of Concrete*, 2d ed., John Wiley & Sons, New York, 1973.

2.18. G. E. Troxell, H. E. Davis, and J. W. Kelly: Composition and Properties of Concrete, 2d ed., McGraw-Hill Book Company, New York, 1968.

2.19. *ACI Manual of Concrete Practice*, Parts I, II, and III, American Concrete Institute, Detroit, 1979.

2.20. S. Popovics: *Concrete Making Materials*, McGraw-Hill Book Company, New York, 1979.

2.21. S. Ahmad: "Properties of Confined Concrete Under Static and Dynamic Loads," Ph.D. thesis submitted to the University of Illinois at Chicago Circle, Department of Materials Engineering, Chicago, March 1981.

2.22. U.S. Bureau of Reclamation: *Concrete Manual*, 8th ed., U.S. Government Printing Office, Denver, Colorado, 1975.

2.23. A. D. Ross: "Creep and Shrinkage of Plain, Reinforced, and Prestressed Concrete: A General Method of Calculation," *Journal Institution of Civil Engineers*, London, November 1943, pp. 38–57.

2.24. "Prediction of Creep, Shrinkage, and Temperature Effects in Concrete Structures," Report by ACI Committee 209, *ACI Special Publication SP-27*, Designing for Effects of Creep, Shrinkage, Temperature in Concrete Structures, 1971, pp. 51–93.

2.25. D. E. Branson and K. M. Kripanarayanan: "Loss of Prestress, Camber, and Deflection of Non-Composite and Composite Prestressed Concrete Structures," *PCI Journal*, vol. 16, no. 5, September/October 1971, pp. 22–52.

2.26. "Recommendations for Estimating Prestress Losses," Report by PCI Committee on Prestress Losses, *PCI Journal*, vol. 20, no. 4, July/August 1975, pp. 108–126.

2.27. R. L'Hermite: *Les Déformations du Béton* (in French), ed. Eyrolles, Paris, 1961.

2.28. A. M. Neville: *Creep of Concrete : Plain, Reinforced, and Prestressed*, North Holland Publishing Co., Amsterdam, 1970.

2.29. "Volume Changes in Precast Prestressed Concrete Structures," PCI Committee on Design Handbook, *PCI Journal*, vol. 22, no. 5, September/October 1977, pp. 38–53.

2.30. Z. P. Bazant and L. Panula: "Creep and Shrinkage Characterization for Analyzing Prestressed Concrete Structures," *PCI Journal*, vol. 25, no. 3, May/June 1980, pp. 86–122.

2.31. V. Jittawait and M. K. Tadros: "Deflection of Partially Prestressed Concrete Members," Final Report, PCI Fellowship, Department of Civil Engineering, West Virginia University, 1979. (Also available as M.S. thesis of first author.)

2.32. W. G. Corley, J. M. Hanson, and T. Helgason: "Design of Reinforced Concrete for Fatigue," *Journal of the Structural Division, ASCE*, vol. 104, no. ST6, June 1978.

PROBLEMS

2.1 (a) Using Eq. (2.3) plot on a semilogarithmic graph the percent loss due to relaxation (that is $\Delta f_{pR}(t)/f_{pi}$) versus time in hours for values of $f_{pi}/f_{py} = 0.6, 0.7, 0.8,$ and 0.9, respectively, assuming f_{pi} remains constant with time.

(b) Compute the percent loss due to relaxation at the following times given in hours: $24, 10^2, 10^3, 10^4, 10^5, 10^6$, assuming $f_{pi}/f_{py} = 0.80$.

2.2 Using the prediction relationship given in Table 2.9 plot two curves showing the strength of moist-cured and steam-cured concrete (in percent of their 28 days strength) versus time up to $t = 90$ days.

2.3 Using the information given in Table 2.9 and assuming $K_{CS} = 1$ and a standard value of the ultimate creep coefficient $C_{CU} = 3$, compute the ultimate creep coefficient for a moist-cured concrete member for the following conditions: $H = 40, 60,$ and 80 percent and $t_A = 7, 28,$ and 90 days, respectively.

ANALYSIS OR INVESTIGATION VERSUS DESIGN

idamental difference exists between analysis and design. In dealing with civil
eering structures, design implies an unknown product, at least in part, while
sis implies investigating or reviewing a finished or proposed product. The
sis process, also called investigation or review process, is concerned with
sing the response of the structure to the application of loadings. It deals in
with the determination of stresses and stress or force resultants and in part
checking if the structure satisfies acceptable design criteria. This is often done
mparing actual findings with corresponding limits and ranges recommended
vailing codes. To analyze is also to compare with what engineering judgment
e state of the art considers acceptable. As it generally involves no unknown,
nalysis process is easier than the design process.

n civil engineering structures, design involves the selection among a large
of possibilities of many particulars, such as the structural layout, the shape of
mber, the structural material, and even the construction process. Within each
the design deals with the actual versus the ideal and at different levels of
ls. Although design does not necessarily imply finding the optimum solution,
tainly aims at being within an appropriate range of the optimum. Because of
herent nature of dealing with unknowns and because infinite combinations of
bilities exist, design is mostly an iterative process. An efficient design is one in
1 the number of iterations is reduced to a minimum. This often depends on the
ience and skills of the designer. A distinction is made between preliminary
n or designs, in which many alternatives can be explored quickly (using
eering judgment, rule of thumb, etc.) and final design which is a more refined
ion, ready for implementation.

More on some detailed steps of what "analysis versus design" implies in
ng with prestressed concrete beams can be found in Chap. 4. Similar compari-
could apply to other structural elements and systems.

DESIGN OBJECTIVES

der to fulfill its purpose, a structure must satisfy a number of design objectives.
basic are feasibility, safety, serviceability, economy, functionality, and aes-
s (Ref. 3.2).

Given available materials and technologies, a structure must be first of all
le. At present, for instance, it is not possible to build a simply supported
ressed concrete bridge with a main span of more than about 500 meters.
bility and economy often go together. To be built, a proposed design must be
nably economical in comparison to other potential alternatives. Assuming
bove objectives are within range, a structure must be safe, that is, it should not
ose under loads foreseeable during its service life. It should also be serviceable,
is, it should perform properly under load and render the service it was built
prestressed beam may be perfectly safe, yet unserviceable if it undergoes very

2.4 Using the relationship given in Table 2.9, plot a curve showing the creep coefficient versus time up to $t = 10,000$ days for a precast rectangular concrete column with cross section of 15×15 in (38×38 cm). The following information is given: $C_{CU} = 3$, $H = 60$ percent, steam-cured concrete, $t_A = 1$ day.

2.5 Compute for the purpose of design the predicted direct tensile strength, modulus of rupture, and modulus of elasticity for normal weight and all lightweight concretes of compressive strength $f'_c = 5000$ psi, 6000 psi, and 7000 psi, respectively. Assume $\gamma_c = 150$ lb/ft^3 for normal weight concrete and 105 lb/ft^3 for lightweight concrete.

Montreal Olympic Stadium where C-shaped posttensioned rigid frames made of segmentally precast elements were used. *(Courtesy Canadian Prestressed Concrete Institute.)*

THE PHILOSOPHY

3.1 WHAT IS DESIGN?

In its most general definition "design is regarded as the pr applying the total spectrum of science and technology to the a result which serves a valuable purpose" (Hill, Ref. 3.1). To de put together something new, to rearrange things in a new way a latest state of the art. Design is an intersection of both art and creativity combined with skills and knowledge.

Many professions have designers, including engineers of c The design process spans a wide range of responsibilities from very detailed level. For instance, the term design applies to desig system (transportation design engineer), designing a bridge gineer), and designing the beam of a bridge or the reinforcemer tural designer).

Good design requires the ability of both analysis and synt should be capable of thinking in relationships and correlations. his product he should be able to see the detail and the whole, the core, the immediate and the ultimate; he should understand i tions and external interfaces. Nature gives us the best example Many phenomena are a matter of design and infinite successio design. In a way we are by design and design within design.

large cambers. Similarly a water tank may be safe but unserviceable if it cracks and leaks substantially.

The safety and serviceability of structures are generally achieved by satisfying a number of code limitations or criteria. To ensure safety, several design approaches are available and accepted by various codes. They are treated in more detail in the next section. Serviceability generally includes aspects of camber and deflections, fatigue, corrosion, cracking, and fire resistance. Serviceability criteria and how to account for them in the design of prestressed concrete structures are addressed in Chap. 7 and Vol. 2.

3.4 DESIGN APPROACHES

In designing for safety, several design approaches can be followed. These approaches are generally based on theory, supported by experimental evidence. Currently encountered design approaches include: working stress design (WSD), ultimate strength design (USD), limit or plastic design, limit state design, nonlinear design, and probabilistic design. Although a single approach is generally sufficient, current practice in prestressed concrete involves the combination of working stress design (WSD) and ultimate strength design (USD).

Except for probabilistic design, design approaches begin with the choice of specified loads to design for. The nature and magnitude of these loads depend on the type of structure. Some are described in Sec. 3.6.

1. Working Stress Design

In this approach the stresses under working loads are limited to permissible values or allowable stresses and the structure is analyzed assuming linear elastic behavior. Safety is ensured by selecting allowable stresses as relatively small fractions of the characteristic strengths of the component materials. Allowable stresses are specified in various codes and may vary from one code to another. Typical values are described in Sec. 3.7. For instance, the maximum permissible compressive stress on concrete flexural members may be taken as $0.45 f'_c$. This implies a safety factor of $(1/0.45) = 2.22$ against concrete failure. Note that in the working stress design all types of loads are treated the same, no matter how different their variability is. The design of prestressed concrete beams by the working stress design approach is covered in detail in Chap. 4.

2. Ultimate Strength Design

In this approach, the design working loads are multiplied by load factors and the structure is designed to resist at its ultimate capacity the factored loads. The load factors associated with a type of loading are adjusted to reflect the degree of variability and uncertainty of that loading. This is more realistic than in the WSD approach where all loads are treated the same.

3. Plastic Design or Limit Design

This is a design based on the formation of plastic hinges, or yielding mechanisms within a structure under loading. It has been mostly applied to statically indeterminate steel structures where it is observed that collapse cannot occur due to a single section undergoing yielding. A large reserve of strength generally exists between first yielding and general collapse. Limit design is recognized in the ACI code but to a very limited extent: Secs. 8.4 and 18.10 of the code allow redistribution of negative moments in reinforced and prestressed concrete continuous members (Sec. 10.14). In effect, limit design for concrete structures is more often synonymous of nonlinear design than plastic design.

4. Limit State Design, Nonlinear Design, Probabilistic Design

These approaches are put under the same heading as they are very much correlated. In the limit state design, a structure may become unsafe or unserviceable when one of the specified limit states is reached. Limit states include the ultimate strength limit state and serviceability limit states such as fatigue, corrosion, cracking, excessive deflection, and the like. The object of the design is to ensure that there is an accepted probability for the structure not to reach any limit state. Limit state design combines some of the merits of ultimate strength design, working stress design, deterministic methods, and reliability theory. In probabilistic design, the applied load and the inherent resistance are assumed random variables. The margin of safety itself, which is defined as the difference between resistance and load, is a random variable. Its mean value is associated with the probability of failure where failure means not only collapse but also unserviceability or violation of any other specified limit state. More on both the probabilistic design of prestressed concrete and nonlinear design can be found in Vol. 2.

3.5 DESIGN CODES

Although basic engineering concepts and judgment can be used to design a structure, most of the guesswork can be reduced and better efficiency achieved if structural requirements set by design codes are satisfied. Design codes (such as Refs. 3.3 to 3.9) are written to protect the user and society as a whole. They provide information on methods of analysis and design, minimum design requirements and minimum expected performance. They represent a summary of the collective opinion or agreed upon state of knowledge of the profession.

In the United States most reinforced and prestressed concrete structures (except bridges) are designed in accordance with the Building Code Requirements for Reinforced Concrete (ACI 318-77) published by the American Concrete Institute (Refs. 3.3 and 3.4). It is regarded as an authoritative statement of current good practice in the field of concrete structures. The ACI code is incorporated entirely or in part in many municipal and regional codes in the United States and is used as a reference in many foreign countries.

Most prestressed concrete bridges for highways or railways in the United States are designed in accordance with two major codes: the AASHTO (American Association of State Highway and Transportation Officials) Standard Specifications for Highway Bridges (Ref. 3.5) and the AREA (American Railway Engineering Association) Manual for Railway Engineering (Ref. 3.6). Except for some subtle differences, the sections of these codes related to prestressed concrete are essentially identical in scope with the corresponding sections of the ACI code.

In this text, reference to these codes will often be made and preferably their latest editions should be used. These codes contain specific information on analysis and design methods, as well as service loads, load factors, and allowable stresses. These last three items will be discussed next.

3.6 LOADS

For the purpose of design, loads are classified into two categories: dead loads and live loads.

Dead loads include primarily the self-weight of the structure and any permanent component such as tiles, false ceiling, and partitions. They are assumed to remain constant during the life of the structure. Self-weight is estimated from the dimensions of the element and the unit weight of the material. Dead loads due to partitions and the like are generally approximated by an equivalent uniform load applied to the surface area of the structure. Typical values of unit weights of some common materials are given in Table 3.1. In general, one can assume a unit weight of 150 lb/ft³ (23.6 kN/m³) for normal weight concrete (including the steel) and between 100 and 120 lb/ft³ (15.7 to 18.9 kN/m³) for structural lightweight concrete.

Contrary to dead loads, live loads are variable in nature and fluctuate with time. They include (1) occupancy loads caused by people, furnishings, or movable objects, (2) vehicle loads such as trucks or trains, (3) snow, rain, water, ice, wind, earth pressure, temperature loads and the like. Occupancy loads are generally prescribed in various codes as uniformly distributed loads. The ACI code does not prescribe occupancy loads. However, the American National Standards Institute (ANSI), as well as local and regional codes, prescribe such values. Typical selected values of occupancy loads recommended by the ANSI code (Ref. 3.7) are reproduced in Table 3.2. A more extensive list can be found in Ref. 3.7 as well as particular details for other live loads. Wind loads of 15 to 30 lb/ft² ($\simeq 0.7$ to 1.4 kN/m²) and snow loads of 10 to 40 lb/ft² ($\simeq 0.5$ to 2 kN/m²) are common design ranges.

Vehicle loadings are prescribed by the AASHTO and AREA specifications respectively, for highway and railway bridges. The live loads recommended by AASHTO belong to two groups: moving vehicle loads and equivalent lane loading simulated by a combination of prescribed uniform and concentrated loads. Both

Table 3.1 Weights and specific gravity of various materials (*Adapted from AISC Manual, Ref. 3.9*)

Substance	Density† lb/ft^3 (average)	Specific gravity (range)
Building materials		
Asphaltum	81	1.1–1.5
Brick (common)	120	1.8–2.0
Cement, Portland, loose	90	....
Cement, Portland, set	183	2.7–3.2
Earth, dry, loose	76	....
Earth, dry, packed	95	....
Earth, moist, loose	78	....
Earth, moist, packed	96	....
Glass (common)	156	2.40–2.60
Lime, gypsum, loose	53–64	....
Lime, mortar, set	103	1.4–1.9
Sand, gravel, dry, loose	90–105	....
Sand, gravel, dry, packed	100–120	....
Sand, gravel, wet	118–120	....
Timber, pine (various types, seasoned, 15 to 20% moisture by weight)	30–44	0.48–0.70
Timber, oak (various types, seasoned, 15 to 20% moisture by weight)	41–59	0.65–0.95
Liquids		
Alcohol, 100%	49	0.79
Oils	58	0.90–0.94
Water, 4°C	62.43	1.0
Water, ice	56	0.88–0.92
Water, snow, fresh fallen	8	0.125
Water, seawater	64	1.02–1.03
Metals		
Aluminum, cast, hammered	165	2.55–2.75
Copper, cast, rolled	556	9.8–9.0
Iron, cast, pig	450	7.2
Lead	710	11.37
Steel, rolled	490	7.85
Minerals		
Asbestos	153	2.1–2.8
Granite, syenite	175	2.5–3.1
Gypsum, alabaster	159	2.3–2.8
Limestone, marble	165	2.5–2.8
Pumice, natural	40	0.37–0.90
Quartz, flint	165	2.5–2.8
Sandstone, bluestone	147	2.2–2.5
Shale slate	175	2.7–2.9
Reinforced, prestressed concrete		
Normal weight	150	2.3–2.5
Lightweight, structural	100–120	1.6–1.9

† To obtain density in kilograms per cubic meter, multiply by 16.03.

[handwritten note:] 1 lb/ft.^{3} = 16.03 kg/m^3

Table 3.2 Typical values of uniformly distributed design live loads (*Adapted from ANSI Code, Ref. 3.7*)

Occupancy or use	Live load	
	lb/ft^2	kPa or kN/m^2
Assembly halls		
Fixed seats	60	2.9
Movable seats	100	4.8
Balcony (exterior)	100	4.8
Dining rooms, restaurants, dance halls	100	4.8
Garages (passenger cars)	100	4.8
Floors shall be designed to carry 150% of the maximum wheel load anywhere on the floor		
Hospitals		
Operating rooms	60	2.9
Private rooms	40	1.9
Hotels		
Guest rooms	40	1.9
Public rooms	100	4.8
Housing		
Private houses and apartments	40	1.9
Public rooms (in multifamily units)	100	4.8
Libraries		
Reading rooms	60	2.9
Stack rooms	150	7.2
Office buildings		
Offices	80	3.8
Lobbies	100	4.8
Schools		
Classrooms	40	1.9
Corridors	100	4.8
Sidewalks, driveways subject to trucking	250	12
Stairs	100	4.8
Storage warehouses		
Light	100	4.8
Heavy	250	12
Yards and terraces, pedestrians	100	4.8

are to be considered and the more critical is to be accounted for in the design. Commonly, vehicle loadings prevail for small spans, while the equivalent lane loadings prevail for longer spans. As many factors, such as the influence of impact, the number of lanes, the width of the lanes, are associated with bridge loadings, their treatment is left to Chap. 14.

3.7 ALLOWABLE STRESSES

In the working stress design approach a number of stress limits, called allowable stresses or permissible stresses, are needed. These allowable stresses are not to be exceeded by actual stresses under the application of service loads. They are in

Table 3.3 Allowable stresses in concrete in prestressed flexural members (*Adapted from ACI 318-77 Code, Ref. 3.3*)

	psi	MPa
1. Stresses immediately after prestress transfer (before prestress losses) shall not exceed the following:		
(a) Extreme fiber stress in compression $\bar{\sigma}_{ci}$	$0.60f'_{ci}$	$0.60f'_{ci}$
(b) Extreme fiber stress in tension except as permitted in c $\bar{\sigma}_{ti}$	$-3\sqrt{f'_{ci}}$	$-0.25\sqrt{f'_{ci}}$
(c) Extreme fiber stress in tension at ends of simply supported members Where computed tensile stresses exceed these values, bonded auxiliary reinforcement (nonprestressed or prestressed) shall be provided in the tensile zone to resist the total tensile force in the concrete computed with the assumption of an uncracked section.	$-6\sqrt{f'_{ci}}$	$-0.50\sqrt{f'_{ci}}$
2. Stresses at service loads (after allowance for all prestress losses) shall not exceed the following:		
(a) Extreme fiber stress in compression $\bar{\sigma}_{cs}$	$0.45f'_c$	$0.45f'_c$
(b) Extreme fiber stress in tension in precompressed tensile zone $\bar{\sigma}_{ts}$	$-6\sqrt{f'_c}$	$-0.50\sqrt{f'_c}$
(c) Extreme fiber stress in tension in precompressed tensile zone of members (except two-way slab systems) where analysis based on transformed cracked sections and on bilinear moment-deflection relationships shows that immediate and long-time deflections comply with requirements stated elsewhere in the code	$-12\sqrt{f'_c}$	$-\sqrt{f'_c}$
3. The permissible stresses of sections 1 and 2 may be exceeded if shown by test or analysis that performance will not be impaired.		

Table 3.4 Typical values of allowable stresses in concrete (psi units)

Allowable stress†	Specified compressive strength, f'_c				
	5000	6000	7000	8000	9000
$0.60f'_{ci}$	2400	2880	3360	3840	4320
$-3\sqrt{f'_{ci}}$	−190	−208	−224	−240	−255
$-6\sqrt{f'_{ci}}$	−380	−416	−449	−480	−509
$0.45f'_c$	2250	2700	3150	3600	4050
$-6\sqrt{f'_c}$	−424	−465	−502	−537	−569

† Assuming $f'_{ci} = 0.80f'_c$.

general prescribed by the code adopted for the design. Typical values of allowable stresses imposed by the ACI code are summarized in Tables 3.3 to 3.8 for concrete, prestressing steel, and reinforcing steel, respectively.

Allowable stresses for concrete (Table 3.3) are separated into two groups: the first one corresponds to initial stresses at time of transfer of prestress (before prestress losses) and the second one corresponds to final or service load stresses (after allowance for prestress losses). The subscript i such as in f'_{ci} is associated with the first group. Tension in concrete has been given a negative sign in accordance with the sign convention adopted in this text. Also given in Table 3.3 are the stresses in the equivalent SI system where factors have been slightly rounded off. Note that a tension of $-6\sqrt{f'_c}$ is generally allowed during service on the concrete precompressed fiber. A fictitious value of $-12\sqrt{f'_c}$ or more is allowed essentially to

Table 3.5 Typical values of allowable stresses in concrete (MPa units)

Allowable stress†	Specified compressive strength, f'_c			
	30	40	50	60
$0.60f'_{ci}$	14.4	19.2	24	28.8
$-0.25\sqrt{f'_{ci}}$	−1.22	−1.41	−1.58	−1.73
$-0.50\sqrt{f'_{ci}}$	−2.45	−2.83	−3.16	−3.46
$0.45f'_c$	13.5	18	22.5	27
$-0.50\sqrt{f'_c}$	−2.74	−3.16	−3.54	−3.87

† Assuming $f'_{ci} = 0.80f'_c$.

Table 3.6 Allowable stresses in prestressing steel (*Adapted from ACI 318-77 Code, Ref. 3.3*)

Tensile stress in prestressing tendons shall not exceed the following:	
1. Due to tendon jacking force whichever is smaller, but not greater than the maximum value recommended by the manufacturer of the prestressing tendons or anchorages	$f_{pe} = 0.80f_{pu}$ or $0.94f_{py}$
2. Pretensioning tendons immediately after prestress transfer	$0.70f_{pu}$
3. Post-tensioning tendons immediately after tendon anchorage	$0.70f_{pu}$

Table 3.7 Typical values of allowable stresses in prestressing tendons

	Specified ultimate strength, f_{pu}									
Allowable stress	145 ksi	1000 MPa	ksi	MPa	ksi	MPa	250 ksi	1723 MPa	270 ksi	1860 MPa
At jacking $0.8f_{pu}$	116	800	128	882	188	1296	200	1378	216	1488
After transfer $0.7f_{pu}$	101.5	700	112	773	164.5	1134	175	1206	189	1302

Table 3.8 Allowable stresses in reinforcing steels (*Adapted from ACI 318-77 Code, Ref. 3.3*)

	Allowable stress, $\bar{f}_s$	
Conditions of application	ksi	MPa
1. Grade 40 or grade 50 reinforcement	20	138
2. Grade 60 or greater and welded wire fabric (smooth or deformed)	24	165
3. For flexural reinforcement $\frac{3}{8}$ in (9.5 mm) or less in diameter, in one-way slabs of not more than 12-ft (3.7-m) span	$0.50f_y \leq 30$	$0.50f_y \leq 207$

accommodate partial prestressing. Numerical values of allowable stresses are given in Tables 3.4 and 3.5 for typical concrete strengths in both U.S. and SI units.

Some of the allowable stresses recommended by the AASHTO specifications for bridges are slightly different from the corresponding ACI values. Mostly, under service loads, a compressive stress of $0.40f'_c$ is prescribed instead of $0.45f'_c$ and tension is not allowed unless bonded reinforcement is used. Allowable stresses by AASHTO are reproduced in App. E and should be used in bridge design.

Allowable stresses in prestressing steels imposed by the ACI code are summarized in Table 3.6 and some practical values appear in Table 3.7. Different values are given depending on whether the stress applies before or after transfer of the force from the steel to the concrete. Previous editions of the code used also to specify a limiting final stress of $0.60f_{pu}$ after all losses. This does not seem to be any longer necessary as, in practice, the actual final stress or effective prestress is seldom larger (commonly of the order of $0.5f_{pu}$ to $0.55f_{pu}$).

Allowable stresses in the reinforcing steel are given in Table 3.8. As the current ACI code deals mostly with ultimate strength design, these stresses are not given directly in the body of the code but in its App. B where working stress design (termed alternate design method) provisions are explained. In principle, they should apply to reinforcing steels used in prestressed concrete. Note that for cases where temporary tension in the concrete is to be resisted (such as in the note of case 1(c) of Table 3.3) a permissible stress of $0.60f_y$ or 30 ksi (207 MPa), whichever is smaller, can be used throughout.

3.8 LOAD FACTORS AND STRENGTH REDUCTION FACTORS

In ultimate strength design (USD), service loads are multiplied by load factors and the member is designed to resist the factored loads. Load factors prescribed by the ACI code are summarized in Table 3.9. Different factors are given for different loadings. Their magnitude is in effect adjusted to reflect the degree of uncertainty associated with the type of loading. Thus, load factors for dead loads are smaller than those for live loads. Several load combinations are to be considered in the design. Reduction factors are used on some combinations because of the low probability of their simultaneous occurrences. One of the combinations that often controls the design is the combination of dead and live loads $(1.4D + 1.7L)$. It always prevails if other combinations are not critical.

A large number of load factors and their combinations is given in the AASHTO specifications for bridges. These factors may be substantially different in magnitude from those recommended by the ACI building code. For flexural design, the combination for group IA, where only dead and live loads are considered, is $1.3D + 2.2(L + I)$, in which I stands for impact. An approach very similar to that of AASHTO is taken by ACI Committee 343 in its report on the analysis and design of reinforced concrete bridge structures (Ref. 3.10).

Table 3.9 Load factors for determining required strength U (*Adapted from ACI 318-77 Code, Ref. 3.3*)

Conditions of application	Required strength U is largest of	
1. Resistance to dead load D and live load L	$1.4D + 1.7L$	(1)
2. If resistance to structural effects of a specified wind load W is included in design, load combinations prescribed include both full value and zero value of L to determine the more severe condition, and the case where gravity counteracts wind-load effects	$0.75(1.4D + 1.7L + 1.7W)$	(2a)
	$0.75(1.4D + 1.7W)$	(2b)
	$0.9D + 1.3W$	(3)
3. If resistance to specified earthquake loads of forces E are included in design, load combinations of section 2 shall apply except that $1.1E$ shall be substituted for W		
4. If resistance to lateral earth pressure H is included in design, load combinations prescribed comprise the additive or counteractive effects of H to D	$1.4D + 1.7L + 1.7H$	(4a)
	$0.9D + 1.7H$	(4b)
5. If resistance to lateral liquid pressure F is included in design, load combinations of section 4 shall apply, except that $1.4F$ shall be substituted for $1.7H$. Vertical liquid pressure shall be considered as dead load D		
6. If resistance to impact effects is taken into account in design, such effects shall be included with live load L		
7. Where structural effects T of differential settlement, creep, shrinkage, or temperature change may be significant in design, estimations of the effects of T shall be based on realistic assessment of such effects occurring in service	$0.75(1.4D + 1.4T + 1.7L)$	(5)
	$1.4(D + T)$	(6)

A concrete structural element is designed to achieve a nominal strength at least equal to the required strength. The nominal strength is generally predicted using accepted analytic procedures. However, in order to account for the degree of accuracy or uncertainty with which the nominal strength can be predicted, a strength reduction factor ϕ is recommended by the ACI code. A safe design will be achieved when the required strength obtained from the factored loadings is less than or equal to the strength obtained as the product of the nominal strength by the reduction factor ϕ. Thus, the following most general condition exists:

$$\text{(Code required strength or design strength} \leq \phi \text{ (nominal strength)} \qquad (3.1)$$

Applied to moment, shear, torque, and axial forces, it leads to:

$$\begin{cases} M_u \leq \phi M_n \\ V_u \leq \phi V_n \\ T_u \leq \phi T_n \\ P_u \leq \phi P_n \end{cases} \qquad (3.2)$$

in which the subscripts n and u stand for nominal and ultimate required strengths, respectively.

Values of the strength reduction factor ϕ, given in the ACI code, are summarized in Table 3.10. Different numerical values apply to different types of load effects. Factors affecting the choice of these values include possible variability in materials properties, the nature, mode, and consequence of failure, should it occur, and dimensional inaccuracies (Ref. 3.11).

The combined application of factored loads and strength reduction factors given by the ACI code is aimed at producing approximate probabilities of understrength of the order of 1/100 and overloads of 1/1000 (Ref. 3.12). This results in a probability of structural failure of the order of 1/100,000.

Table 3.10 Strength reduction factors ϕ (*Adapted from ACI 318-77 Code, Ref. 3.3*)

	Resisting effect	ϕ
Reinforced and prestressed concrete	Flexure	0.90
	Compression	
	(*a*) members with spirals	0.75
	(*b*) others	0.70
	Axial tension	0.90
	Shear	0.85
	Torsion	0.85
	Development length, bond	1
	Any effect using working stress design	1
Plain concrete	Bending	0.65
	Bearing	0.70

3.9 DETAILS OF REINFORCEMENT

The designer should be able to see the whole and the detail. Although this text is not intended to cover detailing procedures, a minimum number of details are necessary in the early stages of the design. Once the required amount of reinforcement is determined, it is the designer's responsibility to make certain that such reinforcement can be properly placed inside the concrete section. Information related to spacing of the tendons and the minimum protection or clear concrete cover of the reinforcement is needed.

In its Sec. 7.6 the ACI code prescribes a clear distance between pretensioning tendons at each end of a member not less than four times the diameter for wires and three times the diameter for strands. Closer vertical spacing and bundling of

Table 3.11 Concrete protection of reinforcement in prestressed concrete
(*Adapted from ACI 318-77 Code, Ref. 3.3*)

Conditions of application	Clear concrete cover for prestressed concrete (ACI Sec. 7.7.3)	
	in	mm†
The following minimum clear concrete covers shall be provided for prestressed and nonprestressed reinforcement, ducts, and end fittings. For prestressed concrete members exposed to earth, weather, or corrosive environment, the minimum concrete cover shall be increased by 50% if service tensile stress exceeds $|6\sqrt{f'_c}|$ in magnitude		
1. Concrete cast against and permanently exposed to earth	3	75
2. Concrete exposed to earth or weather		
(*a*) Wall panels, slabs, joints	1	25
(*b*) Other members	$1\frac{1}{2}$	40
3. Concrete not exposed to weather or in contact with ground		
(*a*) Slabs, walls, joints	$\frac{3}{4}$	20
(*b*) Beams, columns		
primary reinforcement	$1\frac{1}{2}$	40
ties, stirrups, spirals	1	25
(*c*) shells, folded plate members		
No. 5 bar, W31 or D31 wire, and smaller	$\frac{3}{8}$	10
Other reinforcement	$\geq d_b$ or $\frac{3}{4}$	$\geq d_b$ or 20

† Values in millimeters have been rounded off to nearest five or ten.

strands may be permitted in the middle portion of a span. Another requirement can be derived from Sec. 3.3.3 of the code where it is stated that the clear spacing between individual reinforcing bars, wires, bundles of bars, prestressing tendons, and ducts shall not be less than one-and-one-third times the maximum size of the coarse aggregate used in the concrete. Posttensioning ducts may be bundled if it is shown that concrete can be satisfactorily placed and if provision is made to prevent the tendons, when tensioned, from breaking through their ducts. Special precautions must be taken with posttensioned tendons at the ends of a member where the anchors are placed. In general, additional stirrups are recommended near the ends of the member (with or without end blocks) and confining reinforcements in the form of spirals or grids are placed in the concrete directly behind the anchors. A conservative rule of thumb is to place the end anchors in such a way that the average stress in the concrete directly behind the anchor plate is less than $0.80f'_{ci}$ (up to f'_{ci} is often acceptable). A more representative approach can be found in Sec. 18.13 of the Commentary on the ACI code (Ref. 3.4).

Prestressing tendons must be placed inside a concrete section so as to have sufficient protection. A summary of the ACI code provisions for the minimum clear concrete cover of the reinforcement in prestressed concrete is given in Table 3.11.

It has been this author's experience that in dealing with prestressed pretensioned beams where tendons of not more than half-inch diameter are used, a cover of two inches (50 mm) to the centroid of the tendon and a center spacing of two inches (50 mm) between tendons commonly lead to a feasible design (Fig. 3.1). Note that a center spacing of two inches leaves a clear spacing of $1\frac{1}{2}$ in (38 mm), often just sufficient to allow the internal vibration of the concrete during casting.

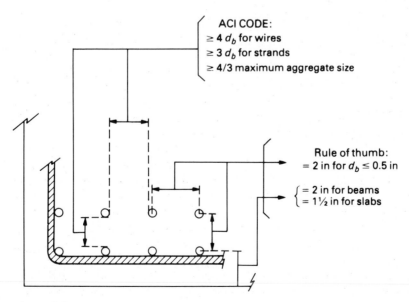

ACI CODE:
$\geq 4\,d_b$ for wires
$\geq 3\,d_b$ for strands
$\geq 4/3$ maximum aggregate size

Rule of thumb:
$= 2$ in for $d_b \leq 0.5$ in

$\begin{cases} = 2 \text{ in for beams} \\ = 1\frac{1}{2} \text{ in for slabs} \end{cases}$

Figure 3.1

When posttensioning cables made of several tendons are used, it is desirable to leave a clear spacing larger than 2 in (50 mm). The prescribed amount should depend on the capacity of the cable.

3.10 PRESTRESS LOSSES

As pointed out in Chap. 2, substantial losses of stress occur in the prestressing steel due to relaxation of the steel and creep and shrinkage of the concrete. Other losses also take place, due to elastic shortening of the concrete at load transfer and due to friction between the tendons and the concrete. Several methods exist to estimate the total loss of prestress and may require different levels of difficulty. A basic and detailed treatment of prestress losses is given in Chap. 8. However, it is important to estimate prestress losses in routine situations *a priori* to proceed with the design. Their values may later be revised if necessary. Lump sum estimates of individual losses are given in many technical documents (see References of Chap. 8). Lump sum values of total losses (35 ksi for pretensioned structures and 25 ksi for posttensioned structures not including friction) were suggested by the ACI-ASCE Joint Committee on Prestressed Concrete (Ref. 3.13). They were recommended in previous editions of the code and its current commentary but are presently considered unconservative when designing for serviceability. Note that misestimating prestress losses has little effect on the ultimate resistance of the member.

The AASHTO specifications (Ref. 3.5) suggest lump sum loss values for bridges where average standard conditions prevail. These values for a specified concrete compressive strength of (5 ± 0.5) ksi are shown in Table 3.12.

Let us define by f_{pi} the initial stress in the prestressing steel at time of load transfer and f_{pe} the effective stress remaining after all losses have taken place. The corresponding prestressing forces are defined as F_i and F. F_i is the force that the concrete experiences under initial loading. Thus f_{pi} may mean the stress just after transfer for a pretensioned member and just before transfer for a posttensioned member. The difference between f_{pi} and f_{pe} essentially represents the total prestress

Table 3.12 Typical lump sum loss of prestress for routine design (*Adapted from AASHTO Sec. 1.6.7, Ref. 3.5*)

Type of prestressing steel	Total loss for $f'_c = 5$ ksi (34.5 MPa)	
	ksi	MPa
Pretensioning strand	45	310
Post-tensioning		
Wire or strand	33†	228†
Bars	23†	159†

† Losses due to friction are not included.

losses for a posttensioned beam while it represents time-dependent losses (excluding initial relaxation) for a pretensioned beam. In the preliminary design and dimensioning of the structure the above difference between f_{pi} and f_{pe} is not as useful as their ratio. Let us define

$$\eta = \frac{f_{pe}}{f_{pi}} = \frac{F}{F_i} \qquad (3.3)$$

$f_{pi} = 0.70 f_{pu}$
$f_{pe} = 0.55 f_{pu}$

where F_i is the initial and F the final or effective prestressing force.

Note, as the stress in the steel generally varies along the length of the member, the coefficient η will also vary. However, η is needed mostly at the critical section of a simply supported beam or at a limited number of critical sections of a continuous beam. For all practical purposes it could be assumed the same for all critical sections. As a first approximation in design the value of η can be taken between 0.75 and 0.85. Assuming a value of $f_{pi} = 0.70 f_{pu}$, the corresponding value of f_{pe}, the effective prestress, will be somewhere between $0.53 f_{pu}$ and $0.58 f_{pu}$. Such approximations will frequently be used in preliminary designs. However, for final designs a more exact assessment of prestress losses is recommended.

REFERENCES

3.1. P. H. Hill: *The Science of Engineering Design*, Holt, Reinhart and Winston, Inc., New York, 1975.

3.2. R. N. White, P. Gergely, and R. G. Sexsmith: *Structural Engineering, Vol. 1: Introduction to Design Concepts and Analysis*, John Wiley & Sons, New York, 1972.

3.3. Building Code Requirements for Reinforced Concrete, ACI 318-77, American Concrete Institute, Detroit, 1977.

3.4. Commentary on Building Code Requirements for Reinforced Concrete, ACI 318-77, American Concrete Institute, Detroit, 1977.

3.5. Standard Specifications for Highway Bridges, 12th ed., American Association of State Highway and Transportation Officials (AASHTO), Washington, D.C., 1977, 496 pp.

3.6. Manual of Railway Engineering, American Railway Engineering Association (AREA), Washington, D.C., 1973.

3.7. Building Code Requirements for Minimum Design Loads in Buildings and Other Structures, American National Standards Institute (ANSI), A 58.1, 1972, New York, 1972.

3.8. British Standard Code of Practice for the Use of Structural Concrete, British Standards Institution, CP-110, London, 1972.

3.9. Manual of Steel Construction, 8th ed., American Institute of Steel Construction (AISC), New York, 1980.

3.10. Analysis and Design of Reinforced Concrete Bridge Structures, Report by ACI Committee 343, ACI 343R-77, American Concrete Institute, Detroit, 1977.

3.11. J. G. MacGregor: "Safety and Limit States Design for Reinforced Concrete," *Canadian Journal of Civil Engineering*, vol. 3, no. 4, 1976.

3.12. G. Winter: "Safety and Serviceability Provisions in the ACI Building Code," *ACI Special Publication*, SP-59, 1979.

3.13. "Tentative Recommendations for Prestressed Concrete," Report by ACI-ASCE Joint Committee 423, formerly 323, *Journal of the American Concrete Institute*, vol. 54, no. 7, January 1958, pp. 548–578.

Typical precast prestressed concrete structural building system. *(Courtesy Pre-stressed Concrete Institute.)*

CHAPTER 4

FLEXURE:
WORKING STRESS ANALYSIS AND DESIGN

4.1 ANALYSIS VERSUS DESIGN

Two categories of problems can generally be identified in dealing with prestressed beams: the first is called *analysis* or *investigation* and the second is described as *design*. The major difference between them is that in the design process a number of unknowns must be determined by the designer (Fig. 4.1).

The analysis or investigation process can be defined as follows: Given problem description and beam characteristics (such as loadings, span, cross-sectional dimensions, materials properties of steel and concrete, etc.) check if specified design criteria are satisfied at every section along the member. Design criteria usually involve many facets such as flexure, shear, cracking, camber, and deflections (Fig. 4.2) but are limited in this chapter to flexural stresses under service conditions as specified by available codes of practice. It is important however to view the overall problem before getting into the details of every step.

Similarly to the analysis process, the design process in its entirety requires also many steps (Fig. 4.3). In flexure, design usually implies the determination of some unknowns to satisfy specified allowable stresses under working load conditions. The complexity of the design increases with the number of unknowns involved, thus covering a whole range of specific problems (Fig. 4.1). For example, a design problem may imply the determination of the shape of the beam cross section, its dimensions as well as the prestressing force and its eccentricity. If a rectangular shape is desired, only four unknowns will have to be determined namely b, h, F, and e_o. The number of unknowns is reduced to three for a slab and can be equal to

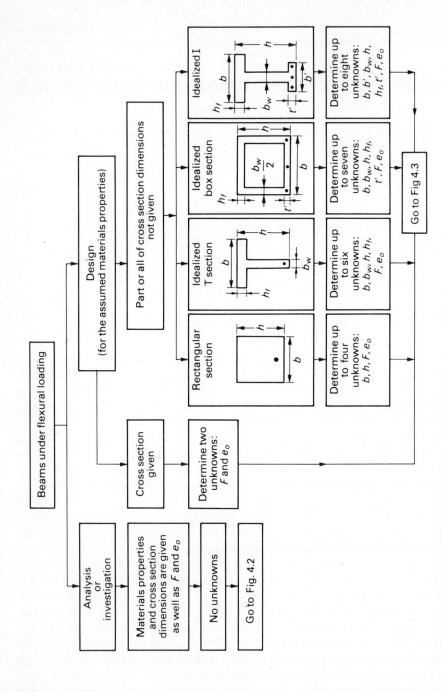

Figure 4.1 Levels of difficulty encountered in the flexural design of beams.

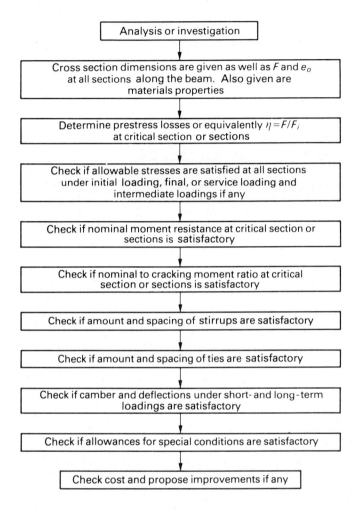

Figure 4.2 Major steps encountered in the analysis or investigation of beams.

eight for an ideal I beam (Fig. 4.1). If a beam cross section is selected a priori, say from a set of standard shapes, then only two unknowns are left, namely the prestressing force F and its eccentricity e_o. Note that knowing the value of F is equivalent to knowing the required area of prestressing steel.

It is important to realize that the design process described in Figs. 4.1 and 4.3 involves generally a repetitive procedure within each step and between steps in order to approach a satisfactory and close-to-optimum solution. Often the designer has to assume a practical value for some of the unknowns in order to start the first cycle (see, for example, Sec. 14.10 on preliminary design of bridges). For instance, in selecting the size of a rectangular beam, the designer may assume a depth equal to four percent of the span and a width equal to half the depth. Once a

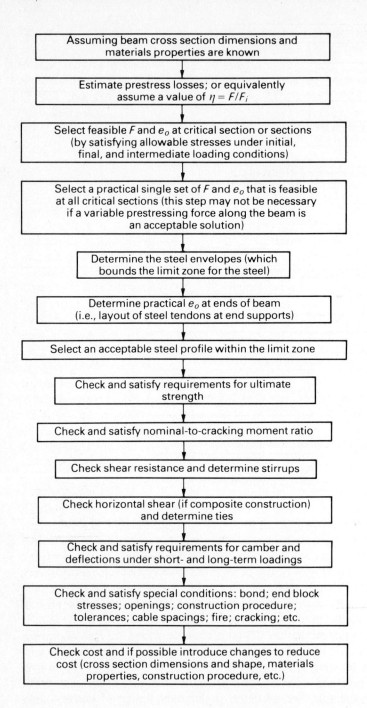

Figure 4.3 Major design steps after a beam cross section is selected.

satisfactory cross section has been arrived at, the problem is reduced to finding the values of F and e_o, and becomes substantially simpler.

Figure 4.4 is provided here in combination with the three others in order to allow the reader to view possible actions that might be taken in an iterative procedure not only for flexure but also for other aspects of the design, such as shear, and deflections. The above described first four figures put in perspective the content of other chapters in relation to this one.

4.2 CONCEPTS OF PRESTRESSING

Three different concepts may be applied to analyze the behavior of prestressed concrete members (Fig. 4.5). The first concept treats prestressed concrete essentially as an elastic composite material with no specific consideration to the two major components, steel and concrete (Fig. 4.5a). For example, a prestressed concrete tie in which the final prestress is 1000 psi may be considered as an elastic material which can sustain tensile stresses of up to 1000 psi without risk of cracking (whether these stresses are due to a tensile load or to a moment). It is the same as saying that a glass rod can sustain a tension of 10,000 psi before failing or that a steel rod can sustain 60,000 psi before yielding. This concept may be advantageous when tensile stresses in the concrete component are not allowed and for simple problems.

The second concept treats prestressed concrete in a way similar to reinforced concrete where, in analyzing the section, the concrete component is seen to carry the compressive force while the steel component carries the tensile force (Fig. 4.5b). Thus the two components, steel and concrete, can be considered separately or in combination, as free bodies in equilibrium under the effects of external forces and reactions. There is, however, a fundamental difference in behavior between a reinforced concrete section and a prestressed concrete section.

A reinforced concrete section is assumed to crack under flexural loading. The resulting compressive force on the uncracked portion of the concrete section and the tensile force on the steel are equal in magnitude, and form an internal couple which resists the applied external moment. If, in the elastic range of behavior, the moment applied to the section increases, the forces of the couple increase while the distance between them remains almost constant (actually it decreases at a very slow rate).

A prestressed concrete section is assumed uncracked under service loads and, if the moment applied to the section increases, the forces of the couple remain essentially constant while the distance between them increases. This is illustrated next.

Let us consider the stress diagram along a cross section of a simply supported beam (Fig. 4.6). If the beam is assumed weightless, the compressive force C on the concrete section considered as a free body is equal in magnitude to the prestressing force F and acts along the same line as F. The stress profile along the cross section due to C is shown in Fig. 4.6a. If a bending moment is applied to the section, it

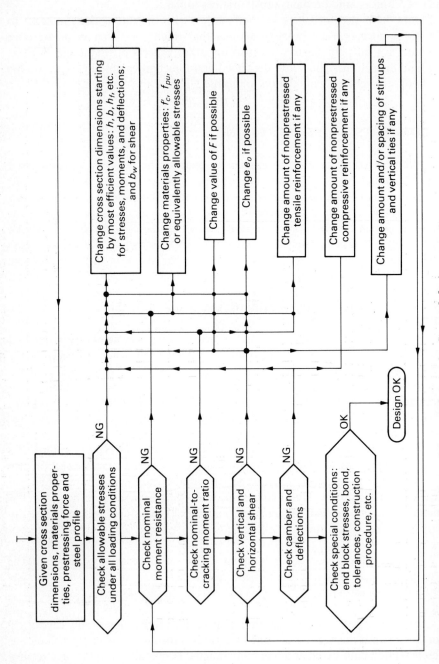

Figure 4.4 Iterative steps and possible remedies in the flexural design process.

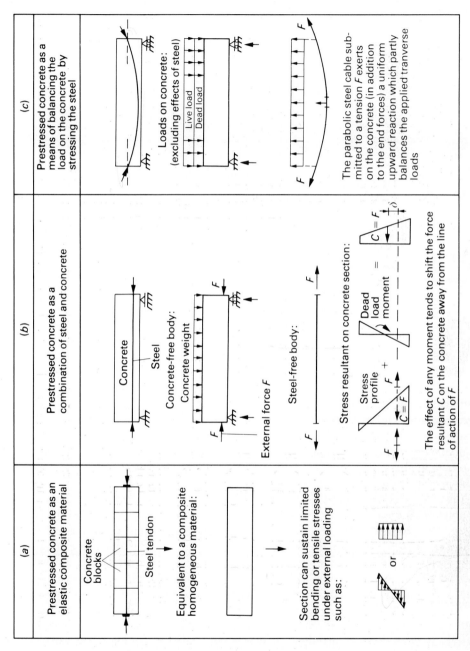

Figure 4.5 Three concepts used to analyze prestressed concrete elements.

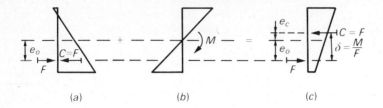

(a)   (b)   (c)

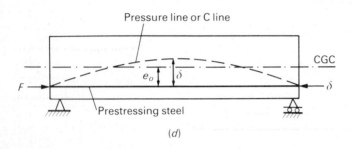

Pressure line or C line

CGC

Prestressing steel

(d)

Figure 4.6 Effect of a moment on resulting stresses and C line in concrete.

produces a stress diagram as shown in Fig. 4.6b. The resulting stress diagram due to the combined action of the prestressing force and the moment is shown in Fig. 4.6c (assuming no tension or cracking). The sum of the forces on the section is still equal to C as any external moment leads to an equal internal couple and the sum of the forces of a couple is zero. Therefore, the stress diagram shown in (c) can be seen as resulting from a compressive force C, same as in (a), but acting along a different line parallel to and located a distance δ from the line of action of F. This is the same as saying that the combined action of a moment and a force is equivalent to the action of the same force displaced a certain distance δ parallel to its line of action. Note that δ is such that $\delta F = \delta C$ = external moment. It can be seen that an external moment on the section does not affect the value of $C = F$ but only changes the distance between them. (In fact because of the change in curvature due to loading F changes slightly.) It is interesting to note that the location of C can be characterized by an eccentricity e_c similar to the eccentricity e_o of the prestressing force. The geometric lieu of C at every section along a given member is called pressure line or C line. For the simply supported beam considered in this example and subjected to a prestressing force F and a bending moment due to its own weight and/or to a uniform load, the C line is schematically represented in Fig. 4.6d; it has a parabolic shape to reflect the bending moment diagram.

The third concept used in analyzing prestressed beams is called the load-balancing concept. Here the entire concrete is considered a free body subjected to externally applied loads including its own weight. The prestressing force and the steel profile are selected to directly balance *part* of these external loads (Fig. 4.5c). The beam is then in a state of uniform compression with zero deflection, and uncracked elastic beam theory is used to analyze the effects of the remaining part of the external loads. The load-balancing technique is an extremely simple and powerful technique for the design of continuous beams, two-way slabs, frames and shells. It has been first introduced by T. Y. Lin and is discussed in more detail in Chap. 10.

4.3 LOADING STAGES

In the design of prestressed concrete members, loading refers not only to externally applied loads such as dead and live loads but often to a combination of these loads acting with the prestressing force on the concrete section. Several loading stages can be identified in the elastic range of behavior among which the *initial* and the *final* loadings are generally most critical.

The *initial loading* refers primarily to the stage where the prestressing force is transferred to the concrete and no external loads are present except the weight of the member. At this time the prestressing force is maximum as prestress losses have not yet taken place and the concrete strength is minimum as the concrete is still young; consequently, the stresses in the concrete can be critical. In pretensioned members in order to speed production, the prestressing tendons are released simultaneously at a time when the strength of the concrete has reached 60 to 80 percent of its specified 28 days strength. By curing it at higher temperatures, these strengths can be achieved in less than 24 hours after the concrete is poured. In posttensioned members, often the prestressing tendons are not tensioned all at the same time but rather in two or three steps to allow the concrete to reach its specified strength before the prestressing force is fully applied. For example, 20 percent of the tendons may be tensioned about three days after the concrete is poured to compensate shrinkage stresses; an additional 40 to 50 percent at about seven days in order to remove the molds, etc., at 28 days. This approach has also the advantage of distributing more uniformly the work of the posttensioning crew in the field. In most cases the initial loading leads to critical stresses and its effect must be carefully assessed.

The *final loading* stage refers here to the most severe loading under service conditions; it is then assumed that all prestress losses have occurred, i.e., the prestressing force has its final and smallest value, and that the most critical combination of external loadings is applied; such a combination includes the weight of the members, superimposed dead loads, live loads, impact, and the like. Load combinations are generally specified in various codes and specifications.

Although the initial and final loadings are often the two most critical loadings, some intermediate loadings may become critical in the design. For example, special conditions during handling, transportation, and erection of precast prestressed members may lead to stresses more critical than those induced by the initial and final loadings. Every particular case must be studied with care and if necessary integrated in the design.

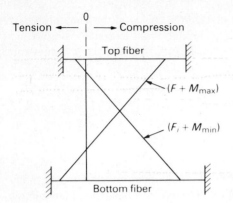

Figure 4.7 Typical stress diagrams under extreme loadings.

Among all possible loadings applied to a prestressed section, two will bound the others in terms of flexural stresses, and will be identified as the *two extreme loading* conditions. In a majority of cases they are due to the effect of $(F_i + M_{min})$ and $(F + M_{max})$ where, under initial and final conditions, M_{min} and M_{max} are the minimum and maximum bending moments, respectively, and F_i and F are the initial and effective values of the prestressing force at the section considered. For example, M_{min} may be the same as the moment due to the weight of the member while M_{max} may be the moment due to weight plus the superimposed dead load and live load. The relation between the initial and the final prestressing force is described by their ratio $\eta = F/F_i$; η can be estimated in a preliminary design and then more carefully assessed in the final design (Sec. 3.10). It may happen that one or both extreme loadings do not necessarily correspond to what is generally described as the initial and/or final loadings. Although they must be clearly identified, their effect is treated similarly to the above approach. Typical stress diagrams on the concrete section of a pretensioned member are shown in Fig. 4.7. The two extreme loading conditions are shown to induce flexural stress which are within allowable stress. Allowable stresses are represented by short vertical lines dashed on one side.

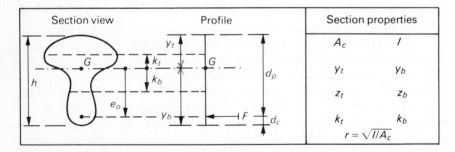

Figure 4.8 Typical characterization of beam cross section.

4.4 NOTATIONS FOR FLEXURE

In the most general case, the uncracked cross section of a prestressed concrete beam can be characterized for design purposes by a number of variables and geometric properties. Geometric properties can be determined directly from the dimensions of the cross section. Referring to Fig. 4.8, the following notations and definitions will be used:

A_c = area of concrete cross section (it may indicate the net or the gross area depending on the problem at hand and whether preliminary or final design are considered; practically, in pretensioned members it is taken as the gross area while in posttensioned members it is often the net or the transformed area)

A_{ps} = area of prestressing steel

d_p = distance from extreme compression fiber to the centroid of prestressing steel (or force)

d_c = concrete cover from the precompressed tensile fiber to the centroid of prestressing steel (d_c can generally be estimated in a preliminary design and revised for the final design)

$(d_c)_{\min}$ = minimum feasible value of d_c

e_o = eccentricity of the prestressing force (or centroid of prestressing steel) with respect to the centroid of the concrete section (e_o varies along the member, thus $e_o(x)$ is used when needed)

F = final prestressing force or effective prestressing force after all losses

h = total height of concrete section ($h = y_t + y_b = d_p + d_c$)

I = moment of inertia of the section with respect to an axis passing by its centroid (it generally implies gross sectional inertia; for a final design it may indicate the transformed inertia)

y_t = distance from the centroid of the concrete section to the extreme top fiber

y_b = distance from the centroid of the concrete section to the extreme bottom fiber

$\rho_p = A_{ps}/bd_p$ prestressed reinforcement ratio (b is the width of a rectangular section or flange width of T section)

σ = stress in concrete in general (it will be used unless a widely standard notation such as f'_c applies)

$Z_t = I/y_t$ = section modulus with respect to the top fiber

$Z_b = I/y_b$ = section modulus with respect to the bottom fiber

$r = \sqrt{I/A_c}$ = radius of gyration of the section

$k_t = -I/A_c y_b = -Z_b/A_c = -r^2/y_b$ = distance from the centroid of the concrete section to the upper limit of the central kern

$k_b = I/A_c y_t = Z_t/A_c = r^2/y_t$ = distance from the centroid of the concrete section to the lower limit of the central kern

$\gamma = (-k_t + k_b)/h = I/A_c y_t y_b = r^2/y_t y_b = Z_t/A_c y_b = Z_b/A_c y_t$ = geometric efficiency of the cross section with respect to bending

Note that $y_b = e_o + d_c$ and $d_p = e_o + y_t$. The central kern defines an area within the concrete section where a compressive force will not lead to tensile stresses on any part of the section. The upper limit of the central kern k_t has been given a negative sign to be consistent with the sign convention given for e_o (Sec. 4.5) and because k_t can be seen as a particular value of e_o. The reinforcement ratio of prestressed concrete members is generally less than one percent and most values in beams and slabs practically fall between 0.1 and 0.5 percent.

The geometric efficiency of the section γ is equal to $\frac{1}{3}$ for rectangular beams and 1 for a fictitious webless I beam having two identical and infinitely thin flanges. Practically the value of γ ranges from about 0.45 to 0.50 for T beams and 0.55 to 0.65 for box beams.

It is important for the designer to rapidly determine the main cross-sectional properties of any given cross section of concrete. Some background information on the moment of inertia is given in Fig. 4.9. An example is given in the next section and can be essentially used as a model.

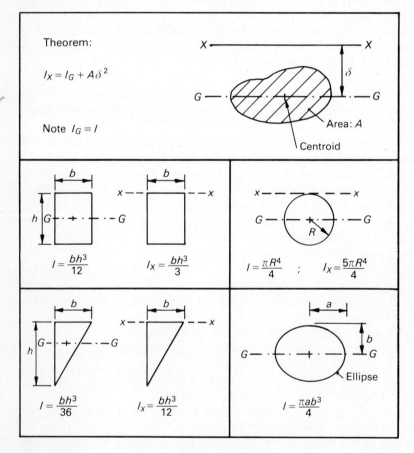

Figure 4.9

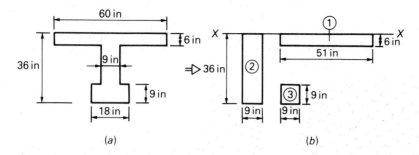

Figure 4.10

1. Example: Computation of Sectional Properties

Determine the section properties of the beam shown in Fig. 4.10a. The cross section is divided into three parts such as shown in Fig. 4.10b. Typical calculations are set in a tabular form below; the method can be used for any section with any number of parts n. The centroid of each part A_i is assumed at a distance δ_i from the X-X axis.

Part	Area	A_i	δ_i	$A_i\delta_i$	$A_i\delta_i^2$	$(I)_i$	$(I_x)_i$
1	51 × 6 = 306	3	918.0	2,754	918	3,672
2	9 × 36 = 324	18	5832.0	104,976	34,992	139,968
3	9 × 9 = 81	31.5	2551.5	80,372.25	546.75	80,919
4						
⋮						
n						

Total: $A_c = 711$	$Q = 9301.5 = \sum A_i\delta_i$		$I_x = 224{,}559$

So: $y_t = Q/A_c = 13.08$ in; $y_b = 36 - y_t = 22.92$ in
  $I = I_x - A_c\, y_t^2 = 102{,}916$ in^4
  $Z_t = 7868$ in^3; $Z_b = 4490$ in^3
  $k_t = -Z_b/A_c = -6.32$ in; $k_b = Z_t/Ac = 11.07$ in

4.5 SIGN CONVENTION

Although in prestressed concrete many effects such as compression or tension can be found by inspection, it is very important to provide a consistent sign convention in order to reduce error in any complex and systematic analysis; of course, this is essential in any computerized design. The following rules of sign convention will be generally followed in this text:

1. Plus (+) for compressive stresses in concrete
2. Minus (−) for tensile stresses in concrete

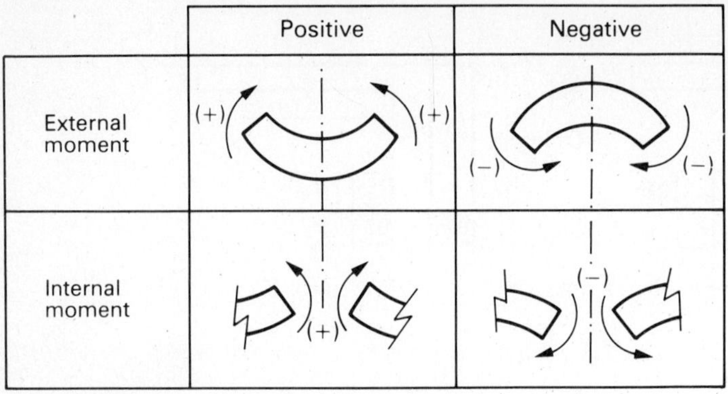

	Positive	Negative
External moment		
Internal moment		

Figure 4.11 Sign convention for moments.

The two extreme fibers of a horizontal member (beam) will be referred to as top and bottom fiber. A vertical member is rendered horizontal by a clockwise rotation of $\pi/2 = 90°$. Thus the left and right fiber of a vertical member are identical from a stress analysis viewpoint respectively to the top and bottom fiber of a horizontal member. In order to evaluate the effect of a bending moment on stresses, the usual sign convention for moments as described in Fig. 4.11 will be used. Thus a moment which tends to bend a simply supported beam so that it will retain water is considered positive. For instance, the moment due to the own weight of a simply supported beam is positive, while the moment at the intermediate support of a two-span continuous beam is negative. Therefore:

3. Use plus (+) the numerical value of the moment for positive moments and minus (−) for negative moments

An extreme fiber stress due to a bending moment of magnitude M can be calculated from the expression M/Z_t or M/Z_b where Z_t and Z_b represent the section moduli with respect to the top and bottom fibers, respectively. In order for flexural stresses to show the correct sign corresponding to a tension or a compression the following rule is set:

4. Multiply the above expression by $+1$ if the stress calculated is on the top fiber and by -1 if it is on the bottom fiber. Note that the same holds for stresses calculated at points " above " and " below " the neutral axis of bending of the section. For vertical members " left " replaces " top " or " above," and right replaces " bottom " or " below."

The prestressing force F acting on a concrete section with eccentricity e_o has the same effect as a concentric force F applied at the centroid of the section thus inducing a uniform compressive stress (positive) and a moment of magnitude Fe_o. If e_o is positive downward such as in Fig. 4.6, the moment is negative (convex) and its value is $(-Fe_o)$. The corresponding flexural stresses are given on the top fiber by:

$$+\left(\frac{-Fe_o}{Z_t}\right) = \frac{-Fe_o}{Z_t}$$

and on the bottom fiber by:

$$-\left(\frac{-Fe_o}{Z_b}\right) = \frac{Fe_o}{Z_b}$$

Another way to present these results is to follow the following rules:

5. Use the absolute value of the moment Fe_o (i.e., F and e_o are positive).

6. Multiply the stress due to Fe_o by $+1$ when it is computed on a fiber on the same side as e_o with respect to the neutral axis and by -1 when it is on the opposite side.

Expressions for the stress on the top and bottom fibers due to a bending moment and to the prestressing force for a typical simply supported prestressed beam are summarized in Table 4.1. For each expression several equivalent forms can be used. However, some may be easier to handle in a particular problem.

Table 4.1 Various expressions of stresses due to a moment or a prestressing force

Effect of	Extreme fiber	Various expressions for stresses
Positive moment of magnitude M	Top	$\sigma_t = \dfrac{My_t}{I} = \dfrac{M}{Z_t} = \dfrac{M}{A_c k_b} = \dfrac{My_t}{A_c r^2}$
	Bottom	$\sigma_b = -\dfrac{My_b}{I} = -\dfrac{M}{Z_b} = +\dfrac{M}{A_c k_t} = -\dfrac{My_b}{A_c r^2}$
Prestressing force F at eccentricity e_o toward bottom fiber of beam	Top	$\sigma_t = \dfrac{F}{A_c} - \dfrac{Fe_o y_t}{I} = \dfrac{F}{A_c}\left(1 - \dfrac{e_o y_t}{r^2}\right)$ $= \dfrac{F}{A_c}\left(1 - \dfrac{e_o A_c}{Z_t}\right) = \dfrac{F}{A_c}\left(1 - \dfrac{e_o}{k_b}\right)$ $= \dfrac{F}{Z_t}(k_b - e_o)$
	Bottom	$\sigma_b = \dfrac{F}{A_c} + \dfrac{Fe_o y_b}{I} = \dfrac{F}{A_c}\left(1 + \dfrac{e_o y_b}{r^2}\right)$ $= \dfrac{F}{A_c}\left(1 + \dfrac{e_o A_c}{Z_b}\right) = \dfrac{F}{A_c}\left(1 - \dfrac{e_o}{k_t}\right)$ $= \dfrac{F}{Z_b}(e_o - k_t)$

The following sign convention will be used for the steel (prestressed or non-prestressed):

7. Plus $(+)$ for tensile stresses

8. Minus $(-)$ for compressive stresses

It may seem awkward to have different sign conventions for stresses in the steel and concrete. However the above recommended sign convention is natural for prestressed concrete as it generally leads to positive stresses in the concrete (compression) and in the steel (tension). When confusion may arise in some equations where compression steel is present, absolute value sign will be used to warn the reader.

The following example illustrates the use of the sign convention in computing the numerical values of stresses in a prestressed concrete section.

1. Example

(a) Compute the stress on the bottom fiber at the midspan section of the simply supported prestressed beam shown in Fig. 4.12 for which $F = 120$ kips, $e_o = 8$ in, and the moment due the beam's own weight $= 60$ kips-ft.

The stress due to the moment is given by: $(-1)(+M)/Z_b$ according to rules 4 and 3 of the sign conventions; i.e., $(-1)(60 \times 12,000)/1152 = -625$ psi (tension). The stress due to the prestressing force is given by: $F/A_c + (+1)(Fe_o)/Z_b$ according to rules 1, 5, and 6 of the sign conventions; i.e., $(120,000/288) + (120,000 \times 8)/1152 = 1250$ psi (compression). Thus the resulting bottom fiber stress is given by: $(-M/Z_b) + (F/A_c + Fe_o/Z_b) = 625$ psi (compression).

(b) Compute the stress on the bottom fiber at the support of a double cantilever simply supported beam (Fig. 4.13) assuming same section properties, same force and eccentricity as for Example **(a)** and a maximum service moment $M = -15$ kips-ft at the supports.

According to rules 3 and 4 of the sign convention, the stress due to the moment is given by: $(-1)(-M)/Z_b = M/Z_b$; i.e., $15 \times 12,000/1152 = 156.2$ psi (compression) and the stress

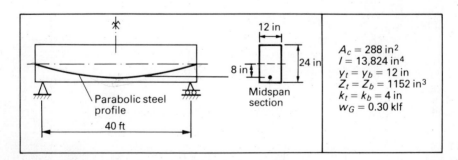

$A_c = 288$ in^2
$I = 13,824$ in^4
$y_t = y_b = 12$ in
$Z_t = Z_b = 1152$ in^3
$k_t = k_b = 4$ in
$w_G = 0.30$ klf

Figure 4.12

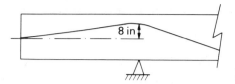

Figure 4.13

due to the prestressing force according to rules 1, 5, and 6 is given by: $F/A_c + (-1)$ $(Fe_o)/Z_b = F/A_c - Fe_o/Z_b = -416.6$ psi (tension). Thus the resulting stress on the bottom fiber is equal to -260.4 psi (tension).

4.6 ALLOWABLE STRESSES

Allowable stresses on the concrete section as well as in the steel are generally provided by the codes of practice or specifications considered for a particular study. Typical values of allowable stresses are given in Sec. 3.7.

As generally two extreme loading conditions bound all others, at least four allowable stresses on the concrete section must be considered in the design: namely two (tension and compression) for the initial loading and two (tension and compression) for the most severe final loading. Without specifically referring to any particular numerical value of allowable stresses, the following notation will be used in describing them:

$\bar{\sigma}_{ti}$ = allowable temporary tensile stress in the concrete (initial most severe loading)

$\bar{\sigma}_{ci}$ = allowable temporary compressive stress in the concrete (initial most severe loading)

$\bar{\sigma}_{ts}$ = allowable service tensile stress in the concrete (final most severe loading)

$\bar{\sigma}_{cs}$ = allowable service compressive stress in the concrete (final most severe loading)

In a composite construction one additional allowable stress must be considered, namely:

$\bar{\sigma}_{c\,slab}$ = allowable service compressive stress in composite slab

Note that in order to identify allowable stresses for concrete a bar is placed on top of the literal value. The same notation is used without the bar for actual stresses under same conditions. For the steel allowable stresses are given directly as a fraction of ultimate strength; thus no literal values are needed.

4.7 MATHEMATICAL BASIS FOR FLEXURAL ANALYSIS

For the analysis, it is assumed that materials behave elastically in the working range of stresses applied. The usual hypothesis of Hooke and Navier-Bresse are assumed valid, namely:

1. The materials (steel and concrete) are elastic and there is a proportional relationship between stresses and strains,
2. Plane sections remain plane after bending, and
3. There is a perfect bond between steel and concrete

This is equivalent to saying that both the stress and strain diagrams along the section of concrete under bending are linear, and that the changes in strains in the steel and in the concrete at the level of the steel are identical. Also the load-deflection or moment-curvature curve are assumed linear for the loadings considered. Typical stress diagrams for the two extreme initial and final loadings have been described in Fig. 4.7. Note that the highest stresses in the section occur at the extreme top and bottom fibers.

As two extreme loadings are generally critical and as for each, two allowable stresses must be specified, at least four allowable stresses must be considered in the analysis. As under flexural loading maximum stresses occur on the two extreme fibers (top and bottom), eight inequality equations comparing actual stresses with allowable stresses can be derived. They are of the form:

$$\text{(Actual stress)} \left\{ \begin{matrix} \geq \\ \text{or} \\ \leq \end{matrix} \right\} \text{(allowable stress)} \tag{4.1}$$

Let us develop one of these equations for a pretensioned simply supported member. The actual stress on the top fiber under initial conditions must be more than or equal to the allowable initial tensile stress. Therefore

$$\sigma_{ti} = \frac{F_i}{A_c} - \frac{F_i e_o}{Z_t} + \frac{M_{\min}}{Z_t} \geq \bar{\sigma}_{ti} \tag{4.2}$$

where $M_{\min}$ represents the dead load moment at the section considered. Equation (4.2) could also be rewritten in several different ways, one of which may be more suitable if a particular variable is to be emphasized, such as, for example:

$$\left\{ \begin{aligned} & F_i \leq (M_{\min} - \bar{\sigma}_{ti} Z_t)/(e_o - k_b) \\ & e_o \leq k_b + (1/F_i)(M_{\min} - \bar{\sigma}_{ti} Z_t) \\ & 1/F_i \geq (e_o - k_b)/(M_{\min} - \bar{\sigma}_{ti} Z_t) \end{aligned} \right. \tag{4.3}$$

As mentioned above eight inequality equations (which will also be described as "stress conditions" or "stress constraints") can be derived and are similar in form to Eq. (4.2). However, in actual design problems, out of the eight conditions four are generally nonbinding. For example, if for a simply supported member the tensile stress on the top fiber for the initial loading is of concern and is checked against allowable (as for Eq. (4.2)), there is certainly no need to check, against allowable, the compressive stress on the same fiber and for the same loading. If for the same loading we were checking the section at the intermediate support of a two-span continuous beam, we would check the stress on the top fiber against initial allowable compression and that eliminates the need to check against initial

allowable tension. Thus the number of inequality equations that must be considered in the analysis at a given section is essentially reduced to four, i.e., four of
them are binding while the four others are not. The four that are binding in a
particular design depend on the sign of the applied moments.

The eight stress inequality equations written in various ways are shown four at
a time in Tables 4.2 and 4.3. They have been numbered in roman notation I to IV
and I′ to IV′. The coefficient η was defined in Sec. 3.10 and is the ratio of the final
prestressing force after all losses to the initial prestressing force. For a cross section
where all applied moments (M_{min} and M_{max}) are positive, only stress inequalities I
to IV need to be considered; similarly, if all applied moments are negative stress
inequalities I′ to IV′ become binding. When a particular section is subject to
moments of different signs it is possible, by inspection, to select out of the eight
inequalities the four that would be binding; on the other hand, one can also check
systematically the eight inequalities against allowable stresses and select the four
that are binding.

Table 4.2 Useful ways of writing the four stress inequality conditions

Way	Stress condition	Inequality equation
1	I	$(F_i/A_c)[1 - (e_o/k_b)] + M_{min}/Z_t \geq \bar{\sigma}_{ti}$
	II	$(F_i/A_c)[1 - (e_o/k_t)] - M_{min}/Z_b \leq \bar{\sigma}_{ci}$
	III	$[(F \text{ or } \eta F_i)/A_c][1 - (e_o/k_b)] + M_{max}/Z_t \leq \bar{\sigma}_{cs}$
	IV	$[(F \text{ or } \eta F_i)/A_c][1 - (e_o/k_t)] - M_{max}/Z_b \geq \bar{\sigma}_{ts}$
2	I	$e_o \leq k_b + (1/F_i)(M_{min} - \bar{\sigma}_{ti} Z_t)$
	II	$e_o \leq k_t + (1/F_i)(M_{min} + \bar{\sigma}_{ci} Z_b)$
	III	$e_o \geq k_b + [1/(F \text{ or } \eta F_i)](M_{max} - \bar{\sigma}_{cs} Z_t)$
	IV	$e_o \geq k_t + [1/(F \text{ or } \eta F_i)](M_{max} + \bar{\sigma}_{ts} Z_b)$
3	I	$F_i \leq (M_{min} - \bar{\sigma}_{ti} Z_t)/(e_o - k_b)$
	II	$F_i \leq (M_{min} + \bar{\sigma}_{ci} Z_b)/(e_o - k_t)$
	III	$F = \eta F_i \geq (M_{max} - \bar{\sigma}_{cs} Z_t)/(e_o - k_b)$
	IV	$F = \eta F_i \geq (M_{max} + \bar{\sigma}_{ts} Z_b)/(e_o - k_t)$
4	I	$1/F_i \geq (e_o - k_b)/(M_{min} - \bar{\sigma}_{ti} Z_t)$
	II	$1/F_i \geq (e_o - k_t)/(M_{min} + \bar{\sigma}_{ci} Z_b)$
	III	$1/F = 1/\eta F_i \leq (e_o - k_b)/(M_{max} - \bar{\sigma}_{cs} Z_t)$
	IV	$1/F = 1/\eta F_i \leq (e_o - k_t)/(M_{max} + \bar{\sigma}_{ts} Z_b)$
All	V	$e_o \leq (e_o)_{mp} = y_b - (d_c)_{min} = \text{maximum practical eccentricity}$

Table 4.3 Useful ways of writing the four complementary stress inequality conditions

Way	Stress condition	Inequality equation
1	I′	$(F_i/A_c)[1 - (e_o/k_b)] + M_{min}/Z_t \le \bar\sigma_{ci}$
	II′	$(F_i/A_c)[1 - (e_o/k_t)] - M_{min}/Z_b \ge \bar\sigma_{ti}$
	III′	$[(F \text{ or } \eta F_i)/A_c][1 - (e_o/k_b)] + M_{max}/Z_t \ge \bar\sigma_{ts}$
	IV′	$[(F \text{ or } \eta F_i)/A_c][1 - (e_o/k_t)] - M_{max}/Z_b \le \bar\sigma_{cs}$
2	I′	$e_o \ge k_b + (1/F_i)(M_{min} - \bar\sigma_{ci} Z_t)$
	II′	$e_o \ge k_t + (1/F_i)(M_{min} + \bar\sigma_{ti} Z_b)$
	III′	$e_o \le k_b + [1/(F \text{ or } \eta F_i)](M_{max} - \bar\sigma_{ts} Z_t)$
	IV′	$e_o \le k_t + [1/(F \text{ or } \eta F_i)](M_{max} + \bar\sigma_{cs} Z_b)$
3	I′	$F_i \ge (M_{min} - \bar\sigma_{ci} Z_t)/(e_o - k_b)$
	II′	$F_i \ge (M_{min} + \bar\sigma_{ti} Z_b)/(e_o - k_t)$
	III′	$F = \eta F_i \le (M_{max} - \bar\sigma_{ts} Z_t)/(e_o - k_b)$
	IV′	$F = \eta F_i \le (M_{max} + \bar\sigma_{cs} Z_b)/(e_o - k_t)$
4	I′	$1/F_i \le (e_o - k_b)/(M_{min} - \bar\sigma_{ci} Z_t)$
	II′	$1/F_i \le (e_o - k_t)/(M_{min} + \bar\sigma_{ti} Z_b)$
	III′	$1/(F \text{ or } \eta F_i) \ge (e_o - k_b)/(M_{max} - \bar\sigma_{ts} Z_t)$
	IV′	$1/(F \text{ or } \eta F_i) \ge (e_o - k_t)/(M_{max} + \bar\sigma_{cs} Z_b)$
All	V	$e_o \le (e_o)_{mp} = y_b - (d_c)_{min} = \text{maximum practical eccentricity}$

Note that the stress inequalities I′ to IV′ given in Table 4.3 are described here as "complementary stress inequalities." This is because often one does not need to use them. It can be shown that if all applied moments are negative, stress conditions I to IV can still be used provided the concrete section is assumed in its inverted position (i.e., use properties of inverted section) and the sign of the moments is changed from negative to positive; the position of F within the cross section remains unchanged. It is because stress conditions I to IV can cover the majority of practical problems that they are often encountered alone in the technical literature and with no reference to the four others.

In Tables 4.2 and 4.3, a fifth condition number V has been included and will be described as the "practicality condition." Essentially it states that the prestressing force must be inside the concrete section with an adequate cover $(d_c)_{min}$. Thus the design eccentricity e_o must be less than or equal to a maximum practical value $(e_o)_{mp} = y_b - (d_c)_{min}$. Although in an analysis or investigation problem, condition V is obviously satisfied, in a design problem condition V can be binding and can be used with advantage in optimizing or simplifying a solution. This is why it has been included in the tables.

In an analysis or investigation problem, the above stress inequality equations can be directly checked as all quantities are known. In a design problem, however, these inequalities can be used to determine exactly or put bounds on some of the unknown variables such as prestressing force F, eccentricity e_o, and/or section properties. For example, if the concrete cross section is given, the stress conditions can be used to determine bounds on all the possible values of F and e_o that would be acceptable for the problem at hand. This is clarified in the next sections.

4.8 GEOMETRIC INTERPRETATION OF THE STRESS INEQUALITY CONDITIONS

The geometric interpretation of the stress conditions has been first explored by Magnel (Ref. 1.7). As emphasized throughout this text, the geometric representation can be a very useful and powerful technique for the solution of many problems where the working stress design approach is used.

Let us assume that the geometric properties of the concrete cross section are given; then (as η is estimated a priori) only two unknown variables remain in equations I to IV, namely e_o and F_i or F. One can plot on a two-dimensional scale the curves corresponding to the four equations at equality; each curve will separate the plane into two parts, one where the inequality is satisfied and the other where it is not. If e_o is plotted versus F_i, the curves will be hyperbola. However, if e_o is plotted versus $1/F_i$, then straight lines are obtained and the geometric representation is much simplified. For this it is better to use the second way of writing the equations in Table 4.2 as they are written in the form: $e_o = a(1/F_i) + b$ where b is the intercept and a the slope of the line. When plotted as shown in Fig. 4.14 the inequality equations delineate a domain of feasibility limited by a quadrilateral A, B, C, D. Essentially any point inside this domain has coordinates F_i and e_o which satisfy the four stress inequality conditions I to IV. The practicality condition V can also be represented at equality on the same graph by a horizontal line parallel to the $1/F_i$ axis. If this line intersects the quadragon A, B, C, D, such as case (b) of Fig. 4.14, then a new reduced feasibility domain is defined such as $EBCDG$. Any point inside this new domain would have satisfactory and practically feasible values of F_i and e_o. If the line representing condition V does not intersect the domain A, B, C, D, such as in cases (a) or (c) of Fig. 4.14, then either there is no practical solution (case (a)) or any point of the domain A, B, C, D represents a practically feasible solution (case (c)). In case (a) a new concrete cross section must be used leading to higher section moduli. In case (c), as any point of the domain A, B, C, D is feasible, one must select the one leading to the smallest prestressing force, i.e., point A, intersection of lines representing conditions I and IV. The corresponding analytical solution is obtained by solving two equations, I and IV, to determine two unknowns, F_i and e_o. In case (b) the smallest value for the prestressing force is obtained at point E intersection of IV and V. The corresponding analytical solution is obtained by solving IV for F_i, after replacing e_o by $(e_o)_{mp} = y_b - (d_c)_{min}$.

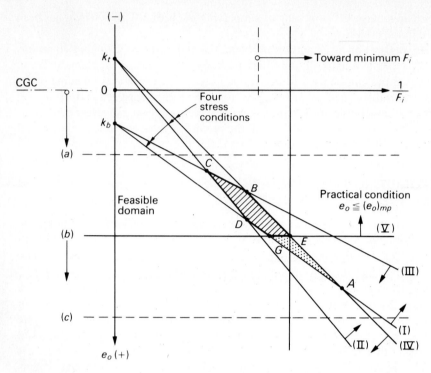

Figure 4.14

Note that the geometric interpretation of the stress conditions gives a very clear picture of what is or what should be done about a particular problem. For example, in a given analysis or design problem, one can plot the feasibility domain and check if the proposed values of F and e_o are represented by a point inside the domain; if it is inside, there is no need to check the stresses; if it is not, one can spot right away the condition or conditions that are not satisfied and devise a corrective action. Other types of practical questions that can be best answered by using the above geometric representation are as follow: (1) Given an eccentricity e_o, what are the maximum and minimum feasible values of prestressing force? (2) Given a prestressing force, what is the range of feasible eccentricities at a given section?

A typical example is treated next, in which the geometric representation is used in both an investigation problem and a design problem where the concrete section is given.

4.9 EXAMPLE: ANALYSIS AND DESIGN OF A PRESTRESSED BEAM

This example is also continued in Secs. 4.12, 4.15, 5.5, 6.10, 6.18, 7.7, and 7.8.

Consider the pretensioned simply supported member shown in Fig. 4.15 with a span length of 70 feet. It is assumed that $f'_c = 5000$ psi, $f'_{ci} = 4000$ psi, $\bar{\sigma}_{ti} = -189$ psi, $\bar{\sigma}_{ci} = 2400$

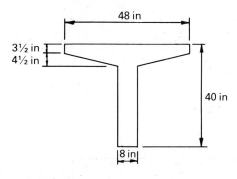

Figure 4.15

SECTION PROPERTIES

$A_c = 550$ in²

$I = 82,065$ in⁴

$y_t = 12.9$ in.; $y_b = 27.1$ in.

$Z_t = 6362$ in³; $Z_b = 3028$ in³

$k_t = -5.51$ in.; $k_b = 11.57$ in.

$w_G = 0.573$ klf

$$\bar{\sigma}_{ti} = -3\sqrt{f'_{ci}}$$
$$\bar{\sigma}_{ci} = 0.60\, f'_{ci}$$
$$\bar{\sigma}_{cs} = 0.45\, f'_c$$
$$\bar{\sigma}_{ts} = -6\sqrt{f'_c}$$

psi, $\bar{\sigma}_{ts} = -424$ psi, $\bar{\sigma}_{cs} = 2250$ psi. Normal weight concrete is used, i.e., $\gamma_c = 150$ pcf, live load $= 100$ psf and superimposed dead load $= 10$ psf. Assume that $\eta = F/F_i = 0.83$ and $(e_o)_{mp} = y_b - 4 = 23.1$ in.

(a) Investigate flexural stresses at midspan given: $F = 229.5$ kips (corresponding to ten $\frac{1}{2}$-in diameter strands) and $e_o = 23.1$ in.
Note : $F_i = F/\eta = 276.5$ kips.

In order to calculate the stresses the geometric properties of the section (given in Fig. 4.15) and the applied moments are needed.

$$\text{Minimum moment } M_{min} = M_G = 0.573(70^2/8) = 350.962 \text{ kips-ft}$$

$$\frac{110 \times 70 \times 4}{1000 \times 70}$$

Additional moment due to superimposed dead load and live load $\Delta M = 0.44(70^2/8) = 269.5$ kips-ft.

$$\text{Maximum moment } M_{max} = M_{min} + \Delta M = 620.462 \text{ kips-ft.}$$

Referring to the four stress inequality equations given in Table 4.2 (way 1) and multiplying the values of moments by 12,000 in order to have units of pounds per square inch leads to:

$$\text{Condition I} \qquad \sigma_{ti} = \frac{F_i}{A_c}\left(1 - \frac{e_o}{k_b}\right) + \frac{M_{min}}{Z_t}$$

$$\sigma_{ti} = \frac{276,500}{550}\left(1 - \frac{23.1}{11.57}\right) + \frac{350.962 \times 12,000}{6362} \simeq 161 \text{ psi} > \bar{\sigma}_{ti} = -189 \text{ psi} \qquad \text{OK}$$

The results for the other conditions are given as follows:

$$\text{Condition II} \qquad \sigma_{ci} = 1219 \text{ psi} < \bar{\sigma}_{ci} = 2400 \text{ psi} \qquad \text{OK}$$

$$\text{Condition III} \qquad \sigma_{cs} = 754 \text{ psi} < \bar{\sigma}_{cs} = 2250 \text{ psi} \qquad \text{OK}$$

$$\text{Condition IV} \qquad \sigma_{ts} = - \Leftrightarrow V \Leftrightarrow \text{psi} > \bar{\sigma}_{ts} = -424 \text{ psi} \qquad \text{OK}$$

Therefore the section is satisfactory with respect to flexural stresses.

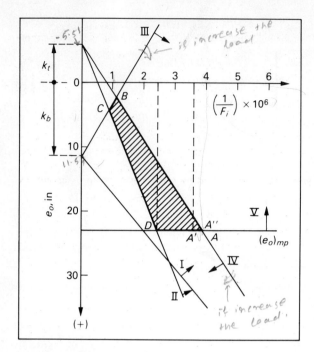

Figure 4.16

(b) Plot the feasibility domain for the above problem and check geometrically if allowable stresses are satisfied.

The equations at equality given in Table 4.2 (way 2) are used to plot linear relationships of e_o versus $1/F_i$ on Fig. 4.16. They are reduced to the following form:

Condition I $\qquad e_o = 11.57 + 5.410\left(\dfrac{10^6}{F_i}\right)$

Condition II $\qquad e_o = -5.51 + 11.4787\left(\dfrac{10^6}{F_i}\right)$

Condition III $\qquad e_o = 11.57 - 8.276\left(\dfrac{10^6}{F_i}\right)$

Condition IV $\qquad e_o = -5.51 + 7.424\left(\dfrac{10^6}{F_i}\right)$

where e_o is in inches and F_i is in pounds. Also equation V showing $(e_o)_{mp} = 23.1$ in is plotted in Fig. 4.16. The five lines delineate a feasibility domain $ABCD$.

Let us check if the given values of F_i and e_o are represented by a point which belongs to the feasible region:

$$\frac{1}{F_i} = \frac{1}{276{,}500} \simeq 3.6 \times 10^{-6}$$

The representative point is shown in Fig. 4.16 as point A'. As it is on line AD, it belongs to the feasible region and therefore all allowable stresses are satisfied. Note that all stresses would still be satisfied if the eccentricity is reduced to approximately 21 in for the same force. This is shown as point A'' on line AB and allows the designer to accept a reasonable tolerance on the value of e_o actually achieved during the construction phase.

(c) Assuming the prestressing force is not given, determine its value.

This is essentially a typical design problem where the concrete cross section is given. It can be solved directly analytically or from the graphical representation of the feasibility domain. Anyway, the graphical representation helps in the analytic solution. It dictates the choice of point A of Fig. 4.16 as the solution that minimizes the prestressing force. Point A corresponds to the intersection of the line representing $(e_o)_{mp}$ with that representing stress condition IV. The corresponding value of F is obtained by replacing e_o by $(e_o)_{mp}$ in Eq. IV (way 3) of Table 4.2; that is:

$$F = \frac{M_{max} + \bar{\sigma}_{ts} Z_b}{(e_o)_{mp} - k_t} = \frac{620.462 \times 12,000 - 424 \times 3028}{23.1 + 5.51} = 215,368 \text{ lb}$$

and

$$F_i = \frac{F}{0.83} = 259,479 \text{ lb} \simeq 259.5 \text{ kips}$$

Graphically the coordinates of A can be read on Fig. 4.16 as $e_o = 23.1$ in and $1/F_i = 3.9 \times 10^{-6}$ which leads to $F_i \simeq 257,000$ lb $= 257$ kips. It can be seen that the graphical solution gives essentially the same answer as the analytical one. Note that the practical value of the prestressing force to use in the design should correspond to an integer number of tendons. In this case, exactly 9.38 strands each with a final force of 22.95 kips would be required. The number is rounded off to 10. The resulting higher prestressing force allows for an acceptable tolerance on the value of e_o.

(d) If the beam is to be used with different values of live loads, what is the maximum value of live load it can sustain?

Referring to the stress conditions, it can be observed that conditions I and II do not change and therefore lines I and II on Fig. 4.16 are fixed. Increasing the value of the live load will increase the value of M_{max} and thus will change the slopes of lines III and IV so as to reduce the size of the feasible domain. Consequently, point A of the feasible domain will move in the direction of AD and line BA tends to rotate toward CD. The maximum value of live load in this case is the one that will make lines II and IV have same slopes. Therefore:

$$\frac{M_{max} + \bar{\sigma}_{ts} Z_b}{\eta} = 11.4787 \times 10^6$$

which leads to $M_{max} = 10,811,193$ lb-in $= 900.93$ kips-ft. Subtracting from M_{max} the values of moments due to dead load and superimposed dead load, leads to the live load moment, from which the live load can be determined as 214.5 psf. The representative point on Fig. 4.16 is D which shows the following coordinates: $e_o = 23.1$ in and $10^6/F_i \simeq 2.5$, i.e., $F_i \simeq 400,000$ lb $= 400$ kips. The reader is encouraged to check numerically in this case that the two allowable stresses $\bar{\sigma}_{ci}$ and $\bar{\sigma}_{ts}$ are attained while the two others are satisfied, as the geometric representation indicates. Note that such a design may have to be revised if the assumed value of e_o cannot be practically achieved.

4.10 USE OF THE STRESS INEQUALITY CONDITIONS
FOR THE DESIGN OF SECTION PROPERTIES

It has been shown in the preceding two sections that to determine feasible values of F and e_o the stress conditions can be used (1) in an investigation problem to check if allowable stresses are satisfied and (2) in a design problem where the concrete cross section is known. As there are four stress inequality conditions, one can attempt to use them at equality to determine four unknowns, namely F, e_o, and two others related to the dimensions or geometric properties of the section. Generally, the two section moduli Z_t and Z_b are sought as they do not necessarily relate to any particular shape of the section.

It is important to understand that the main objective of the design here is to use the most efficient beam cross section, i.e., the smallest possible section moduli. Once a satisfactory section has been selected, the objective becomes to minimize the required prestressing force (or maximize the eccentricity).

Let us assume that we have a cross section such that at the critical (say midspan) section of the beam the two allowable stresses $\bar{\sigma}_{ti}$ and $\bar{\sigma}_{ci}$ are attained under initial loading. Referring to Fig. 4.17, it means that under initial loading $(F_i + M_{min})$ the flexural stress diagram along the section is represented by line ab.

Let us assume that prestress losses occur suddenly (or, equivalently, that time is allowed to pass so that prestress losses take place) and that the loading becomes $(F + M_{min})$, thus leading to stress diagram $a'b'$. If an additional moment is applied, the stress diagram tends to rotate around point Ω which represents the neutral axis for bending. Let us assume that such an additional moment is the moment amplitude $\Delta M = M_{max} - M_{min}$. If under the action of ΔM the two allowable stresses $\bar{\sigma}_{ts}$ and $\bar{\sigma}_{cs}$ are not attained, it indicates that the section moduli Z_t and/or Z_b of the

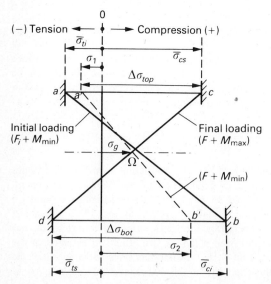

Figure 4.17

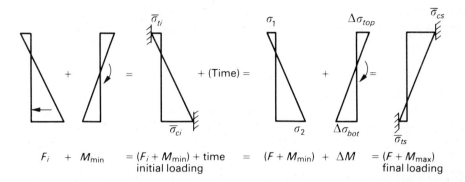

$$F_i + M_{min} = (F_i + M_{min}) + \text{time} = (F + M_{min}) + \Delta M = (F + M_{max})$$
$$\text{initial loading} \qquad\qquad\qquad\qquad\qquad \text{final loading}$$

Figure 4.18

section are larger than required (as $\Delta\sigma = \Delta M/Z$). Similarly, if the two allowable stresses are exceeded, it indicates that the section moduli are smaller than required. One can therefore attempt to determine the minimum required values of Z_t and Z_b for which allowable stresses would be just attained.

Referring to Fig. 4.18 where the flexural stress diagrams of the steps assumed above are represented separately, let us, for example, determine the stresses on the top fiber.

For the initial loading $(F_i + M_{min})$, we have:

$$\frac{F_i}{A_c}\left(1 - \frac{e_o A_c}{Z_t}\right) + \frac{M_{min}}{Z_t} = \bar\sigma_{ti} \tag{4.4}$$

Assuming that all prestress losses have occurred leads to a loading $(F + M_{min})$ for which:

$$\frac{\eta F_i}{A_c}\left(1 - \frac{e_o A_c}{Z_t}\right) + \frac{M_{min}}{Z_t} = \sigma_1 \tag{4.5}$$

which can also be written as:

$$\eta\left[\frac{F_i}{A_c}\left(1 - \frac{e_o A_c}{Z_t}\right) + \frac{M_{min}}{Z_t}\right] - \frac{\eta M_{min}}{Z_t} + \frac{M_{min}}{Z_t} = \sigma_1 \tag{4.6}$$

Using Eq. (4.4) into Eq. (4.6) gives:

$$\eta\bar\sigma_{ti} + \frac{M_{min}(1 - \eta)}{Z_t} = \sigma_1 \tag{4.7}$$

If we add a moment amplitude ΔM to the section, the corresponding additional stress on the top fiber will be

$$\Delta\sigma_{top} = \frac{\Delta M}{Z_t} \tag{4.8}$$

The resulting stress due to the combined effect of $(F + M_{min})$ and ΔM must be less than or equal to the allowable compressive stress $\bar{\sigma}_{cs}$, i.e.,

$$\sigma_1 + \Delta\sigma_{top} \leq \bar{\sigma}_{cs} \tag{4.9}$$

Using Eqs. (4.7) and (4.8) in Eq. (4.9) gives:

$$\eta\bar{\sigma}_{ti} + \frac{M_{min}(1-\eta)}{Z_t} + \frac{\Delta M}{Z_t} \leq \bar{\sigma}_{cs} \tag{4.10}$$

Noting that $M_{min} + \Delta M = M_{max}$, Eq. (4.10) leads to:

$$\boxed{Z_t \geq \frac{M_{max} - \eta M_{min}}{\bar{\sigma}_{cs} - \eta\bar{\sigma}_{ti}}} \tag{4.11}$$

By similarly examining the state of stress on the bottom fiber, it can be shown that:

$$\boxed{Z_b \geq \frac{M_{max} - \eta M_{min}}{\eta\bar{\sigma}_{ci} - \bar{\sigma}_{ts}}} \tag{4.12}$$

Equations (4.11) and (4.12) have been first derived by Guyon (Ref. 4.1) and expanded thereafter by Nilson (Ref. 4.2) and the author (Ref. 4.3).

Equations (4.11) and (4.12) can also be rewritten in functions of ΔM and M_{min} as follows:

$$\boxed{Z_t \geq \frac{\Delta M + (1-\eta)M_{min}}{\bar{\sigma}_{cs} - \eta\bar{\sigma}_{ti}}} \tag{4.13}$$

$$\boxed{Z_b \geq \frac{\Delta M + (1-\eta)M_{min}}{\eta\bar{\sigma}_{ci} - \bar{\sigma}_{ts}}} \tag{4.14}$$

The advantage of this form is that it separates ΔM, which is independent of the beam cross section, from M_{min} which is generally equal to the deadweight moment of the beam. Equations (4.11) and (4.12) or, equivalently, Eqs. (4.13) and (4.14) can be used in combination with up to two other stress inequality conditions to determine Z_t, Z_b, F, and e_o.

Note that the required values of Z_t and Z_b given by Eqs. (4.11) and (4.12) are functions of M_{min} (i.e., the dead load moment in this case) which itself depends on the weight of the section; knowing the weight of the section implies the knowledge of section dimensions, thus the values of Z_t and Z_b. In general a solution for the required values of Z_t and Z_b is obtained by assuming, to start the design, a cross section (and corresponding M_{min}) and through a number of rapidly converging iterations. As observed earlier, the required values of Z_t and Z_b apply whatever the shape of the cross section of concrete. It can be shown, however, that the simpler the shape of the section, such as a slab, the easier it is to obtain a satisfactory solution (Ref. 4.3). This will be discussed in the chapter on optimum design in Vol. 2.

The determination of a cross section of concrete having the minimum required values of Z_t and Z_b guarantees a feasibility domain for the prestressing force and its eccentricity (Fig. 4.14). This domain may be as small as a single point for which the four stress conditions are satisfied. However, the presence of a feasible domain satisfying the four allowable stresses does not guarantee that the domain or part of it falls inside the cross section so as to allow the prestressing steel to be placed inside the beam.

One additional requirement must therefore be satisfied. Generally, once a beam cross section has been selected the practicality condition V (Table 4.2) is checked; if it is not satisfied such as in case (a) of Fig. 4.14 the cross section dimensions must be accordingly changed. This will invariably lead to increases in section properties.

Once a satisfactory cross section of concrete has been selected either by direct solution or from a list of standard sections, the determination of F and e_o is achieved according to Secs. 4.8 and 4.9.

Note that finding a concrete cross section with section moduli both exactly equal to those required may not be possible if a specific shape of cross section is desired. This happens, for example, if both required values of Z_t and Z_b are equal and if a T section is desired, or if both values are unequal and a rectangular section is desired. Generally, it is possible to determine a section with one of the section moduli exactly equal to that required while the other is larger than required.

The determination of section moduli Z_t and Z_b of a trial section, nonrectangular in shape, can be time-consuming. Charts may be prepared to expedite the design at least during the trial-and-error phase. Such a chart (Fig. 4.19) has been developed for use with idealized I and box sections and applies (at the limit) to T and inverted T sections as well. It can be used (1) to estimate the section moduli of a concrete cross section with given dimensions or (2) to estimate the dimensions of a potential concrete section given required values of section moduli.

Although the methodology described in this chapter aims at using the most efficient cross section (or equivalently the least weight beam), it may not necessarily lead to a least cost design. However, for currently prevailing unit costs of materials, it seems to lead to a near minimum cost solution. More on minimum weight versus minimum cost designs can be found in Vol. 2.

4.11 EXAMPLE OF DETERMINATION OF SECTION DIMENSIONS WITH REQUIRED MINIMUM SECTION PROPERTIES

As mentioned in the previous section, the determination of a concrete cross section with section moduli equal to those required leads to a minimum weight beam. Two examples are treated next and illustrate the methodology used and the level of difficulties that might be encountered.

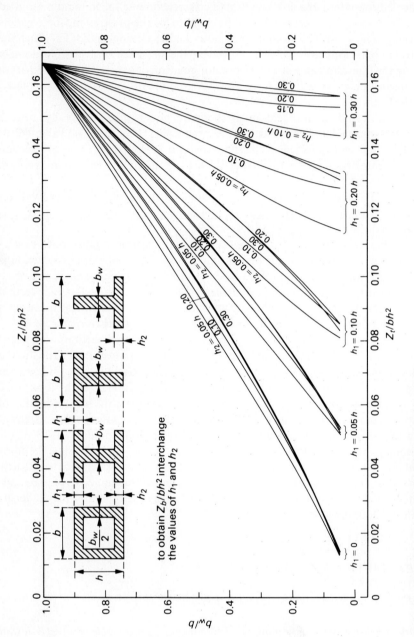

Figure 4.19 Chart to determine section moduli.

For the following examples use a simply supported prestressed element with a span of $l = 40$ ft and:

$$\bar{\sigma}_{ti} = -189 \text{ psi} \qquad \bar{\sigma}_{ci} = 2400 \text{ psi}$$

$$\bar{\sigma}_{ts} = 0 \qquad \bar{\sigma}_{cs} = 2250 \text{ psi}$$

$$\eta = F/F_i = 0.80$$

$$(e_o)_{mp} = y_b - 3$$

$$\gamma_c = 150 \text{ pcf}$$

(a) Slab. Design the least weight slab section assuming a live load of 500 psf (Fig. 4.20a).

Referring to the required section moduli given by Eqs. (4.11) and (4.12) it can be observed that both have the same numerator. As a slab by symmetry must have same values of section moduli, the higher value obtained from Eqs. (4.11) and (4.12) will control; the controlling value corresponds to the equation with the smaller denominator. Let us call:

$$\Delta\bar{\sigma} = \text{the smaller of } \left. \begin{matrix} \bar{\sigma}_{cs} - \eta\bar{\sigma}_{ti} = 2401 \text{ psi} \\ \text{and} \quad \eta\bar{\sigma}_{ci} - \bar{\sigma}_{ts} = 1920 \text{ psi} \end{matrix} \right\} = 1920 \text{ psi}$$

thus the minimum required values of section moduli for the slab will be given by:

$$Z = \frac{M_{max} - \eta M_{min}}{\Delta\bar{\sigma}}$$

Let us assume that $M_{min} = M_G$ at the midspan of the slab, and that $M_{max} = M_G + \Delta M =$ moment due to dead load plus live load. The controlling value of Z can be written as:

$$Z = \frac{\Delta M}{\Delta\bar{\sigma}} + \frac{(1 - \eta)M_G}{\Delta\bar{\sigma}}$$

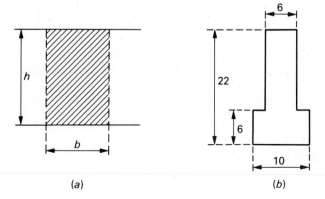

(a)

(b)

Figure 4.20

But for a simply supported slab section of dimensions b and h we have the following:

$$M_G = \gamma_c \frac{bhl^2}{96}$$

$$Z = \frac{bh^2}{6}$$

where γ_c is in pounds per cubic foot, b and h in inches, l in feet, and M in pound-inches. Replacing M_G and Z by their values in the above equation leads to:

$$\frac{bh^2}{6} = \frac{\Delta M}{\Delta\bar{\sigma}} + \frac{(1-\eta)}{\Delta\bar{\sigma}} \gamma_c \frac{bhl^2}{96}$$

which could be written as a quadratic equation in h:

$$h^2 \frac{b}{6} - h \frac{(1-\eta)\gamma_c \, bl^2}{96 \, \Delta\bar{\sigma}} - \frac{\Delta M}{\Delta\bar{\sigma}} = 0$$

Note that this equation has two roots, one of them positive.

If we use a unit width for the slab, say $b = 12$ in, the positive root is given by:

$$h = \frac{1}{4}\left(\frac{(1-\eta)\gamma_c \, l^2}{8 \, \Delta\bar{\sigma}} + \sqrt{\left[\frac{(1-\eta)\gamma_c \, l^2}{8 \, \Delta\bar{\sigma}} \right]^2 + \frac{8 \, \Delta M}{\Delta\bar{\sigma}}} \right)$$

which for our example with the values of l in feet and $\Delta M = $ live load moment $= 1,200,000$ lb-in, leads to

$$h = 18.47 \simeq 18.5 \text{ in}$$

The corresponding values of M_G, M_{max} and Z are:

$$M_G = 555,000 \text{ lb-in}$$

$$M_{max} = 1,755,000 \text{ lb-in}$$

$$Z = 684.5 \text{ in}^3$$

One can check that the controlling required value of Z from Eq. (4.11) leads to about the same result as above.

It can be shown that the feasible domain for this slab section is similar to case (c) of Fig. 4.14; the corresponding minimum value of F is obtained from the coordinates of point A, intersection of lines I and IV for which $F = 195,797$ lb and $e_o = 5.88$ in.

Note that the determination of h as shown in this example does not guarantee that the feasible domain or part of it falls inside the concrete section so as to allow the prestressing steel to be placed inside the beam. Such a case may occur when the magnitude of the live load moment is small compared to that of the dead load moment. The general solution of this problem, for slabs and rectangular beams, is provided in Vol. 2.

(b) **Beam.** Design a least weight beam section assuming the same live load moment as in case (a), that is $\Delta M = 1,200,000$ lb-in.

The required values of Z_b from Eq. (4.14) can be written as:

$$Z_b \geq \frac{\Delta M}{\eta\bar{\sigma}_{ci} - \bar{\sigma}_{ts}} + \frac{(1-\eta)M_G}{\eta\bar{\sigma}_{ci} - \bar{\sigma}_{ts}} = \frac{1,200,000}{1920} + \frac{0.2M_G}{1920}$$

or

$$Z_b \geq 625 + \frac{M_G}{9600}$$

and similarly from Eq. (4.13) we get:

$$Z_t \geq 500 + \frac{M_G}{12{,}005}$$

As the required value of Z_b is larger than that of Z_t, an inverted T section can be considered for the design.

The actual determination of Z_b and Z_t requires trial-and-error procedures and a few iterations as the value of M_G is unknown.

In this case it is desirable to assume a depth h for the cross section and a web width b_w so as to achieve values of section moduli for a rectangular section ($h \times b_w$) smaller than those required. The addition of a suitable flange will then increase the values of Z_b and Z_t to within the requirements. A few iterations are necessary. Figure 4.19 can be used.

For the present example the following values were found adequate: $h = 22$ in, $b_w = 6$ in, to which a net bottom flange of 4×6 in was added; the cross section obtained has the following geometric properties (Fig. 4.20b):

$$A_c = 156 \text{ in}^2 \qquad I = 6693 \text{ in}^4$$

$$y_b = 9.77 \text{ in} \qquad y_t = 12.23 \text{ in}$$

$$Z_t = 547 \text{ in}^3 \qquad Z_b = 685 \text{ in}^3$$

$$k_t = -3.51 \text{ in} \qquad k_b = 4.39 \text{ in}$$

$$w_G = 0.1625 \text{ klf}$$

The corresponding dead load moment at midspan is: $M_G = 390{,}000$ lb-in.

It can be checked that the required values of Z_b and Z_t as derived from Eqs. (4.14) and (4.13) are equal to 665.6 in^3 and 532.5 in^3, respectively, and are slightly smaller than those provided; thus the proposed beam section seems adequate.

It can be shown that the feasible domain for this section is similar to case (c) of Fig. 4.14; the corresponding minimum value of F is equal to 151,284 lb for an eccentricity $e_o \simeq 7$ in. Here, too, note that the determination of cross section dimensions as shown above does not always guarantee that the feasible domain or part of it falls inside the concrete section so as to place the prestressing force inside the beam.

4.12 LIMITING THE ECCENTRICITY ALONG THE SPAN

Once the required prestressing force and its eccentricity have been determined at the critical section, it is generally assumed that the same force will be used throughout the span. Thus there is a need to determine the limiting eccentricities of a known force at any section along the span so that none of the allowable stresses are violated. One way to do this is to build the feasibility domain (such as in Fig. 4.16) at several sections and select the upper and lower limits of eccentricities acceptable for a given prestressing force. Of course, this procedure is too tedious. A more elegant method is developed next. It requires first defining the limit kern.

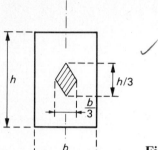

Figure 4.21 Central kern.

1. Limit Kern Versus Central Kern

The central kern area of a section is a region within which an axial compressive force of any magnitude will not produce any tension in the section. The central kern of a rectangular section has the shape of a parallelepiped and spans the middle third of the section in either of the principal directions (Fig. 4.21). The upper and lower limits of the central kern in the y direction have been defined in Sec. 4.4 as k_t and k_b, and can be determined for any type of cross section. For the same cross section throughout the span, they remain constant. It is observed that the central kern depends on the geometry of the section but is independent of the applied compressive force or the allowable stresses. The limit kern, however, accounts for these variables.

The limit kern is the area of the section within which an axial compressive force of a given magnitude can be placed while none of the allowable stresses (tension or compression) are violated. For the same cross section, force, and allowable stresses, the limit kern remains the same throughout the span. The limit kern can be considered as bounding the geometric lieu of the force (C line or pressure line) along the beam.

It was stated in Sec. 4.2 that the combined effect of an axial force $C = F$ and an externally applied moment M is equivalent to that of a force C displaced a distance $\delta = M/F$ from the line of action of F. The position of C with respect to the centroid of the section can be characterized by an eccentricity e_c. It is easily shown (Fig. 4.6c) that:

$$e_c = e_o - \frac{M}{F} \tag{4.15}$$

Note that as e_o and M generally vary along the span e_c also varies. In statically determinate structures e_c becomes equal to e_o when no external loading is applied. This is not the case for statically indeterminate structures (Chap. 10).

Similarly to the central kern, the limit kern is bound respectively by an upper and a lower limit, k'_t and k'_b. Their values can be determined as shown next from the four stress inequality conditions given in Table 4.2.

Let us first define the stresses at the centroid of the concrete section under the effect of initial force F_i and final or effective force F:

$$\begin{cases} \sigma_{gi} = \dfrac{F_i}{A_c} \\[2ex] \sigma_g = \dfrac{F}{A_c} = \dfrac{\eta F_i}{A_c} \end{cases} \qquad (4.16)$$

The first stress inequality condition of Table 4.2 can be written as:

$$e_o - \frac{M_{min}}{F_i} \le k_b - \frac{\bar{\sigma}_{ti} Z_t}{F_i} \qquad (4.17)$$

Replacing Z_t by $k_b A_c$ and using the definition of σ_{gi} from Eq. (4.16) in Eq. (4.17) gives:

$$e_o - \frac{M_{min}}{F_i} \le k_b \left(1 - \frac{\bar{\sigma}_{ti}}{\sigma_{gi}} \right) \qquad (4.18)$$

Similarly, the three other stress inequality conditions lead to:

$$e_o - \frac{M_{min}}{F_i} \le k_t \left(1 - \frac{\bar{\sigma}_{ci}}{\sigma_{gi}} \right) \qquad (4.19)$$

$$e_o - \frac{M_{max}}{\eta F_i} \ge k_b \left(1 - \frac{\bar{\sigma}_{cs}}{\sigma_g} \right) \qquad (4.20)$$

$$e_o - \frac{M_{max}}{\eta F_i} \ge k_t \left(1 - \frac{\bar{\sigma}_{ts}}{\sigma_g} \right) \qquad (4.21)$$

The left-hand sides of Eqs. (4.18) to (4.21) represent the eccentricity (Eq. (4.15)) of the C force in the concrete for the two extreme loading conditions. At equality four limiting eccentricities are obtained, two of which will control (Fig. 4.22a). These last two represent the upper and lower limits of the limit kern, k'_t and k'_b. Thus:

$$k'_t = \text{the larger (algebraically) of} \begin{cases} k_b \left(1 - \dfrac{\bar{\sigma}_{cs}}{\sigma_g} \right) \\[2ex] \text{and} \\[2ex] k_t \left(1 - \dfrac{\bar{\sigma}_{ts}}{\sigma_g} \right) \end{cases} \qquad (4.22)$$

$$k'_b = \text{the smaller (algebraically) of} \begin{cases} k_b \left(1 - \dfrac{\bar{\sigma}_{ti}}{\sigma_{gi}} \right) \\[2ex] \text{and} \\[2ex] k_t \left(1 - \dfrac{\bar{\sigma}_{ci}}{\sigma_{gi}} \right) \end{cases} \qquad (4.23)$$

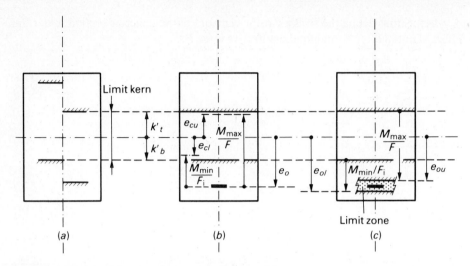

Figure 4.22 (*a*) **Determination of limit kern.** (*b*) **Upper and lower limit of C line for given** e_o. (*c*) **Limit zone for the prestressing force.**

and the eccentricity of the C force shall satisfy the following condition:

$$k_t' \le e_c \le k_b' \tag{4.24}$$

Generally for a given force F and eccentricity e_o, the actual upper and lower limits of e_c, e_{cu} and e_{cl}, for the two extreme loadings will be within the limit kern (Fig. 4.22*b*).

The equations developed above imply the same concrete section, the same prestressing force, and the same allowable stresses throughout the span. If these vary, the equation remains the same but the numerical results will be different depending on the section considered.

The values of k_t' and k_b' given in Eqs. (4.22) and (4.23) were derived from the four stress inequality conditions of Table 4.2. The latter were derived assuming both M_{min} and M_{max} are positive. If both moments were negative, the four complementary conditions given in Table 4.3 should be used and different expressions for k_t' and k_b' would be obtained. However, one can bypass this additional computation and still use the expression given in Eqs. (4.22) and (4.23) provided these steps are followed: (*a*) temporarily assume the section is inverted; (*b*) change the signs of the moments from negative to positive; (*c*) compute $(k_t')_{in}$ and $(k_b')_{in}$ from Eqs. (4.22) and (4.23) for the inverted section and (*d*) compute, for the noninverted section, $k_b' = (-k_t')_{in}$ and $k_t' = (-k_b')_{in}$.

If both positive and negative moments exist at a given section, the upper and lower limits of the limit kern should be determined from the four stress inequality conditions that control the design. These conditions are taken from the eight conditions described in Tables 4.2 and 4.3.

2. Steel Envelopes and Limit Zone

The required prestressing force and its eccentricity are first determined at the critical section of the beam. As F is assumed constant and as the applied external moment varies along the span, the eccentricity of the prestressing force should in general also vary. The question becomes: *Given a prestressing force and an external moment that varies along the span, find the limiting eccentricities of the force at each section so that none of the allowable stresses are violated.* The geometric lieu of the upper and lower limits of these eccentricities along the span are called the *steel envelopes* and the zone between them is called the *limit zone*.

It was shown in the preceding section that the C force in the concrete section is bound by the upper and lower limits of the limit kern k'_t and k'_b. As the eccentricity of the C force e_c is directly related to the eccentricity of the prestressing force e_o (Eq. (4.15)), the limiting eccentricities of the prestressing force can be directly derived from k'_t and k'_b. For the two extreme loadings applied we have (algebraically):

$$e_o - \frac{M_{max}}{F} \geq k'_t \tag{4.25}$$

$$e_o - \frac{M_{min}}{F_i} \leq k'_b \tag{4.26}$$

from which the following condition on e_o is derived

$$k'_t + \frac{M_{max}}{F} \leq e_o \leq k'_b + \frac{M_{min}}{F_i} \tag{4.27}$$

Thus the upper and lower limits of e_o, the eccentricity of the prestressing force, with respect to the centroid of the section, are:

$$e_{ou} = k'_t + \frac{M_{max}}{F} \tag{4.28}$$

$$e_{ol} = k'_b + \frac{M_{min}}{F_i} \tag{4.29}$$

They vary along the span and define the limit zone for the prestressing steel. They can be easily calculated as, from structural analysis, the moments are already known throughout the beam.

In summary, in order to build the limit zone the following steps are recommended (Fig. 4.22c):

1. Determine from Eqs. (4.22) and (4.23) the values of k'_t and k'_b.
2. Determine from Eqs. (4.28) and (4.29) the values of e_{ou} and e_{ol}.

The above steps apply at any section along the span (Fig. 4.23). Once the limit zone is determined, the controlling prestressing force is placed inside this zone by adjusting the eccentricities of various tendons.

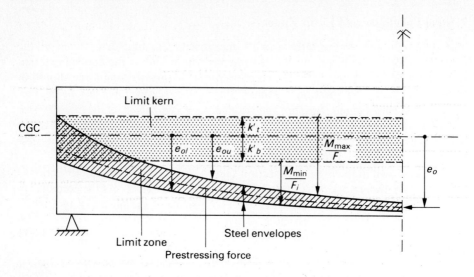

Figure 4.23 Relationship between limit kern and limit zone.

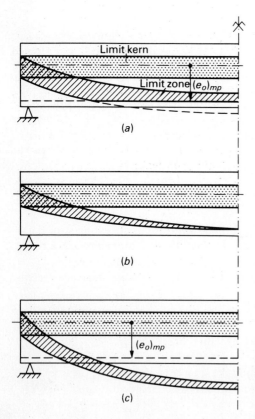

(a)

(b)

(c)

Figure 4.24 Typical shapes of limit zone. (a) Most common design (lower limit controlled by maximum practical eccentricity). (b) Optimum design (only one set of F and e_o is feasible). (c) Inadequate concrete section (preliminary design).

Typical shapes of limit zones are shown in Fig. 4.24. Case (a) is most common and occurs when an adequate concrete section is selected; the lower steel envelope extends beyond the reach of the maximum practical eccentricity, but sufficient area remains to place the prestressing force. Case (b) of Fig. 4.24 applies to optimum design whereas a single set of values of F and e_o at midspan is feasible. Case (c) is typical of preliminary design in which an insufficient cross section of concrete is provided; the limit zone extends outside the beam, thus the prestressing force cannot be placed inside the section. To overcome such difficulty, higher concrete cross-sectional properties are needed.

Often in a preliminary design it is not necessary to determine the limit zone; the profile of the prestressing steel can be approximated by a parabola with a vertex ordinate equal to e_o at midspan and passing by the centroid of the concrete section at supports; another appropriate profile for pretensioned members is a profile with two draping points (Fig. 1.9) where the distance between the draping points is taken about equal to one-third the span length.

Note that the determination of the limit zone is often left to a later stage in the design after the requirements for ultimate and cracking moments are verified.

3. Example

The beam of example Sec. 4.9a is used to illustrate how the limit kern and the limit zone can be determined. First some detailing is needed. The beam is assumed to be pretensioned and to have two draping points for the steel profile. The location of the draping points will be determined from the limit zone. The configuration shown in Fig. 4.25a is selected for the midspan section and comprises six draped strands and four straight ones (such detailing is in agreement with ACI code recommendations (Sec. 3.9)). The resulting eccentricity of the centroid of the strands at midspan is within the feasible range (Fig. 4.16) and thus is acceptable. A detail of how the draped strands can be placed at the ends of the beam is shown in Fig. 4.25b. The centroid of the prestressing steel is brought close to that of the concrete section and within the central kern (also within the limit kern).

In order to determine the upper and lower limits of the limit kern, the stress at the centroid of the concrete section is needed; thus:

$$\begin{cases} \sigma_g = \dfrac{F}{A_c} = \dfrac{229{,}500}{550} = 417.27 \text{ psi} \\[2ex] \sigma_{gi} = \dfrac{F_i}{A_c} = 502.74 \text{ psi} \end{cases}$$

The upper limit of the limit kern is given by Eq. (4.22):

$$k_t' = \text{the larger (algebraically) of} \begin{cases} k_b\left(1 - \dfrac{\bar{\sigma}_{cs}}{\sigma_g}\right) \\[2ex] \text{and} \\[2ex] k_t\left(1 - \dfrac{\bar{\sigma}_{ts}}{\sigma_g}\right) \end{cases}$$

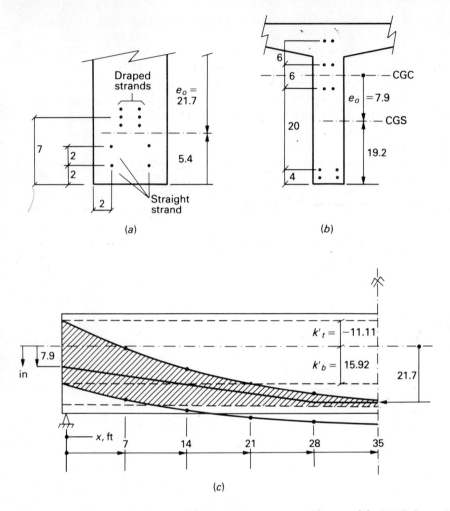

Figure 4.25 Example beam. (*a*) **Steel layout at midspan.** (*b*) **Steel layout at support.** (*c*) **Determination of the limit zone.**

that is:

$$k'_t = \text{the larger of} \begin{cases} 11.57\left(1 - \dfrac{2250}{417.27}\right) = -50.82 \\ \text{and} \\ -5.51\left(1 - \dfrac{-424}{417.27}\right) = -11.11 \end{cases}$$

thus $k'_t = -11.11$ in.

Similarly, the lower limit of the limit kern is given by Eq. (4.23):

$$k'_b = \text{the smaller (algebraically) of} \begin{cases} k_b\left(1 - \dfrac{\bar{\sigma}_{ti}}{\sigma_{gi}}\right) \\ \text{and} \\ k_t\left(1 - \dfrac{\bar{\sigma}_{ci}}{\sigma_{gi}}\right) \end{cases}$$

that is:

$$k'_b = \text{the smaller of} \begin{cases} 11.57\left(1 + \dfrac{189}{502.74}\right) = 15.92 \\ \text{and} \\ -5.51\left(1 - \dfrac{2400}{502.74}\right) = 20.79 \end{cases}$$

thus $k'_b = 15.92$ in.

The upper and lower limits of the limit zone, or the steel envelopes, are determined from Eqs. (4.28) and (4.29). To proceed with the calculations, the minimum and maximum external moments are needed at each section considered. Their value at any section distant x from the support is given by

$$(M_{\min}(x) \text{ or } M_{\max}(x)) = (w_{\min} \text{ or } w_{\max}) \frac{x(l - x)}{2}$$

Table 4.4 Results for example beam

	Distance x from support, ft					
	0	7	14	21	28	35
$M_{\min}$, k-in	0	1516	2695	3538	4043	4211
$M_{\max}$, k-in	0	2680	4765	6254	7148	7445
$\dfrac{M_{\min}}{F_i}$, in	0	5.48	9.75	12.80	14.62	15.23
$\dfrac{M_{\max}}{F}$, in	0	11.68	20.76	27.25	31.14	32.44
$e_{ou} = k'_t + \dfrac{M_{\max}}{F}$	−11.11	0.57	9.65	16.14	20.03	21.33
$e_{ol} = k'_b + \dfrac{M_{\min}}{F_i}$	15.92	21.40	25.67	28.72	30.54	31.15
e_o, in	7.9	11.35	14.80	18.25	21.70	21.70

It is advisable to run the computations at every tenth of the span. The results are summarized in Table 4.4 for the half-span because of symmetry. The following values were used: $F = 229.5$ kips; $F_i = 276.5$ kips; $w_{min} = w_G = 0.573$ klf; $w_{max} = 1.013$ klf. As the eccentricities are in inches and the prestressing force is given in kips, the moments were calculated in kip-inch units.

The upper and lower limits of the steel envelopes are plotted in Fig. 4.25c. As the lower envelope extends below the concrete section, the limit zone (shaded area) is also limited by the maximum practical eccentricity. A steel profile is selected to fit within the limit zone; it has a draping point at a distance from midspan equal to one-tenth the span. It can be easily shown that the corresponding value of the eccentricity of the resultant prestressing force is given by:

$$
\begin{cases}
e_0(x) = 21.7 \text{ in} & \text{for } (28 \leq x \leq 35 \text{ ft}) \\[2mm]
e_o(x) = 7.9 + 13.8\,\dfrac{x}{28} & \text{for } x < 28 \text{ ft}
\end{cases}
$$

where $e_o(x)$ is in inches and x in feet. The numerical values of $e_o(x)$ are summarized in the last line of Table 4.4.

4.13 SOME PRELIMINARY DESIGN HINTS

If the concrete cross section is properly selected for the problem at hand (such as from a handbook of standard beams), it is very likely that the maximum practical eccentricity (Fig. 4.14, case (b)) will control the design. It becomes clear that the prestressing force can then be determined from point E which represents the inter-section of stress condition IV with the line corresponding to $(e_o)_{mp}$. Referring to Table 4.2, the prestressing force can be determined from equation IV at equality in which e_o is replaced by $(e_o)_{mp}$, that is:

$$
F = \eta F_i = \frac{M_{max} + \bar{\sigma}_{ts} Z_b}{(e_o)_{mp} - k_t}
\tag{4.30}
$$

The designer can then check if the other three stress inequality conditions are satisfied for the values of F and e_o. A similar approach was used in the example Sec. 4.9c. If the other three equations are not satisfied, it is better then to build the feasibility domain such as in Fig. 4.14 to see what needs to be done.

A very approximate method can often be used in cost estimate studies to estimate globally the value of the prestressing force. It is derived from observing that in order to have a feasible limit zone the upper eccentricity limit e_{ou} must be less than or equal to the lower eccentricity limit e_{ol}; thus using Eqs. (4.28) and (4.29):

$$
k'_t + \frac{M_{max}}{F} \leq k'_b + \frac{M_{min}}{F_i}
\tag{4.31}
$$

Replacing F_i by F/η and solving for F leads to:

$$
F \geq \frac{M_{max} - \eta M_{min}}{k'_b - k'_t}
\tag{4.32}
$$

which indicates that the minimum value of F is given by

$$(F)_{\min} = \frac{M_{\max} - \eta M_{\min}}{k'_b - k'_t} = \frac{\Delta M + (1 - \eta)M_{\min}}{k'_b - k'_t} \tag{4.33}$$

Because (1) η is generally around 0.80, (2) the dead load moment is often of the same order as the live load moment, and (3) often no tension is allowed in the section, the following approximations can be made:

$$\begin{cases} M_{\max} - \eta M_{\min} \simeq 1.2\,\Delta M & (4.34) \\ k'_b - k'_t \simeq k_b - k_t = \gamma h & (4.35) \end{cases}$$

ΔM is independent of the beam cross section and γ, the geometric efficiency of the section, can be estimated within a reasonable range depending on the section shape as described in Sec. 4.4. Thus Eq. (4.33) becomes:

$$(F)_{\min} \simeq \frac{1.2\,\Delta M}{\gamma h} \tag{4.36}$$

Applied to example Sec. 4.9a, Eq. (4.36) gives:

$$(F)_{\min} \simeq \frac{1.2 \times 269.5 \times 12}{0.427 \times 40} = 227.2 \text{ kips}$$

which is very close to the answer obtained in the exact analysis. If we did not know the value of γ, we could have estimated $\gamma \simeq 0.45$ for a T section and obtained $F \simeq 215.6$ kips.

4.14 CRACKING MOMENT

Although this chapter deals essentially with uncracked prestressed concrete beams, the load or moment at which cracking occurs is needed in the design (Chap. 5). The cracking moment is the moment for which the tensile stress on the extreme fiber of the concrete section reaches a value equal to the modulus of rupture of the concrete. For the bottom fiber of a prestressed concrete section, the cracking moment can be determined from satisfying the following equation:

$$\frac{F}{A_c}\left(1 - \frac{e_o}{k_t}\right) - \frac{M_{cr}}{Z_b} = f_r \tag{4.37}$$

where the first term of the left-hand side of the equation is the stress due to the prestressing force and the second term is the stress due to the cracking moment just before cracking. Solving for M_{cr} gives:

$$M_{cr} = Z_b\left[\frac{F}{A_c}\left(1 - \frac{e_o}{k_t}\right) - f_r\right] = F\left(e_o + \frac{Z_b}{A_c}\right) - f_r Z_b$$

$$= F(e_o - k_t) - f_r Z_b \tag{4.38}$$

Note that the values of k_t and the modulus of rupture are negative.

Let us assume $f_r = -7.5\sqrt{f'_c}$; the cracking moment of the beam treated in Example 4.9a is given by:

$$M_{cr} = 229,500\ (23.1 + 5.51) + 530 \times 3028 = 8,170,835 \text{ lb-in}$$

or

$$M_{cr} = 680.9 \text{ k-ft}$$

4.15 LIMITING THE AMOUNT OF PRESTRESSED REINFORCEMENT

The working stress design procedure followed in this chapter leads to the determination of the prestressing force or equivalently the amount of prestressing steel often described by its reinforcement ratio ρ_p.

In order to ensure that the member is not too much reinforced to lose its ductility and not too little reinforced to collapse should cracking occur, the ACI code contains some provisions that would limit the amount of reinforcement used in the design. As these requirements have to do with the mode of failure of prestressed beams, they are covered in Chap. 5. In the majority of cases, however, the working stress design procedure leads to an amount of reinforcement that satisfies the code limitations.

4.16 END ZONE: PRETENSIONED MEMBERS

1. Transfer Length and Development Length

The prestressing force has to be transferred from the steel to the concrete. For pretensioned members, this is achieved gradually by bond between the two materials mainly in the end part of the member, called end zone. The distance over which the effective prestressing force is transferred to the concrete is called transfer length l_t. After transfer, the tendons have zero stress at the end of the member and an effective stress f_{pe} far away from the end. Hence, the transfer length can be seen as the distance needed to develop the effective prestress f_{pe} in the tendons. Test results of the transfer length of prestressing tendons commonly show large scatters.

It is generally agreed that the transfer length is influenced by many factors that depend on both the steel and the concrete. They include the size and type of tendons (wires versus strands), their surface conditions (smooth, deformed, rusted, and the like), the tendon stress, the method of transfer (sudden versus gentle release), the concrete strength, the concrete compaction, and the state of strain in the transfer region. Observed values of transfer length range from about 50 to 160 times the diameter of the tendon tested. The lower limit is more characteristic of prestressing strands which, because of their twisted shape, have a good mechanical bond.

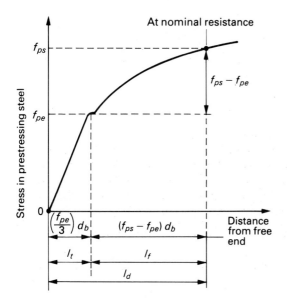

Figure 4.26 ACI assumed variation of steel stress with distance from free end for pretensioned strands.

In order to develop the full design strength of the tendons f_{ps} to resist flexural stresses at nominal moment resistance of the member, an additional bond length is required. It is called the flexural bond length l_f. The flexural bond length l_f added to the transfer length l_t leads to a value called the development length l_d. The following minimum value of development length is prescribed by the ACI code for prestressing strands:

$$l_d = l_t + l_f = \left(\frac{f_{pe}}{3}\right) d_b + (f_{ps} - f_{pe}) d_b = (f_{ps} - \tfrac{2}{3}f_{pe}) d_b \qquad (4.39)$$

where f_{pe} and f_{ps} are in kips per square inch and d_b is the diameter of the strand in inches. Equation (4.39) is graphically illustrated in Fig. 4.26. Using megapascals for stresses and meters for d_b and l_d leads to:

$$l_d = \frac{1}{6.895} (f_{ps} - \tfrac{2}{3}f_{pe}) d_b \qquad (4.40)$$

Equation (4.39) is primarily based on tests of prestressing strands and, hence, does not apply to other types of tendons (Refs. 4.4 to 4.6). An extensive survey of data related to the transfer and development lengths can be found in Ref. 4.7 where regression equations are proposed to also predict l_t and l_d.

The above prescribed minimum value of development length applies in the end zone of a member nearest each support. When partly sheathed (or blanketed) tendons are used, no bond will develop along the sheathed portion of the tendon. The bond will become effective only at the end of the sheathed portion which generally falls outside the end zone. There, the stress transfer between steel and concrete is less effective as the compression, due to the vertical reaction, cannot be counted on and as tensile stresses, due to bending, may exist. The ACI code prescribes the doubling of the development length value given by Eq. (4.39) for cases where the bond does not extend to the end of a member, such as for blanketed strands.

2. End Zone Reinforcement

Because of high tensile splitting stresses that exist at the end of pretensioned members, their end zones must be additionally reinforced by vertical strirrups to limit the openings of cracks. The ACI code does not give any particular guidance on how to determine such reinforcement. However, the AASHTO specifications prescribe the use of stirrups acting at a stress of 20 ksi (138 MPa) to resist at least four percent of the total initial prestressing force F_i. They shall be placed within a distance $d_p/4$ of the end of the beam, the first stirrup being as close to the end as practicable. For at least a distance d_p from the end of the beam, nominal reinforcement shall be placed to enclose the prestressing steel in the bottom flange.

The necessary area of stirrup reinforcement in the end zone of a pretensioned member can also be estimated using a simple semirational approach proposed by Marshall and Mattock (Ref. 4.8), which leads to the following equation:

$$A_v = 0.021 \frac{F_i h}{\bar{f}_s l_t} \tag{4.41}$$

where A_v is the area of vertical stirrups to be uniformly distributed over a distance $h/5$ from the end of the beam and $\bar{f}_s$ is the allowable stress in the stirrups. A value of $\bar{f}_s = 20$ ksi (138 MPa) as recommended by the AASHTO specifications is considered appropriate. The transfer length l_t used in Eq. (4.41) can be taken either equal to 50 tendon diameters or it can be obtained from Eq. (4.39).

Applying Eq. (4.41) to the beam of Example 4.9 leads to the following values:
For

$$h = 40 \text{ in}$$

and

$$l_t \simeq 50 \times \tfrac{1}{2} = 25 \text{ in}$$

$$A_v = 0.021 \frac{276.5 \times 40}{20 \times 25} = 0.465 \text{ in}^2$$

Three #3 closed stirrups will be used within the first eight-inch distance from the end providing a total area of 0.66 in². The stirrups are provided in addition to other stirrups required to resist shear or torsional forces. If, instead of Eq. (4.41), we follow the AASHTO recommendations, the required area of stirrups will be given by:

$$A_v = 0.04 \frac{F_i}{\bar{f}_s} = 0.553 \text{ in}^2$$

which is also achieved practically using three #3 closed stirrups, as found earlier.

4.17 END ZONE: POSTTENSIONED MEMBERS

1. Analysis of Stresses

The prestressing force in a posttensioned member is transferred from the tendon to the concrete essentially by direct bearing through the anchorage. In the immediate vicinity of the bearing area very high compressive stresses exist and transverse tensile stresses develop in the nearby concrete. This is illustrated in Fig. 4.27a for a member of depth h assumed loaded uniformly at its center through a bearing plate of depth approximately equal to $h/4$. The high compressive stresses immediately behind the plate can cause bursting of the concrete (high strains) which generates

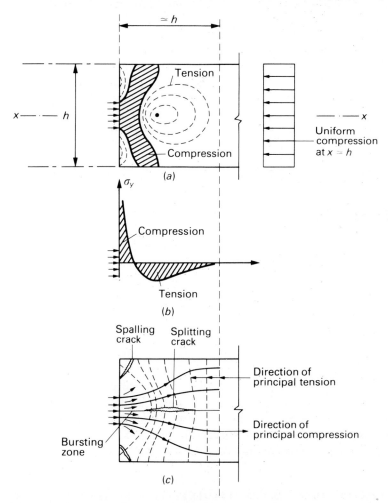

Figure 4.25 Example beam. (a) Steel layout at midspan. (b) Steel layout at support. (c) Determination of the limit zone.

high tensile stresses in the transverse direction away from the plate (Fig. 4.27b). The tensile stress field is such that splitting cracks tend to appear in the horizontal x-x plane accompanied by spalling cracks at the corners of the member (Fig. 4.27c). Such cracks must be contained by addition of proper reinforcement. The effects described in Fig. 4.27 for a single tendon are cumulated and superimposed when many tendons are used. Hence, a zone with high stress concentrations and potentially dangerous cracking develops at the end of a posttensioned member. It is called "lead-in zone," "anchorage zone," or simply "end zone" as for pretensioned members. It extends for a distance about equal to the depth of the member after which splitting tensile stresses become negligible. Elastic analysis indicates that for a loading case such as shown in Fig. 4.27a, the transverse stress (normal to the horizontal x plane) changes from compression to tension at about 0.1 h and the maximum tensile stress occurs at about 0.3 h from the end (Fig. 4.27b).

The accurate determination of stresses in the anchorage zones of posttensioned members is complex and may require more time than the flexural design itself. Several methods essentially based on elastic analysis are available to estimate these stresses. They include in particular, the methods suggested by Guyon (Ref. 4.9), Magnel (Ref. 4.10), and Zielinski and Rowe (Ref. 4.11). Most of these methods are also reviewed at length in traditional texts such as Refs. 4.12 and 4.13. Although these methods lead to a sufficient understanding of the state of stress in the anchorage zone, they do not accurately represent actual conditions of the anchorage. This is because compressive stresses in the immediate vicinity of the anchorage can be of the same order as the strength of the concrete and, hence, are accompanied by high inelastic strains and deformations. Current computational methods, such as nonlinear finite elements, allow for a three-dimensional analysis that should closely correlate with actual behavior. However, for economical reasons such techniques are beyond the scope of most applications. Current design provisions are based on a cumulation of past experience, semirational analysis, and engineering judgment. They are summarized next.

2. Anchorage Zone Design

At the ends of posttensioned members it has been customary to provide zones, called "end blocks," of larger widths than the web widths of the members (Fig. 4.28). The end block is meant to allow for a practical distribution of the anchorages or bearing plates in the end zone and to reduce the magnitude of transverse tensile stresses. However, an investigation by Gergeley and Sozen has concluded that actual tensile stresses are not reduced, but increased instead (Ref. 4.14). Nevertheless, end blocks may still be necessary as suggested by the ACI code to accommodate the anchorages and for support bearing. However, end blocks are specifically required by AASHTO for all posttensioned members. According to the AASHTO specifications end blocks shall have a length at least equal to three-fourths the depth of the member, and in any case not less than 24 in (0.61 m). Preferably they should be as wide as the narrower flange of the member.

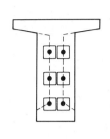

Grid   Stirrups

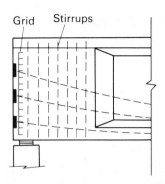

Figure 4.28 Typical end block and end block reinforcement.

A closely spaced grid of both horizontal and vertical bars is required near the face of the end block to resist bursting stresses. When specific recommendations by the supplier of the end anchorages are not available, the grid shall consist of at least 3 bars on 3-in (75-mm) centers in each direction, placed not more than $1\frac{1}{2}$ in (38 mm) from the inside face of the anchor-bearing plate. In addition, closely spaced reinforcement shall be placed both vertically and horizontally throughout the length of the end block in accordance with accepted methods of end block stress analysis.

The design of the anchorage zone (or end block) in posttensioned members is essentially reduced to (1) sizing the anchorages to limit the stresses in the concrete and (2) determining the transverse reinforcement to control splitting tensile cracking in the end zone.

The ACI commentary of the code suggests the following procedure to size end anchorages where either experimental data or a more refined analysis are not available. The average bearing stress in the concrete just behind the anchorage should not exceed the following allowable values:

1. Immediately after tendon anchorage

$$\bar{\sigma}_{bi} = 0.8f'_{ci}\sqrt{A_2/A_1 - 0.2} \leq 1.25f'_{ci} \qquad (4.42)$$

2. After allowance of prestress losses

$$\bar{\sigma}_{bi} = 0.6f'_c\sqrt{A_2/A_1} \leq f'_c \qquad (4.43)$$

where A_1 = bearing area of anchor plate of post-tensioning tendon

A_2 = maximum area of the portion of the anchorage surface that is geometrically similar to and concentric with the area of the anchor plate of the posttensioning tendon

$\bar{\sigma}_{bi}, \bar{\sigma}_b$ = allowable concrete bearing stresses under the anchor plate of posttensioning tendons with the end anchorage zone adequately reinforced

Consideration of the strength reduction factor ϕ (see Sec. 3.8) is already incorporated in Eqs. (4.42) and (4.43). The limiting values on $\bar{\sigma}_{bi}$ and $\bar{\sigma}_b$ suggest that A_2/A_1 cannot be taken larger than about 2.7. The determination of A_2 is illustrated in

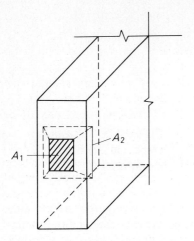

Figure 4.29 Assumed effective bearing area of anchorage.

Fig. 4.29. If the ACI code commentary (Sec. 10.16) is followed, A_2 can be determined by plotting from the periphery of A_1 planes sloping at a ratio of two to one (hence forming a truncated pyramid or, if A_1 is circular, a truncated cone). Experience has shown that good control of end zone cracking can be achieved by keeping the bearing area of all anchorages to less than one-third the area of the end section of concrete and by providing grid or spiral reinforcement in the concrete immediately behind the anchorage or bearing plate.

Although, according to the ACI code, reinforcement shall be provided where required in anchorage zones to resist bursting, splitting, and spalling forces, no

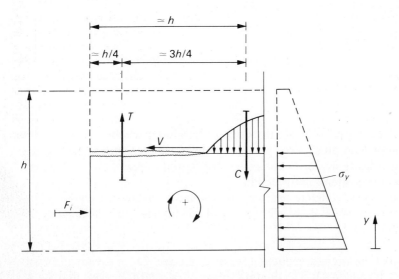

Figure 4.30 Free body model for end zone design *(Ref. 4.14).*

particular guidance is provided on how to determine such reinforcement. A simple method proposed by Gergeley and Sozen can be used (Ref. 4.14). It is based on considering the free body diagram of the end block in which a splitting horizontal crack (in any possible position) has occurred and the crack opening is resisted by vertical stirrups (Fig. 4.30). Moment equilibrium of the free body leads to the required area of stirrups. As a first approximation, the tensile force resisted by the stirrups can be placed at $h/4$ from the end and the counteracting compressive force in the plane of the crack can be placed at a distance h from the end.

In analyzing the free body of concrete end zone below the crack, the vertical component of the prestress at the support is neglected, and undisturbed bonding stresses induced by the prestressing moment are assumed to act at the other end of the end block. Moments are then calculated in the transverse direction (along the horizontal plane) of the free body and reinforcement is provided to resist the maximum moment obtained. Note that the method assumes that there is a longitudinal crack in the end zone and the role of the reinforcement is to confine the crack. The likely position of such crack is at the level of the maximum moment obtained. When moments of different signs are obtained, they may indicate a splitting crack away from the anchorage. Such cracks should also be confined by reinforcement. The procedure is best illustrated by an example.

3. Example: Design of End Zone Reinforcement

Let us consider the beam of Examples 4.9 and 4.12.3 and assume that it will be post-tensioned with the same initial prestressing force $F_i = 276.5$ kips at the same end eccentricity $e_o = 7.9$ in. According to ACI, no end block is needed as there will be very few anchorages in the end zone. For instance, two VSL-type cables (App. C), one with seven strands (and especially designed bearing plate) and one with three strands, can be used. The free body diagram of the end zone, assuming its length is equal to the depth of the beam, that is 40 in, and assuming a constant eccentricity for the prestressing force, is shown in Fig. 4.31a. The vertical component of the prestressing force is neglected. The prestressing force acts on the free end while elastic bending stresses, due to prestressing only, act at the other end. The moments on any horizontal plane of ordinate y can be computed from the contribution of both the prestressing force and the stresses. The moments due to the stresses can be computed by dividing the section vertically into several parts (say 10) and determining the equivalent force in each. Computations for this example are summarized in Table 4.5 and the net moment diagram is shown in Fig. 4.31b. Referring to the stress diagram of Fig. 4.31a, the moment due to concrete stresses can be determined from:

$$(1.224 - \sigma_y) \frac{8y^2}{3} + 4\sigma_y y^2$$

where σ_y is the stress at level y. The above equation is obtained by dividing the trapezoidal stress diagram into a rectangle and a triangle. For instance, at $y = 27.1$ in, which corresponds to the centroid of the concrete section, the moment due to concrete stresses is obtained from:

$$(1.224 - 0.503) \frac{8 \times 27.1^2}{3} + 4 \times 0.503 \times 27.1^2 = 2889.66 \text{ kips-in}$$

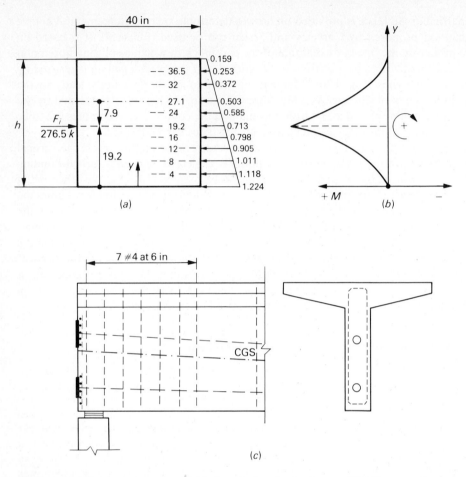

Figure 4.31 Design example. (*a*) End zone free body forces. (*b*) Moment diagram. (*c*) Reinforcement.

The moment due to the prestressing at the same level y is given by:

$$-F_i(y - 19.2) = -2184.35 \text{ kips-in}$$

and the net moment at level y is given by:

$$2889.66 - 2184.35 = 705.31 \text{ kips-in}$$

Referring to Table 4.5, it can be observed that the maximum moment occurs at $y = 19.2$ in. Hence, this will be the preferential level at which a splitting tensile crack will occur. The tensile force T contributed by the needed stirrups can be determined assuming that T acts at $h/4$ from the end of the beam, the corresponding compressive force C in the concrete acts at h from the end, and the couple produced by T and C is equal to the maximum moment,

Table 4.5 Moments at horizontal sections of end zone

Distance y from bottom fiber, in	Moment due to concrete stresses, kips-in	Moment due to F_i, kips-in	Net moment, kips-in
4	76.07	....	76.07
8	295.17	....	295.17
12	643.78	....	643.78
16	1107.97	....	1107.97
19.2	1553.69	....	1553.69
24	2329.34	−1327.2	1002.14
27.1	2889.66	−2184.35	705.31
32	3850.24	−3539.20	311.04
36.5†	4831.62	−4783.45	48.17
40†	5714.5	−5751.2	−36.7 ≃ 0

† The beam width at this level is larger than eight inches.

that is:

$$T\left(h - \frac{h}{4}\right) = 1553.69 \text{ kips-in}$$

from which:

$$T = \frac{1553.69}{30} \simeq 51.79 \text{ kips}$$

Using an allowable stress $\bar{f}_s = 20$ ksi leads to the following required area of stirrups

$$A_v = \frac{51.79}{20} = 2.59 \text{ in}^2$$

For this, seven #4 closed stirrups at approximately six-inch spacing can be used (Fig. 4.31c). These stirrups are in addition to those required for shear.

REFERENCES

4.1. Y. Guyon: *Prestressed Concrete, Vol. 1*, John Wiley & Sons, New York, 1960.
4.2. A. H. Nilson: "Flexural Design Equations for Prestressed Concrete Members," *PCI Journal*, vol. 14, no. 1, February 1969, pp. 62–71.
4.3. A. E. Naaman: "Minimum Cost Versus Minimum Weight of Prestressed Slabs," *Journal of the Structural Division, ASCE*, vol. 102, no. ST7, July 1976, pp. 1493–1505.
4.4. N. W. Hanson and P. H. Kaar: "Flexural Bond Tests of Pretensioned Prestressed Beams," *ACI Journal*, vol. 30, no. 7, January 1959, pp. 783–802.
4.5. P. H. Kaar and D. D. Magura: "Effect of Strand Blanketing on Performance of Pretensioned Girders," *PCI Journal*, vol. 10, no. 6, December 1965, pp. 20–34.
4.6. N. W. Hanson: "Influence of Surface Roughness of Prestressing Strands on Bond Performance," *PCI Journal*, vol. 14, no. 1, February 1969, pp. 32–45.

4.7. P. Zia and T. Mostafa: "Development Length of Prestressing Strands," *PCI Journal,* vol. 22, no. 5, September/October 1977, pp. 54–65.

4.8. W. T. Marshall and A. H. Mattock: "Control of Horizontal Cracking in the Ends of Pretensioned Prestressed Concrete Girders," *PCI Journal,* vol. 7, no. 5, October 1962, pp. 56–74.

4.9. Y. Guyon: *Prestressed Concrete, Vol. 2,* John Wiley and Sons, New York, 1960, 741 pp.

4.10. G. Magnel: *Prestressed Concrete,* McGraw-Hill Book Company, New York, 1954, 345 pp.

4.11. J. Zielinski and R. E. Rowe: "An Investigation of the Stress Distribution in Anchorage Zones of Post-Tensioned Concrete Members," Research Report No. 9, September 1960, Publication 41.009, Cement and Concrete Association, Wexham Springs, Slough, England, 32 pp.

4.12. F. Leonhardt: *Prestressed Concrete,* Wilhelm Ernst and Sohn, Berlin, Germany, 1964, 677 pp.

4.13. P. W. Abeles, B. K. Bardhan-Roy, and F. H. Turner: *Prestressed Concrete Designer's Handbook,* 2d ed., Viewpoint Publishers, Cement and Concrete Association, Wexham Springs, Slough, England, 1976, 548 pp.

4.14. P. Gergeley and M. A. Sozen: "Design of Anchorage Zone Reinforcement in Prestressed Concrete Beams," *PCI Journal,* vol. 12, no. 2, April 1967, pp. 63–75.

PROBLEMS

4.1 A plain concrete beam 12×18 inches in cross section supports, in addition to its own weight, a uniform live load of 100 plf on a simple span of 40 ft. (Fig. P4.1). Assume $\gamma_c = 150$ pcf and $f'_c = 5000$ psi.

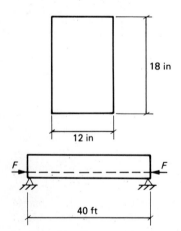

18 in

12 in

$F \rightarrow$ $\leftarrow F$

40 ft

Figure P4.1

(a) Determine the bending stresses at midspan. If the modulus of rupture of the concrete is $-7.5\sqrt{f'_c}$ are such stresses allowable?

(b) Find the magnitude of the smallest longitudinal force which, if applied at the centroid of the cross section, would reduce the tensile stress at the bottom of the beam to zero under full load at midspan.

(c) Find the magnitude of the smallest longitudinal force which, if applied at the bottom of the central kern at midspan, would also lead to zero stress at the bottom of the beam under full load.

(d) Find the magnitude of the smallest longitudinal force which, if applied at 6 in from the center line, would also produce the same result.

4.2 Consider the simply supported rectangular beam (Fig. P4.2) for which we have the following information: $\bar{\sigma}_{ti} = \bar{\sigma}_{ts} = 0$; $\bar{\sigma}_{ci} = 2400$ psi; $\bar{\sigma}_{cs} = 2000$ psi; span $= 40$ ft; $M_G = 720,000$ lb-in; $(e_o)_{mp} = 9$ in; $\eta = 0.80$. Determine the live load ($w_L =$ plf) for which the point of intersection of the two lines representing stress conditions I and IV leads to a value of eccentricity equal $(e_o)_{mp}$.

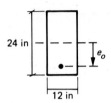

24 in

12 in

Figure P4.2

4.3 Given the rectangular simple span beam (Fig. P4.3) and the following information: span $= 30$ ft; live load $= 0.64$ klf; $w_G = 0.225$ klf; $f'_c = 5000$ psi; $f'_{ci} = 4000$ psi; $\bar{\sigma}_{ti} = -189$ psi; $\bar{\sigma}_{ci} = 2400$ psi; $\bar{\sigma}_{ts} = 0$; $\bar{\sigma}_{cs} = 2000$ psi; $\eta = 0.80$.

(a) Determine the magnitude and eccentricity of the minimum prestressing force at midspan. Build geometrically the feasible domain for F and e_o.

(b) Based on the answer found in (a) determine the limits of the limit kern.

(c) Using the information provided by (a) and (b) determine the envelopes of the prestressing force at every tenth of the span from support to midspan.

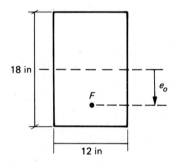

18 in

F

e_o

12 in

Figure P4.3

4.4 You are exploring the feasibility of posttensioning a double cantilever slab using straight prestressing bars (Fig. P4.4). Information on section properties, allowable stresses, loading and bending moments are given below: $h = 15$ in; $b = 12$ in; $\eta = 0.8$; $(e_o)_{mp} = 5$ in; $\bar{\sigma}_{ti} = \bar{\sigma}_{ts} = 0$; $\bar{\sigma}_{ci} = 2400$ psi; $\bar{\sigma}_{cs} = 2000$ psi; $w_G = 187.5$ plf; $w_L = 100$ psf; moments at support A, $M_{max} = -172.5$ kips-in and $M_{min} = 112.5$ kips-in; moments at midspan, $M_{max} = 577.4$ kips-in and $M_{min} = 337.4$ kips-in.

(a) Find graphically a prestressing force F and its eccentricity e_o suitable for the elastic solution of the problem (i.e., do not check ultimate moment, shear, etc.).

(b) If you were told that the minimum prestressing force for both sections A and B corresponds to satisfying stress condition IV, derive the analytical solution for question (a).

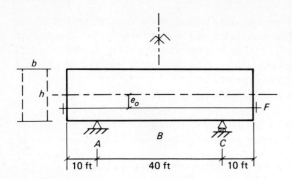

Figure P4.4

4.5 A slab deck is composed of simply supported standard precast pretensioned double T concrete beams to be obtained from a local supplier (Fig. P4.5). Typical cross section and available dimensions are given in Table P4.5 and Fig. (P4.5).

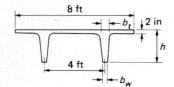

Figure P4.5

Table P4.5

h in	b_t in	b_w in	A in^2	I in^4	y_t in	y_b in	Z_t in^3	Z_b in^3	W_G plf
16	8.00	6.00	388	8,944	4.87	11.13	1837	804	310
18	9.75	7.75	472	14,623	6.16	11.84	2374	1235	377
20	9.75	7.50	503	19,354	6.94	13.06	2789	1482	401
24	9.75	7.00	560	31,192	8.49	15.51	3674	2011	447
32	8.00	4.00	549	51,286	10.29	21.71	4984	2362	439

Other information is given as follows: $f'_c = 5000$ psi; $f'_{ci} = 3750$ psi; $\bar{\sigma}_{ti} = -184$ psi; $\bar{\sigma}_{ci} = 2250$ psi; $\bar{\sigma}_{ts} = -424$ psi; $\bar{\sigma}_{cs} = 2250$ psi; $\eta = 0.85$; $(e_o)_{mp} = y_b - 3.5$ in; span = 60 ft; super-imposed dead load plus live load = 55 psf or 75 psf. For each of the above two loadings, select the least weight beam which satisfies the working stress design requirements in flexure and determine the corresponding values of F and e_o at midspan.

4.6 A T beam (Fig. P4.6) supports in addition to its own weight a live load of 60 psf. The following information is provided: $f'_c = 6000$ psi; $f'_{ci} = 4500$ psi; $\bar{\sigma}_{ti} = -201$ psi; $\bar{\sigma}_{ci} =$

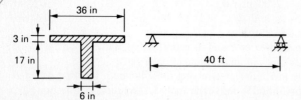

Figure P4.6

2700 psi; $\bar{\sigma}_{ts} = -465$ psi; $\bar{\sigma}_{cs} = 2700$ psi; $\gamma_c = 150$ pcf; $\eta = 0.80$; $(d_c)_{min} = 3$ in; $f_{pu} = 270$ ksi; $f_{pe} = 151$ ksi; final effective force of 1 strand $= 23.1$ kips.

(a) Assuming you are told there is a domain of feasibility for F and e_o, determine the value of F necessary at midspan. Round off its value to the nearest integer number of strands. Check that all stresses are within allowable limits.

(b) Determine graphically the feasibility domain for the beam and find graphically the value of F. (Use graph paper.)

(c) Assuming e_o is fixed, what is the maximum value of F that the beam can be subjected to, without any of the allowable stresses being exceeded?

(d) Let us assume that the live load is not specified. Assuming the same $(d_c)_{min}$, what is the maximum value of live load and corresponding F that can be applied to the beam (from a working stress design approach in flexure only).

Going back to question (a):

(e) Determine the two limits of the limit kern.

(f) Determine the upper and lower limits of the steel envelopes at every tenth of the span.

(g) Suggest a profile for the center of gravity of the prestressing steel along the beam. Show midspan as well as end cross section details.

4.7 A double cantilever simply supported beam (Fig. P4.7) supports in addition to its weight a live load of 80 psf and a concentrated load at its ends (as shown) of 750 lb. This

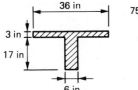 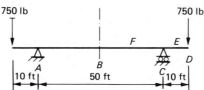

Figure P4.7

concentrated load can be considered as a dead load due to the weight of a wall. The following information is given: $f'_c = 5000$ psi; $f'_{ci} = 4000$ psi; $\bar{\sigma}_{ti} = -189$ psi; $\bar{\sigma}_{ci} = 2400$ psi; $\bar{\sigma}_{ts} = -213$ psi; $\bar{\sigma}_{cs} = 2250$ psi; $f_r = -7.5\sqrt{f'_c}$; $\gamma_c = 150$ pcf; $\eta = 0.80$; $(d_c)_{min} = 3$ in; $f_{pu} = 270$ ksi; $f_{pe} = 150$ ksi; area of one strand $= 0.153$ in^2.

(a) Determine the required value of the prestressing force to be used throughout the length of the beam (i.e., check F at A, F at B, and select the largest value; then check that all stresses are satisfied). Think about using the feasibility domain for solution. Select F corresponding to an integer number of strands.

(b) Determine the upper and lower limits of the steel envelopes at support, midspan, quarter span, end D, and 5 ft from D (i.e., E).

(c) Plot graphically the limit zone and show an acceptable profile for the prestressing force.

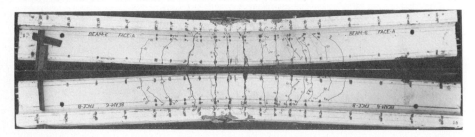

Cracking pattern at failure of test prestressed pretionsioned beams. *(Courtesy Edward Nawy, Rutgers University.)*

CHAPTER 5

FLEXURE:
ULTIMATE STRENGTH ANALYSIS
AND DESIGN

Similarly to reinforced concrete, prestressed concrete offers a great versatility in its flexural behavior. Depending on the values of the design variables and parameters, a prestressed concrete member can be made either to exhibit a great ductility after cracking and before failure, or to fail altogether in a sudden manner. It can be designed to carry a relatively small or large load before failure; thus in order to achieve a good design, it is essential to understand the causal effects of important variables on the behavior at ultimate.

5.1 LOAD-DEFLECTION RESPONSE

The overall behavior of a simply supported prestressed concrete beam subjected to a monotonically increasing load can be well described by its load-deflection curve. Such a typical curve is shown in Fig. 5.1 for an underreinforced beam with bonded tendons. Several points are marked on the curve and correspond to a particular state of behavior. Points 1 and 2 correspond to the theoretically predicted camber of the beam, assumed weightless, when the initial or the effective prestresses are applied respectively. However, when the prestress is applied, self-weight acts automatically. Point 3 represents the camber due to the combined effects of self-weight and the effective prestressing force assuming all prestress losses have taken place. Typical stress diagrams along the cross section of maximum moment corresponding to points 3 to 9 are also shown in Fig. 5.1. If additional load beyond self-weight

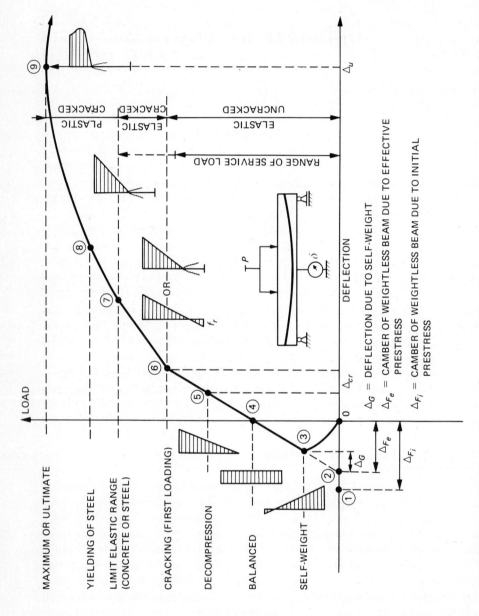

Figure 5.1 Typical load-deflection curve of a prestressed beam (underreinforced; bonded tendons; first loading).

is applied, several points of interest can be identified until failure: point 4 represents the point of zero deflection and corresponds to a uniform state of stress in the section (also called balanced state); point 5 represents decompression or zero stress at the bottom fiber; if cracking has already occurred due to prior loading, or if the tensile strength of the concrete is assumed nil, point 5 would represent the boundary between cracked and uncracked section behavior and thus would take the place of point 6; point 6 represents the onset of cracking in the concrete under first loading; beyond point 6 the prestressed concrete section behaves similarly to a cracked reinforced concrete section subjected to combined bending and compression; if the applied load on the beam keeps increasing the stresses in the steel and the concrete extreme compressive fiber would continue to increase until either material reaches its nonelastic characteristics; this limit is represented by point 7 of Fig. 5.1. For increasing loads, the steel would first reach its yielding strength (bonded tendons) represented by point 8, and finally the maximum capacity of the beam is attained at point 9. Note that point 9 represents the point of maximum load which is described as the ultimate load; generally beyond this point the beam still provides some resistance to increasing deflections but at values of loads less than the ultimate load.

Figure 5.1 shows the load-deflection curve for a prestressed concrete beam using bonded tendons. Everything else being equal, if the prestressed tendons were unbonded the load-deflection curve would, in comparison, fall under the first curve as shown in Fig. 5.2, and very likely failure at ultimate would occur before yielding of the steel. This is because under loading the increase in strain in the unbonded tendon is averaged (as no bond exists) over the whole length of the tendon, and thus is much smaller than the strain increase in the bonded tendon taken at the section of maximum moment. More on the behavior of beams with unbonded tendons can be found in Vol. 2.

It is informative to understand how the stresses in the prestressing steel and in the concrete top fiber vary when the load applied on the beam increases to its maximum or ultimate value. This is illustrated in Fig. 5.3 where the stress values at the various stages described in Fig. 5.1 are qualitatively shown directly on the

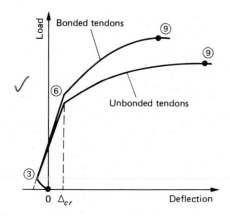

Figure 5.2 Effect of bonded versus unbonded tendons on load-deflection curve.

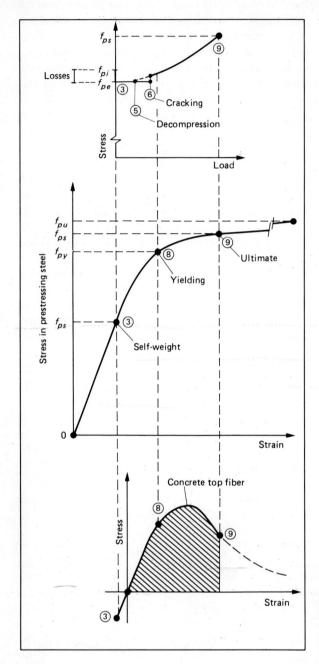

Figure 5.3 Variations of stresses in the concrete and the prestressing steel with applied load.

stress-strain curve of the steel and the concrete top fiber (assuming an under-reinforced beam with bonded tendons). Note that at the ultimate point 9 the stress in the prestressing steel f_{ps} is not equal to its ultimate tensile strength f_{pu}; it is in general smaller because at maximum load, while the concrete reaches its maximum capacity, the stress in the steel increases to the level needed to maintain the force equilibrium in the section. Note that f_{ps} here is larger than the yield strength f_{py}; this is mostly the case for underreinforced beams and leads us to explain the different types of failure that might be observed under flexural loading.

5.2 FLEXURAL TYPES OF FAILURES

The typical load-deflection response of a prestressed concrete beam as described in Fig. 5.1 is a desirable type of behavior; design limitations in various codes tend in general to ensure such behavior. Other types of behavior can, however, be observed. As the load is progressively increased on a simply supported prestresed concrete beam, the following types of flexural failures might occur depending on the amount of steel reinforcement provided:

1. Fracture of the steel immediately after concrete cracking and thus sudden failure
2. Crushing of the concrete compressive zone, preceded by yielding and plastic extension of the steel
3. Crushing of the concrete compressive zone before yielding of the steel

Figure 5.4 qualitatively illustrates the above types of failure, assuming an identical beam reinforced with increasing amounts of reinforcement.

1. Minimum Reinforcement Ratio $\left(\rho_{min} \right)$

When the amount of reinforcement is very small, the steel is not capable of resisting the increase in tensile force generated immediately after cracking to maintain moment equilibrium in the cracked section, and thus a brittle failure occurs by fracture of the steel. To avoid such undesirable failure, codes of practice recommend minimum amounts of reinforcement by specifying a minimum value of the reinforcement ratio. The ACI code does not give a specific limit for prestressed concrete. However, for reinforced concrete it recommends a minimum value $\rho_{min} = 200/f_y$ where f_y is in pounds per square inch. Using megapascals for f_y leads to $\rho_{min} = 1.4/f_y$. For prestressed concrete the British code CP.110 recommends a value $\rho_{min} = 2.5/f_{pu}$ where f_{pu} is in megapascals; it becomes $\rho_{min} = 357/f_{pu}$ where f_{pu} is expressed in pounds per square inch. The above minimum ratios apply for rectangular sections. According to the ACI code, ρ_{min} shall be computed using the width of the web only for T sections. In a general case, it should be quite simple to determine the minimum amount of reinforcement which corresponds to the fracture of the steel immediately after cracking by writing that the ultimate moment is

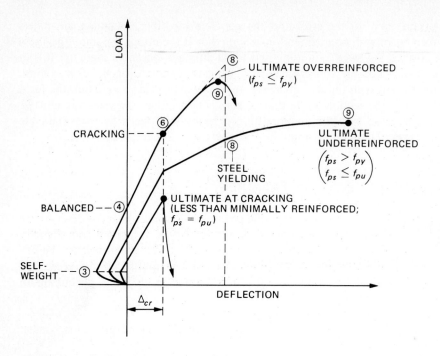

Figure 5.4 Typical change in load-deflection curve with an increase in the amount of reinforcement.

equal to the cracking moment as illustrated below. The cracking moment is given by Eq. (4.38):

$$M_{cr} = A_{ps} f_{pe}\left(e_o + \frac{Z_b}{A_c}\right) - f_r Z_b \tag{5.1}$$

where f_r is the modulus of rupture of the concrete. The moment resistance at ultimate (termed nominal moment M_n by the code) can be determined as a first approximation assuming a lever arm equal to $0.95d_p$. For the purpose of design, ϕM_n should be used, thus:

$$\phi M_n \simeq \phi \times 0.95 A_{ps} f_{pu} d_p \tag{5.2}$$

where $\phi = 0.9$ for flexure. Equating M_{cr} to ϕM_n leads to:

$$A_{ps} = \frac{-f_r Z_b}{0.86 f_{pu} d_p - f_{pe}(e_o + Z_b/A_c)} \tag{5.3}$$

Dividing both sides by bd_p leads to the minimum reinforcement ratio required:

$$\rho_{min} = \frac{A_{ps}}{bd_p} = \frac{-f_r Z_b}{bd_p[0.86 f_{pu} d_p - f_{pe}(e_o + Z_b/A_c)]} \tag{5.4}$$

To avoid brittle failure immediately after cracking, the ACI code recommends that the nominal moment be at least 20 percent higher than the cracking moment. The corresponding minimum reinforcement ratio, if determined as above using the same approximation of the nominal moment as in Eq. (5.2), would be:

$$\rho_{min} = \frac{A_{ps}}{bd_p} = \frac{-1.2f_r Z_b}{bd_p[0.86f_{pu}d_p - 1.2f_{pe}(e_o + Z_b/A_c)]} \qquad (5.5)$$

For the example beam described in Sec. 4.9, the value of ρ_{min} given by Eq. (5.5) is:

$$\rho_{min} = \frac{1.2 \times 530 \times 3028}{48 \times 34.6[0.86 \times 270,000 \times 34.6 - 1.2 \times 150,000(21.7 + 5.51)]} = 0.00037$$

Note that this value is much smaller than that recommended by the British code: $357/270,000 = 0.00132$ because the section is a T section. If the section was rectangular, with the same flange width, the corresponding ρ_{min} given by Eq. (5.5) would be equal to 0.00167 (quite close to 0.00132).

2. Maximum Reinforcement Ratio

When the amount of reinforcement in a given beam is too large, the beam behaves as an overreinforced beam in which the stress in the steel does not reach yielding and little ductility is observed before failure. Going back to Figs. 5.3 and 5.4, point 9 would appear before point 8 on the load-deflection curve and the stress-strain curve of the steel. On the diagram showing the stress change in the concrete top fiber (Fig. 5.3) point 8 would appear after point 9. Thus failure occurs by crushing of the concrete before yielding of the steel. Because of this "brittle" or "nonductile" type of failure, overreinforced beams are normally avoided in design. To achieve this goal in reinforced concrete beams, the ACI code limits the maximum value of the reinforcement ratio to 75 percent of the balanced reinforcement ratio, i.e., $\rho_{max} = 0.75\rho_b$; the balanced ratio corresponds to a failure for which the strain in the steel is exactly equal to the yield strain at the time of concrete crushing. For reinforced concrete the balanced ratio ρ_b can be easily estimated. For instance if the ACI assumptions for the stress and strain distributions in the concrete at ultimate are used, the value of ρ_b would be:

$$\rho_b = 0.85\beta_1 \frac{f'_c}{f_y} \frac{\varepsilon_{cu}}{\varepsilon_{cu} + \varepsilon_y} = 0.85\beta_1 \frac{f'_c}{f_y} \frac{0.003}{0.003 + \varepsilon_y} \qquad (5.6)$$

where $\begin{cases} \beta_1 = 0.85 \text{ for } f'_c \leq 4000 \text{ psi} \\ \beta_1 = 0.65 \text{ for } f'_c \geq 8000 \text{ psi} \\ \beta_1 = 0.85 - 5 \times 10^{-5}(f'_c - 4000) \text{ for } 4000 \leq f'_c \leq 8000 \text{ psi} \end{cases} \qquad (5.7)$

In prestressed concrete, the yield strength and hence the yield strain of the prestressing steel are not very well defined. However, to achieve ductile failure, the ACI code provides a maximum limit of 0.30 for a parameter called the reinforcing index which is defined, for a fully prestressed beam, by $\omega_p = (A_{ps}/bd_p)(f_{ps}/f'_c)$. If the reinforcing index is less than 0.30, the member is considered underreinforced. Note that, although for reinforced concrete members the ACI code does not permit an overreinforced design, the code allows the design of an overreinforced prestressed beam, in spite of the fact that it is undesirable. The author's opinion on this question and the limiting values of the reinforcement ratio and the reinforcing index are expanded upon in Secs. 5.4 and 5.5.

Because of the extensive technical literature available on *ultimate strength design*, only a few references, combining some classical papers, codes, and recent developments, are suggested at the end of this chapter. (Refs. 5.1 to 5.15).

5.3 ANALYSIS OF THE SECTION AT ULTIMATE

The purpose of the analysis at ultimate or maximum load is generally to determine the nominal moment resistance (moment at ultimate behavior) of the section, assuming that cross-sectional dimensions, materials properties, and amounts of reinforcement are given. As pointed out earlier, the stress in the steel and in the concrete at ultimate are well outside their linear range of behavior. An exact analysis which takes into account the actual stress and strain distributions in the section and the constitutive equations of the concrete and the reinforcement is presented in Vol. 2. Such an accurate analysis, however, is not very efficient in everyday situations, and codes of practice tend to propose simplifying assumptions which allow a quick but sufficiently accurate evaluation of nominal moment resistance. Widely accepted assumptions are (for most codes) as follows:

1. Plane sections remain plane under loading. Consequently it is assumed that a linear strain distribution exists along the concrete section up to ultimate load.
2. Perfect bond exists between steel and concrete. (The case of unbonded tendons must be treated separately and is discussed in Vol. 2.) A direct result of this assumption is that any strain change in the steel due to an applied load is equal to the strain change in the concrete at the level of the steel due to the same load.

The ACI code proposes the following additional assumptions (Fig. 5.5):

3. The limiting compressive strain of the concrete ε_{cu} is equal to 0.003, regardless of the strength of the concrete, the shape of the section, and the amount of reinforcement. It is also considered the same for normal weight and structural lightweight concrete.
4. The tensile strength of concrete is neglected. Thus the point of zero stress represents the boundary between the cracked and the uncracked part of the section.
5. The total force in the concrete compressive zone can be well approximated by considering a uniform stress of magnitude $0.85f'_c$ over a rectangular block of width b and depth $a = \beta_1 c$, where c represents the depth of the neutral axis and

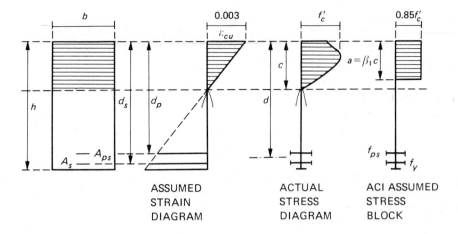

ASSUMED   ACTUAL   ACI ASSUMED
STRAIN   STRESS   STRESS
DIAGRAM   DIAGRAM   BLOCK

Figure 5.5 Stresses and strains at ultimate behavior as assumed by the ACI code.

β_1 a factor defined in Eq. (5.7). The concept of the rectangular stress block was first proposed by Whitney (Ref. 5.1). Other shapes of stress distribution such as the trapezoidal and parabolic shapes or the exact shape have been used and are acceptable by the ACI code (Sec. 10.2.6). However, the rectangular distribution leads to the simplest prediction equations with little loss in accuracy.

In addition to the above assumptions the ACI code introduces in the design a factor called strength reduction factor ϕ, which accounts for the possible loss of resistance due to unintended defects in materials and construction. The ϕ factor applies only when ultimate strength analysis or design are undertaken and its use is not limited to flexure but extends to shear, torsion, compression, and other effects. It is most generally used in the following two relations:

$$\begin{aligned} &\text{(Code acceptable resistance} \\ &\text{or design nominal resistance)} = \phi \times \text{(nominal resistance)} \qquad (5.8) \\ &\text{Design strength} \qquad\qquad\qquad \leq \phi \times \text{(nominal resistance)} \end{aligned}$$

where the nominal resistance is the resistance directly derived from the theoretical equations associated with the problem being studied. Values of the ϕ factor as given in the ACI code are summarized in Table 3.10. The design strength is the minimum strength required by the code and corresponds to the effect of factored loads given in Table 3.9. It is assumed positive throughout.

Whatever the assumptions and the particular approaches taken by different codes, two equations of statics pertaining to the equilibrium (force and moment) of the section at ultimate can be written. They have a different form depending on the shape of section and are developed below for rectangular and T sections with prestressed and nonprestressed conventional reinforcement. The nonprestressed reinforcement is assumed to have an elastic perfectly plastic stress-strain relation, and to be in a yielding state at ultimate behavior of the section. When only prestressing is used, set $A_s = 0$; when only reinforced concrete is used set $A_{ps} = 0$; the equations provided below remain valid.

1. Rectangular Section (or Rectangular Section Behavior)

Referring to Figs. 5.5 and 5.6, the force equilibrium equation can be set by writing that the compressive force in the concrete compression block is equal to the total tensile force in the steel, thus:

$$0.85f'_c ba = A_{ps} f_{ps} + A_s f_y \tag{5.9}$$

The moment equation states that the nominal moment resistance of the section is equal to the internal couple or moment of the above equal and opposite forces, thus:

$$M_n = 0.85f'_c ba \left(d - \frac{a}{2} \right) \tag{5.10}$$

where d is the distance from the extreme compressive fiber of concrete to the centroid of the tensile force; its value can be estimated somewhere in between d_p and d_s or calculated exactly from the following equation (Fig. 5.6):

$$d = \frac{A_{ps} f_{ps} d_p + A_s f_y d_s}{A_{ps} f_{ps} + A_s f_y} \tag{5.11}$$

In the above three equations, the following four variables are considered unknowns: $a, f_{ps}, d,$ and M_n. In order to determine their values, a fourth equation is

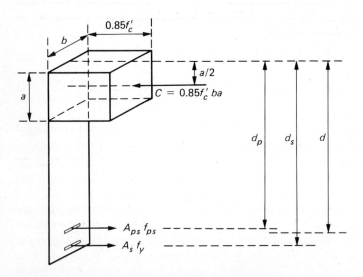

Figure 5.6 Rectangular section: forces at ultimate.

needed. Such an equation is given, for instance, in the ACI code which recommends an expression to estimate the value of f_{ps}, the stress in the prestressing steel at ultimate capacity of the section. Once f_{ps} is known, a can be computed from Eq. (5.9), d from Eq. (5.11), and M_n from Eq. (5.10).

In lieu of a more accurate analysis based on strain compatibility (described in Vol. 2), the ACI code recommends the use of the following approximate values of f_{ps} if f_{pe}, the effective prestress, is not less than $0.5f_{pu}$:

1. For members with bonded prestressing tendons:

$$f_{ps} = f_{pu}\left(1 - 0.5\rho_p \frac{f_{pu}}{f_c'}\right)$$

(5.12)

where f_{pu} is the ultimate strength and ρ_p the reinforcement ratio of the prestressing steel given by A_{ps}/bd_p.

2. For members with unbonded prestressing tendons:

$$
\left|
\begin{array}{l}
f_{ps} = f_{pe} + 10,000 + \dfrac{f_c'}{100\rho_p} \quad \text{in psi} \\[2ex]
f_{ps} = f_{pe} + 69 + \dfrac{f_c'}{100\rho_p} \quad \text{in MPa}
\end{array}
\right.
$$

(5.13)

but f_{ps} in Eq. (5.13) shall not be taken greater than f_{py} or $(f_{pe} + 60,000)$ psi or $(f_{pe} + 414)$ MPa. Note that Eq. (5.12) is valid in all systems of units. For slabs with unbonded tendons, an equation different from Eq. (5.13) is suggested in Secs. 11.2 and 11.9.

Once the nominal moment resistance at ultimate M_n is determined, the design moment resistance acceptable by the ACI code is obtained by multiplying M_n by the strength reduction factor ϕ, thus:

$$\text{Design moment resistance} = \phi M_n$$

(5.14)

ϕM_n will be considered as the nominal moment capacity for the purpose of design and is often made equal to the strength design moment required for the given loading and termed M_u.

It is worth noting that Eq. (5.10), which leads to the value of the nominal moment resistance, is not unique in its form. It could be written in different ways; for instance, if the force in the steel is used instead of using the force in the concrete, Eq. (5.10) becomes:

$$M_n = (A_{ps} f_{ps} + A_s f_y)\left(d - \frac{a}{2}\right)$$

(5.15)

or

$$M_n = A_{ps} f_{ps}\left(d_p - \frac{a}{2}\right) + A_s f_y\left(d_s - \frac{a}{2}\right)$$

(5.16)

Another way to write M_n is in function of the reinforcing index q as shown in Eq. (5.37) of Sec. 5.4, that is:

$$M_n = f_c' \, bd^2 q(1 - 0.59q) \qquad (5.37)$$

This form is convenient because it shows only one unknown, q, instead of two, namely a and f_{ps}. However, it may be misleading because it suggests a direct relation between M_n and f_c' while in reality M_n is generally only slightly sensitive to f_c' because f_c' is in the denominator of q.

To illustrate the use of the above equations, the following example of a rectangular section is developed.

Note also that a general flow chart illustrating the computations of nominal moments according to the steps described in this chapter and the ACI code is given later in Fig. 5.13. It applies to rectangular and T sections, reinforced, prestressed, and partially prestressed with and without compressive steel, and allows for bonded or unbonded tendons. Its content will be fully understood only after the entire treatment of this chapter. It could also be used as an alternative in solving the following example.

2. Example: Analysis at Ultimate of a Rectangular Section

(a) Determine the nominal moment resistance of a partially prestressed concrete rectangular section with dimensions given in Fig. 5.7, for which $f_c' = 5000$ psi, $f_{pu} = 270$ ksi, $f_{pe} = 0.55 f_{pu} = 148.5$ ksi, $f_y = 60$ ksi, $A_{ps} = 0.92$ in^2 or six $\frac{1}{2}$-in-diameter prestressing strands, $A_s = 1.2$ in^2 or two #7 reinforcing bars, $d_p = 20.75$ in, $d_s = 21.5$ in. The prestressing tendons are assumed bonded. The following steps are followed according to the procedure described above and later expanded in the flow chart (Fig. 5.13).

Compute:

$$\rho_p = \frac{A_{ps}}{bd_p} = 0.00369$$

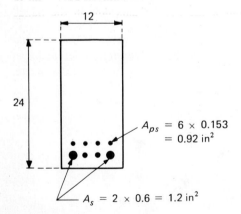

$$A_{ps} = 6 \times 0.153 = 0.92 \text{ in}^2$$

$$A_s = 2 \times 0.6 = 1.2 \text{ in}^2$$

Figure 5.7

From Eq. (5.12) compute

$$f_{ps} = 270\left(1 - 0.5 \times 0.00369 \times \frac{270}{5}\right) = 243 \text{ ksi}$$

From Eq. (5.11) estimate

$$d = \frac{0.92 \times 243 \times 20.75 + 1.2 \times 60 \times 21.5}{0.92 \times 243 + 1.2 \times 60} = 20.93 \text{ in}$$

From Eq. (5.9) compute

$$a = \frac{A_{ps}f_{ps} + A_s f_y}{0.85 f'_c b} = \frac{0.92 \times 243 + 1.2 \times 60}{0.85 \times 5 \times 12} = 5.8 \text{ in}$$

From Eq. (5.10) compute the nominal moment resistance

$$M_n = 0.85 \times 5 \times 12 \times 5.8(20.93 - 5.8/2) = 5333.3 \text{ kips-in} \simeq 444.4 \text{ kips-ft}$$

The moment resistance acceptable by the code is given by Eq. (5.14):

$$\phi M_n = 0.9 M_n = 400 \text{ kips-ft}$$

(b) The reader may want to check that if $A_s = 0$ and $d_p = 21.33$ in, everything else being the same, the design nominal moment resistance of the corresponding fully prestressed section would be $\phi M_n = 321.8$ kips-ft.

(c) Similarly, if, for the same partially prestressed section, unbonded tendons were used, everything else being the same, the following results would have been obtained: $f_{ps} = 172.050$ ksi; $d = 20.98$ in; $a = 4.51$ in; $M_n = 359.3$ kips-ft; $\phi M_n = 323.3$ kips-ft.

3. T Section Behavior

According to the assumption that the tensile strength of concrete is neglected in the analysis, the portion of the concrete section that falls below the neutral axis is considered to offer no resistance and thus is, in effect, ignored in the computation of nominal moment. It could be of any shape, one of which is the shape corresponding to that of a rectangular section. In approaching the analysis particular to T sections, it is important to realize that if the neutral axis falls in the flange (Fig. 5.8), the section is treated exactly similarly to a rectangular section of same

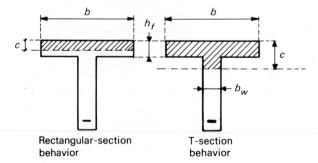

Rectangular-section behavior

T-section behavior

Figure 5.8

width b and the corresponding equations developed above apply without any modification. Thus, depending on the various variables and parameters, a T section does not necessarily imply T section behavior at ultimate. In fact, as the depth of the neutral axis c at ultimate moment capacity of reinforced and prestressed concrete members is generally small in comparison to the depth of the member, T sections behave often as rectangular sections. This is particularly true for small values of the reinforcement ratio or the reinforcing index. If, however, the neutral axis falls outside the flange (i.e., in the web), a special analysis must be undertaken.

It is clear that the first step would be to determine the depth of the neutral axis c. For this, the T section is first assumed to behave as a rectangular section for which c at ultimate capacity is computed from $c = a/\beta_1$ where a is given by Eq. (5.9); c is then compared to the flange depth h_f of the T section and if c is larger than h_f, the section is considered to behave as a T section and the following approach is proposed according to the ACI code: the T section is divided into two fictitious parts, one corresponding to the overhanging portion of the flange and the other to the web alone (Fig. 5.9). The analysis is then pursued whereas the two parts are treated separately as two rectangular sections and their nominal capacities are summed up to give the nominal capacity of the T section.

In order to reduce the computational effort and clarify the approach used it is essential to estimate first the value of d (Eq. (5.11)) when prestressed and non-prestressed reinforcements are used. The design may be revised if d is found inadequate at the end of the first iteration.

Assuming f_{ps} can be estimated from Eq. (5.12) or (5.13), the equation of force equilibrium in the section, which states that the total tensile force in the steel is equal to the sum of compressive forces in the concrete, can be written for $c \geq h_f$, as:

$$A_{ps} f_{ps} + A_s f_y = 0.85 f'_c (b - b_w) h_f + 0.85 f'_c b_w a \qquad (5.17)$$

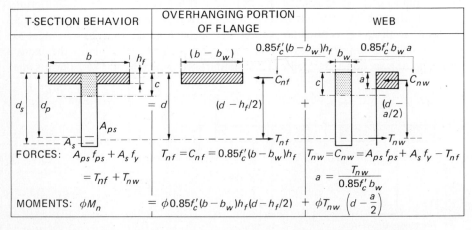

Figure 5.9 Partially prestressed T section: forces and moments at ultimate.

Note that the total tensile force at ultimate can be separated into two parts, one, T_{nf}, balancing the compressive force in the overhanging portion of the flange and the other, T_{nw}, balancing the compressive force in the web, thus:

$$T_{nf} = 0.85f'_c(b - b_w)h_f \tag{5.18}$$

$$T_{nw} = 0.85f'_c\, b_w\, a \tag{5.19}$$

The depth of the stress block a can be computed from the force equilibrium Eq. (5.17) as:

$$a = \frac{A_{ps}\, f_{ps} + A_s\, f_y - 0.85f'_c(b - b_w)h_f}{0.85f'_c\, b_w} \tag{5.20}$$

Thus the nominal moment resistance is given by:

$$M_n = T_{nf}\left(d - \frac{h_f}{2}\right) + T_{nw}\left(d - \frac{a}{2}\right) \tag{5.21}$$

The corresponding design moment resistance acceptable by the ACI code is:

$$\phi M_n = \phi T_{nf}\left(d - \frac{h_f}{2}\right) + \phi T_{nw}\left(d - \frac{a}{2}\right) \tag{5.22}$$

Note that the approach followed above to analyze T section behavior, although slightly different from that usually followed by the code, leads to identical results. When T section behavior exists, the usual code approach is first to compute the areas of steel reinforcement needed to develop the ultimate strength of the overhanging portion of the flange and the web and then to determine corresponding forces and moments. This is illustrated in Fig. 5.10 for a fully prestressed section where most relevant equations are also given. However, in the previous

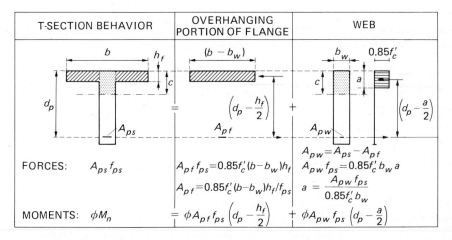

Figure 5.10 Fully prestressed T section: ACI code approach.

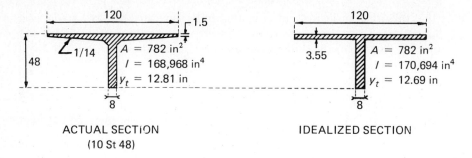

ACTUAL SECTION
(10 St 48)

IDEALIZED SECTION

Figure 5.11 Actual and idealized T section.

approach (Fig. 5.9), the forces generated by the reinforcement instead of the steel areas were used, because if both prestressed and nonprestressed reinforcement are present, it becomes tedious to find in what proportion, if any, each type of reinforcement is associated with the flange or the web. A more general and rational procedure is proposed in Vol. 2 where the same proportion is taken for prestressed and nonprestressed steel and set a priori equal to a parameter called *partial prestressing ratio.*

The above equations have been developed assuming an idealized T section where the thickness of the flange h_f was assumed constant and no fillets were considered. However, in practice, T beams, mainly those precast prestressed, have tapering flanges and fillets at the web-flange junctions. In such a case, although no theoretical difficulties exist, the calculations leading to the exact depth of the neutral axis c and the force in the concrete compressive zone become tedious. Fortunately, little accuracy in the results is lost when an equivalent idealized section is considered. Such an equivalent section is shown, for example, in Fig. 5.11 where the flange thickness was selected to keep the total cross-sectional area of concrete the same as for the actual section. Note that the idealized section is needed only for the analysis at ultimate; the actual section and corresponding geometric properties should be used in the analysis under service loads.

A remark is in order here. In treating the overhanging portion of the flange according to the ACI code assumptions an inconsistency may develop leading to some confusing results. For instance, let us assume that in a particular problem the value of c is about equal to h_f; let us define c^+ as a small epsilon higher than h_f and c^- as a small epsilon smaller than h_f. If c^- is used, the T section is designed as a rectangular section, and in computing the compressive force in the concrete the factor β_1 used in $a = \beta_1 c$ applies throughout; thus Eq. (5.18) becomes:

$$T_{nf} = 0.85f'_c(b - b_w)\beta_1 c^- = 0.85f'_c(b - b_w)\beta_1 h_f \qquad (5.23)$$

However, if c^+ is used, the section behaves as a T section and Eq. (5.18), in which β_1 is not included, applies. Obviously the two results are substantially different due to the value of β_1. It is the author's opinion that in order to provide continuity, Eq. (5.23) should be used instead of Eq. (5.18). However, one can also argue that

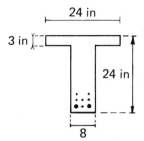

Figure 5.12

when c is substantially larger than h_f an in-between approach may be more representative. Note that the use of Eq. (5.23) would lead to a safer design than the use of Eq. (5.18).

The following example illustrates the steps described above. An expanded computational chart for the calculation of nominal moments is also given in Fig. 5.13 and may be used as an alternative.

4. Example: Analysis at Ultimate of a T Section

(a) Determine the nominal moment capacity of a partially prestressed concrete T section with dimensions described in Fig. 5.12. The amounts and location of reinforcement and the materials properties are exactly the same as for the first Example in Sec. 5.3. Assume bonded tendons are used.

Compute

$$\rho_p = \frac{A_{ps}}{bd_p} = 0.00185$$

Estimate

$$f_{ps} = 270\left(1 - 0.5 \times 0.00185 \times \frac{270}{5}\right) = 256.5 \text{ ksi}$$

Estimate

$$d = \frac{0.92 \times 256.5 \times 20.75 + 1.2 \times 60 \times 21.5}{0.92 \times 256.5 + 1.2 \times 60} = 20.92 \text{ in}$$

First assume rectangular section behavior and determine neutral axis c; using Eq. (5.9):

$$a = \frac{0.92 \times 256.5 + 1.2 \times 60}{0.85 \times 5 \times 24} = 3.02 \text{ in}$$

and

$$c = a/\beta_1 = \frac{3.02}{0.80} = 3.77 \text{ in}$$

where β_1 is computed from Eq. (5.7). As c is larger than $h_f = 3$ in, the section should be treated as a T section.

Determine the depth of the stress block a from Eq. (5.20):

$$a = \frac{0.92 \times 256.5 + 1.2 \times 60 - 0.85 \times 5(24 - 8)3}{0.85 \times 5 \times 8} = 3.06 \text{ in}$$

$$c = a/\beta_1 = 3.06/0.80 = 3.82 \text{ in}$$

Equations (5.18), (5.19), (5.21), and (5.22) give:

$$T_{nf} = 0.85 \times 5 \times (24 - 8)3 = 204 \text{ kips}$$

$$T_{nw} = 0.85 \times 5 \times 8 \times 3.06 = 104.04 \text{ kips}$$

$$M_n = 204,000(20.92 - 3/2) + 104,040\left(20.92 - \frac{3.06}{2}\right)$$

$$= 5,979,015 \text{ lb-in} = 498.25 \text{ kips-ft}$$

$$\phi M_n = 0.9 \times M_n = 448.42 \text{ kips-ft}$$

The reader may want to check the results obtained for the following two particular cases:

(b) If $A_s = 0$ and everything else is kept the same, the following results would be obtained: $\rho_p = 0.00185$; $f_{ps} = 256.5$ ksi; $d = d_p = 20.75$ in; the section behaves as a rectangular section at ultimate with $a = 2.31$ in; $c = 2.89$ in, smaller than $h_f = 3$ in; $M_n = 385.3$ kips-ft, and $\phi M_n = 346.8$ kips-ft.

(c) If unbonded tendons are used instead of bonded tendons, everything else being same as in (a) above, the following results would be observed: $\rho_p = 0.00185$; $f_{ps} = 185.527$ ksi; $d = 20.97$ in; the section behaves as a rectangular section with $a = 2.38$ in; $c = 2.97$ in, smaller than $h_f = 3$ in; $M_n = 400.15$ kips-ft; $\phi M_n = 360.13$ kips-ft.

(d) Example cases (a), (b), and (c) did not present any difficulty or noticeable inconsistency because in all three cases both values of a and c were either simultaneously smaller or simultaneously larger than the flange thickness h_f. It may happen, however, that a is smaller and c is larger than h_f. In such a case the strict application of the ACI procedure may lead to a seemingly inconsistent result as pointed out at the end of Sec. 5.3. To illustrate what happens, let us assume that we have exactly the same problem as case (a) except that h_f is 3.5 in instead of 3 in. The computed values of a and c are: $a = 3.02$ in and $c = 3.77$ in. As c is larger than $h_f = 3.5$ in, the section behaves as a T section. If we attempt to determine again the depth of the stress block in the web using Eq. (5.20), we would get:

$$a = \frac{0.92 \times 256.5 + 1.2 \times 60 - 0.85 \times 5(24 - 8)3.5}{0.85 \times 5 \times 8} = 2.06 \text{ in}$$

$$c = a/\beta_1 = 2.06/0.8 = 2.57 \text{ in}$$

Contrary to previous results the above values would indicate that the neutral axis is now in the flange and a rectangular section behavior is at hand. This inconsistency can be avoided by using Eq. (5.23) instead of Eq. (5.20) in determining a. If this is done, Eq. (5.20) would read:

$$a = \frac{A_{ps} f_{ps} + A_s f_y - 0.85 f'_c(b - b_w)h_f \beta_1}{0.85 f'_c b_w} \tag{5.24}$$

which, for our example, with $\beta_1 = 0.80$, leads to $a = 3.46$ in and $c = 4.32$ in, both—as expected—larger than those assuming rectangular section behavior.

5.4 CONCEPT OF REINFORCING INDEX

In order to identify between underreinforced and overreinforced sections, the ACI code uses a parameter called here the combined reinforcing index q, or simply the reinforcing index, and defined as follows:

$$q = \omega_p + \omega - \omega' \tag{5.25}$$

where the partial reinforcing indexes ω_p, ω and ω' are given by:

$$\omega_p = \frac{A_{ps} f_{ps}}{bdf'_c} = \rho_p \frac{f_{ps}}{f'_c} \tag{5.26}$$

$$\omega = \frac{A_s f_y}{bdf'_c} = \rho \frac{f_y}{f'_c} \tag{5.27}$$

$$\omega' = \frac{A'_s |f'_y|}{bdf'_c} = \rho' \frac{|f'_y|}{f'_c} \tag{5.28}$$

The absolute value of f'_y is used as it is a compressive stress in the steel and is negative. A_s and A'_s are the areas of the nonprestressed tensile and compressive reinforcement and d is the distance to the centroid of the tensile force as defined in Eq. (5.11). Note that for a purely prestressed section $d = d_p$ while for a purely reinforced section $d = d_s$. If we replace the partial reinforcing indexes by their values in Eq. (5.25), the combined index becomes:

$$q = \frac{A_{ps} f_{ps}}{bdf'_c} + \frac{A_s f_y}{bdf'_c} - \left| \frac{A'_s f'_y}{bdf'_c} \right| \tag{5.29}$$

or

$$q = \rho_p \frac{f_{ps}}{f'_c} + \rho \frac{f_y}{f'_c} - \rho' \frac{|f'_y|}{f'_c} \tag{5.30}$$

where ρ_p, ρ, and ρ' are the reinforcing ratios of the prestressing steel, the non-prestressed tensile steel, and the compressive steel, respectively.

It is important to fully understand the meaning of the reinforcing index q. It is a unifying parameter between reinforced, prestressed, and partially prestressed sections and can be very useful in design. As shown below, q is directly proportional to the forces in the section at ultimate behavior whether they come from the reinforcing steel, the prestressing steel, or their combination.

Let us assume that we have a rectangular section reinforced with prestressed and nonprestressed tensile and compressive reinforcement. Let us also assume that at ultimate capacity, the nonprestressed tensile and compressive steels yield. Equilibrium of the internal forces in the section leads to:

$$0.85 f'_c ba = A_{ps} f_{ps} + A_s f_y - |A'_s f'_y| \tag{5.31}$$

If we divide both sides by bdf'_c, the second term of the equation becomes q and Eq. (5.31) is reduced to:

$$0.85 \frac{a}{d} = q \tag{5.32}$$

Thus, q can also be expressed as:

$$q = \frac{1}{bdf'_c} \times \begin{matrix} \text{(compressive force} \\ \text{in concrete)} \end{matrix} = \frac{1}{bdf'_c} \times \begin{matrix} \text{(net force in all} \\ \text{the steel)} \end{matrix} \tag{5.33}$$

Note that, generally, the assumption that the nonprestressed tensile steel yields at ultimate is valid. However, the compressive steel may not always yield at ultimate. When this is the case, a small difference will appear between the definition of q (Eq. (5.29)) and that derived from the force equilibrium Eqs. (5.31) and (5.32) above. By assuming yielding of the compressive steel in the definition of q, the ACI code allows for a small error in some cases, but the design procedure is greatly simplified.

Many of the equations derived in Sec. 5.3 can be put in function of q. For instance, let us consider a partially prestressed rectangular section without compressive reinforcement. The force equilibrium Eq. (5.31) can be written:

$$0.85f'_c \, ba = q \times bdf'_c \tag{5.34}$$

from which the depth of the stress block a can be obtained:

$$a = \frac{1}{0.85} qd = 1.18qd \tag{5.35}$$

Thus the depth of the neutral axis becomes:

$$c = \frac{a}{\beta_1} = \frac{1.18qd}{\beta_1} \tag{5.36}$$

Replacing the value of a, Eq. (5.35) in Eq. (5.10) leads to the nominal moment resistance

$$M_n = f'_c \, bd^2 q(1 - 0.59q) \tag{5.37}$$

The corresponding design nominal moment resistance is given by:

$$\phi M_n = \phi f'_c \, bd^2 q(1 - 0.59q) \tag{5.38}$$

When T section instead of rectangular section behavior exists, the nominal moment can be divided into two parts, one due to the resistance of the flange and one to that of the web; thus:

$$\phi M_n = \phi(M_{nf} + M_{nw}) \tag{5.39}$$

As M_{nw} is the moment of a rectangular section, it can be expressed in function of q_w, similarly to what was achieved in Eq. (5.38). Thus, using Eqs. (5.18), (5.22), and (5.38) leads to the design nominal moment resistance for T section behavior:

$$\phi M_n = \phi[0.85f'_c(b - b_w)h_f(d - h_f/2) + f'_c \, b_w d^2 q_w(1 - 0.59q_w)] \tag{5.40}$$

Equation (5.40) illustrates the use of the reinforcing index in computing the nominal moment resistance, assuming underreinforced T section behavior. A simpler equation is given in the flow chart, Fig. 5.13.

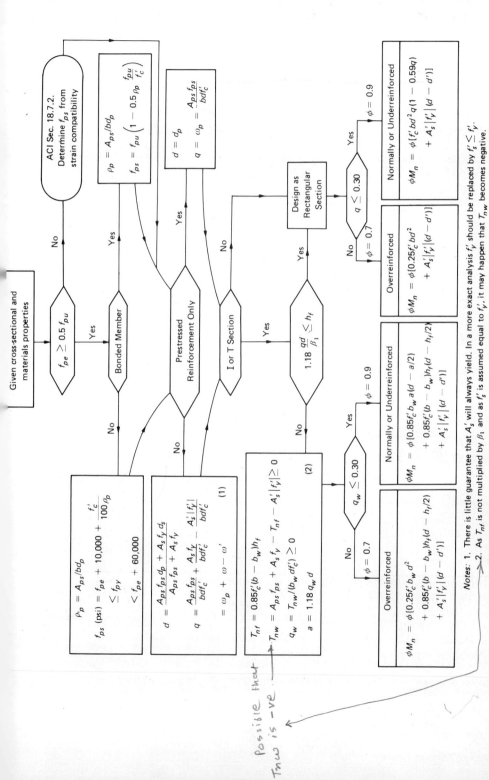

Figure 5.13 Flow chart to determine the nominal moment resistance of prestressed and partially prestressed sections in accordance with the ACI code.

Notes: 1. There is little guarantee that A'_s will always yield. In a more exact analysis f'_y should be replaced by $f'_s \leq f'_y$.

2. As T_{nf} is not multiplied by β_1 and as f'_s is assumed equal to f'_y, it may happen that T_{nw} becomes negative.

Possible that T_{nw} is -ve.

The earlier comment that, in the expression of M_{nf}, the factor β_1 should theoretically be used as a multiplier, applies here also. However, to avoid confusion with the code approach, β_1 is not included in the expression of M_{nf} in Eq. (5.40) and Fig. 5.13.

5.5 LIMITING VALUES OF THE REINFORCING INDEX

1. Underreinforced and Overreinforced Sections

The ACI code defines underreinforced rectangular sections as sections for which the value of q is less than or equal to $q_{max} = 0.30$. If q is larger than q_{max}, it is then assumed that the section is overreinforced and the ultimate moment capacity is limited by that of the concrete compressive block. For such a case, the following approximate but conservative relationship, suggested in the commentary of the ACI code, is proposed to determine the nominal moment resistance, assuming rectangular section or rectangular section behavior:

$$\phi M_n = \phi(0.25 f'_c bd^2) \tag{5.41}$$

where ϕ is taken equal to 0.7 (crushing compression failure). The above equation can be arrived at assuming that the maximum possible value of a in Eq. (5.10) is about equal to $0.36d$.

For a T section behavior the limiting value of the reinforcing index q should correspond to $q_w = 0.30$ where q_w is the reinforcing index of the web. It can be computed from Eq. (5.33) noting that q is equal to the force in the concrete divided by bdf'_c thus:

$$(q)_T = \frac{1}{bdf'_c} [\overset{(flange)}{0.85f'_c(b - b_w)h_f} + \overset{(web)}{0.85f'_c b_w a}] \tag{5.42}$$

or

$$(q)_T = \frac{0.85(b - b_w)h_f}{bd} + \frac{0.85a}{d}\frac{b_w}{b} \tag{5.43}$$

However, for a rectangular section Eq. (5.32) indicated that

$$(q)_{rect} = \frac{0.85a}{d} \tag{5.44}$$

Thus using Eq. (5.44) in Eq. (5.43) leads to:

$$(q)_T = \frac{0.85(b - b_w)h_f}{bd} + (q)_{rect}\frac{b_w}{b} \tag{5.45}$$

and its limiting value for $(q)_{rect} = 0.30$ is given by:

$$(q_{max})_T = 0.30 \frac{b_w}{b} + \frac{0.85(b - b_w)h_f}{bd} \leq 0.30$$ (5.46)

If the actual reinforcing index of a T section is larger than its corresponding limit $(q_{max})_T$, the section is assumed to behave as an overreinforced section and the corresponding ultimate resisting moment is given by:

$$\phi M_n = \phi \left[0.25 f'_c b_w d^2 + 0.85 f'_c (b - b_w) h_f \left(d - \frac{h_f}{2} \right) \right]$$ (5.47)

where $\phi = 0.7$.

It is important to understand the parallel that exists between the limiting value of the reinforcing index q in prestressed concrete and the limiting value of the reinforcement ratio ρ in reinforced concrete. For reinforced concrete members the ACI code limits the value of the reinforcement ratio to 75 percent of the balanced reinforcement ratio. Thus

$$\rho_{max} = 0.75 \rho_b$$ (5.48)

If both sides of the equation are multiplied by (f_y/f'_c), Eq. (5.48) becomes (for reinforced concrete):

$$q_{max} = 0.75 \rho_b \frac{f_y}{f'_c} = 0.75 q_b$$ (5.49)

Computing the values of q_{max} using Eq. (5.6) for ρ_b, and assuming several values of f_y and f'_c, lead to results close to 0.30 in most cases. This limiting value of 0.30 has been confirmed also by results derived from an extensive nonlinear analysis of reinforced concrete members with strengths up to 14 ksi. (Ref. 5.15.)

It seems, therefore, that a limiting value of $q_{max} = 0.30$ should apply not only to prestressed concrete but also to reinforced concrete and partially prestressed concrete. It is a unifying parameter that presents the advantage of remaining about constant for various f_y and f'_c while the reinforcement ratio ρ_{max} would substantially change depending on the values of these variables.

2. Minimally Reinforced Sections

It was pointed out earlier in Sec. 5.2 that in order to avoid sudden failure immediately after cracking the ACI code recommends a minimum value of the reinforcement ratio for reinforced concrete members while, for prestressed concrete, such a limit was indirectly set by requiring that the ultimate moment be 20 percent higher than the cracking moment. An attempt is made below to show that a lower value of the reinforcing index q can be set as a common lower limit for reinforced concrete, prestressed concrete, and partially prestressed concrete. The design moment resistance of the above three classes of beams can be written as a function of q (assuming rectangular section behavior) as:

$$\phi M_n = \phi f'_c bd^2 q(1 - 0.59q)$$ (5.50)

Furthermore, their cracking moment can be most generally written as:

$$M_{cr} = A_{ps} f_{pe} \left(e_o + \frac{Z_b}{A_c} \right) - f_r Z_b \tag{5.51}$$

where for a purely reinforced section $A_{ps} = 0$. To avoid sudden collapse after cracking, we can write the condition that:

$$\phi M_n > \alpha M_{cr} \tag{5.52}$$

where α is a factor larger than or equal to 1. (For prestressed concrete ACI recommends $\alpha = 1.2$.)

Replacing Eqs. (5.50) and (5.51) in Eq. (5.52) leads to a quadratic equation in q with two positive roots; the smaller root gives the minimum required value of q.

The solution is different whether a fully reinforced or a fully prestressed concrete section is used because the value of M_{cr} for a fully prestressed section is itself dependent on q. The reader may want to check the following results. For reinforced concrete:

$$q_{min} = 0.85 \left[1 - \sqrt{1 + \frac{2.36\alpha f_r Z_b}{\phi f'_c b d_s^2}} \right] \tag{5.53}$$

and for prestressed concrete:

$$q_{min} = 0.85 \left[(1 - \lambda) - \sqrt{(1 - \lambda)^2 + \frac{2.36\alpha f_r Z_b}{\phi f'_c b d_p^2}} \right] \tag{5.54}$$

where

$$\lambda = \frac{\alpha}{\phi d_p} \frac{f_{pe}}{f_{pu}} \left(e_o + \frac{Z_b}{A_c} \right) \tag{5.55}$$

For partially prestressed concrete the value of d_p in Eqs. (5.54) and (5.55) should be replaced by d, where d is the distance to the centroid of the tensile force. As a first approximation the average of d_p and d_s can be used.

When the section has a minimum amount of reinforcement, it can be shown that f_{ps} is very close to f_{pu}; in developing the above equation of q_{min} for prestressed concrete, it was assumed that $f_{ps} \simeq f_{pu}$.

Note also that in most cases a T section with a low reinforcement ratio or reinforcing index behaves as a rectangular section at ultimate. Thus, the above values of q_{min} should in general apply to T sections as well as to rectangular sections.

The following example illustrates the computations of the upper and lower limits of the reinforcing index.

3. Example: q_{max} and q_{min} for a T section

Determine the values of q_{max} and q_{min} for the T beam shown in Fig. 5.12 and used in the second Example in Sec. 5.3, assuming $f'_c = 5000$ psi and $f_{pu} = 270$ ksi.

(a) Assume prestressed reinforcement only with $d_p = 20$ inches in computing q_{max}. Equation (5.46) gives:

$$(q_{max})_T = 0.30 \frac{b_w}{b} + \frac{0.85(b - b_w)h_f}{bd_p}$$

$$= 0.30 \frac{8}{24} + \frac{0.85(24 - 8)3}{24 \times 20} = 0.185$$

The above value is less than 0.30, the maximum index for a rectangular section. Note that given q, one can easily determine the amount of reinforcement A_{ps} as shown in the two Examples Sec. 5.7.

(b) Assume $d_p = d_s = 22$ inches in computing q_{min}. The following values of the variables are used:

$$\alpha = 1.2$$

$$A_c = 240 \text{ in}^2$$

$$Z_b = 956 \text{ in}^3$$

$$e_o = d_p - y_t = 12.1 \text{ in}$$

$$\frac{f_{pe}}{f_{pu}} = 0.55$$

$$f_r = -530 \text{ psi}$$

If the section is fully reinforced, Eq. (5.53) leads to

$$q_{min} = 0.85\left(1 - \sqrt{1 - \frac{2.36 \times 1.2 \times 530 \times 956}{0.9 \times 5000 \times 24 \times 22^2}}\right) = 0.01175$$

If the section is fully prestressed, Eq. (5.55) leads to:

$$\lambda = \frac{1.2}{0.9 \times 22} 0.55(12.1 + 3.98) = 0.536$$

and Eq. (5.54) leads to:

$$q_{min} = 0.85[(1 - 0.536) - \sqrt{(1 - 0.536)^2 - 0.02745}] = 0.026.$$

Note that, if the ACI code is used to estimate the value of q_{min} for reinforced concrete, the following results would have been obtained:

$$\rho_{min} = \frac{200}{f_y} = \frac{200}{60,000} = 0.0033$$

The above ρ_{min} is based on the web part of the T section. Transformed to the flange, it leads to:

$$0.0033 \frac{b_w}{b} = 0.0033 \frac{8}{24} = 0.0011$$

The corresponding reinforcing index is given by:

$$q_{min} = 0.0011 \frac{f_y}{f'_c} = 0.0133$$

which is not too different from the result obtained earlier.

4. Example: Ultimate Moment Requirements for Example Beam of Secs. 4.9a and 4.12

The beam is reinforced with ten strands at a midspan eccentricity of 21.7 in. Thus:

$$A_{ps} = 10 \times 0.153 = 1.53 \text{ in}^2$$

$$d_p = e_o + y_t = 21.7 + 12.9 = 34.6 \text{ in}$$

Following the design flow chart Fig. 5.13, we have:

$$\rho_p = \frac{A_{ps}}{bd_p} = \frac{1.53}{48 \times 34.6} = 0.00092$$

$$f_{ps} = 270 \left(1 - 0.5\rho_p \frac{270}{5} \right) = 263.28 \text{ ksi}$$

$$q = \omega_p = 0.0485$$

As we have a T section, check neutral axis:

$$\frac{1.18qd_p}{\beta_1} = \frac{1.18 \times 0.0485 \times 34.6}{0.80} = 2.475 \text{ in} < h_f$$

An equivalent h_f value of 5.75 in can be taken for this beam as it has a tapering flange.

Thus, design as a rectangular section; as $q < 0.30$, the ultimate resisting moment is given by:

$$\phi M_n = \phi f'_c bd_p^2 q(1 - 0.59q) = 12{,}184.8 \text{ kips-in} = 1015.4 \text{ kips-ft}$$

The cracking moment is given by Eq. (4.38):

$$M_{cr} = F \left(\frac{Z_b}{A_c} + e_o \right) - f_r Z_b = 229{,}500 \left(\frac{3028}{550} + 21.7 \right) + 530.3 \times 3028$$

$$= 7{,}849{,}400 \text{ lb-in} \simeq 654 \text{ kips-ft}$$

Check required ratio of ultimate to cracking moment:

$$\frac{\phi M_n}{M_{cr}} = \frac{1015.4}{654} = 1.55 > 1.2 \qquad\qquad\qquad \text{OK}$$

Note that:

$$\phi M_n > M_u = 1.4M_D + 1.7M_L \simeq 942 \text{ kips-ft} \qquad\qquad \text{OK}$$

5.6 SATISFYING ULTIMATE STRENGTH REQUIREMENTS

A general flow chart summarizing the computations of the nominal moment resistance and taking into account the case of an underreinforced or an overreinforced section is given in Fig. 5.13. It applies to rectangular and T sections, reinforced, prestressed, and partially prestressed with bonded or unbonded tendons. It should be used in everyday routine computations where there is no need to go back to the theory.

Once the design moment resistance ϕM_n is determined, its value must be compared to the strength moment required or specified by the code which is also called factored moment or strength design moment M_u. Ultimate strength requirements in flexure are considered satisfied when the following condition is valid:

$$\phi M_n \geq M_u \qquad (5.56)$$

where M_u is assumed positive.

According to the ACI code M_u may represent any of the factored moments obtained using the factors described in Sec. 3.8. However, if the most commonly encountered set of factors is used, Eq. (5.56) becomes:

$$\phi M_n \geq 1.4 M_D + 1.7 M_L \qquad (5.57)$$

If it is found that the design nominal resistance is not sufficient, i.e., the above condition is not satisfied, several corrective actions can be taken. They are listed below in order of their effectiveness and simple implementation:

1. Add nonprestressed reinforcement.
2. Increase the amount of prestressing steel. However, make sure that none of the allowable stresses are then violated under working stress conditions.
3. Increase the eccentricity of the steel or equivalently d_p if possible without violating allowable stresses.
4. Change materials properties and/or sectional dimensions.

The above remedies can also be used in part when the nominal moment resistance is to be increased to satisfy the ACI requirement that the nominal moment resistance should be at least 20 percent higher than the cracking moment. In order of effectiveness it is recommended to use first 1; then use 2 noting that the cracking moment will also increase; do not consider 3; and, finally, use 4.

5.7 DESIGN FOR ULTIMATE STRENGTH

The design of prestressed concrete beams in flexure starts generally by the working stress design procedure described in Chap. 4, where the prestressing force and the corresponding area of prestressing steel are determined. In order to satisfy ultimate strength requirements, the nominal moment resistance is then determined as

shown in Sec. 5.3 and its value is compared to the strength design moment specified by the code. In this step, it is assumed that materials properties and cross-
sectional dimensions are known.

However, it is possible to design prestressed concrete sections strictly on the
basis of ultimate strength, without reference to the working stress approach. This is
in effect similar to reinforced concrete design currently used in the ACI code. As
only two equations of equilibrium are available at ultimate, two unknowns can be
determined. They are generally a, the depth of the stress block and A_{ps}, the area of
the prestressing steel. Thus, for such calculations, one must assume all the other
unknowns given, i.e., the cross-sectional dimensions, the materials properties, and
the loading conditions or the value of the strength design moment M_u.

The purpose of the design is then to determine the area of prestressed reinforcement A_{ps}, which would lead to a design nominal moment resistance ϕM_n
equal to the specified strength design moment M_u.

Going back to Eq. (5.37) it can be observed that, for a given M_u, the required
value of q can be determined from satisfying the above quadratic relation:

$$q(1 - 0.59q) = \frac{M_u}{\phi f'_c \, bd_p^2} \qquad (5.58)$$

As for a fully prestressed section:

$$q = \frac{A_{ps} f_{ps}}{bd_p f'_c} \quad \text{as per eqn. 5-33} \qquad (5.59)$$

and as f_{ps} is given in function of A_{ps} in Eqs. (5.12) or (5.13), the particular values of
the two unknowns f_{ps} and A_{ps} can be determined by solving simultaneously
Eq. (5.59) and Eq. (5.12) or (5.13). The solution is given directly in the flow
chart, Fig. 5.14 for fully prestressed rectangular and T sections with bonded and
unbonded reinforcement and in accordance with the ACI code. The reader is
encouraged to verify the equations given in Fig. 5.14.

A more general solution is given in Vol. 2 for the case where prestressed and
nonprestressed reinforcements are used.

Note that Eq. (5.58) is general and is very useful; its numerical solution is
provided in Table 5.1, where for a given value of $M_u/\phi f'_c \, bd^2$ one can get the
corresponding value of q and vice versa.

The following example illustrates the design procedure.

1. Example: Design for Nominal Resistance; Rectangular Section

Determine the area of prestressed reinforcement required to develop a design moment
resistance $\phi M_n = M_u = 300$ kips-ft in the rectangular section shown in Fig. 5.15 assuming
the following properties: $f'_c = 5000$ psi; $f_{pu} = 270$ ksi; $f_{pe} = 0.55 f_{pu}$; $d_p = 21$ in; bonded
tendons.

Referring to the flow chart Fig. 5.14 for the steps followed, compute q by solving the

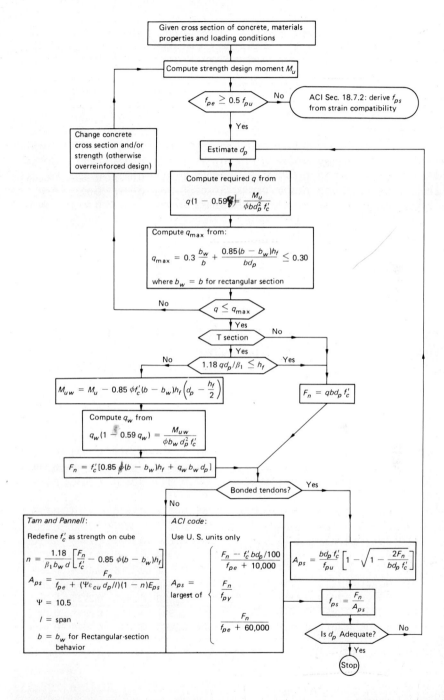

Figure 5.14 Flow chart to determine the prestressing steel to satisfy ultimate strength requirements.

Table 5.1 Numerical values of the reinforcing index q versus $M_u/(\phi f_c' bd^2)$ and vice versa

q	0.000	0.001	0.002	0.003	0.004	0.005	0.006	0.007	0.008	0.009
0.0	0	0.0010	0.0020	0.0030	0.0040	0.0050	0.0060	0.0070	0.0080	0.0090
0.01	0.0099	0.0109	0.0119	0.0129	0.0139	0.0149	0.0159	0.0168	0.0178	0.0188
0.02	0.0197	0.0207	0.0217	0.0226	0.0236	0.0246	0.0256	0.0266	0.0275	0.0285
0.03	0.0295	0.0304	0.0314	0.0324	0.0333	0.0343	0.0352	0.0362	0.0372	0.0381
0.04	0.0391	0.0400	0.0410	0.0420	0.0429	0.0438	0.0448	0.0457	0.0467	0.0476
0.05	0.0485	0.0495	0.0504	0.0513	0.0523	0.0532	0.0541	0.0551	0.0560	0.0569
0.06	0.0579	0.0588	0.0597	0.0607	0.0616	0.0625	0.0634	0.0643	0.0653	0.0662
0.07	0.0671	0.0680	0.0689	0.0699	0.0708	0.0717	0.0726	0.0735	0.0744	0.0753
0.08	0.0762	0.0771	0.0780	0.0789	0.0798	0.0807	0.0816	0.0825	0.0834	0.0843
0.09	0.0852	0.0861	0.0870	0.0879	0.0888	0.0897	0.0906	0.0915	0.0923	0.0932
0.10	0.0941	0.0950	0.0959	0.0967	0.0976	0.0985	0.0994	0.1002	0.1011	0.1020
0.11	0.1029	0.1037	0.1046	0.1055	0.1063	0.1072	0.1081	0.1089	0.1098	0.1106
0.12	0.1115	0.1124	0.1133	0.1141	0.1149	0.1158	0.1166	0.1175	0.1183	0.1192
0.13	0.1200	0.1209	0.1217	0.1226	0.1234	0.1243	0.1251	0.1259	0.1268	0.1276
0.14	0.1284	0.1293	0.1301	0.1309	0.1318	0.1326	0.1334	0.1342	0.1351	0.1359
0.15	0.1367	0.1375	0.1384	0.1392	0.1400	0.1408	0.1416	0.1425	0.1433	0.1441
0.16	0.1449	0.1457	0.1465	0.1473	0.1481	0.1489	0.1497	0.1506	0.1514	0.1522
0.17	0.1529	0.1537	0.1545	0.1553	0.1561	0.1569	0.1577	0.1585	0.1593	0.1601
0.18	0.1609	0.1617	0.1624	0.1632	0.1640	0.1648	0.1656	0.1664	0.1671	0.1679
0.19	0.1687	0.1695	0.1703	0.1710	0.1718	0.1726	0.1733	0.1741	0.1749	0.1756
0.20	0.1764	0.1772	0.1779	0.1787	0.1794	0.1802	0.1810	0.1817	0.1825	0.1832
0.21	0.1840	0.1847	0.1855	0.1862	0.1870	0.1877	0.1885	0.1892	0.1900	0.1907
0.22	0.1914	0.1922	0.1929	0.1937	0.1944	0.1951	0.1959	0.1966	0.1973	0.1981
0.23	0.1988	0.1995	0.2002	0.2010	0.2017	0.2024	0.2031	0.2039	0.2046	0.2053
0.24	0.2060	0.2067	0.2075	0.2082	0.2089	0.2096	0.2103	0.2110	0.2117	0.2124
0.25	0.2131	0.2138	0.2145	0.2152	0.2159	0.2166	0.2173	0.2180	0.2187	0.2194
0.26	0.2201	0.2208	0.2215	0.2222	0.2229	0.2236	0.2243	0.2249	0.2256	0.2263
0.27	0.2270	0.2277	0.2284	0.2290	0.2297	0.2304	0.2311	0.2317	0.2324	0.2331
0.28	0.2337	0.2344	0.2351	0.2357	0.2364	0.2371	0.2377	0.2384	0.2391	0.2397
0.29	0.2404	0.2410	0.2417	0.2423	0.2430	0.2437	0.2443	0.2450	0.2456	0.2463
0.30	0.2469	0.2475	0.2482	0.2488	0.2495	0.2501	0.2508	0.2514	0.2520	0.2527
0.31	0.2533	0.2539	0.2546	0.2552	0.2558	0.2565	0.2571	0.2577	0.2583	0.2590
0.32	0.2596	0.2602	0.2608	0.2614	0.2621	0.2627	0.2633	0.2639	0.2645	0.2651
0.33	0.2657	0.2664	0.2670	0.2676	0.2682	0.2688	0.2694	0.2700	0.2706	0.2712
0.34	0.2718	0.2724	0.2730	0.2736	0.2742	0.2748	0.2754	0.2760	0.2766	0.2771
0.35	0.2777	0.2783	0.2789	0.2795	0.2801	0.2807	0.2812	0.2818	0.2824	0.2830
0.36	0.2835	0.2841	0.2847	0.2853	0.2858	0.2864	0.2870	0.2875	0.2881	0.2887
0.37	0.2892	0.2898	0.2904	0.2909	0.2915	0.2920	0.2926	0.2931	0.2937	0.2943
0.38	0.2948	0.2954	0.2959	0.2965	0.2970	0.2975	0.2981	0.2986	0.2992	0.2997
0.39	0.3003	0.3008	0.3013	0.3019	0.3024	0.3029	0.3035	0.3040	0.3045	0.3051

Notes: 1. Enter $\dfrac{M_u}{\phi f_c' bd^2}$ and get q or vice versa.

2. $\dfrac{M_u}{\phi f_c' bd^2} = q(1 - 0.59q)$.

3. $q = \dfrac{A_{ps} f_{ps} + A_s f_y - A_s' f_y'}{bd f_c'}$

178

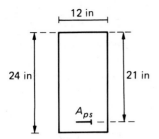

Figure 5.15

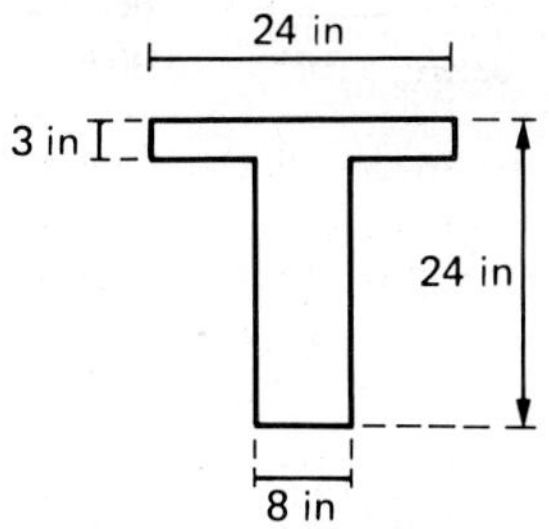

24 in

12 in

21 in

A_{ps}

24 in

24 in

3 in

8 in

Figure 5.16

following quadratic equation:

$$q(1 - 0.59q) = \frac{M_u}{\phi f'_c bd_p^2} = \frac{300 \times 12}{0.9 \times 5 \times 12 \times 21^2} = 0.1512 \quad \llap{\textit{↙}}$$

for which $q \simeq 0.168$. (See also Table 5.1.)

As we have a rectangular section, the total tensile force taken by the prestressing steel at nominal resistance is given by:

$$F_n = qbd_p f'_c = 211.680 \text{ kips} \qquad \textit{From chart 5·14}$$

Thus, as we have bonded tendons:

$$A_{ps} = \frac{bd_p f'_c}{f_{pu}} \left(1 - \sqrt{1 - \frac{2F_n}{bd_p f'_c}}\right) = 0.864 \text{ in}^2 \qquad \textit{chart 5·14}$$

and

$$f_{ps} = \frac{F_n}{A_{ps}} = 245 \text{ ksi} \qquad \textit{chart 5·14}$$

2. Example: Design for Nominal Resistance; T Section

Aps = 9

chart 5·14 and Table 5·1

Determine the area of prestressed reinforcement required to develop a design moment resistance $\phi M_n = M_u = 450$ kips-ft in the T section shown in Fig. 5.16, assuming the same materials properties as for the preceding Example and $d_p = 20$ in.

Referring to the flow chart Fig. 5.14, the following steps are pursued. Compute q from the following equation:

$$q(1 - 0.59q) = \frac{M_u}{\phi f'_c bd_p^2} = \frac{450 \times 12}{0.9 \times 5 \times 24 \times 20^2} = 0.1250 \quad \llap{\textit{↙}}$$

which leads to a value of $q \simeq 0.136$. Note that q is less than $(q_{max})_T = 0.185$ found in the first Example in Sec. 5.5, indicating that we will have an underreinforced section.

To check if T section behavior exists, compute:

$$a = 1.18qd_p = 1.18 \times 0.136 \times 20 = 3.21 \text{ in} > h_f = 3$$

Thus, we have T section behavior. The design strength required from the web is defined as M_{uw} and computed from:

$$M_{uw} = M_u - 0.85\phi f'_c(b - b_w)h_f\left(d_p - \frac{h_f}{2}\right)$$

$$= 450 - \frac{3,396,600}{12,000} = 450 - 283.05 = 166.95 \text{ kips-ft}$$

Compute q_w by solving the following equation:

$$q_w(1 - 0.59q_w) = \frac{M_{uw}}{\phi f'_c b_w d_p^2} = \frac{166.95 \times 12}{0.9 \times 5 \times 8 \times 20^2} = 0.139$$

for which $q_w \simeq 0.153$. (See also Table 5.1.) Note, q_w is less than 0.30. Compute the total tensile force taken by the prestressing steel at ultimate behavior and the required area of prestressing steel from:

$$F_n = f'_c[0.85\phi(b - b_w)h_f + q_w b_w d_p]$$

$$= 5[0.85 \times 0.9 \times 16 \times 3 + 0.153 \times 8 \times 20] = 306 \text{ kips}$$

$$A_{ps} = \frac{bd_p f'_c}{f_{pu}}\left(1 - \sqrt{1 - \frac{2F_n}{bd_p f'_c}}\right) \simeq 1.2166 \text{ in}^2$$

Note that for the above results the stress in the prestressing steel at ultimate is given by:

$$f_{ps} = \frac{F_n}{A_{ps}} = 251.52 \text{ ksi}$$

The reader may want to go back to the analysis flow chart Fig. 5.13 and check backward if for the reinforcement found here the designed ultimate moment resistance is arrived at.

5.8 CONTINUOUS AND COMPOSITE BEAMS

The above computations have addressed specifically the ultimate strength of a prestressed or partially prestressed concrete section. If the section considered is the critical section or section of maximum moment in a simply supported beam, then its ultimate resistance is also the ultimate resistance of the beam. The computation of the ultimate resisting moment at a given section of a continuous beam is similar to that of a simply supported beam. However, the value obtained at one critical section does not correspond in general to the failure moment of the beam itself, as many critical sections exist and plastic hinges may form at these sections. In such a case the plastic hinge analysis approach may be applied to determine the maximum load at which collapse will occur (see Sec. 10.14).

For a composite section made out of a cast-in-place concrete slab on top of a prestressed concrete girder or beam the procedure leading to the ultimate moment resistance is exactly similar to that of a T section, with one important modification. Because in general the cast-in-place slab has a compressive strength different from that of the prestressed girder, the effective width of the slab associated with the girder is to be multiplied by a factor equal to the ratio of their respective strengths, i.e., $(f'_c)_{slab}/(f'_c)_{girder}$. Then the composite section is analyzed as a noncomposite T section assuming a uniform strength equal to that of the prestressed girder. Note that in Chap. 9 on composite beams, the transformed section was obtained by multiplying the effective width of the slab by the ratio of moduli of elasticity, i.e., $(E_c)_{slab}/(E_c)_{girder}$ instead of the strength ratio. However, as the moduli ratio is higher than the strength ratio, the author feels that, for ultimate strength computations where f'_c is the important variable, it is safer to use the strength ratio. Note that the above remark on composite beams applies mostly if T section behavior is observed. If a rectangular section behavior exists, the ultimate strength can be directly computed (similarly to a noncomposite section) using the effective width of the slab and $(f'_c)_{slab}$ as, under the neutral axis, the section is assumed cracked and the prestressed girder does not contribute (see Sec. 9.8).

5.9 CONCLUDING REMARKS

The problems and procedures addressed in this chapter cover the majority of design situations where flexure of prestressed beams at ultimate behavior is involved. Although an attempt was made to follow closely the recommendations and equations given by the ACI code, diverging approaches were pointed out where necessary. The reader may want to supplement the present chapter by referring to Vol. 2 where design at ultimate using partial prestressing is systematically explored and where nonlinear analysis at ultimate and strain compatibility analysis are covered.

REFERENCES

5.1. C. S. Whitney: "Plastic Theory in Reinforced Concrete Design," *Transactions, ASCE*, vol. 107, 1942, pp. 251–326.

5.2. E. Hognestad, N. W. Hanson, and D. McHenry: "Concrete Stress Distribution in Ultimate Strength Design," *ACI Journal, Proceedings*, vol. 52, no. 44, December 1955, pp. 455–479.

5.3. J. R. Janney, E. Hognestad, and D. McHenry: "Ultimate Flexural Strength of Prestressed and Conventionally Reinforced Concrete Beams," *ACI Journal, Proceedings*, vol. 52, February 1956, pp. 601–617.

5.4. C. S. Whitney and E. Cohen: "Guide for Ultimate Strength Design of Reinforced Concrete," *ACI Journal, Proceedings*, vol. 53, November 1956, pp. 455–475.

5.5. A. H. Mattock, L. B. Kriz, and E. Hognestad: "Rectangular Concrete Stress Distribution in Ultimate Strength Design," *ACI Journal, Proceedings*, vol. 57, February 1961, pp. 877–928.

5.6. J. Warwaruk, M. Sozen, and C. P. Siess: "Investigation of Prestressed Reinforced Concrete for Highway Bridges, Part 3: Strength and Behavior in Flexure of Prestressed Concrete Beams," *University of Illinois, Urbana*, Engineering Experiment Station, Bulletin 464, 1962, 105 pp.

5.7. G. Gurfinkel and N. Khachaturian: "Ultimate Design of Prestressed Concrete Beams," University of Illinois, Urbana, Engineering Experiment Station, Bulletin 478, April 1965.

5.8. F. N. Pannell: "The Ultimate Moment Resistance of Unbonded Prestressed Concrete Beams," *Magazine of Concrete Research*, vol. 21, no. 66, March 1969, pp. 43–54.

5.9. *British Standard Code of Practice for the Use of Structural Concrete*, CP-110, British Standards Institution, London, 1972.

5.10. A. Tam and F. N. Pannell: "The Ultimate Moment of Resistance of Unbonded Partially Prestressed Reinforced Concrete Beams," *Magazine of Concrete Research*, vol. 28, no. 97, December 1976, pp. 203–208.

5.11. ACI Committee 318: "Building Code Requirements for Reinforced Concrete (ACI 318-77)," American Concrete Institute, Detroit, 1977, chap. 18.

5.12. ACI Committee 318: "Commentary on Building Code Requirements for Reinforced Concrete (ACI 318-77)," American Concrete Institute, Detroit, 1977, chap. 18.

5.13. *PCI Design Handbook—Precast, Prestressed Concrete*, 2d ed., Prestressed Concrete Institute, Chicago, Illinois, 1978.

5.14. A. E. Naaman: "Ultimate Analysis of Prestressed and Partially Prestressed Sections by Strain Compatibility," *PCI Journal*, vol. 22, no. 1, January/February 1977, pp. 32–51.

5.15. ———: "A Proposal to Extend some Code Provisions on Reinforcement to Partial Prestressing," *PCI Journal*, vol. 26, no. 2, March/April 1981, pp. 74–91.

PROBLEMS

5.1 Investigate the prestressed concrete simply supported box beam (described in Fig. P5.1) according to the ACI specifications given the following information: span 80 ft, $f'_c = 5000$ psi, $f'_{ci} = 4000$ psi, live load = 100 psf, $A_{ps} = 2.45$ in^2, $F = 370$ kips, $e_o = 15$ in at midspan, $\gamma_c = 150$ pcf, $\eta = 0.80$, $f_{pu} = 270$ ksi. Check stresses at initial and service loads and ultimate moment requirements at midspan.

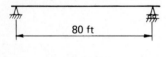

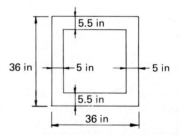

Figure P5.1

5.2 Consider the same cross section of the beam given in Prob. 5.1 (Fig. P5.1) and the same corresponding materials properties and e_o. Determine (a) the minimum value of reinforcement ratio or reinforcing index that can be used without violating the ACI code, (b) the maximum value of the reinforcing index if an underreinforced section is desired, and (c) in each case the corresponding ultimate moment resistance of the section.

5.3 Determine the ultimate resisting moment of the prestressed concrete beam section shown in Fig. P5.3 given the following: bonded member, $A_{ps} = 1.53$ in^2, $f'_c = 5000$ psi,

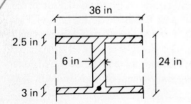

Figure P5.3

$e_o = 8.15$ in, $f_{pu} = 270$ ksi, $f_{pe} = 0.567 f_{pu}$, $A_c = 309$ in^2, $I = 25,500$ in^4, $Z_b = 2170$ in^3, $Z_t = 2080$ in^3, $y_t = 12.25$ in, $y_b = 11.75$ in, $d_p = 20.4$ in. Check ultimate to cracking moment ratio. Repeat assuming nonprestressed reinforcement is added in the lower flange having: $A_s = 1.24$ in^2, $f_y = 60$ ksi, $d_s = 22$ in.

5.4 Refer back to Prob. 4.6 after question (a) where it is assumed that the final prestressing force is achieved using three strands at $d_c = 3$ in from bottom fiber. (a) Check ultimate moment requirements at midspan according to the ACI specifications. Check also nominal to cracking moment requirements assuming a modulus of rupture $f_r = -7.5\sqrt{f_c'}$. (b) Repeat assuming unbonded strands are used. (If some requirements are then not satisfied, propose a solution to overcome them.)

5.5 Referring back to Prob. 4.7, check ultimate moment requirements and ratio of nominal to cracking moments at support and at midspan. Discuss the results observed at the support and provide some possible solutions.

5.6 The beam shown in Fig. P5.6 is to be used for an elevated structure designed for a transit system. The following information is given:

$\bar{\sigma}_{ti} = -189$ psi	$f_c' = 5000$ psi	$A_c = 576$ in^2	
$\bar{\sigma}_{ci} = 2400$ psi	$f_{ci}' = 4000$ psi	$y_t = 10.33$ in	$y_b = 13.67$ in
$\bar{\sigma}_{ts} = -213$ psi	$f_{pu} = 270$ ksi	$Z_t = 3990$ in^3	$Z_b = 3020$ in^3
$\bar{\sigma}_{cs} = 2250$ psi	$f_{pe} = 150$ ksi	$k_t = -5.24$ in	$k_b = 6.94$ in
$\eta = 0.80$	live load $= 1$ klf	$I = 41,300$ in^4	
$(d_c)_{min} = 3$ in		$w_G = 0.60$ klf	
$(e_o)_{mp} = 10.67$ in			

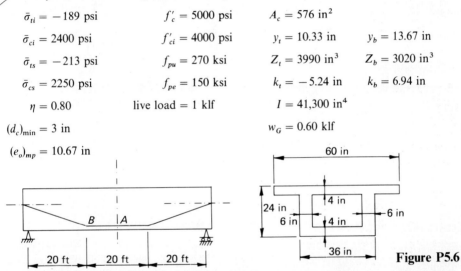

Figure P5.6

(a) Find the minimum value of F which satisfies all allowable stresses at midspan given the maximum and minimum moments:

$$M_{max} = 8640 \text{ kips-in} \qquad M_{min} = 3240 \text{ kips-in}$$

Round off the value of F to correspond to the nearest integer number of strands assuming the final force per strand is 22.95 kips. (b) Check ultimate moment requirements at midspan according to the ACI specifications. (c) Check the requirement for nominal-to-cracking moment ratio at midspan.

5.7 Determine the nominal moment resistance of a rectangular section given the following information: $b = 12$ in, $h = 24$ in, $f_c' = 5000$ psi, $f_{pu} = 270$ ksi, $f_{pe} = 148.5$ ksi, $f_y = f_y' = 60$ ksi, $A_{ps} = 1.07$ in^2, $A_s = 2$ in^2, $A_s' = 0.62$ in^2, $d_p = d_s = 21$ in, $d_s' = 2$ in. Check if, according to the assumptions used by the ACI code, the compressive steel yields at ultimate.

5.8 Determine (strictly based on ultimate strength requirements) the area of prestressed reinforcement and its eccentricity needed to provide an ultimate moment resistance of 4000 kips-in to a rectangular section with the following properties: $b = 12$ in, $h = 24$ in, $f_c' = 6000$ psi, $f_{pu} = 270$ ksi.

Prestressed concrete monorail structure at Disney World, Florida. *(Courtesy Prestressed Concrete Institute.)*

CHAPTER 6

DESIGN FOR SHEAR AND TORSION

6.1 INTRODUCTION

The effects of external loadings on structures seldom lead to pure flexure alone. Other actions are simultaneously induced and include shear, torsion, and axial forces. Shear is most commonly encountered in combination with flexure. Its consideration in design logically follows that of flexure and represents a major step as shown in Figs. 4.1 to 4.3. Torsion is dealt with only occasionally in everyday design.

Basically, shear and torsion are different in nature: shear is a force and torsion is a twisting moment. However, they both lead to similar shearing stresses in the structure. Such stresses can be resolved into a principal tensile stress, also called diagonal tension, and a principal compressive stress. The diagonal tension is of the greatest concern, as it induces cracking in the concrete. To ensure that such cracking does not lead to failure, transverse reinforcement resisting shear and/or torsion is generally provided in the form of stirrups or ties. Additional longitudinal reinforcement is also needed for torsion.

In spite of considerable research on, and analysis of, the behavior of reinforced concrete in shear (Refs. 6.1 to 6.15), no general theory has been established for all types of concrete structures. Special provisions exist in various codes, including the ACI code for the design of reinforced or prestressed concrete subjected to shear forces. No such coverage is provided in the ACI code for the torsion design of prestressed concrete, although reinforced concrete is addressed. Some recent investigations have indicated, however, that an approach similar to reinforced concrete applies to prestressed concrete, provided some minor modifications are introduced.

Sufficient information is given in this chapter to cover most common design problems involving shear and torsion of prestressed concrete. No attempt is made to use a sign convention for shear or torsional stresses. They are assumed positive throughout as treated in the code and various publications. Also assumed positive are values of specified factored strength and nominal resistance.

6.2 SHEAR DESIGN

In earlier codes the design for shear was based on limiting the magnitude of diagonal tension under working loads, thus providing a safety factor against cracking. However, in prestressed concrete an overload may induce substantial changes in compressive stresses, thus leading to disproportionately high increases in diagonal tension at some points of the section and seriously jeopardizing the corresponding margin of safety. In effect, the magnitude and direction of principal tensile stresses theoretically derived lose their applicability once cracking occurs. After cracking considerable changes in stresses and stress distribution take place. A more meaningful approach to shear design is reflected in the ACI code, where shear is considered at the ultimate limit state and shear reinforcement design for factored loads. Because after cracking concrete retains a significant shear resistance (crack roughness and aggregate interlocking), the full contribution of transverse reinforcement takes place long after cracking.

6.3 PRESTRESSED VERSUS REINFORCED CONCRETE IN SHEAR

It is often said that, with respect to shear, prestressed concrete offers two major advantages over reinforced concrete, namely:

1. For the same external loading, everything else being equal, the shear force in prestressed concrete is smaller than that in reinforced concrete. This is illustrated in Fig. 6.1 where the sign convention for shear adopted in this text is also given. It is observed that at any section x, the difference in shear between a prestressed concrete beam and an otherwise identical reinforced concrete beam is essentially due to the vertical component of the prestressing force $V_F = F \sin \alpha$. V_F acts in a direction opposite to the external loads, thus reducing the shear force in prestressed concrete which at any section x is given by:

$$V(x) = V_{\text{loads}}(x) - F \sin (\alpha_x) \tag{6.1}$$

2. Due to the compression induced by prestressing, the diagonal tension is smaller in prestressed concrete than in reinforced concrete. Furthermore, its angle of inclination with respect to the beam's axis is reduced. This implies that if cracking occurs and if, for safe design, the inclined crack is assumed to cross at least one stirrup, the stirrups' spacing in prestressed concrete would be larger and

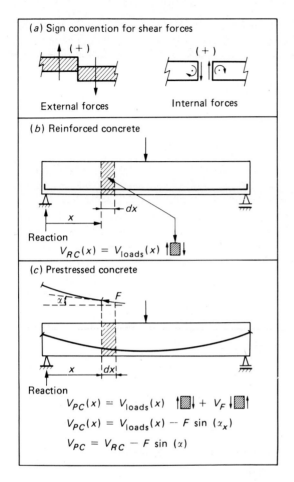

(a) Sign convention for shear forces

(+)   (+)

External forces   Internal forces

(b) Reinforced concrete

Reaction

$$V_{RC}(x) = V_{loads}(x)$$

(c) Prestressed concrete

Reaction

$$V_{PC}(x) = V_{loads}(x) + V_F$$

$$V_{PC}(x) = V_{loads}(x) - F \sin (\alpha_x)$$

$$V_{PC} = V_{RC} - F \sin (\alpha)$$

Figure 6.1 Shear in reinforced and prestressed concrete.

their required area smaller than in reinforced concrete. Thus, a more economical solution (for shear) is obtained. The determination of principal tension is clarified in the next section.

6.4 DIAGONAL TENSION IN UNCRACKED SECTIONS

Let us analyze, assuming elastic behavior, the stresses on a square element of main axis XX, YY taken along a concrete beam subjected to transverse loading. If the beam is plain or reinforced, an element taken along the neutral axis such as element A (Fig. 6.2) will not be subjected to any axial stresses, only to shear stresses. If the beam is prestressed, an axial stress will be exerted on A in addition to the shear stress. The analysis, in this case, is similar to that of an element B taken above the neutral axis of a plain or reinforced concrete beam.

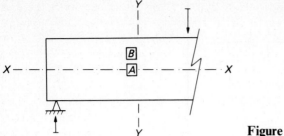

Figure 6.2

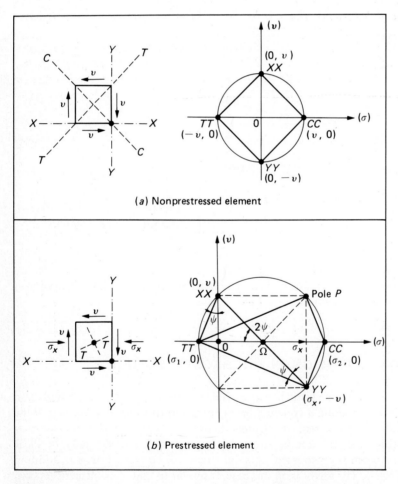

(a) Nonprestressed element

(b) Prestressed element

Figure 6.3 Mohr's circle for an element taken along the neutral axis.

To find the magnitude of principal stresses along an element of beam, it is convenient to construct the Mohr's circle of the element. The following general notation is adopted: σ for axial stress, v for shear stress, σ_1 for principal tensile stress, σ_2 for principal compressive stress, XX for the horizontal plane, YY for the vertical plane, CC for the plane of principal compression, and TT for the plane of principal tension.

The Mohr's circle of an element taken along the neutral axis of a non-prestressed beam is shown in Fig. 6.3a. The point of the circle representing each plane of interest is indicated. Note that XX and YY are diametrically opposed, thus bounding an arc of angle 180° on the circle (that is, twice their actual angle of 90°). It can be observed that the magnitudes of principal stresses (points with zero shear ordinate) represented by points CC and TT on the circle are equal to the magnitude of the shear stress and their angles of inclination are at 45° with respect to the horizontal and vertical axes. (The actual angle is half the angle on the Mohr's circle.) Also, the radius of the circle is equal in magnitude to the shear stress v.

Figure 6.3b gives the Mohr's circle for an element prestressed along its x axis. It can be observed that the magnitude of principal tension represented by the distance from the center of the circle to point TT is substantially smaller than in the previous case. Assuming v and σ_x given, the radius of the circle can be determined from:

$$R = \sqrt{v^2 + \left(\frac{\sigma_x}{2}\right)^2} \qquad (6.2)$$

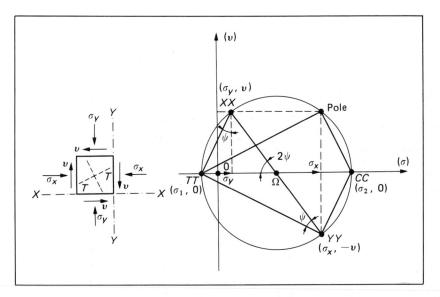

Figure 6.4 Mohr's circle for an element biaxially prestressed.

The principal stresses are then computed from:

$$\sigma_1 = -R + \frac{\sigma_x}{2} = -\sqrt{v^2 + \frac{\sigma_x^2}{4}} + \frac{\sigma_x}{2} \tag{6.3}$$

$$\sigma_2 = R + \frac{\sigma_x}{2} = \sqrt{v^2 + \frac{\sigma_x^2}{4}} + \frac{\sigma_x}{2} \tag{6.4}$$

The angle of inclination φ of the plane of principal tension with respect to the XX plane is obtained from (Fig. 6.3b):

$$\tan \varphi = \frac{\sigma_1}{v} = \frac{\sigma_x}{2v} - \sqrt{1 + \left(\frac{\sigma_x}{2v}\right)^2} \qquad \leftarrow \text{dividing by} \tag{6.5}$$

dividing by v² in eqn 6.2

It could also be obtained from:

$$\tan 2\varphi = \frac{2v}{\sigma_x} \qquad \tan 2\varphi = \frac{v}{\frac{\sigma_x}{2}} = \frac{2v}{\sigma_x} \tag{6.6}$$

Note that, as $\tan \varphi$ from Eq. (6.5) is smaller than 1, φ is smaller than 45°.

There is one convenient method to determine the inclination of different planes from the Mohr's circle. First, a pole point is determined: it is a point of the circle representing the intersection of a line parallel to a known plane and passing by its representative point on the circle. For instance, a line parallel to the horizontal plane and passing by XX intersects the circle at the pole. The pole has the following property: any line passing by the pole and another point of the circle is parallel to the plane of the element represented by that point. For instance, the line joining P to TT in Fig. 6.3b gives the actual direction of the TT plane.

In comparing the results obtained in Figs. 6.3a and b, it can be observed that both the principal tension and its inclination are smaller for a prestressed element than for a nonprestressed element. For the same shear stress v we have

$$(\sigma_1)_{\text{prestressed}} = (\tan \varphi)(\sigma_1)_{\text{nonprestressed}} \tag{6.7}$$

Mohr's circle representation can also be very useful when biaxial prestressing is applied. This occurs, for instance, when, in addition to horizontal prestressing, vertical prestressing is used near the supports of long bridges. Such technique can substantially reduce and/or eliminate diagonal tension. A typical representation is shown in Fig. 6.4 for an element subjected to shear stresses, and compressive stresses σ_x and σ_y with $\sigma_x > \sigma_y$. The following expressions can easily be derived:

$$R = \sqrt{v^2 + \left(\frac{\sigma_x - \sigma_y}{2}\right)^2} \tag{6.8}$$

$$0\Omega = \frac{\sigma_x + \sigma_y}{2} \tag{6.9}$$

$$\sigma_1 = \frac{\sigma_x + \sigma_y}{2} - R \tag{6.10}$$

$$\sigma_2 = \frac{\sigma_x + \sigma_y}{2} + R \qquad (6.11)$$

$$\tan 2\varphi = \frac{2v}{\sigma_x - \sigma_y} \qquad (6.12)$$

It is interesting to point out that one can eliminate the principal tension by properly selecting the value of σ_y, the vertical prestress. Setting $\sigma_1 = 0$ in Eq. (6.10) and replacing R by its value from Eq. (6.8) leads to:

$$\sigma_y = \frac{v^2}{\sigma_x} \qquad (6.13)$$

Note that, in order to compute the principal stresses, the values of v and σ or σ_x (shear and axial stresses) are needed. The determination of σ under the combined effects of prestressing and external loads was covered in Chap. 4. The determination of shear stress v is treated next.

6.5 SHEAR STRESSES IN UNCRACKED SECTIONS

The shear stress along any section of a flexural beam, assumed elastic uncracked, is given by (Fig. 6.5):

$$v_y = \left(\frac{Q}{b_y I}\right) V = \left(\frac{A_y \bar{y}}{b_y I}\right) V \qquad (6.14)$$

where v_y = shear stress at a distance y from the neutral axis (y within section)
Q = first static moment about the neutral axis of the portion of section outside the shear plane considered
A_y = area of the portion of section outside the shear plane
$\bar{y}$ = distance from neutral axis to the centroid of A_y
b_y = width of section at level y of shear plane considered
I = moment of inertia of the cross section.

Note that V is the shear force at the section of beam considered and, therefore, generally varies along the span. For a prestressed beam, V is the net shear (Eq. (6.1)) due to the combination of applied load and prestressing.

For a given value of V and section dimensions, the shear stress v_y varies along the section. Typical shear stress diagrams for common sections are shown in Fig. 6.5. It is observed that the maximum value of v_y occurs when the shear plane considered is taken along the neutral axis of the beam. The magnitude of maximum shear stress, also given in Fig. 6.5, can be put in the following form for rectangular and flanged sections:

$$v_{max} = \lambda \frac{V}{b_w h} \qquad (6.15)$$

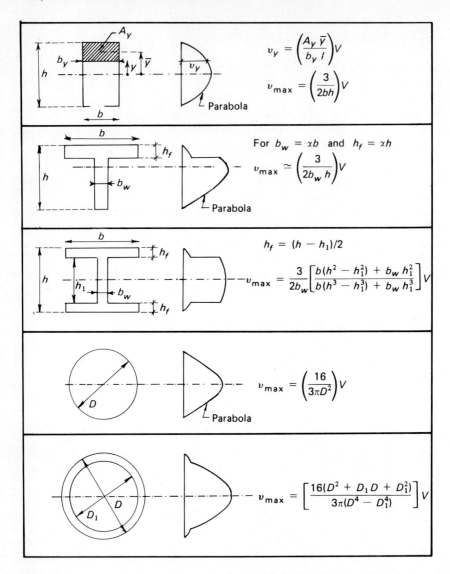

Figure 6.5 Typical shear stress diagrams and maximum shear stresses for various sections.

where λ is a factor larger than one. In computing shear stress from the shear force, the ACI code essentially adopts a value of λ equal to one and uses d (d_p or d_s) instead of h.

If cracking occurs, the shear stress distribution along the section will change. According to the analysis, it appears that the shear stress will be constant in the cracked region of the section.

6.6 SHEAR CRACKING BEHAVIOR

Structural cracking occurs in concrete when its tensile strength is exceeded. As any external loading generally leads to combined effects such as flexure and shear, it is likely that more than one type of cracking will be critical depending on the variables at hand.

An extensive number of investigations (Refs. 6.1 to 6.7) have shown or confirmed that two types of shear related cracks can develop in reinforced and prestressed concrete beams: *flexure-shear cracks and web-shear cracks* (Fig. 6.6a). The manner in which these cracks develop and grow strongly depends on the relative magnitude of shearing and flexural stresses.

Flexure-shear cracking is due to a combined effect of flexure and shear. The corresponding cracks start as flexural cracks (normal to the beam's axis). Then, due to the increased effect of diagonal tension at the tip of the crack, they deviate and propagate at an inclined direction corresponding essentially to the inclination of the diagonal tension plane. Typical flexure-shear cracks are shown in Fig. 6.6a.

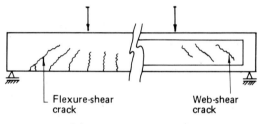

Flexure-shear crack   Web-shear crack

(a) Types of shear cracking

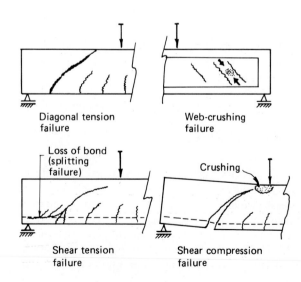

Diagonal tension failure   Web-crushing failure

Loss of bond (splitting failure)   Crushing

Shear tension failure   Shear compression failure

(b) Major types of shear failures

Figure 6.6 Shear cracking and shear failures.

Flexure-shear cracking can lead to several types of failures (Refs. 6.4, 6.5, 6.12), schematically illustrated in Fig. 6.6b. Very slender beams generally fail in flexure either by their tensile reinforcement or by the concrete compressive zone.

However, in beams with smaller shear span-to-depth ratio, failure may occur by flexure-shear cracking before the flexural capacity is developed. In moderately slender beams one of the cracks may continue to propagate until it becomes unstable, reaching throughout the depth of the beam and leading to what is described as diagonal tension failure. In relatively deep beams a secondary crack triggered by a flexure-shear crack may propagate horizontally along the longitudinal reinforcement, leading to a loss of bond followed by a loss of anchorage near the support and subsequent failure described as shear tension failure. Alternatively, the concrete at the upper end of the crack may fail by crushing in what is described as shear compression failure.

Web-shear cracking (Fig. 6.6a) occurs when the magnitude of principal tension is relatively high in comparison to flexural stresses. It is characteristic of beams with narrow webs, such as I beams, where cracking due to diagonal tension develops before flexural cracking. Web-shear cracking may lead to the same types of failures described for flexure-shear cracking, namely, diagonal tension, shear tension, and shear compression failures. In addition (Fig. 6.6b), crushing under diagonal compression may occur within the web, leading to web-crushing failure.

A photograph of typical shear cracking in prestressed concrete beams is shown in Fig. 6.7.

In an attempt to better understand shear failure mechanisms, several models have been proposed and include limit analysis mechanisms or analogies with arches, trusses, or frames. The arch and truss analogies are illustrated in Fig. 6.8. A more comprehensive treatment of each mechanism and their comparison can be found in Refs. 6.7, 6.8, and 6.13.

It is preferable to avoid shear failure as it is substantially more brittle than flexural failure. To supplement the shear resistance of concrete members and to ensure flexural failure prior to shear failure, shear reinforcement (also called web or transverse reinforcement) in form of stirrups is generally provided. The function of shear reinforcement is expanded upon in Ref. 6.9.

6.7 SHEAR REINFORCEMENT AFTER CRACKING

The approach used to determine shear reinforcement in prestressed concrete is identical to that used for reinforced concrete. It is assumed that a shear crack extends at 45° to the beam's axis, as shown in Fig. 6.9. The force carried by the web reinforcement is assumed to balance a principal tension across the crack equal in magnitude to the shear stress v. According to the ACI code, the value of v, balanced by the shear reinforcement, is the shear stress in excess of the shear strength contribution of the concrete. Let us call it v_s. It is also assumed that, at the onset of shear failure, the shear reinforcement will be stressed to its yield strength f_y.

Figure 6.7 Typical shear failure in prestressed beams without web reinforcement. *(Courtesy Prestressed Concrete Institute.)*

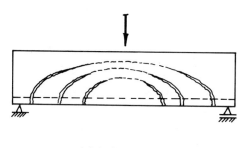

(*a*) Arch analogy

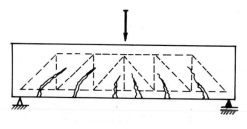

(*b*) Truss analogy

Figure 6.8 Typical analogies for shear failure mechanisms.

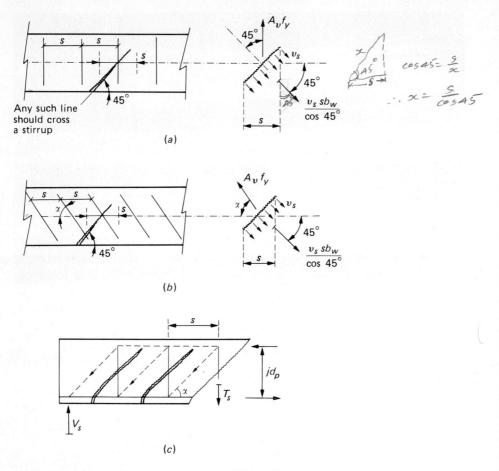

Figure 6.9 Shear resistance provided by (a) Vertical stirrups, (b) Inclined stirrups, (c) Truss analogy model.

Let us first consider the case of shear reinforcement placed normally to the beam's axis (vertical stirrups) with spacing s (Fig. 6.9a). The crack can be divided in equal segments, each traversed by a single stirrup. A free body diagram of a typical segment is also shown in Fig. 6.9a. The length of the segment is given by $(s/\cos 45°)$. Projecting the forces on the vertical axis leads to the following equilibrium equation:

$$A_v f_y = \left(\frac{b_w v_s s}{\cos 45°}\right) \cos 45° = b_w v_s s \qquad (6.16)$$

from which the required area of vertical reinforcement (stirrups) is obtained:

$$A_v = \frac{b_w v_s s}{f_y} \qquad (6.17)$$

If the stirrups were placed at an angle α to the beam's axis (inclined stirrups), an approach similar to the above would lead to the area of inclined reinforcement given by

$$A_v = \frac{b_w v_s s}{f_y (\sin \alpha + \cos \alpha)} \qquad (6.18)$$

with inclined stirrups

Although inclined stirrups are more efficient than vertical stirrups, the increased labor cost associated with their placing seldom justifies their use.

Note that in designing the shear reinforcement according to the ACI code, the value of v_s in Eqs. (6.17) and (6.18) is to be replaced by $(v_u/\phi - v_c)$ as explained in Sec. 6.8.

Equations (6.17) and (6.18) were derived assuming a 45° shear crack. It is expected that a smaller angle will develop in prestressed beams, thus leading to a conservative design.

The above results can also be simply derived using the truss analogy model of the web reinforced cracked beam in which the stirrups act as tension members of the analogous truss and the concrete struts running parallel to the diagonal cracks act as compression members (Fig. 6.9c). Assuming vertical stirrups, the shear at the section shown is given by:

$$V_s = T_s \qquad (6.19)$$

where T_s is the tensile force in the stirrup. The web steel force per unit length is given by T/s in which $s = jd_p \cot \alpha$. Hence,

$$\frac{T_s}{s} = \frac{V_s}{s} = \frac{V_s}{jd_p \cot \alpha} \qquad (6.20)$$

Assuming, as a first approximation, $jd_p \simeq d_p$ and $\alpha = 45°$ leads to:

$$\frac{T_s}{s} = \frac{V_s}{d_p} \qquad (6.21)$$

As stirrups are assumed to yield at ultimate, T_s is replaced by $A_v f_y$, leading to:

$$A_v = \frac{V_s s}{f_y d_p} \qquad (6.22)$$

in which

$$V_s = v_s b_w d_p \qquad (6.23)$$

Eq. (6.22) is identical to Eq. (6.17).

Shear reinforcement in prestressed beams consists mostly of stirrups. A stirrup has generally two legs, leading to an area of vertical reinforcement twice the cross-sectional area of the stirrup bar. Typical shapes of stirrups are shown in Fig. 6.10a. Stirrups used for shear can be "open" stirrups but must be appropriately anchored in the compressive zone of the members. Typical hooks and ties for stirrups, as recommended by the ACI code, are shown in Fig. 6.10b. Because of

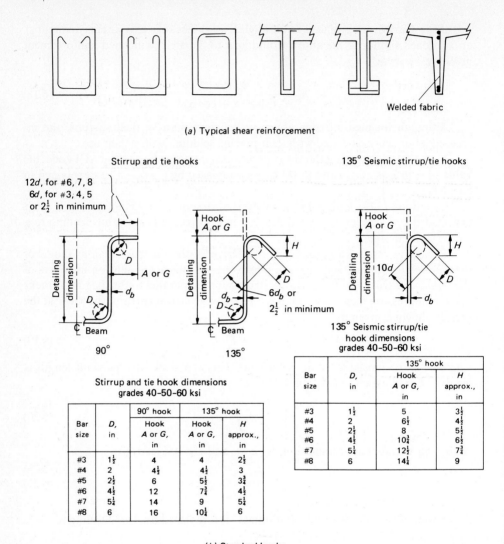

(a) Typical shear reinforcement

Stirrup and tie hooks

135° Seismic stirrup/tie hooks

90°

135°

135° Seismic stirrup/tie
hook dimensions
grades 40–50–60 ksi

Bar size	D, in	135° hook	
		Hook A or G, in	H approx., in
#3	$1\frac{1}{2}$	5	$3\frac{1}{2}$
#4	2	$6\frac{1}{2}$	$4\frac{1}{2}$
#5	$2\frac{1}{2}$	8	$5\frac{1}{2}$
#6	$4\frac{1}{2}$	$10\frac{3}{4}$	$6\frac{1}{2}$
#7	$5\frac{1}{4}$	$12\frac{1}{2}$	$7\frac{3}{4}$
#8	6	$14\frac{1}{4}$	9

Stirrup and tie hook dimensions
grades 40–50–60 ksi

Bar size	D, in	90° hook	135° hook	
		Hook A or G, in	Hook A or G, in	H approx., in
#3	$1\frac{1}{2}$	4	4	$2\frac{1}{2}$
#4	2	$4\frac{1}{2}$	$4\frac{1}{2}$	3
#5	$2\frac{1}{2}$	6	$5\frac{1}{2}$	$3\frac{3}{4}$
#6	$4\frac{1}{2}$	12	$7\frac{3}{4}$	$4\frac{1}{2}$
#7	$5\frac{1}{4}$	14	9	$5\frac{1}{4}$
#8	6	16	$10\frac{1}{4}$	6

(b) Standard hooks

Figure 6.10 (*a*) **Typical shear reinforcement.** (*b*) **Standard hooks for stirrups.** (*Courtesy American Concrete Institute.*)

relatively high labor costs associated with the placing of stirrups, an alternative solution consisting of welded wire fabric is often used with precast prestressed products (Fig. 6.10a). Commercial welded wire fabrics used for shear reinforcement have few longitudinal wires and selected sets of transverse wire spacings covering a practical range. It is common practice to assume that the vertical wires are properly anchored in the tensile zone if two welded junctions are provided.

6.8 ACI CODE DESIGN CRITERIA FOR SHEAR

1. Basic Approach

As mentioned earlier, the ACI design approach is based on ultimate strength requirements (factored loads). A design shear strength at each section of a member is required. It is achieved by the combined contributions of concrete and shear reinforcement. The concrete is assumed to provide a prescribed shear resistance; the shear reinforcement (stirrups) provides the excess resistance, if any, needed to satisfy the required strength. In the 1971 edition of the code, shear related equations were given in stress units. In the 1977 edition, they are given in force units where the force is obtained by multiplying the shear stress by $b_w d_p$. Although the approaches used and the results obtained are identical, the author prefers the use of stresses, as they are independent of the size of the member and, thus, can be easily correlated when different solutions are compared.

The required design shear strength at each section of the member should be less than or equal to its shear resistance, that is:

$$v_u \leq \phi v_n \tag{6.24}$$

where v_u = design shear stress resulting from factored loads
$\quad\ v_n$ = nominal shear strength of the section
$\quad\ \phi$ = strength reduction factor (equal to 0.85 for shear)

Equation (6.24) can be conveniently rewritten as:

$$\frac{v_u}{\phi} \leq v_n \tag{6.25}$$

The design shear stress v_u is obtained from

$$v_u = \frac{V_u}{b_w d_p} \tag{6.26}$$

where V_u = design shear force at section considered due to factored loads
$\quad\ b_w$ = width of rectangular section or web width of flanged section
$\quad\ d_p = d_p$ or $0.8h$ whichever is larger.

The provision to use $0.8h$ in case the actual value of d_p is smaller, is substantiated by test data. It is valid because the section is uncracked and the web area available to compute nominal shear stress is not a direct function of the location of the prestressing steel (Ref. 6.5).

The nominal shear strength v_n is defined as the sum of the shear resistance of the concrete and the shear contribution of the web reinforcement at the section considered. That is:

$$v_n = v_c + v_s \tag{6.27}$$

where v_c = nominal shear strength provided by the concrete
$\quad\ v_s$ = nominal shear strength provided by the shear reinforcement

The values of v_c and v_s are defined in the next sections.

2. Shear Strength Provided by Concrete ✓

According to the ACI code, the value of v_c is to be taken as the smaller of v_{ci}, the resistance to flexure-shear cracking, and v_{cw}, the resistance to web-shear cracking. The equations given by the code for v_{ci} and v_{cw} are simplified approximations of equations derived either theoretically from analyzing the section (v_{cw}) or from semirational analysis supported by experimental evidence (v_{ci}).

In order not to interrupt the design process, the derivations of v_{ci} and v_{cw} are given at the end of this chapter in Sec. 6.20. It is found that the flexure-shear cracking stress v_{ci} is made out of three components: (1) the shear stress needed to transform a flexural crack into an inclined crack, $0.6\sqrt{f'_c}$, (2) the shear stress due to the self-weight of the member, V_G/bd_p, and (3) the additional factored shear stress that initially will cause a flexural crack to occur, $\Delta M_{cr}/(\Delta M_u/\Delta V_u)b_w d_p$. Hence, the shear resistance of the concrete, corresponding to flexure-shear cracking, can be taken as (using the notation adopted in this text):

$$\checkmark \quad v_{ci} = 0.6\sqrt{f'_c} + \frac{V_G}{b_w d_p} + \left(\frac{\Delta V_u \times \Delta M_{cr}}{\Delta M_u}\right)\frac{1}{b_w d_p} \tag{6.28}$$

but not less than $1.7\sqrt{f'_c}$, where

b_w = width of rectangular section or web width of flanged section
$d_p = d_p$ or $0.80h$ whichever is larger at section considered
V_G = shear force due to self-weight of member at section considered
ΔV_u = factored shear force due to superimposed dead load plus live load at section considered under same loading as ΔM_u
ΔM_u = factored bending moment due to superimposed dead load plus live load at section considered
ΔM_{cr} = moment in excess of self-weight moment, causing flexural cracking in the precompressed tensile fiber at section considered $(= M_{cr} - M_G)$

The effect of self-weight shear V_G is considered separately for two reasons:

1. Self-weight is generally uniformly distributed, whereas live loads can have any distribution and can be concentrated loads. The ratio $\Delta M_u/\Delta V_u$ which appears in Eq. (6.28), remains constant at a given section when external loads increase proportionately and are independent of load factors. Corresponding calculations are easier to handle.
2. By separating the effect of self-weight, Eq. (6.28) remains applicable to composite members. For such members, V_G should include not only the weight of the precast unit but also that of the cast-in-place slab. However, note that the construction sequence in composite members (i.e., shored versus unshored) influences the value of ΔM_{cr}, as part of the load is resisted by the precast section alone and part by the composite section. Appropriate section properties should be used to determine ΔM_{cr}.

Note that the cracking moment increment ΔM_{cr} used in Eq. (6.28) is to be computed assuming a modulus of rupture of $-6\sqrt{f'_c}$ instead of $-7.5\sqrt{f'_c}$. Using Eq. (4.38) and assuming the beam is statically determinate, leads to:

$$\Delta M_{cr} = Z_b\left[6\sqrt{f'_c} + \frac{F}{A_c}\left(1 + \frac{e_o A_c}{Z_b}\right)\right] - M_G \qquad (6.29)$$

The nominal resistance of the concrete, corresponding to web-shear cracking at the section considered, is to be taken as:

$$v_{cw} = 3.5\sqrt{f'_c} + 0.3\sigma_g + \frac{V_p}{b_w d_p} \qquad (6.30)$$

where σ_g = compressive stress in the concrete at the centroid of the cross section or at the junction of web and flange if the centroid is in the flange, due to effective prestress

V_p = vertical component of prestressing force at section considered ($= F \sin \alpha$)

$d_p = d_p$ or $0.8h$ whichever is larger

Alternatively, v_{cw} may be theoretically computed as the shear stress corresponding to a combination of dead load and live load that results in a principal tensile stress of $-4\sqrt{f'_c}$ at the centroid of the section or at the junction of web and flange when the centroid is in the flange. Its derivation is given in Sec. 6.20.

The computation of the values of v_{ci} and v_{cw}, given by the above expressions, is time-consuming, especially as several sections along the span must be considered. An alternative, more conservative method for beams with uniform loads is suggested in the code and can be used when the effective prestress f_{pe} is not less than $0.40 f_{pu}$ (Ref. 6.9). The shear resistance of the concrete can then be taken as:

$$v_c = 0.6\sqrt{f'_c} + 700\frac{V_u d_p}{M_u} \qquad (6.31)$$

but not less than $2\sqrt{f'_c}$ and not more than $5\sqrt{f'_c}$. V_u and M_u are the factored shear and moment and d_p is the depth to the centroid of the prestressing force at the section considered. The ratio $V_u d_p/M_u$ is not to be taken greater than 1 and the lower bound value of $0.8h$ on d_p, used elsewhere, does not apply. Once the value of v_c has been determined, the shear force resistance of the concrete can be obtained from:

$$V_c = v_c b_w d_p \qquad (6.32)$$

3. Required Area of Shear Reinforcement

In order to determine the required area of shear reinforcement, the value of v_s is needed. The minimum design value of v_s is obtained from Eqs. (6.24) and (6.27) at equality. That is:

$$\frac{v_u}{\phi} = v_n = v_c + v_s \qquad (6.33)$$

from which

$$v_s = \left(\frac{v_u}{\phi} - v_c\right) \tag{6.34}$$

where v_s is the required nominal shear resistance of the shear reinforcement. The area of shear reinforcement can then be obtained from Eqs. (6.17) and (6.18) in which v_s is replaced by its value from Eq. (6.34):

For vertical stirrups:

$$A_v = \frac{(v_u/\phi - v_c)b_w\, s}{f_y} = \frac{(V_u/\phi - V_c)s}{f_y\, d_p} \tag{6.35}$$

For stirrups inclined at an angle α to the beam's axis:

$$A_v = \frac{(v_u/\phi - v_c)b_w\, s}{f_y\,(\sin \alpha + \cos \alpha)} = \frac{(V_u/\phi - V_c)s}{f_y\, d_p\,(\sin \alpha + \cos \alpha)} \tag{6.36}$$

The above two equations can also be used to find the required spacing s for a given type of stirrup A_v:

For vertical stirrups:

$$s = \frac{A_v\, f_y}{(v_u/\phi - v_c)b_w} = \frac{A_v\, f_y\, d_p}{V_u/\phi - V_c} \tag{6.37}$$

For inclined stirrups:

$$s = \frac{A_v\, f_y\,(\sin \alpha + \cos \alpha)}{(v_u/\phi - v_c)b_w} = \frac{A_v\, f_y\,(\sin \alpha + \cos \alpha)d_p}{V_u/\phi - V_c} \tag{6.38}$$

4. Limitations and Special Cases

A number of design limitations relative to shear reinforcement are given in the ACI code:

1. *Maximum spacing.* The spacing s of shear reinforcement measured parallel to the axis of the member shall not exceed $3h/4$ or 24 in (61 cm). In common practice the spacing is taken not less than 3 in (75 mm).
2. *Maximum shear.* The value of $(v_u/\phi - v_c)$ shall not exceed $8\sqrt{f'_c}$. If it does, the section is insufficient for shear resistance and its dimensions should be changed (namely, b_w and/or d_p). If $(v_u/\phi - v_c)$ exceeds $4\sqrt{f'_c}$, the spacing obtained from Eqs. (6.35) to (6.38) should be halved and should not exceed $3h/8$ or 12 in (30 cm).
3. *Minimum shear reinforcement.* Shear reinforcement restrains the growth of inclined cracking and, hence, increases ductility and provides a warning of failure. Otherwise, in an unreinforced web, the sudden formation of inclined cracking

might lead directly to sudden failure. Accordingly, the code requires a minimum area of shear reinforcement to be provided in all reinforced and prestressed concrete structural members when the factored shear stress v_u exceeds half the shear strength of concrete ϕv_c, except in (1) slabs and footings, (2) concrete joist construction for which the code allows a 10 percent increase in v_c, and (3) beams with total depth not greater than 10 in (25 cm), $2\frac{1}{2}$ times the thickness of the flange or one-half the web width, whichever is greater.

When the effective prestress f_{pe} exceeds $0.40f_{pu}$ (which is common in prestressed concrete), the minimum required area of shear reinforcement may be computed by:

when $f_{pe} > 0.40 f_{pu}$

$$(A_v)_{min} = \frac{A_{ps}}{80} \frac{f_{pu}}{f_y} \frac{s}{d_p} \sqrt{\frac{d_p}{b_w}} \qquad (6.39)$$

where d_p has a lower bound value of $0.80h$ and other notations are standard.

A more general equation for the minimum required area of shear reinforcement is given in the code and applies to reinforced as well as prestressed concrete (with no limitation on the effective prestress). It is given by:

$$(A_v)_{min} = 50 \frac{b_w s}{f_y} \qquad (6.40)$$

where b_w and s are in inches and f_y in pounds per square inch. Equation (6.40) will generally require greater minimum shear reinforcement in typical building members than will Eq. (6.39).

4. *Lightweight concrete.* When prestressed lightweight concrete is used instead of normal weight concrete, the same procedure to determine the shear reinforcement is used with the following modifications prescribed by the code:

(a) When the split cylinder strength f'_{tc} is specified for lightweight concrete and when $f'_{tc}/6.7$ is less than $\sqrt{f'_c}$, replace $\sqrt{f'_c}$ by $f'_{tc}/6.7$ in the equations giving v_c.

(b) When f'_{tc} is not specified, the values of $\sqrt{f'_c}$ used in the expressions of v_c and ΔM_{cr} shall be multiplied by 0.75 for "all-lightweight" concrete and 0.85 for "sand-lightweight" concrete.

5. Critical Sections for Shear $\left(\frac{h}{2}\right)$

Contrary to reinforced concrete where commonly the most critical section is near the support, in prestressed concrete several sections along the beam may be more critical than the support section. This is why shear is to be investigated at several sections along the span.

The first critical section for shear is to be taken at $h/2$ from the face of the support and the same shear reinforcement required at $h/2$ is to be used at sections located at a distance less than $h/2$. However, in some particular cases the first critical section must be taken at less than $h/2$, as illustrated in Fig. 6.11 (Ref. 6.15).

In the design of slabs and footings, shear reinforcement is not required when $v_u < \phi v_c$. Similarly to what is done for beams, the first critical section in each

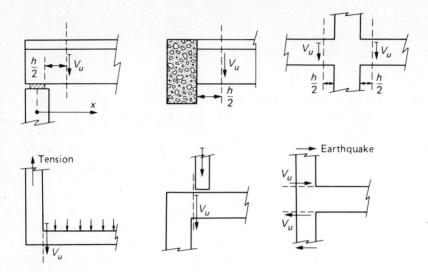

Figure 6.11 Typical support conditions for locating first critical section in shear design.

direction is to be taken at $h/2$ from the face of the support. Simultaneously, punching shear must be checked along a critical section perpendicular to the plane of the slab with perimeter b_o (Fig. 6.12) located a distance $d_p/2$ from the perimeter of the concentrated load or reaction area. The value of v_c for punching shear is taken as:

$$v_c = \left(2 + \frac{4}{\beta_c}\right)\sqrt{f'_c} \qquad (6.41)$$

but not greater than $4\sqrt{f'_c}$, where β_c is the ratio of long side to short side of concentrated load or reaction.

Punching shear resistance in slabs is covered more extensively in Chap. 11.

6.9 DESIGN EXPEDIENTS

Designing for shear can be very time-consuming, as one does not know a priori the locations of the critical sections which require more than the minimum amount of

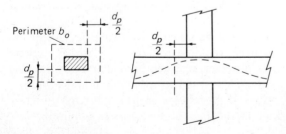

Figure 6.12 Critical section for punching shear in slabs.

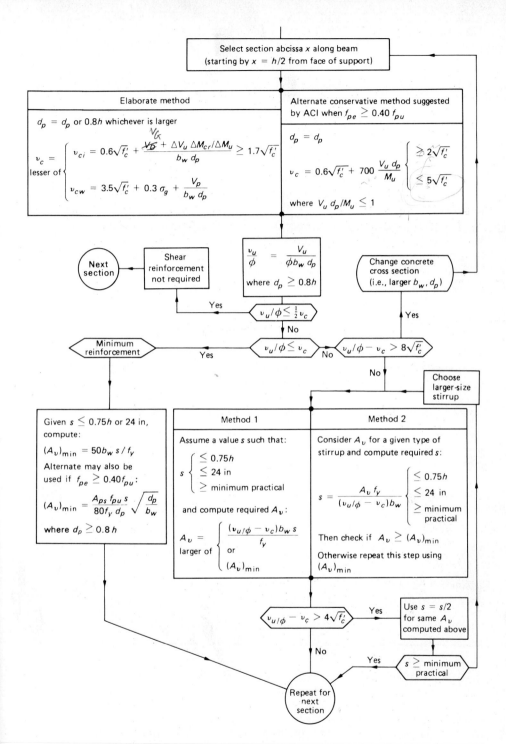

Figure 6.13 Flow chart for the design of shear reinforcement in beams.

reinforcement. It is not uncommon to find out (assuming uniform loading) that excess reinforcement is needed at two sections and is not needed at a section in between.

In most common design problems involving beams, it is very likely that the minimum amount of shear reinforcement $(A_v)_{min}$ at the maximum allowable spacing of $0.75h$ or 24 in ($\simeq$ 60 cm) will prevail throughout the span. If the alternative method to determine v_c is considered, its lower bound value of $2\sqrt{f'_c}$ can be used as a first approximation in design and checked against the value of v_u/ϕ at several sections. Only if significant differences exist, the more accurate analysis will be pursued.

In order to ease the design procedure, a design flow chart for shear is proposed in Fig. 6.13. It applies to normal weight prestressed concrete beams and synthesizes most of the steps encountered if the ACI code approach is followed. As in each complete iteration only one section is considered, it is appropriate to set the numerical results in a table, such as in Tables 6.1 and 6.2.

Once the computations are completed, the results can be directly used to determine stirrups' areas and spacings. However, a graphical representation of the variation of shear stresses and strengths may give a better visual representation of stirrup distribution along the span. Such a typical graph is shown in Fig. 6.17. As only a discrete number of sections is analyzed, the graph can help isolate graphically the segments of the span where same shear reinforcement and spacing prevail.

Often, in practice, a type of stirrup is selected for a given beam, say a #3 U stirrup, and the spacing is selected at each section according to the design requirements. However, only a discrete number of spacings is considered, such as 4, 6, 8, 12, 18, and 24 in (10, 15, 20, 30, 45, and 60 cm). Thus one can go backward and, given A_v and a spacing, find graphically the value of $(v_u/\phi - v_c)$ and the corresponding segment of beam where such spacing applies. Charts can also be developed to determine the value of $(v_u/\phi - v_c)$ that can be resisted by a certain stirrup at a certain spacing. Using Eq. (6.35) for vertical stirrups we have:

$$\left(\frac{v_u}{\phi} - v_c\right) = \frac{A_v f_y}{b_w s} \qquad \qquad s \text{ varies.} \qquad (6.42)$$

which can be plotted versus b_w for different values of s, assuming A_v and f_y given. Such a chart for a #3 U stirrup with $f_y = 60$ ksi is given in Fig. 6.14.

6.10 EXAMPLE: DESIGN OF SHEAR REINFORCEMENT

The T beam described in Sec. 4.9 and the Example of Sec. 4.12 is considered (Fig. 6.15). The design characteristics of interest to this example are: $h = 40$ in, $b_w = 8$ in, $A_c = 550$ in^2, $A_{ps} = 1.53$ in^2, $y_t = 12.9$ in, $Z_b = 3028$ in^3, $f'_c = 5000$ psi (normal weight concrete), $F = 229.5$ kips, $f_{pe} > 0.4f_{pu}$, $l = 70$ ft, live load = 0.4 klf, superimposed dead load = 0.04 klf, self-weight $w_G = 0.573$ klf.

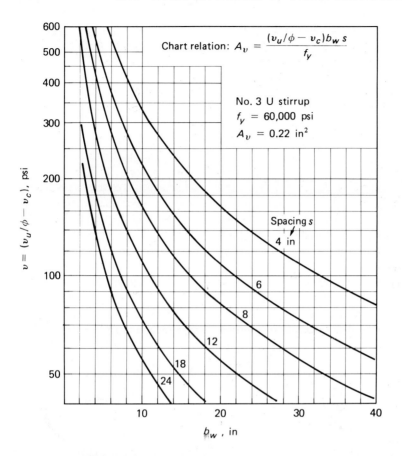

Chart relation: $A_v = \dfrac{(v_u/\phi - v_c)b_w\, s}{f_y}$

No. 3 U stirrup
$f_y = 60{,}000$ psi
$A_v = 0.22$ in^2

Spacing s

4 in

6

8

12

18

24

$v = (v_u/\phi - v_c)$, psi

b_w , in

Figure 6.14 Design chart for stirrup spacings.

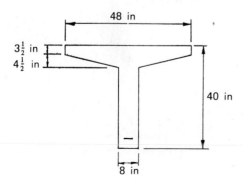

48 in

$3\frac{1}{2}$ in

$4\frac{1}{2}$ in

40 in

8 in

Figure 6.15

The final eccentricity of the prestressing force along the half span was arrived at in Sec. 4.12 and is given by:

$$\begin{cases} e_o = 21.7 \text{ in} & \text{for } x \geq 28 \text{ ft} \\ e_o = 7.9 + 13.8 \dfrac{x}{28} & \text{for } 0 \leq x < 28 \text{ ft} \end{cases}$$

Let us define the following values of factored loads:

$$w_u = 1.4(0.573 + 0.04) + 1.7 \times 0.4 = 1.538 \text{ klf}$$

$$\Delta w_u = 1.4(0.04) + 1.7 \times 0.4 = 0.736 \text{ klf}$$

The first critical section to be analyzed is taken at $h/2 = 20$ in from the face of the support. If a support pad eight inches wide is used, the location of the section from the center of the support becomes $x = 20 + 8/2 = 24$ in or 2 ft.

Two approaches will be followed, the more elaborate approach and the alternative conservative approach, according to the steps described in the flow chart Fig. 6.13. The computations for the first critical section will be covered in detail. Results obtained for this section and four others are summarized in Tables 6.1 and 6.2.

1. Elaborate approach. The eccentricity of the prestressing force at $x = 2$ ft is given by:

$$e_o = 7.9 + 13.8 \frac{2}{28} = 8.9 \text{ in}$$

Thus

$$d_p = e_o + y_t = 8.9 + 12.9 = 21.8 \text{ in}$$

For the equations used in the elaborate approach, d_p is limited by $0.8h = 32$ in. Thus, use $d_p = 32$ in. Following the design flow chart Fig. 6.13, let us first compute v_{ci}. For this we need the following values:

$$V_G = w_G \left(\frac{l}{2} - x\right) = 0.573 \times 33 = 18.91 \text{ kips}$$

$$\Delta V_u = \Delta w_u \left(\frac{l}{2} - x\right) = 0.736 \times 33 = 24.288 \text{ kips}$$

$$\Delta M_u = \Delta w_u \frac{x(l - x)}{2} = \frac{0.736 \times 2 \times 68}{2} = 50.05 \text{ kips-ft} = 600.57 \text{ kips-in}$$

$$M_D = w_G \frac{x(l - x)}{2} = \frac{0.573 \times 2 \times 68}{2} = 38.964 \text{ kips-ft} = 467.57 \text{ kips-in}$$

$$\Delta M_{cr} = Z_b \left[6\sqrt{f_c'} + \frac{F}{A_c}\left(1 + \frac{e_o A_c}{Z_b}\right)\right] - M_G$$

$$= 3028 \left[6\sqrt{5000} + \frac{229,500}{550}\left(1 + \frac{8.9 \times 550}{3028}\right)\right] - 467,570$$

$$= 4,123,153 \text{ lb-in} = 4123 \text{ kips-in}$$

Table 6.1 Summary of shear design computations for Example 6.10, case *a*

Section, x	V_D, kips	ΔV_u, kips	ΔM_u, kips-in	ΔM_{cr}, kips-in	d_p or $0.8h$ in	v_{ci}, psi	V_p, kips	v_{cw}, psi	v_c, psi	V_u, kips	v_u/ϕ, psi	$v_u/\phi - v_c$	A_v at s or s for A_v	$(A_v)_{min}$ at s_{max}
0.5h	18.91	24.288	600.57	4123	32	767.6	←	←	409.5	50.75	233.2	negative	use next column	(0.13 in² required) (0.22 in² provided)
1/20	18.05	23.184	1027.82	3958	32	461.3			409.5	48.45	222.6			
2/20	16.044	20.608	1947.45	3638	32	255.3		409.5	255.3	43.06	197.9			
3/20	14.039	18.032	2758.9	3402	32	184	9.426		183	37.68	173.2			
4/20	12.033	15.456	3462.14	3250	32	146	→	→	146	32.30	148.4			
⋮ 1/2	⋯	⋯	⋯	⋯	⋯	⋯			⋯	⋯	⋯			

Table 6.2 Summary of shear design computations using the conservative method; Example 6.10, case *b*

Section	Distance, x ft	d_p, in	V_u, kips	M_u, kips-in	$\dfrac{V_u d_p}{M_u} \le 1$	$v_c \begin{cases} \le 5\sqrt{f'_c} \\ \ge 2\sqrt{f'_c} \end{cases}$ psi	$v_u = \dfrac{V_u\dagger}{\phi b_w d_p}$ psi	$\left(\dfrac{v_u}{\phi} - v_c\right)$ psi	A_v for s or s for A_v	$(A_v)_{min}$ at s_{max}
0.5h	2	21.8	50.75	1255.2	0.88	~~658~~ 353.5 / ~~399~~	233.2	....	use next column	(0.13 in² required) (0.22 in² provided)
1/20	3.5	22.52	48.45	2148.1	0.51	~~399~~ 353.5	222.6	....		
2/20	7	24.25	43.06	4070	0.26	224.4	197.9	⋯		
3/20	10.5	25.8	37.68	5765.9q	0.17	161.4	173.2	11.8		
4/20	14	27.7	32.30	7235.7	0.12	~~126~~ 141.4	148.4	7		
⋮ 1/2	⋯	⋯	⋯	⋯	⋯	⋯	⋯	⋯		

† $d_p = d_p$ or $0.8h$ whichever is larger

Thus:

$$v_{ci} = 0.6\sqrt{f'_c} + \frac{V_G + (\Delta V_u \times \Delta M_{cr})/\Delta M_u}{b_w d_p} \qquad \text{eqn. } 6.28$$

$$= 0.6\sqrt{5000} + 1000\left(\frac{18.91 + (24.288 \times 4123)/600.57}{8 \times 32}\right)$$

$$= 767.6 \text{ psi} > 1.7\sqrt{f'_c} = 120 \text{ psi} \qquad\qquad \text{OK}$$

In order to determine v_{cw}, we need the following:

$$\sigma_g = \frac{F}{A_c} = \frac{229{,}500}{550} = 417.3 \text{ psi}$$

(*Note:* The centroid is in the web.)

$$\sin \alpha \simeq \tan \alpha = \frac{(e_o)_{28} - (e_o)_{support}}{28 \times 12} = 0.04107$$

$$V_p = F \sin \alpha = 229{,}500 \times 0.04107 = 9426 \text{ lb}$$

Thus:

$$v_{cw} = 3.5\sqrt{f'_c} + 0.3\sigma_g + \frac{V_p}{b_w d_p} \qquad \longrightarrow \quad \text{eqn. } 6.30$$

$$= 3.5\sqrt{5000} + 0.3 \times 417.3 + \frac{9426}{8 \times 32} = 409.5 \text{ psi}$$

The shear resistance v_c is the smaller of v_{ci} and v_{cw}. Thus at $x = 2$ ft:

$$v_c = 409.5 \text{ psi}$$

In order to determine v_u, we need V_u:

$$V_u = w_u\left(\frac{l}{2} - x\right) = 1.538(35 - 2) = 50.75 \text{ kips}$$

thus

$$\frac{v_u}{\phi} = \frac{V_u}{\phi b_w d_p} = \frac{50{,}750}{0.85 \times 8 \times 32} = 233.2 \text{ psi}$$

The value of v_u/ϕ is to be compared to v_c. As $v_c = 409.5 > v_u/\phi$, only minimum reinforcement is needed at this section.

Let us select a stirrup with $f_y = 60$ ksi, at a maximum spacing $s = 24$ in $< 0.75h$. As $f_{pe} > 0.4f_{pu}$, the minimum area of shear reinforcement is obtained from:

$$(A_v)_{min} = \frac{A_{ps} f_{pu} s}{80 f_y d_p}\sqrt{\frac{d_p}{b_w}} = \frac{1.53 \times 270 \times 24}{80 \times 60 \times 32}\sqrt{\frac{32}{8}} \simeq 0.13 \text{ in}^2$$

Note, a #3 U stirrup would provide 0.22 in² and thus would be adequate. The other sections covered in Table 6.1 lead to the same minimum reinforcement.

If, instead of selecting the spacing in the above equation, we would have selected the area of a #3 U stirrup, the required spacing would have been $s = 40.9$ in. Of course, then the maximum allowable spacing of 24 in would control, leading to the same result.

Providing a given area of reinforcement at a given spacing allows the determination of the maximum shear stress to be resisted by the reinforcement. In this case, we have:

$$v_s = \left(\frac{v_u}{\phi} - v_c\right) \leq \frac{A_v f_y}{b_w s} = \frac{0.22 \times 60,000}{8 \times 24} \simeq 69 \text{ psi}$$

2. **Alternate conservative approach.** In order to compute v_c we need:

$$V_u = 50.75 \text{ kips} \qquad \text{(calculated above)}$$

$$M_u = w_u \frac{x(l - x)}{2} = \frac{1.538 \times 2 \times 68}{2} = 104.58 \text{ kips-ft} = 1255 \text{ kips-in}$$

$$d_p = 21.8 \text{ in}$$

$$\frac{V_u d_p}{M_u} = 0.88 < 1 \qquad\qquad\qquad\qquad \text{OK}$$

Thus:

$$v_c = 0.6\sqrt{f'_c} + 700 \frac{V_u d_p}{M_u} = 658 \text{ psi}$$

v_c is larger than $2\sqrt{f'_c} = 141.4$ psi but should be taken less than $5\sqrt{f'_c} = 353.5$ psi. Thus, use $v_c = 353.5$ psi. Results for the other sections are summarized in Table 6.2.

If we compare v_u/ϕ and v_c we find out that, similar to the previous case, a minimum reinforcement is sufficient.

3. **Design for increased live load.** To illustrate a case where more than the minimum reinforcement is needed, let us assume that the specified live load for the previous example is increased from 0.4 klf to 1 klf. Let us also assume that the beam is designed as a partially prestressed beam, whereas the amount and profile of the prestressing steel are the same as

Table 6.3 Summary of shear design computations for Example 6.10, case c

Section, x	ΔV_u, kips	ΔM_u, kips-in	v_{ci}, psi	v_{cw}, psi	v_c, psi	V_u, kips	v_u/ϕ, psi	$v_u/\phi - v_c$, psi	Required s for A_v†
0.5h	57.948	1,432.9	767.6	↑	409.5	84.42	388	....	24
$l/20$	55.314	2,452.3	418.9		409.5	80.58	370	....	24
$2l/20$	49.168	4,646.4	255		255	71.63	329	74	22.3
$3l/20$	43.022	6,582.4	183		183	62.68	288	105	15.7
$4l/20$	36.876	8,260.2	146	409.5	146	53.72	247	101	16.3
$5l/20$	30.73	9,680	121		121	44.77	206	85	19.4
$6l/20$	24.584	10,841.5	120		120	35.81	165	45	24
$7l/20$	18.438	11,745	120	↓	120	26.86	120	0	24
$8l/20$	12.292	12,390	120	373	120	17.91	76	....	24
$9l/20$	6.146	12,778	120	373	120	8.95	38	....	24
$l/2$	0	12,907	120	373	120	0	0	....	24

† For $A_v = 0.22 \text{ in}^2$ and $s \leq 24$ in.

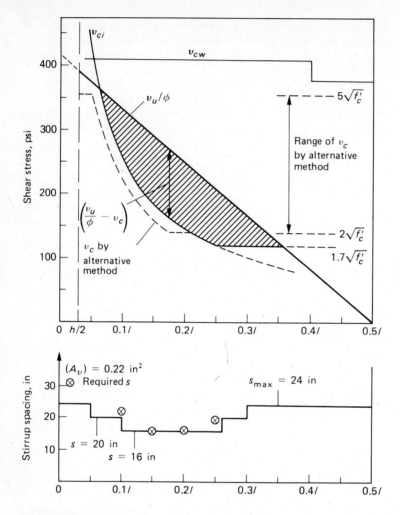

Figure 6.16 Shear stress distribution and stirrups spacing for example beam.

described above but, in order to balance the increase in the required resistance, nonpre-stressed reinforcement is provided. It amounts to four #9 bars with $A_s = 4$ in², $f_y = 60$ ksi, and $d_s = 36$ in at midspan.

For this case, the values of V_G, d_p, M_{cr}, and V_p are the same as in case 1 and are given in Table 6.1. Due to the increase in live load, the following values of factored loads apply:

$$w_u = 1.4(0.573 + 0.04) + 1.7 \times 1 = 2.5582 \text{ klf}$$

$$\Delta w_u = 1.4 \times 0.04 + 1.7 \times 1 = 1.756 \text{ klf}$$

The corresponding computations for ΔV_u, ΔM_u, v_{ci}, v_{cw}, v_c, and the required spacing (Eq. (6.37)) of #3 U stirrups are summarized in Table 6.3. It can be observed that, although the live load has been substantially increased, the required amount of shear reinforcement is still relatively small. If the alternative conservative approach was used, the value of v_c would have been the same as shown in Table 6.2. This is because the ratio V_u/M_u remains the same with an increase in uniform live load. The values of v_{ci}, v_{cw}, and v_c, as well as the required stirrups spacing, are plotted versus x in Fig. 6.16. The shaded area corresponds to the part of the beam where stirrups are theoretically needed. It can be seen that the greatest need for stirrups is not near the support but instead in a region where flexure-shear cracking may control. However note that, although shear may not be critical near the support, additional stirrups are needed to contain tensile splitting cracks in the end zone (see Sec. 4.16).

6.11 TORSION AND TORSION DESIGN

Torsion in form of moment or torque causes twisting of a member along its axis. Common structural concrete elements are often subjected to torsional moments due to unsymmetrical loading. However, the magnitude of these moments is generally small in comparison to other effects and, thus, torsion is not explicitly considered in their design. There are, however, structural members and/or loading conditions for which torsional effects may be significant and must be designed for. This is the case for spandrel beams, cantilevers, unsymmetrically loaded bridges, curved members, spiral stairways, and the like (Fig. 6.17).

An attempt is made in this chapter to unify the torsion design procedure for reinforced and prestressed concrete, using essentially the same notation to avoid confusion. It is stressed, however, that, while the criteria for reinforced concrete are already included in the ACI code (Refs. 6.16 to 6.23), the criteria for prestressed

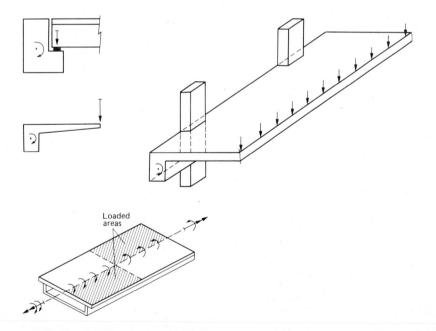

Figure 6.17 Typical members subjected to torsion.

concrete have only been prescribed in the technical literature (Refs. 6.24 to 6.34) and may still be subjected to further simplifications, refinements, and developments.

The design philosophy for torsion is the same as for shear and bending where ultimate strength requirements are sought. As at ultimate cracking prevails, an attempt to guarantee equilibrium (not compatibility) is made. The basic approach consists in considering the concrete contribution to the ultimate torsional strength of a member and then determining the reinforcement, if needed, to balance the remaining portion of the required strength.

Torsion reinforcement comprises both transverse reinforcement in the form of closed stirrups and longitudinal reinforcement (in excess of that needed for flexure) to be distributed along the periphery of the section. Limitations (similar to those encountered for shear and bending) on the minimum and maximum amounts of reinforcement also exist for torsion.

6.12 BEHAVIOR IN PURE TORSION

According to St. Venant's torsion theory for elastic materials, plain concrete members subjected to an increasing torsional moment would eventually fail by developing a spiral type of cracking. However, the actual failure behavior of plain concrete is different. A tension crack generally develops in a plane inclined at about 45° to the axis of the twist and failure is sudden (brittle-like). There is evidence (Refs. 6.19 and 6.20) that, with the exception of thin specimens, failure is mostly caused by bending induced by the bending component of the torsional moment (Fig. 6.18).

If the concrete element is reinforced against torsion, its behavior before cracking is essentially identical to a plain concrete element. Cracking would occur at about the same torque as for the unreinforced element. Prior to cracking, the contribution of the torsional reinforcement is small, as evidenced by the low

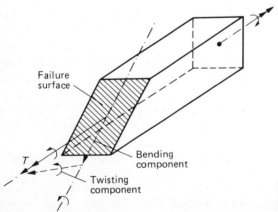

Figure 6.18 Torsional cracking model of plain concrete beam.

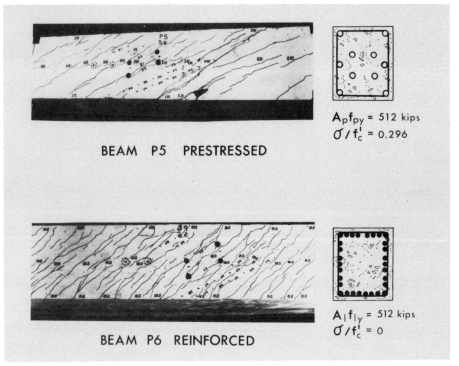

$A_p f_{py}$ = 512 kips
σ / f_c' = 0.296

BEAM P5 PRESTRESSED

$A_l f_{ly}$ = 512 kips
σ / f_c' = 0

BEAM P6 REINFORCED

Figure 6.19 Typical torsion cracking of reinforced members. *(Ref. 6.31, courtesy Prestressed Concrete Institute.)*

stresses recorded during testing. After cracking, the stiffness of the member drops suddenly to a fraction of its precracking value and the stress in the reinforcement increases sharply to maintain equilibrium. With increasing torque, multiple inclined cracks develop at regular spacing and take the form of S shapes (Fig. 6.19). The principal compressive strain measured along elements of concrete bound by two cracks seems to increase at a rate faster than predicted by the theory.

In general, three types of failure by torsion can be observed (Ref. 6.21) depending on the amount of reinforcement, and they are comparable to the flexural types of failure: beams underreinforced for torsion will fail by yielding of the steel while overreinforced beams will fail by crushing of the concrete, mostly on the wider face of the member; a third mixed mode type of failure occurs in partially overreinforced members in which a combination of stirrup yielding or longitudinal steel yielding and concrete crushing occurs.

The functions after cracking of the different elements of a cracked concrete member reinforced for torsion can be visualized by referring to Fig. 6.20 taken from Ref. 6.30. The figure is also useful in summarizing the information on the types of torsional reinforcement and the details that must be addressed in designing them.

Several early theories and models have been proposed to explain the behavior of concrete and reinforced concrete in torsion (Ref. 6.18). A more recent model

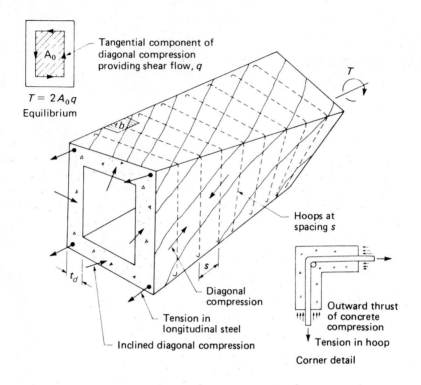

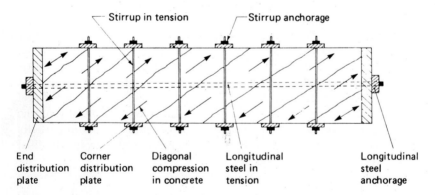

Figure 6.20 Idealized functions of concrete and reinforcement after torsional cracking. *(Ref. 6.30, courtesy American Concrete Institute.)*

developed by Lampert (Ref. 6.23) proposes a tridimensional space truss analogy in which the peripheral longitudinal reinforcement represents the booms of the truss, the closed stirrups form its vertical members, and the inclined concrete elements between cracks play the role of the diagonal compressive members. This model gave realistic predictions of actual behavior and seems to have gained acceptance with many researchers in the field.

6.13 TORSIONAL STRESSES

Using either the mathematical theory of elasticity, the plastic theory, or the skew bending theory, it can be shown that the magnitude of the maximum torsional shear stress in a rectangular section subjected to a torque T is given by:

$$t = \frac{T}{\eta x^2 y} \tag{6.43}$$

where x = shorter overall dimension of rectangular section
y = longer overall dimension of rectangular section
η = torsional coefficient

The value of η varies with y/x from 0.208 to 1/3 in the elastic theory and from 1/3 to 1/2 in the plastic theory. For simplicity, an overall value of 1/3 is adopted by the ACI code for nonprestressed members.

For a plain concrete section at the onset of cracking, Eq. (6.43) can be written as:

$$t_{cr} = \frac{T_{cr}}{\eta x^2 y} \tag{6.44}$$

where t_{cr} = torsional shear stress at cracking
T_{cr} = cracking torque

For flanged sections such as T, I, or L sections, the ACI code suggests the use of $\sum \eta x^2 y$ for the denominator of Eqs. (6.43) and (6.44) in which $\sum$ applies to the various rectangular parts of the section:

$$t_{cr} = \frac{T_{cr}}{\sum \eta x^2 y} \tag{6.45}$$

where x = shorter overall dimension of rectangular part of cross section
y = longer overall dimension of rectangular part of cross section
$\sum \eta x^2 y$ = the sum to be chosen as the largest value of several alternatives if any (safer design)

The above equation is not exactly in accordance with the theory but simplifies the results enormously.

For rectangular box sections, the code prescribes that an equivalent solid section may be taken provided the wall thickness h' is at least $x/4$. If the wall

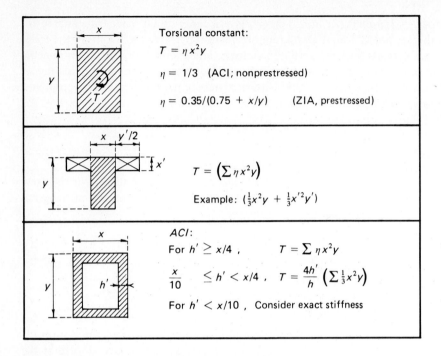

Figure 6.21 Summary of design torsional constants.

thickness h' is less than $x/4$ but greater than $x/10$, an equivalent solid section may be taken, provided the term x^2y is multiplied $4h'/x$. When h' is less than $x/10$, the stiffness of the wall must be considered, as possible buckling and crushing may occur.

The denominator of Eqs. (6.43) to (6.45) is generally called the torsional constant. A summary of the torsional constants recommended by the ACI code for various sections is given in Fig. 6.21.

The ACI code does not cover the torsion design for prestressed concrete. It is accepted, however, that Eqs. (6.43) to (6.45) would remain the same, except for the value of η. Based on 218 test results, Zia and McGee (Ref. 6.26) have shown that the torsion coefficient η can be estimated as a first approximation by:

$$\eta = \frac{0.35}{0.75 + x/y} \qquad (6.46)$$

One can generalize Eq. (6.45) for plain reinforced and prestressed concrete, using $\eta = 1/3$ for plain and reinforced concrete as per ACI code, and η from Eq. (6.40) for prestressed concrete. Note that, according to the discussion of Ref. 6.26, Eq. (6.46) is applicable to reinforced concrete as well as to prestressed concrete.

6.14 TORSIONAL CRACKING STRENGTH

Theoretically and experimentally it has been shown (Refs. 6.19 and 6.20) that the cracking strength t_{cr} of a plain concrete beam subjected to pure torsion is approximately equal to $0.85 | f_r |$ where f_r is the modulus of rupture of the concrete. As a first approximation $0.85 | f_r |$ can be taken as $0.85 (7.5\sqrt{f'_c}) \simeq 6\sqrt{f'_c}$. The stress at cracking in a reinforced concrete beam is considered the same as for plain concrete.

Hsu has shown (Ref. 6.22) that, if the beam is prestressed, its torsional cracking strength is increased with the average prestress and, for all practical purposes, can be estimated by:

$$(t_{cr})_{\text{prestressed}} = \left(\sqrt{1 + \frac{10\sigma_g}{f'_c}} \right) (t_{cr})_{\text{nonprestressed}} \tag{6.47}$$

or equivalently:

$$t_{cr} = 6\sqrt{f'_c} \left(\sqrt{1 + \frac{10\sigma_g}{f'_c}} \right) \tag{6.48}$$

where σ_g = average prestress = F/A_c

Note that Eq. (6.48) also applies to plain and reinforced concrete for which $\sigma_g = 0$.

Knowing the cracking stress, the torque at cracking can be estimated from Eq. (6.45) in which $\eta = 1/3$ for nonprestressed members and η given by Eq. (6.46) for prestressed members.

6.15 TORSIONAL RESISTANCE AFTER CRACKING

A plain concrete section subjected to torsion loses all its resistance after cracking. However, if it is reinforced with torsional reinforcement, the torsional strength contribution by concrete after cracking, t_c, is a significant portion of its precracking strength t_{cr}. Hsu (Ref. 6.21) has shown experimentally that the after-cracking strength t_c is about 40 percent of the precracking strength t_{cr} in reinforced members. He also found that the magnitude of strength loss after cracking is the same for prestressed members as for a nonprestressed (reinforced) member. Thus:

$$(t_{cr} - t_c)_{\text{prestressed}} \simeq (t_{cr} - t_c)_{\text{nonprestressed}} \tag{6.49}$$

where t_c = torsional strength of concrete after cracking (assuming torsional reinforcement is provided)
    t_{cr} = torsional cracking strength

The same notation is used for prestressed and nonprestressed members to minimize confusion. As the design procedure is later unified, the same symbols will apply to both.

The result of Eq. (6.49) is illustrated in Fig. 6.22, adapted from Ref. 6.26, where the concrete contribution to the torsional strength is plotted against the contribution of the torsional reinforcement. The two linear relations are approximate fits

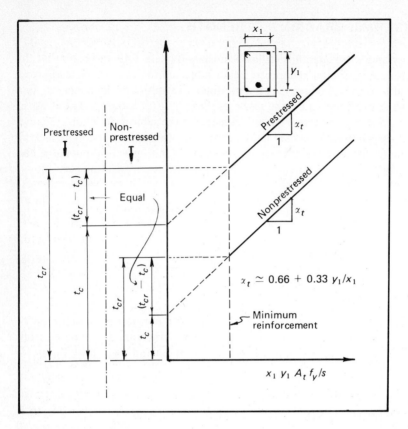

Figure 6.22 Graphical representation of the assumed torsion resistance after cracking.

to actual data. Note that the two lines have the same slope α_t (given by Eq. (6.59)), indicating that the contribution of torsional reinforcement after cracking is the same, whether or not the member is prestressed. However, the concrete contribution t_c after cracking is higher for a prestressed member.

For a nonprestressed member, the strength loss due to cracking is given by:

$$t_{cr} - t_c = t_{cr} - \mu t_{cr} = (1 - \mu)t_{cr} = kt_{cr} \qquad (6.50)$$

where $\mu \simeq 0.4$, as pointed out above. Thus, $k = 0.60 = 1 - t_c/t_{cr}$.

If the member is prestressed, we will have:

$$(t_{cr} - t_c)_{\text{prestressed}} = k(t_{cr})_{\text{nonprestressed}} \qquad (6.51)$$

and using Eqs. (6.48) and (6.49):

$$t_c = \left(\sqrt{1 + \frac{10\sigma_g}{f_c'}} - k\right)(t_{cr})_{\text{nonprestressed}} \qquad (6.52)$$

The above equation can be generalized for prestressed and nonprestressed members with torsional reinforcement, assuming the cracking strength of the nonprestressed member is $6\sqrt{f'_c}$, thus:

$$t_c = 6\sqrt{f'_c}\left(\sqrt{1 + \frac{10\sigma_g}{f'_c}} - k\right)$$ (6.53)

where $k = 0.6$ for non-prestressed members in which case $\sigma_g = 0$
$\quad\quad = 1 - 0.133/\eta$ for prestressed members (Ref. 6.26) in which η is given by Eq. (6.46). Hence $k = (0.75 - 0.4x/y)/1.05$. The value of x/y is determined for the largest component rectangle where the web reinforcement is usually placed.

Note: For nonprestressed members $t_c = 2.4\sqrt{f'_c}$ and $\eta = 1/3$. These values are indirectly used in Eq. (11-22) of the ACI code, where the influence of combined loading with shear is also included.

6.16 DESIGN CRITERIA FOR PURE TORSION

As pointed out in Sec. 6.11, the torsion design procedure for prestressed and nonprestressed members is combined here (although the ACI code does not cover prestressed concrete). The reader will recognize many of the ACI code requirements when nonprestressed members are addressed. The ACI code uses forces and moments in most equations. Stresses will be used here instead, whenever convenient.

The design procedure is summarized in the following steps:

1. Consideration of torsion. Torsion effects shall be included with shear and flexure where the factored torsional moment T_u exceeds $\phi(0.25t_{cr}\sum \eta x^2y)$. Otherwise, torsion effects may be neglected. *Note:* For a nonprestressed member $t_{cr} = 6\sqrt{f'_c}$ and $\eta = 1/3$, leading to $\phi(0.5\sqrt{f'_c}\sum x^2y)$, as given in the code. t_{cr} and η are given by Eqs. (6.48) and (6.46) for prestressed members.

2. Design criterion. The design of cross sections subjected to torsion shall be based on the following condition:

$$T_u \leq \phi T_n$$ (6.54)

where T_u is the factored torsional moment at the section considered and T_n is the nominal torsional moment strength computed from:

$$T_n = T_c + T_s$$ (6.55)

T_c is the nominal torsional moment strength provided by concrete and T_s is the nominal torsional moment strength provided by torsion reinforcement. If T_u exceeds ϕT_c, torsion reinforcement shall be provided. T_c and T_s are given next.

3. Concrete contribution. The contribution of concrete to the torsional resistance is given by the torque

$$T_c = (\sum \eta x^2 y)t_c \tag{6.56}$$

where $\eta = 1/3$ for nonprestressed concrete

$$\eta = \frac{0.35}{0.75 + x/y} \text{ for prestressed concrete}$$

and the torsional strength of the concrete is taken as:

$$t_c = 6\sqrt{f'_c} \left(\sqrt{1 + \frac{10\sigma_g}{f'_c}} - k \right) \tag{6.57}$$

in which $k = 0.60$ for non-prestressed concrete (in which case $\sigma_g = 0$)

$$k = 1 - \frac{0.133}{\eta} = \frac{0.75 - 0.4x/y}{1.05} \text{ for prestressed concrete}$$

4. Reinforcement contribution. The contribution of the torsional reinforcement is given by the nominal torsional moment strength:

$$T_s = \frac{A_t \alpha_t x_1 y_1 f_y}{s} \tag{6.58}$$

where A_t = area of one leg of closed stirrup resisting torsion within a distance s
   s = spacing of torsion reinforcement in direction parallel to longitudinal reinforcement
   x_1 = shorter center-to-center dimension of closed rectangular stirrup
   y_1 = longer center-to-center dimension of closed rectangular stirrup
   f_y = yield strength of the stirrup reinforcement, not to exceed 60 ksi (414 MPa).

The value of α_t corresponds to the slope of the line in Fig. 6.22 and is the same for prestressed and nonprestressed members. Its approximate value, as initially proposed by Hsu (Ref. 6.22), is given in the ACI code by:

$$\alpha_t = 0.66 + 0.33 \frac{y_1}{x_1} \tag{6.59}$$

but should not exceed 1.5.

Typical layouts of closed stirrups used as torsion reinforcement in various sections are shown in Fig. 6.23.

According to the ACI code, the value of T_s for reinforced concrete shall not exceed $4T_c$. Note that an equivalent torsional stress t_s can be derived from T_s using:

$$t_s = \frac{T_s}{\sum \eta x^2 y} \tag{6.60}$$

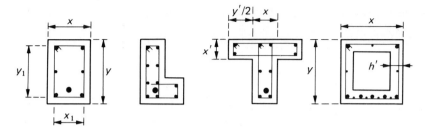

Figure 6.23 Typical layouts of torsion reinforcement.

When using Eq. (6.52) for design to determine torsion reinforcement, the value of T_s can be taken as:

$$T_s = \frac{T_u}{\phi} - T_c \tag{6.61}$$

5. Torsional stirrups. The required area of one leg of a closed stirrup to resist torsion is computed from:

$$A_t = \frac{(T_u/\phi - T_c)s}{\alpha_t f_y x_1 y_1} \tag{6.62}$$

As prescribed in the code, the minimum area of closed stirrups for reinforced concrete shall be such that:

$$A_v + 2A_t = \frac{50 b_w s}{f_y} \tag{6.63}$$

where A_v is the area of web reinforcement required for shear and f_y is in pounds per square inch. The minimum amount of reinforcement reduces the risk of brittle failure due to web reinforcement failure immediately after cracking.

For prestressed concrete, Zia and Hsu recommend the following minimum area of closed stirrups for torsion and shear (Ref. 6.33):

$$A_v + 2A_t = 50 \frac{b_w s}{f_y} \left(1 + \frac{12\sigma_g}{f_c'} \right) \le 200 \frac{b_w s}{f_y} \tag{6.64}$$

The spacing of closed stirrups shall not exceed the smaller of $(x_1 + y_1)/4$ or 12 in (30 cm). Torsional reinforcement shall be provided at least a distance $(d + b)$ beyond the point where it is theoretically required, where d is equal d_s or d_p (not less than $0.8h$), and b is the width of the compression face of the member. The first critical section can be taken the same as for shear design.

Torsional reinforcement comprises longitudinal reinforcement A_l in addition to transverse reinforcement. Longitudinal bars are provided to resist the longitudinal component of the diagonal tension induced by torsion. Their volume should be about equal to that of stirrups for torsion. According to the ACI code, the value of

A_l for nonprestressed members can be taken as the larger of:

$$A_l = 2A_t\left(\frac{x_1 + y_1}{s}\right) \tag{6.65}$$

or

$$A_l = \left[\frac{400xs}{f_y}\left(\frac{T_u}{T_u + V_u/3C_t}\right) - 2A_t\right]\left(\frac{x_1 + y_1}{s}\right) \tag{6.66}$$

in which

$$C_t = \frac{b_w d}{\sum x^2 y} \tag{6.67}$$

It is the author's opinion that, for prestressed members, Eq. (6.65) can be used alone, as it controls the design in most practical cases. In any event, the value of A_l shall not exceed that obtained by substituting $50b_w s\,(1 + 12\sigma_g/f_c')/f_y \leq 200b_w s/f_y$ for $2A_t$ in Eqs. (6.65) and (6.66) (Ref. 6.33). The spacing of longitudinal bars of a size not less than #3 shall be distributed around the periphery of the closed stirrups and shall not exceed 12 in (30 cm). At least one longitudinal bar should be placed at each corner of the section.

6. Maximum torsional moment. In order to reduce the danger of compression failure of the concrete before yielding of the steel due to overreinforcing, a maximum limit of the factored torsional moment must be established. For reinforced concrete, the ACI code limits the value of T_s to $4T_c$. Referring to Eq. (6.61), this is equivalent to limiting the factored torsional moment T_u to:

$$(T_u)_{max} = 5\phi T_c \tag{6.68}$$

Based on the results of Zia and McGee (Ref. 6.26) and using a value of $\eta = 1/3$, Zia and Hsu (Ref. 6.33) have proposed the following limiting value of factored torsional moment for prestressed concrete:

$$(T_u)_{max} = \frac{1/3 \sum x^2 y C(\sqrt{1 + 10\sigma_g/f_c'})\sqrt{f_c'}}{\sqrt{1 + \left(\dfrac{CV_u\sqrt{1 + 10\sigma_g/f_c'}}{30C_t T_u}\right)^2}} \tag{6.69}$$

where

$$C = 12 - 10\sigma_g/f_c' \tag{6.70}$$

and C_t is given by Eq. (6.67).

6.17 COMBINED LOADINGS

Seldom do pure torsional loadings apply in structures. It is more likely that a combination of shear torsion and bending exists. Their interaction is generally represented by an interaction surface and, when taken two at a time, by an interaction curve. Circular, parabolic, or linear interaction curves have been used to

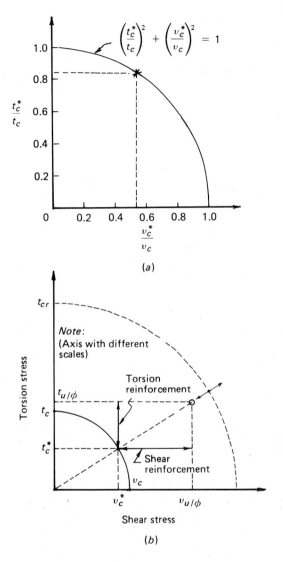

Figure 6.24. (*a*) **Typical interaction curve for combined
torsion and shear.** (*b*) **Typical
representation of concrete and
reinforcement contribution after
cracking.**

model the interaction of bending and torsion or torsion and shear (Refs. 6.23, 6.24,
6.25, 6.27, and 6.29). In the ACI code approach, design for bending is achieved
separately (thus it does not need to be reconsidered) while that of torsion and shear
is combined.

Tests have shown (Refs. 6.27 and 6.29) that the interaction between torsion and
shear for prestressed and nonprestressed members can be adequately represented
by a circular curve (Fig. 6.24*a*) of the form:

$$\left(\frac{t_c^*}{t_c}\right)^2 + \left(\frac{v_c^*}{v_c}\right)^2 = 1 \tag{6.71}$$

in which $t_c^* =$ torsional strength in presence of flexural shear

$\quad\quad t_c =$ torsional strength when member is subjected to torsion alone

$\quad\quad v_c^* =$ shear strength in presence of torsion

$\quad\quad v_c =$ shear strength when member is subjected to flexural shear alone (v_c for prestressed members is lesser of v_{ci} and v_{cw} given by Eqs. (6.28) and (6.30))

Equation (6.71) also applies to plain concrete for which the subscript cr (for cracking) replaces c. Note that, because of the circular shape of the curve, little reduction in strength is observed for one effect, unless the other effect is significant. For instance, about 90 percent of the shear resistance acting alone can be counted on if the torsion is less than 45 percent of the torsion resistance acting alone, and vice versa.

Dividing Eq. (6.71) by either $(v_c^*/v_c)^2$ or $(t_c^*/t_c)^2$ and solving it for t_c^* or v_c^*, leads to:

$$t_c^* = \frac{t_c}{\sqrt{1 + (t_c/v_c)^2(v_c^*/t_c^*)^2}} \tag{6.72}$$

and

$$v_c^* = \frac{v_c}{\sqrt{1 + (v_c/t_c)^2(t_c^*/v_c^*)^2}} \tag{6.73}$$

In nonprestressed members it is generally assumed that for design purposes the ratio t_c^*/v_c^* can be approximated by t_u/v_u, (or, equivalently, by T_u/V_u multiplied by an appropriate factor) and that the ratio between v_c and t_c is equal to the ratio of their respective lower bound values, which are $2\sqrt{f_c'}$ and $2.4\sqrt{f_c'}$. This leads essentially to the ACI code Eq. (11-22) for nonprestressed members as:

$$t_c^* = \frac{2.4\sqrt{f_c'}}{\sqrt{1 + (1.2v_u/t_u)^2}} \tag{6.74}$$

which is written here in function of stresses instead of forces.

For prestressed members, Zia and Hsu have proposed the following approximations to Eqs. (6.72) and (6.73) (Ref. 6.33):

$$t_c^* = \frac{t_c}{\sqrt{1 + \left(\dfrac{t_c}{v_c}\dfrac{v_u}{t_u}\right)^2}} \tag{6.75}$$

and

$$v_c^* = \frac{v_c}{\sqrt{1 + \left(\dfrac{v_c}{t_c}\dfrac{t_u}{v_u}\right)^2}} \tag{6.76}$$

where

$$t_c = 6\sqrt{f_c'}\left(\sqrt{1 + \frac{10\sigma_g}{v_c'}} - 0.6\right) \tag{6.77}$$

and v_c = the lesser of v_{ci} and v_{cw} given by Eqs. (6.28) and (6.30). Hence, in comparing Eq. (6.77) with Eq. (6.57), a value of $k = 0.6$ for both reinforced and prestressed concrete is adopted for combined loadings (Ref. 6.33). Note that v_u/t_u can also be replaced by V_u/T_u multiplied by an appropriate factor.

In an attempt to generalize the use of Eqs. (6.75) and (6.76) for the design of prestressed and nonprestressed members with web reinforcement, the following notation can be used:

t_c^* = torsional stress carried by concrete under combined loading
v_c^* = shear stress carried by concrete under combined loading
t_c = torsional stress carried by concrete subjected to torsion alone:
$t_c = 6\sqrt{f_c'}\,(\sqrt{1 + 10\sigma_g/f_c'} - k)$
$k = 0.6$ for prestressed and nonprestressed members under pure torsion and combined loadings. Using the same k in all cases is suggested in Ref. 6.33
$\eta = 1/3$ for nonprestressed members; according to Ref. 6.33, a value of $1/3$ can also be used as a first approximation for prestressed members
$\eta = 0.35/(0.75 + x/y)$ for prestressed members if a more accurate evaluation is sought
v_c = shear stress carried by concrete subjected to flexural shear alone
$v_c = 2\sqrt{f_c'}$ for nonprestressed members
v_c = the lesser of v_{ci} and v_{cw} given by Eqs. (6.28) and (6.30) for prestressed members
$T_c = (\sum \eta x^2 y)t_c$
$T_c^* = (\sum \eta x^2 y)t_c^*$
v_u = design shear strength = $V_u/b_w d_p$ where $d_p > 0.8h$
t_u = design torsion strength = $T_u/\sum \eta x^2 y$
V_u = factored design shear force
T_u = factored design torsional moment

The design approach for combined loading is the same as for single loading, that is, the torsional strength of concrete in combined loading is considered first and then the web reinforcement, if needed, is proportioned to carry the excess torsion necessary to balance the ultimate required strength (Fig. 6.24b). Thus, the following equations apply:

$$T_u \leq \phi T_n \tag{6.78}$$

$$T_n = T_c^* + T_s \tag{6.79}$$

For design purposes T_s is taken as:

$$T_s = \frac{T_u}{\phi} - T_c^* \tag{6.80}$$

and the torsional reinforcement can be determined from Eqs. (6.62) and (6.65).

A flow chart summarizing the torsion design procedure for prestressed and nonprestressed members for pure or combined loadings is given in Fig. 6.25. It

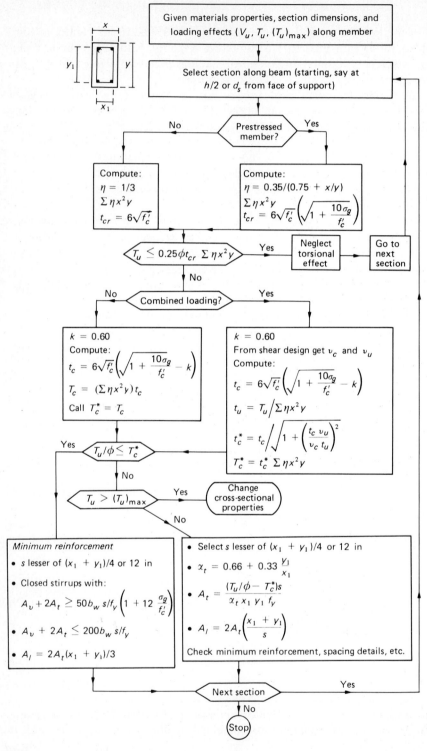

Given materials properties, section dimensions, and loading effects (V_u, T_u, $(T_u)_{max}$) along member

Select section along beam (starting, say at $h/2$ or d_s from face of support)

Prestressed member?

No

Compute:
$$\eta = 1/3$$
$$\Sigma \eta x^2 y$$
$$t_{cr} = 6\sqrt{f'_c}$$

Yes

Compute:
$$\eta = 0.35/(0.75 + x/y)$$
$$\Sigma \eta x^2 y$$
$$t_{cr} = 6\sqrt{f'_c}\left(\sqrt{1 + \frac{10\sigma_g}{f'_c}}\right)$$

$$T_u \leq 0.25\phi t_{cr} \ \Sigma \eta x^2 y$$

Yes → Neglect torsional effect → Go to next section

No

Combined loading?

No

$k = 0.60$
Compute:
$$t_c = 6\sqrt{f'_c}\left(\sqrt{1 + \frac{10\sigma_g}{f'_c}} - k\right)$$
$$T_c = (\Sigma \eta x^2 y)\, t_c$$
Call $T_c^* = T_c$

Yes

$k = 0.60$
From shear design get v_c and v_u
Compute:
$$t_c = 6\sqrt{f'_c}\left(\sqrt{1 + \frac{10\sigma_g}{f'_c}} - k\right)$$
$$t_u = T_u\Big/ \Sigma \eta x^2 y$$
$$t_c^* = t_c\Big/\sqrt{1 + \left(\frac{t_c \, v_u}{v_c \, t_u}\right)^2}$$
$$T_c^* = t_c^* \ \Sigma \eta x^2 y$$

Yes ← $$T_u/\phi \leq T_c^*$$

No

$$T_u > (T_u)_{max}$$

Yes → Change cross-sectional properties

No

Minimum reinforcement

- s lesser of $(x_1 + y_1)/4$ or 12 in
- Closed stirrups with:
$$A_v + 2A_t \geq 50 b_w \, s/f_y \left(1 + 12 \frac{\sigma_g}{f'_c}\right)$$
- $A_v + 2A_t \leq 200 b_w \, s/f_y$
- $A_l = 2A_t(x_1 + y_1)/3$

- Select s lesser of $(x_1 + y_1)/4$ or 12 in
- $\alpha_t = 0.66 + 0.33 \dfrac{y_1}{x_1}$
- $A_t = \dfrac{(T_u/\phi - T_c^*)s}{\alpha_t x_1 y_1 f_y}$
- $A_l = 2A_t\left(\dfrac{x_1 + y_1}{s}\right)$

Check minimum reinforcement, spacing details, etc.

Next section

Yes

No

Stop

covers most common design cases. It is based on the current state of the art which is closest to the ACI code. A value of $k = 0.6$ was used throughout as proposed in Ref. 6.33 and a value of η, given by Eq. (6.46), is suggested for prestressed members. Note that, according to Ref. 6.33, a value of $\eta = 1/3$ can also be used as a first approximation in all cases.

Other design approaches can be used for torsion. It is worth mentioning, in particular, a design procedure for prestressed and nonprestressed beams based on the compression field theory and developed by Collins and Mitchell (Ref. 6.34). The compression field theory, which is an extension of the traditional truss model for shear and torsion, considers geometric compatibility conditions and stress-strain relationships in addition to the truss equilibrium conditions. It offers a better means for understanding the behavior of beams in shear and torsion and is based on a rational analysis. It has one basic drawback because it assumes, according to the truss model, zero shear resistance for the concrete. However, with some modifications, it may very well find wide acceptance in future codes.

6.18 EXAMPLE: TORSION DESIGN OF A PRESTRESSED BEAM

Because the setting of an example involving a combination of bending, shear, and torsion can be very lengthy, let us consider, for simplicity, the same beam (Fig. 6.26a) treated in the Examples of Secs. 4.9, 4.12, and 6.10. Let us also assume that the beam is subjected to the

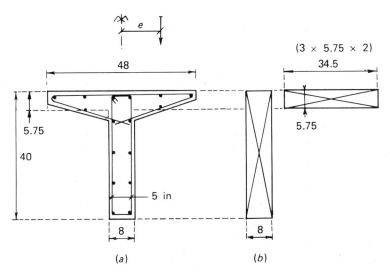

Figure 6.26

Figure 6.25 Proposed design flow chart for torsion (pure or combined) for prestressed and nonprestressed members.

same magnitude of dead and live loads. However, the design calls for a special loading condition in which a line load of magnitude equal to the live load of 0.40 klf can be applied at an eccentricity $e = 1.5$ ft from the axis of the beam. Restraint against rotation is provided at the ends of the beam by a diaphragm linking several such beams transversely. Thus a torsional moment results at each section of the beam in addition to the bending moment and shear.

According to the ACI code procedure, the design for bending remains unchanged while special design considerations must be taken for combined shear and torsion. The first critical section, at a distance of 2 ft from the face of the support, will be covered in detail. A similar procedure can be followed for the other sections along the span, if needed.

The ultimate design torsional moment at the first critical section is given by:

$$T_u = (1.7 w_L)\left(\frac{l}{2} - x\right)e = 0.68 \times 33 \times 1.5 = 33.66 \text{ kips-ft}$$

The design ultimate shear was determined in Example 6.10 (Tables 6.1 and 6.2). Its value is $V_u = 50.75$ kips.

The design flow chart for torsion (Fig. 6.25) will be followed. However, the number of each equation used will also be given lest in the event the reader wants to refer back to the derivations.

Following the flow chart (Fig. 6.25), let us compute first the torsional constant of the section. The beam cross section is idealized to an equivalent T with a constant depth flange of 5.75 in. Then it is separated into two rectangular parts, the web of dimensions 8 by 40 in and the overhanging flange of dimensions 5.75 by 40 in. According to the ACI code, the overhanging flange width used in computing the torsional constant shall not exceed three times the flange thickness, thus, $3 \times 2 \times 5.75 = 34.50$ (Fig. 6.26b).

Using Eq. (6.46), we have:

For the flange:

$$\eta = \frac{0.35}{0.75 + 5.75/34.5} = 0.382$$

For the web:

$$\eta = \frac{0.35}{0.75 + 8/40} = 0.368$$

And the torsional constant is given by:

$$\sum \eta x^2 y = 0.382 \times 5.75^2 \times 34.5 + 0.368 \times 8^2 \times 40 \simeq 1378 \text{ in}^4$$

The torsional cracking strength for the section prestressed with $\sigma_g = 417.3$ psi is given by Eq. (6.48):

$$t_{cr} = 6\sqrt{f_c'}\left(\sqrt{1 + \frac{10\sigma_g}{f_c'}}\right) = 6\sqrt{5000}\left(\sqrt{1 + \frac{10 \times 417.3}{5000}}\right)$$

$$t_{cr} \simeq 575 \text{ psi}$$

To find if torsional effects must be considered in design, $T_u = 33.66$ kips-ft is compared with:

$$0.25\phi t_{cr} \sum \eta x^2 y = (0.25 \times 0.85 \times 575 \times 1378)/12,000 = 14.03 \text{ kips-ft}$$

As $T_u > 14.03$ kips-ft, torsion is considered in combination with shear. Using $k = 0.60$, the torsional concrete resistance after cracking is given by Eq. (6.77):

$$t_c = 6\sqrt{f'_c}\left(\sqrt{1 + \frac{10\sigma_g}{f'_c}} - k\right)$$

$$= 6\sqrt{5000}\left(\sqrt{1 + \frac{10 \times 417.3}{5000}} - 0.60\right) = 320.1 \text{ psi}$$

As we have combined loading, we need the following values:

$$v_c = 409.5 \text{ psi} \qquad \text{from Table 6.1}$$

$$v_u = \frac{V_u}{b_w d_p} = \frac{50,750}{8 \times 32} = 198.2 \text{ psi}$$

$$t_u = \frac{T_u}{\sum \eta x^2 y} = \frac{33.66 \times 12,000}{1378} = 293.1 \text{ psi}$$

The torsional resistance in combined loading is given by Eq. (6.75):

$$t^*_c = \frac{t_c}{\sqrt{1 + \left(\dfrac{t_c v_u}{v_c t_u}\right)^2}} = \frac{320.1}{\sqrt{1 + \left(\dfrac{320.1 \times 198.2}{409.5 \times 293.1}\right)^2}}$$

$$t^*_c = 283 \text{ psi}$$

Thus,

$$T^*_c = t^*_c \sum \eta x^2 y = \frac{283 \times 1378}{12,000} = 32{:}5 \text{ kips-ft}$$

As T^*_c is less than $T_u/\phi = 33.66/0.85 = 39.6$ kips-ft, torsion reinforcement is needed. Note that it can be shown that T_u is less than $(T_u)_{max} = 81.07$ kips-ft given by Eq. (6.69). Thus, the beam will not be overreinforced in torsion. Similarly, it can be shown that $v^*_c = 191.3$ psi and as $v_u/\phi = 233.1 > v^*_c$, shear reinforcement is required.

Closed stirrups will be used with a cover of 1.5 in to the center of the bar on all sides. Their maximum spacing is given by:

$$s = \text{the smaller of} \begin{cases} \dfrac{x_1 + y_1}{4} = \dfrac{5 + 37}{4} = 10.5 \text{ in} \\[2mm] \text{and} \\[2mm] 12 \text{ in} \end{cases}$$

As 10.5 in controls, round it off to $s = 10$ in. To determine the required amount of torsion reinforcement, we need:

$$\alpha_t = 0.66 + 0.33\left(\frac{y_1}{x_1}\right) = 0.66 + 0.33\frac{37}{5} = 3.10$$

but α_t is not to exceed 1.5, thus $\alpha_t = 1.5$. The required area of *one leg* of the closed stirrup, assuming $s = 10$ in, is given by Eq. (6.62):

$$A_t = \frac{(T_u/\phi - T_c^*)s}{\alpha_t x_1 y_1 f_y} = \frac{(39.6 - 32.5)10 \times 12}{1.5 \times 5 \times 37 \times 60} \simeq 0.051 \text{ in}^2$$

Similarly, the area of stirrups required for shear is given by Eq. (6.35):

$$A_v = \frac{(v_u/\phi - v_c^*)b_w s}{f_y} = \frac{(233.1 - 191.32)8 \times 10}{60,000} = 0.056 \text{ in}^2$$

Select a #3 closed stirrup at $s = 10$ in. The area provided per stirrup is 0.22 in^2. It must be larger than the minimum area required for shear and torsion given by:

$$A_v + 2A_t > \frac{50 b_w s}{f_y}\left(1 + \frac{12\sigma_g}{f_c'}\right)$$

or

$$0.056 + 2 \times 0.051 = 0.158 > \frac{50 \times 8 \times 10}{60,000}\left(1 + \frac{12 \times 417.3}{5000}\right) = 0.133$$

Hence, a #3 closed stirrup at $s = 10$ in is satisfactory.

The required area of longitudinal nonprestressed reinforcement for torsion is given by:

$$A_l = 2A_t\left(\frac{x_1 + y_1}{s}\right) = 2 \times 0.051\left(\frac{5 + 37}{10}\right) \simeq 0.428 \text{ in}^2$$

Keeping in mind that the maximum spacing is 12 in between bars, sixteen No. 3 bars ($A_l \simeq 1.76$ in^2) are provided at the periphery of the section, as shown in Fig. 6.26a.

6.19 SHEAR AND TORSION IN PARTIALLY PRESTRESSED MEMBERS

A distinction was made throughout this chapter between a prestressed concrete member and a nonprestressed or reinforced concrete member. A review of the design approaches used for shear and torsion strongly suggests that partially prestressed members, which contain both prestressed and nonprestressed flexural reinforcement, can be designed essentially as prestressed members. The only difference between them is a lower value of the average prestress. In designing for shear and in order to be consistent with the procedures developed earlier, the following value of d can be used:

$$d = \text{the larger of} \begin{cases} PPRd_p + (1 - PPR)d_s \\ \text{or} \\ PPR(0.8h) + (1 - PPR)d_s \end{cases} \tag{6.81}$$

where d = distance from extreme compressive fiber to centroid of tensile force in the reinforcement

d_p = distance from extreme compressive fiber to centroid of prestressing force

d_s = distance from extreme compressive fiber to centroid of tensile reinforcing steel

PPR = Partial Prestressing Ratio defined in Vol. 2

As a first approximation the value of PPR for Eq. (6.81) can be estimated from:

$$PPR \simeq \frac{A_s f_y}{A_s f_y + A_{ps} f_{pu}} \tag{6.82}$$

6.20 DERIVATION OF CONCRETE NOMINAL SHEAR STRENGTH EQUATIONS

The derivation of Eqs. (6.28) and (6.30), on which the ACI shear design provisions for prestressed concrete are based, is given in the commentary of the 1963 ACI code (Ref. 6.5). It is essentially based on the work of MacGregor, Sozen, and Siess (Refs. 6.2 and 6.4). For flexure-shear cracking, their tests have indicated that, in order to reduce the capacity of a beam, a diagonal crack must have a projection on the longitudinal axis of the beam at least equal to the depth of the beam d. (d is used here to comply with the original notation (Ref. 6.5); use d_p for fully prestressed beams.) Let us consider a section along the beam of abcissa x with respect to the support (Fig. 6.27). A flexural crack distant d from x in the direction of decreasing moments may lead to a diagonal crack which could be critical for section x. Tests have indicated that failure becomes imminent when a second flexural crack occurs at a distance $d/2$ from x and extends to the centroid of the section where the increase in the principal tensile stress triggers diagonal cracking. Collapse occurs for an additional load, the effects of which are described later. Let us evaluate the load that led to the second flexural crack. The change in moment

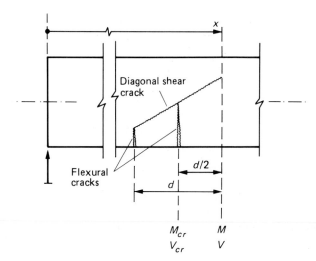

Figure 6.27 Flexure-Shear cracking model.

from section x to section $(x - d/2)$ is equal to the area under the shear diagram between these two sections. That is:

$$M - M_{cr} = \left(\frac{V + V_{cr}}{2}\right)\frac{d}{2} \tag{6.83}$$

where M and V are the moment and shear at section x and M_{cr} and V_{cr} are the cracking moment and corresponding shear at section $(x - d/2)$. The difference between V and V_{cr} over $d/2$ is generally small. Thus, as a first approximation:

$$M - M_{cr} = \frac{Vd}{2} \tag{6.84}$$

or

$$\frac{M}{V} - \frac{M_{cr}}{V} = \frac{d}{2} \tag{6.85}$$

from which:

$$V = \frac{M_{cr}}{M/V - d/2} \tag{6.86}$$

where V is the shear at section x. Note that V appears on the two sides of the equation. However, V can still be computed, because the ratio M/V remains constant at a given section when the applied load increases proportionately.

It can be assumed that, in a beam test, the shear given by Eq. (6.86) is generated by external loads. In design, these loads are represented by superimposed dead loads and live loads. The corresponding shear acts in addition to the dead load shear V_G. Moreover, as mentioned earlier, tests have indicated that, in order for collapse to occur after occurrence of the second flexural crack, a shear force increment estimated at $0.6\sqrt{f'_c}\,b_w d$ is needed. Hence, the total shear force V_{ci}, which will produce a flexure-shear failure, can be estimated from:

$$V_{ci} = 0.6\sqrt{f'_c}\,b_w d + \frac{M_{cr}}{M/V - d/2} + V_G \tag{6.87}$$

The above relationship is compared to actual data in Fig. 6.28 (Ref. 6.5). To simplify the above equation, the term $d/2$ is deleted, leading to a safer design limit. As M, V, and M_{cr} are due to additional external loads (superimposed dead load and live load), and as nominal shear resistance at ultimate is sought, they are replaced, using the notation of this text, by ΔM_u, ΔV_u, and ΔM_{cr}. Hence, Eq. (6.87) becomes:

$$V_{ci} = 0.6\sqrt{f'_c}\,b_w d + \frac{\Delta M_{cr}}{\Delta M_u/\Delta V_u} + V_G \tag{6.88}$$

Dividing the two sides of Eq. (6.88) by $b_w d$ leads essentially to the flexure-shear strength (stress) v_{ci}, given by Eq. (6.28).

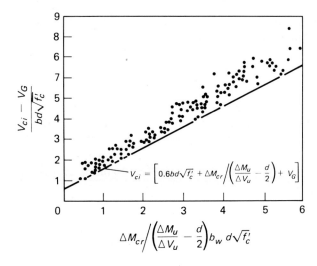

Figure 6.28 **Comparison of Eq. (6.87) with experimental data (Ref. 6.5).**

Web-shear cracking occurs when the maximum principal tensile stress becomes equal to the tensile strength of the concrete, f'_{tc}. The principal tensile stress is assumed maximum at the centroid of the section. Using Eq. (6.3) and assuming all stresses are positive (to keep up with the original derivation), the following relationship is derived to represent the onset of web-shear cracking:

$$f'_{tc} = \sqrt{v_{cw} + \left(\frac{\sigma_g}{2}\right)^2} - \frac{\sigma_g}{2} \qquad (6.89)$$

where σ_g is the compressive stress at the concrete centroid due to the effective prestressing force and v_{cw} is the critical shear stress at web-shear cracking. Solving

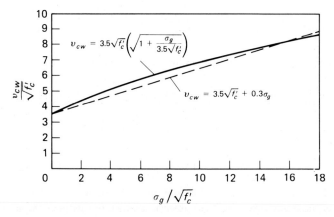

Figure 6.29 **Nominal shear stress at web-shear cracking (Ref. 6.5).**

for v_{cw} leads to:

$$v_{cw} = f'_{tc} \sqrt{1 + \frac{\sigma_g}{f'_{tc}}} \tag{6.90}$$

The magnitude of f'_{tc} indicated by tests appears to be about $4\sqrt{f'_c}$. It is conservatively set equal to $3.5\sqrt{f'_c}$. Hence, Eq. (6.90) becomes:

$$v_{cw} = 3.5\sqrt{f'_c} \sqrt{1 + \frac{\sigma_g}{3.5\sqrt{f'_c}}} \tag{6.91}$$

The curve representing Eq. (6.91) is plotted in Fig. 6.29 and compared to that representing a simplified form given by:

$$v_{cw} = 3.5\sqrt{f'_c} + 0.3\sigma_g \tag{6.92}$$

As the two curves are close, the simplified form is selected for design. In prestressed concrete, the shear stress value given by Eq. (6.92) is augmented by the shear stress generated by the vertical component of the prestressing force. It essentially leads to the nominal web-shear cracking resistance v_{cw} given by Eq. (6.30).

REFERENCES

6.1. M. A. Sozen, E. M. Swoyer, and C. P. Siess: "Strength in Shear of Beams without Web Reinforcement," Engineering Experiment Station Bulletin No. 452, University of Illinois, Urbana, April, 1959.

6.2. J. G. MacGregor, M. A. Sozen, and C. P. Siess: "Strength and Behavior of Prestressed Concrete Beams with Web Reinforcement," University of Illinois Civil Engineering Studies, Structural Research Series 210, Urbana, August, 1960.

6.3. M. Lorentsen: "Theory of the Combined Action of Bending Moment and Shear in Reinforced and Prestressed Concrete Beams," *ACI Journal*, vol. 62, no. 4, April 1965.

6.4. J. G. MacGregor, M. A. Sozen, and C. P. Siess: "Strength of Concrete Beams with Web Reinforcement," *ACI Journal*, vol. 62, no. 12, December 1965, pp. 1503–1519.

6.5. ACI Special Publication SP-10, Commentary on Building Code Requirements for Reinforced Concrete (ACI 318-63), by ACI Committee 318, American Concrete Institute, Detroit, 1965, 91 pp.

6.6. G. N. J. Kani: "Basic Facts Concerning Shear Failure," *ACI Journal*, vol. 63, no. 6, June 1966, pp. 675–692.

6.7. B. Bresler and J. G. MacGregor: "Review of Concrete Beams Failing in Shear," *Journal of the Structural Division, ASCE*, vol. 93, no. 55.2, February 1967.

6.8. G. N. J. Kani: "A Rational Theory for the Function of Web Reinforcement," *ACI Journal*, vol. 66, no. 3, March 1969, pp. 185–197.

6.9. J. G. MacGregor and J. M. Hanson: "Proposed Changes in Shear Provisions for Reinforced and Prestressed Concrete Beams," *ACI Journal*, vol. 66, no. 4, April 1969, pp. 276–288.

6.10. H. P. J. Taylor: "The Fundamental Behavior of Reinforced Concrete Beams in Bending and Shear," Shear in Reinforced Concrete, Special Publication SP-42, vol. 1, American Concrete Institute, Detroit, 1972, pp. 43–77.

6.11. A. H. Mattock and N. M. Hawkins: "Research on Shear Transfer in Reinforced Concrete," *PCI Journal*, vol. 17, no. 2, March/April 1972, pp. 55–75.

6.12. ASCE-ACI Joint Committee 426 Report, "The Shear Strength of Reinforced Concrete Members," *Journal of the Structural Division, ASCE*, vol. 99, no. ST6, June 1973, Chaps. 1–4, pp. 1091–1197. (Also reproduced in ACI Manual of Concrete Practice.)

6.13. O. Gonzales Cuevas, F. Robles, and R. Diaz de Cosio: "Strength and Deformation of Reinforced Concrete Elements," Chap. 5 in Reinforced Concrete Engineering, vol. 1, edited by Boris Bresler, John Wiley & Sons, New York, 1974, pp. 194–301.

6.14. R. Park and T. Paulay: *Reinforced Concrete Structures*, Wiley-Interscience, New York, 1975, 769 pp.

6.15. PCA, Notes on ACI 318-77 Building Code Requirements for Reinforced Concrete with Design Applications, 2d ed., Portland Cement Association, Skokie, Illinois, 1978.

6.16. ACI Special Publication SP-18, "Torsion of Structural Concrete," American Concrete Institute, Detroit, 1968.

6.17. ACI Special Publication SP-35, "Analysis of Structural Systems for Torsion," American Concrete Institute, Detroit, 1973.

6.18. P. Zia: "Torsion Theories for Concrete Members," in "Torsion of Structural Concrete," SP-18, American Concrete Institute, Detroit, 1968, pp. 103–132.

6.19. T. T. C. Hsu: "Torsion of Structural Concrete—A Summary on Pure Torsion," in Ref. 6.16, pp. 165–178.

6.20. ———: "Torsion of Structural Concrete—Plain Concrete Rectangular Sections," in Ref. 6.16, pp. 203–238.

6.21. ———: "Torsion of Structural Concrete—Behavior of Reinforced Concrete Rectangular Members," in Ref. 6.16, pp. 261, 306.

6.22. ———: "Torsion of Structural Concrete—Uniformly Prestressed Rectangular Sections without Web Reinforcement," *PCI Journal*, vol. 13, no. 2, April 1968, pp. 34–44.

6.23. P. Lampert: "Torsion and Bending in Reinforced Concrete and Prestressed Concrete Members," Proceedings of the Institution of Civil Engineers, vol. 50, December 1971, pp. 487–505.

6.24. A. H. Mattock and A. N. Wyss: "Full Scale Torsion, Shear and Bending Tests of Prestressed I-Girders," *PCI Journal*, vol. 23, no. 2, March/April 1978, pp. 22–40.

6.25. H. V. S. Gangarao and P. Zia: "Rectangular Prestressed Beams in Torsion and Bending," *Journal of the Structural Division, ASCE*, vol. 99, no. ST.1, January 1973, pp. 183–198.

6.26. P. Zia and W. D. McGee: "Torsion Design of Prestressed Concrete," *PCI Journal*, vol. 19, no. 2, March/April 1974, pp. 46–65, also discussion in *PCI Journal*, November/December 1974.

6.27. R. L. Henry and P. Zia: "Prestressed Beams in Torsion, Bending, and Shear," *Journal of the Structural Division, ASCE*, vol. 100, no. ST.5, May 1974, pp. 933–952.

6.28. D. Mitchell and M. P. Collins: "Diagonal Compression Field Theory—A Rational Model for Structural Concrete in Pure Torsion," *ACI Journal*, vol. 71, no. 8, August 1974, pp. 396–408.

6.29. D. McGee and P. Zia: "Prestressed Concrete under Torsion, Shear, and Bending," *ACI Journal*, vol. 73, no. 1, January 1976, pp. 26–32.

6.30. D. Mitchell and M. P. Collins: "Detailing for Torsion," *ACI Journal*, vol. 73, no. 9, September 1976, pp. 506–511.

6.31. ——— ———: "Influence of Prestressing on Torsional Response of Concrete Beams," *PCI Journal*, vol. 23, no. 3, May/June 1978, pp. 54–73.

6.32. B. Vijaya Rangan and A. S. Hall: "Strength of Prestressed Concrete I Beams in

Combined Torsion and Bending," *ACI Journal*, vol. 75, no. 11, November 1978, pp. 612–618.

6.33. P. Zia and T. T. C. Hsu: "Design for Torsion and Shear in Prestressed Concrete," ASCE Annual Convention, Chicago, 1978, Preprint No. 3423, 17 pp.

6.34. M. P. Collins and D. Mitchell: "Shear and Torsion Design of Prestressed and Non-Prestressed Concrete Beams," *PCI Journal*, vol. 25, no. 5, September/October 1980, pp. 32–100.

PROBLEMS

6.1 Because of high shear stresses (or principal tension) in the web of a beam near the supports, you propose to use vertical prestressing in combination with horizontal prestressing. Assuming $\sigma_x = \sigma_g = 700$ psi and $v = 400$ psi, determine the magnitude of vertical prestress σ_y so that the principal tension is reduced to either -100 psi or to zero at the centroid of the section.

6.2 Go back to Prob. 4.6g, where a steel profile has been selected, and check shear requirements along the span (Fig. P6.2). What can you conclude if you are told that the beam is part of a joist slab system?

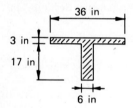

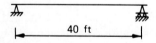

Figure P6.2

6.3 Go back to Prob. 4.7 and check shear requirements near the supports and at sections located 5 ft from the support on the cantilever side, and 5 and 10 ft on the span side.

6.4 The precast prestressed beam of Fig. P6.4 has been selected for the roof of a stadium. The beam is simply supported with, on one side, a cantilever of 20 ft. The following information is given:

Live load 40 psf

Steel

Strands:
$f_{pu} = 270$ ksi
$f_{py} = 45$ ksi
diameter $= \frac{1}{2}$ in
area $= 0.153$ in^2

Stress: f_{pi} at transfer $= 175$ ksi
$f_{pe} = 150$ ksi

Section properties

$A_c = 615$ in^2
$I = 59,720$ in^4
$y_b = 21.98$ in
$y_t = 10.02$ in
$Z_b = 2717$ in^3
$Z_t = 5960$ in^3
$w_G = 0.641$ klf

$k_b = \dfrac{Z_t}{A_c} = 9.69$ in

$k_t = -\dfrac{Z_b}{A_c} = A.42$ in

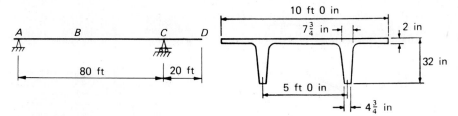

Figure P6.4

Concrete

$f'_c = 5000$ psi $\quad f'_{ci} = 4000$ psi
$\bar{\sigma}_{ti} = -190$ psi $\quad \bar{\sigma}_{ci} = 2400$ psi $\quad \bar{\sigma}_{ts} = -424$ psi $\quad \bar{\sigma}_{cs} = 2250$ psi

(a) Determine the position of point B at which maximum positive moment occurs. Build for sections B and C the two feasible domains of e_o versus $1/F_i$ and select a common satisfactory value of F corresponding to an even integer number of strands. (Note that at C a negative moment exists, that e_o is upward, and that the properties of the section such as k_t, k_b, Z_t, Z_b must be reversed accordingly in order to use the four stress inequality conditions.)

(b) Determine the strands' layout and their centroid at sections B and C and check the ultimate moment requirements according to ACI specifications at both sections. Is the ratio of ultimate to cracking moment at section C satisfactory?

(c) Assuming F constant along the beam, construct the limit zone and the steel envelopes. Suggest a satisfactory steel profile along the beam.

(d) Check shear requirements along the beam and determine the required stirrups. Plot graphically shear stresses versus abcissa and show the selected stirrup spacings.

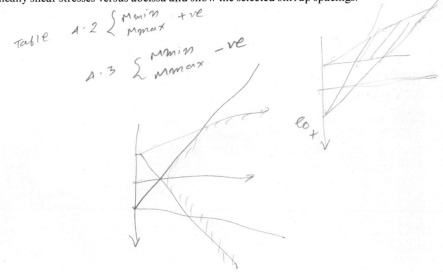

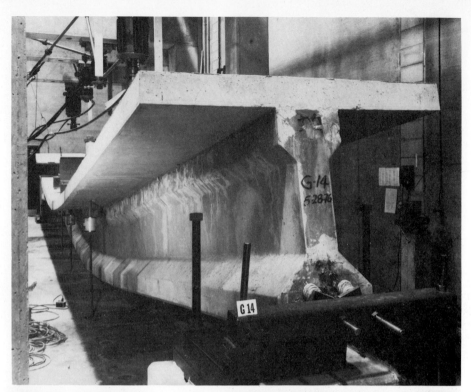

Full-scale testing of prestressed concrete bridge beams. *(Courtesy Portland Cement Association.)*

CHAPTER 7

DEFLECTION COMPUTATION
AND CONTROL

7.1 SERVICEABILITY

Most prestressed concrete structures are first designed and dimensioned on the basis of allowable stresses or strength. If a sound design approach is followed and if code requirements with respect to permissible stresses, strength, shear, and torsion are satisfied, it is very likely that the design can be finalized without any further modification. However, there are increasing situations where it is essential to check if the serviceability of the structure is satisfactory. Serviceability refers to the performance of the structure in service. The most frequently considered serviceability criteria in prestressed concrete are related to short- and long-term camber or deflection, fatigue, cracking (primarily in partial prestressing), corrosion resistance, and durability. Other criteria, such as vibration characteristics, can also be set in the design. In designing for serviceability, loading does not necessarily imply full service load, but should be specified for each criterion. For instance, in evaluating fatigue, the repetitive load for building members may be taken as only 50 percent of the specified live load, while for railway bridges, the full live load should be considered. Similarly, in computing long-term deflection, in addition to dead load, the part of live load that can be considered sustained must be carefully assessed.

In this chapter only the deflection criterion will be addressed. Fatigue and cracking are more characteristic of partially prestressed members than of fully prestressed ones and are covered in Vol. 2. Some aspects of fatigue and corrosion of the component materials can be found in Chap. 2.

7.2 DEFLECTION: TYPES AND CHARACTERISTICS

Deflection is defined as the total movement induced at a point of a member from the position before application of the load to the position after application of the load. The maximum deflection which, in uniformly loaded simply supported beams, occurs at midspan, is generally of main interest in design. A distinction is often made between "camber," which is the deflection caused by prestressing, and "deflection," which is that produced by external loads. They are identical in nature but generally opposite in sign. Typically, prestressing produces *upward* camber in a simply supported beam, while self-weight produces *downward* deflection. Their combination may produce an upward or a downward movement. In order to avoid confusion, the term "deflection" will be used in its most general form, unless the separate effect of camber is addressed.

The following sign convention will be followed: plus (+) for downward deflection and minus (−) for upward deflection.

In reinforced concrete beams, deflection is due to external loads and is always downward. In prestressed concrete, deflection depends on the combined effect of prestressing and external loading. It can easily be controlled by changing the magnitude and profile of the prestressing force. It is not uncommon, in partial prestressing, to achieve a zero deflection design. In both reinforced and prestressed concrete members, deflection under sustained loading continues to increase with time, mainly due to the effects of creep and shrinkage of concrete and relaxation of prestressing steel. Excessive deflections, especially those developing with time, are common causes of trouble and must be limited.

In computing deflection, one may differentiate between camber and deflection, short-term or immediate instantaneous deflection, and long-term or time-dependent deflection. The total deflection itself can be separated into two parts, an instantaneous part and an additional, time-dependent part. Furthermore, a different approach is followed, whether the member is uncracked, such as in fully prestressed members, or cracked, such as in reinforced and partially prestressed members. Most of these differences and how to accommodate them are clarified in the following sections.

7.3 THEORETICAL DEFLECTION DERIVATIONS

Based on general principles of mechanics and assuming linear elastic behavior, several methods can be used to compute the deflections or displacements of structures. One of the most convenient methods in structural design is the moment-area method, because it relies on the knowledge of the moments along the members, and these moments are generally known at this stage of the design. The moment-area method was first developed by Mohr. It is based on the relationship between bending moment and curvature at any point of a flexural member, given by:

$$\Phi = \frac{d\theta}{dx} = \frac{M}{EI} \tag{7.1}$$

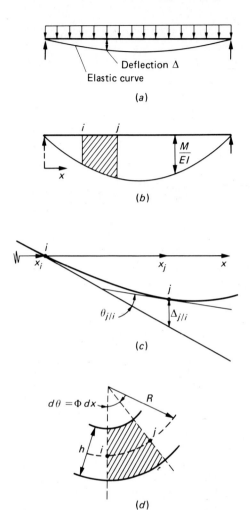

(a)

(b)

(c)

(d)

Figure 7.1 Illustration of deflection, rotation, and curvature.

where Φ is the curvature or angle change per unit length of a deflected flexural member, θ the angle between the tangents at two points of the deflected elastic curve, M the applied bending moment, E the elastic modulus of the beam material, and I the moment of inertia in general. M is a function of x, and I can vary along the member. Two basic theorems are derived and are stated next without proof. The notation is reported in Fig. 7.1.

1. **First moment-area theorem (or Rotation Theorem).** *The change in angle be-tween points i and j on the deflected elastic curve of a flexural member, or the slope at point j relative to the slope at point i, is equal to the area under the M/EI diagram between points i and j, that is:*

$$\theta_{j/i} = \int_{x_i}^{x_j} \left(\frac{M}{EI}\right) dx \qquad (7.2)$$

The first moment-area theorem is essentially used to determine the rotation of one section j with respect to another section i.

2. **Second moment-area theorem (or Deflection Theorem).** *The deflection of point j of a flexural member measured with respect to the tangent at another point i of the member is equal to the first static moment taken about point j of the area under the M/EI diagram along the member between points i and j, that is:*

$$\Delta_{j/i} = \int_{x_i}^{x_j} x\left(\frac{M}{EI}\right) dx \tag{7.3}$$

Note that the change in slope or deflection is taken with respect to a tangent to the elastic curve. The appropriate choice of a base or reference tangent will substantially reduce the computations. For instance, although the midspan section deflects with respect to the support in a uniformly loaded beam, it is preferably selected as the reference section i because the elastic curve has zero slope or horizontal tangent at that section. The computations are run as if the supports are deflecting with respect to midspan. In most common design cases, the moment diagram is either parabolic or linear. Thus, the area under the moment diagram and its centroid with respect to a reference point can be easily determined. Useful expressions for several common cases are shown in Fig. 7.2.

Note that the moment-area theorems are essentially geometric relationships. They hold for any situation where the distribution of curvature can be determined. An example illustrating the use of the second moment-area theorem to compute deflection is given next.

1. Example

Determine the deflection (camber) for a simply supported beam with uniform cross section due to a prestressing force having a profile with two draping points, as shown in Fig. 7.3a. No external loading is considered.

The corresponding M/EI diagram is shown in Fig. 7.3b. The moment is negative and the corresponding deflection will be negative (i.e., camber). However, the sign will be shown only at the end of the computations in order not to carry negative signs with areas. The elastic curve is shown in Fig. 7.3c. The reference point i will be taken at midspan, as it has a horizontal tangent to the elastic curve, and point j will be taken at the support. The area under the M/EI diagram is divided into three parts: A_1, A_2, and A_3. According to the second moment-area theorem, the deflection is equal to the moment of these areas with respect to j. Thus:

$$\Delta_{j/i} = \frac{l}{4} A_1 + \frac{(l/2 + a)}{2} A_2 + \frac{2a}{3} A_3$$

$$= \frac{l}{4} \frac{Fe_2}{EI} \frac{l}{2} + \left(\frac{l/2 + a}{2}\right)\left(\frac{l}{2} - a\right)\left(\frac{Fe_1 - Fe_2}{EI}\right) + \frac{2a}{3} a \left(\frac{Fe_1 - Fe_2}{2EI}\right)$$

It can be reduced to:

$$\frac{Fl^2}{8EI}\left[e_1 + (e_2 - e_1)\frac{4a^2}{3l^2}\right]$$

M/EI Diagram	$\bar{x}_1$ and A_1	$\bar{x}_2$ and A_2
	$x_1 = b/3$	$\bar{x}_2 = \frac{2}{3}b$
	$\bar{y}_1 = h/3$	$\bar{y}_2 = \frac{2}{3}h$
	$A_1 = bh/2$	$A_2 = \frac{1}{2}bh$
	$\bar{x}_1 = \dfrac{b_1(2h + h_1)}{3(h + h_1)}$	$\bar{x}_2 = \dfrac{(2b_1 + b)}{3}$
	$\bar{y}_1 = \dfrac{h_1^2 + hh_1 + h_2^2}{3(h + h_1)}$	$\bar{y}_2 = \dfrac{h_1}{3}$
	$A_1 = \dfrac{b_1(h + h_1)}{2}$	$A_2 = (b - b_1)\dfrac{h_1}{2}$
	$\bar{x}_1 = 3b/8$	$\bar{x}_2 = 3b/4$
	$\bar{y}_1 = 2h/5$	$\bar{y}_2 = 7h/10$
	$A_1 = 2bh/3$	$A_2 = bh/3$
	$\bar{x}_1 = \dfrac{3b_1}{4}\left(\dfrac{h_1 + h}{2h + h_1}\right)$	$\bar{x}_2 = \dfrac{3}{4}\left(\dfrac{b^2h - b_1^2h}{2bh - 2b_1h - b_1h_1}\right)$
	$A_1 = \dfrac{b_1}{3}(2h + h_1)$	$A_2 = \frac{2}{3}bh - A_1$

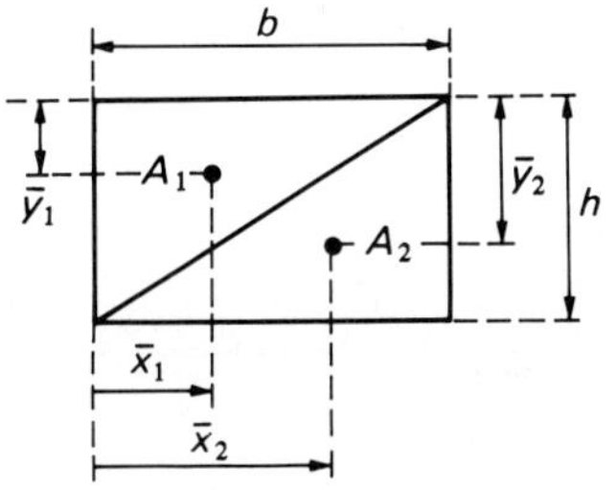

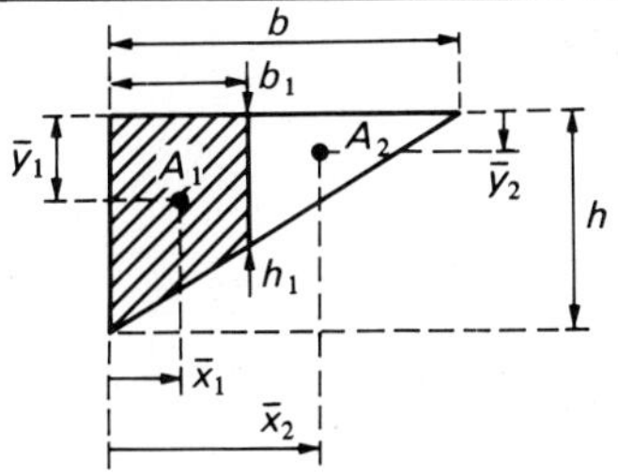

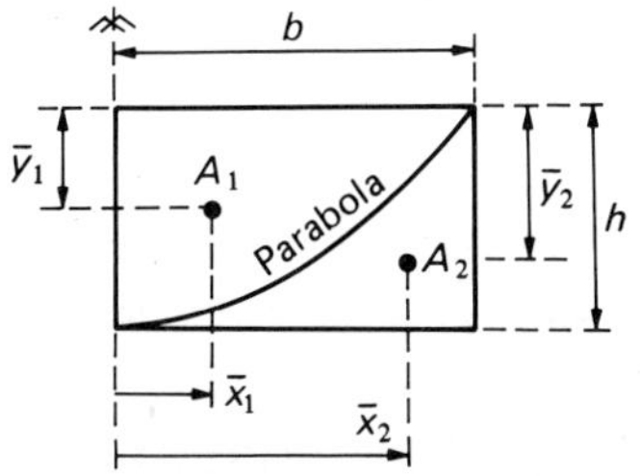

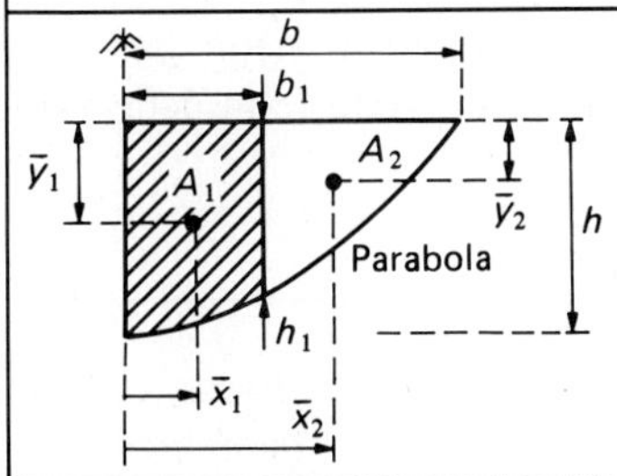

Figure 7.2

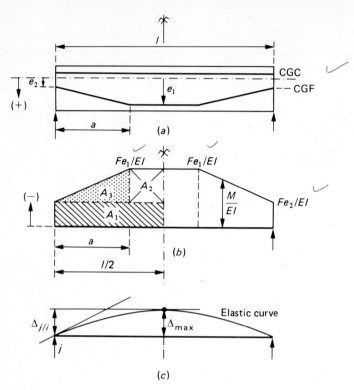

Figure 7.3 Deflection by the second moment area theorem.

to which a minus sign should be applied. The result is the same as the expression shown in Fig. 7.4 for the same case.

7.4 SHORT-TERM DEFLECTIONS IN PRESTRESSED MEMBERS

The approaches treated next apply to the computation of instantaneous or short-term deflections. They also provide the basis for computing long-term deflections by the approximate methods described in Sec. 7.6.

1. Uncracked Members

Fully prestressed concrete members are uncracked under service loads and are assumed linear elastic. Their deflections can be determined using theoretical derivations identical to those described in the previous sections.

Typical formulas for deflections are given in Fig. 7.4 for several profiles of the prestressing force and several types of loading. Simply supported beams with constant prestressing force and constant sectional properties are assumed. Because superposition is valid in computing deflections for uncracked members, many

$\frac{\varepsilon_{ct}}{\varepsilon_{cb}}$

Camber due to prestressing force	Deflection due to loading
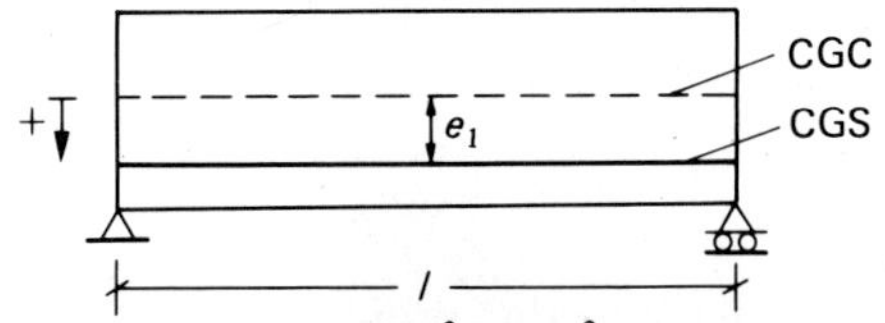 CGC, CGS, e_1 $$\Delta = -\frac{Fe_1 l^2}{8EI} = \Phi_1 \frac{l^2}{8}$$	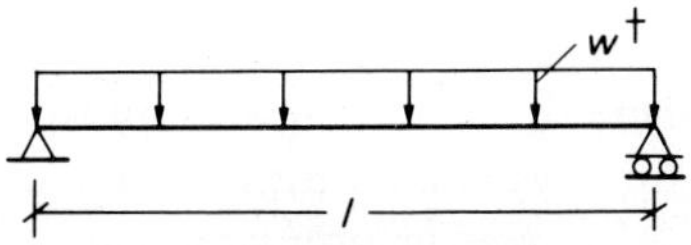 w $$\Delta = \frac{5wl^4}{384EI} = \Phi_1 \frac{5l^2}{48}$$
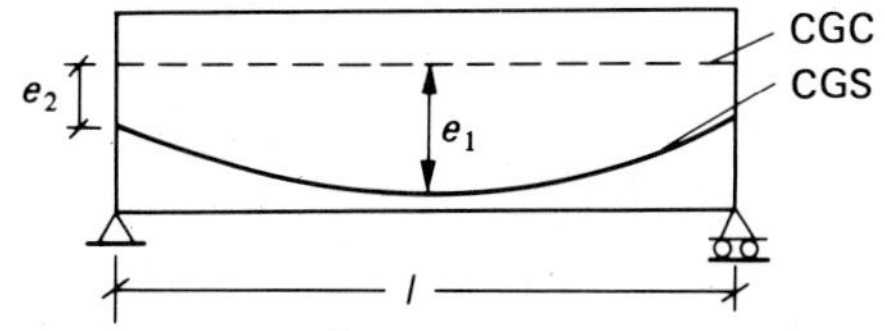 e_2, CGC, CGS, e_1 $$\Delta = -\frac{Fl^2}{8EI}\left[e_2 + \tfrac{5}{6}(e_1 - e_2)\right]$$ $$= \Phi_1 \frac{l^2}{8} + (\Phi_2 - \Phi_1)\frac{l^2}{48}$$	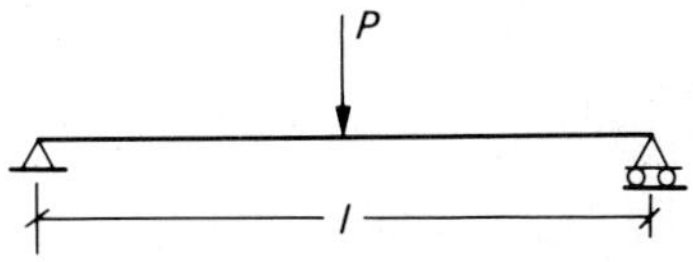 P $$\Delta = \frac{Pl^3}{48EI} = \Phi_1 \frac{l^2}{12}$$
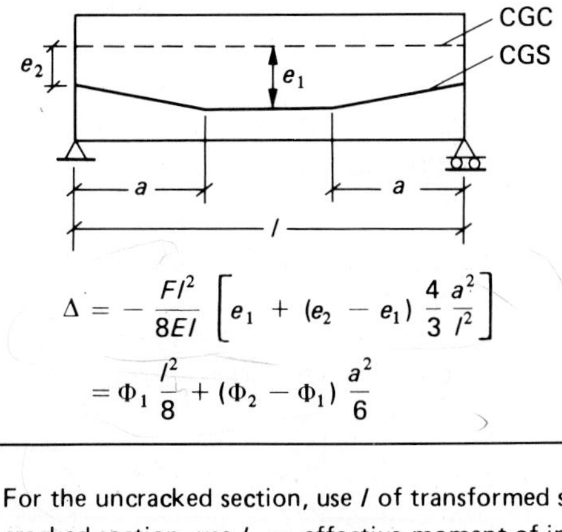 e_2, CGC, CGS, e_1, a, a $$\Delta = -\frac{Fl^2}{8EI}\left[e_1 + (e_2 - e_1)\frac{4}{3}\frac{a^2}{l^2}\right]$$ $$= \Phi_1 \frac{l^2}{8} + (\Phi_2 - \Phi_1)\frac{a^2}{6}$$	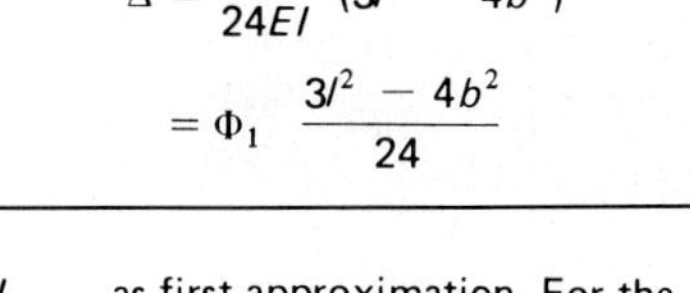 b, P, P, b $$\Delta = \frac{Pb}{24EI}(3l^2 - 4b^2)$$ $$= \Phi_1 \frac{3l^2 - 4b^2}{24}$$

For the uncracked section, use I of transformed section or I_{gross} as first approximation. For the cracked section, use I_e = effective moment of inertia

† Assumed uniform per unit length.

Figure 7.4 Typical midspan deflections for simply supported beams.

combinations are practically covered. Note that two expressions are given for each case, one in function of the prestressing force and the other in function of the curvatures at the midspan section and at the support. The deflection expressed in function of the curvatures (Ref. 7.1) has the following most general form:

$$\Delta = \Phi_1 \frac{l^2}{8} + (\Phi_2 - \Phi_1) \frac{a^2}{6} \qquad (7.4)$$

where Φ_1 = curvature at midspan

Φ_2 = curvature at the support

a = length parameter that is a function of the tendon profile used

The values of a have been directly integrated in the expressions shown in Fig. 7.4. The curvature at any section can be computed from Eq. (7.1) or from the strain distribution along the section as:

$$\Phi = \frac{\varepsilon_{ct} - \varepsilon_{cb}}{h} \qquad (7.5)$$

where ε_{ct} = strain on top fiber of section

ε_{cb} = strain on bottom fiber of section

h = depth of section

Equation (7.5) is illustrated in Fig. 7.5a for uncracked sections and in Fig. 7.5b for cracked sections where ε_{cb} vanishes and h is replaced by c.

Typical formulas for deflections in cantilever beams are given in Fig. 7.6.

In using all deflection expressions such as those given in Figs. 7.4 and 7.6, two important properties must be defined, namely, the modulus of elasticity of the concrete and the moment of inertia of the section (or of each section if variable depth is used).

The modulus of elasticity of the concrete material (secant modulus at $0.45f'_c$) can be estimated from the expression recommended in the ACI code and given in Table 2.8.

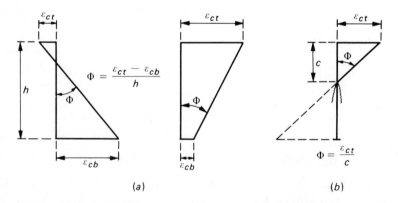

Figure 7.5 Representation of curvature. (a) Uncracked section. (b) Cracked section.

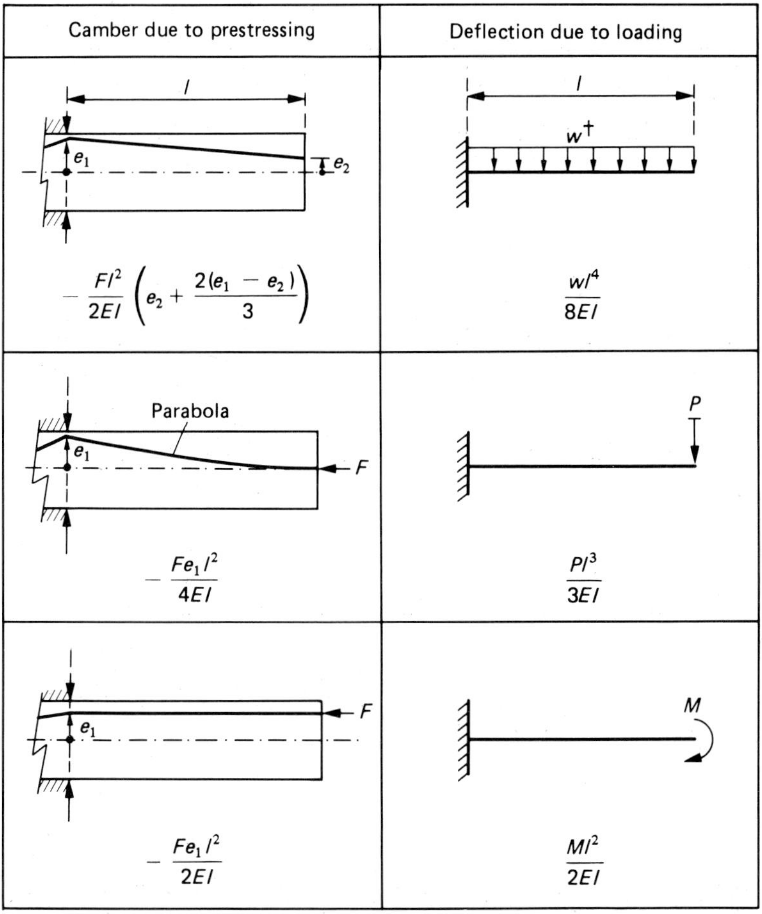

Camber due to prestressing	Deflection due to loading
$$-\frac{Fl^2}{2EI}\left(e_2 + \frac{2(e_1 - e_2)}{3}\right)$$	$$\frac{wl^4}{8EI}$$
Parabola $$-\frac{Fe_1 l^2}{4EI}$$	$$\frac{Pl^3}{3EI}$$
$$-\frac{Fe_1 l^2}{2EI}$$	$$\frac{Ml^2}{2EI}$$

† Assumed uniform per unit length.

Figure 7.6 Typical end deflections for fixed end cantilever beams.

However, the strength of concrete varies with age, and its modulus too. In computing initial camber or deflection, it is common to use the initial modulus E_{ci} while E_c is considered for service load deflections.

The moment of inertia of the section depends on whether the section is cracked or uncracked. When the section is uncracked, it is customary to use the gross moment of inertia I_g for pretensioned members and the net moment of inertia I_n for posttensioned members with unbonded tendons. In all cases with bonded tendons, the moment of inertia of the transformed section can be used. However, often the corresponding lengthier calculations do not result in a significant gain in accuracy.

2. Cracked Members

When cracking occurs in prestressed concrete members, cracks develop at several sections along the span. Theoretically, the cracked moment of inertia I_{cr} applies at

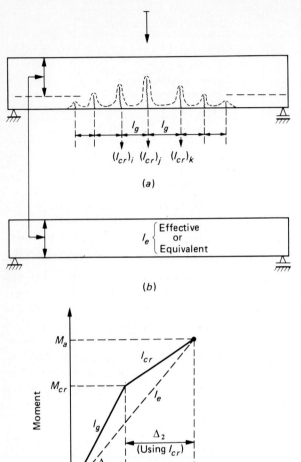

Figure 7.7 (*a*) **Moment of inertia of cracked member.** (*b*) **Effective moment of inertia.** (*c*) **Conceptual representation of bilinear moment deflection relation and effective moment of inertia.**

cracked sections while the gross moment of inertia applies in between cracks (Fig. 7.7*a*). It is generally accepted that methods used in reinforced concrete, where cracking prevails, can be applied to prestressed concrete as well (Ref. 7.6).

The ACI code requires that a bilinear moment deflection relationship (Fig. 7.7*c*) be used to calculate instantaneous deflections when the magnitude of tensile stress in service exceeds $|-6\sqrt{f'_c}|$. This means that for the portion of moment leading to $-6\sqrt{f'_c}$, I_g is used, while for the remaining portion of moment I_{cr} is used.

An alternative method is also allowed by the code for cracked members. An effective moment of inertia is first determined and the deflection is then calculated by substituting I_e for I_g in the deflection calculations. The effective moment of inertia is an average value to be used throughout the span and is weighted depending on the extent of probable cracking under moment. It is an equivalent moment of inertia for a beam that is partly cracked (Figs. 7.7a and b). The difference between the bilinear moment-deflection approach and the I_e approach is illustrated in Fig. 7.7c.

The value of I_e, as originally suggested by Branson (Refs. 7.2 and 7.3) is given by:

$$I_e = I_{cr} + \left(\frac{M_{cr}}{M_a}\right)^3 (I_g - I_{cr}) \le I_g \tag{7.6}$$

where I_{cr} = moment of inertia of the cracked section
I_g = gross moment of inertia of concrete section
M_{cr} = cracking moment for the beam at section of maximum moment M_a
M_a = maximum moment acting on the span at stage for which deflection is computed (essentially same as M_{max} at section of maximum moment)

Several studies have shown good agreement between measured deflections and deflections using I_e. Branson has also recommended the use of I_e for prestressed and partially prestressed concrete cracked members using bonded tendons (Ref. 7.4). In such a case it is suggested that both the cracking moment M_{cr} and the maximum moment M_a be decreased by an amount equal to the decompression moment M_{dec}. The decompression moment is the moment leading to zero stress on the precompressed concrete extreme fiber. For a simply supported prestressed or partially prestressed beam, the cracking moment can be obtained from Eq. (4.38), and the decompression moment from Eq. (4.38) in which f_r is set equal to zero. Hence, Eq. (7.6) becomes:

$$I_e = I_{cr} + \left(\frac{M_{cr} - M_{dec}}{M_a - M_{dec}}\right)^3 (I_g - I_{cr}) \le I_g \tag{7.7}$$

where M_{dec} = decompression moment at section of maximum moment.

Equations (7.6) and (7.7) apply to a simply supported beam or to a continuous beam assumed simply supported at its inflection points (Ref. 7.5). For a cantilever beam, the value of I_e is taken at the face of the support.

In order to compute I_e, the cracked moment of inertia is needed. The determination of I_{cr} for prestressed and partially prestressed concrete members is not as simple as for reinforced concrete because of the following:

The neutral axis of bending in reinforced concrete is the same as the centroid of the cracked transformed section. Thus, the point of zero stress (neutral axis) along the section coincides with its centroid. This is not true for cracked prestressed concrete sections, as their point of zero stress can vary along the section depending on the magnitude of applied moment and/or the prestressing force. The moment of inertia of the section should be determined theoretically with respect to the centroid of the section. For prestressed members, I_{cr} varies with the location of neutral axis. As the neutral axis varies with the applied moment, the centroid of the cracked section varies and thus, I_{cr} varies too. The value of I_{cr} should be in between the gross moment of inertia and the transformed moment of inertia of the cracked section assuming the steel is not prestressed (equivalent to reinforced concrete).

Several methods are proposed in the technical literature to compute I_{cr}. The following empirical expression is suggested in the PCI handbook (Ref. 7.6) for fully prestressed members in a cracked state:

$$I_{cr} = n_p A_{ps} d_p^2 (1 - \sqrt{\rho_p}) \tag{7.8}$$

where $n_p = E_{ps}/E_c$ and the other notations are standard.

Although the above expression does not always lead to results consistent with more accurate methods, it is extremely fast and convenient. Note that a certain variation in I_{cr} generates a much smaller variation in the value of I_e and thus in the computed deflection. Equation (7.8) applies to fully prestressed members. It can be easily extended to partially prestressed members as follows:

$$I_{cr} = (n_p A_{ps} d_p^2 + n_s A_s d_s^2)(1 - \sqrt{\rho_p + \rho_s}) \tag{7.9}$$

where $n_s = E_s/E_c$.

The author (Ref. 7.7) has suggested the use, as a first approximation, of a cracked moment of inertia to be determined with respect to the neutral axis instead of the centroid of the cracked section. Its value for cracked prestressed and partially prestressed rectangular and T sections is given by:

$$I_{cr} = \frac{c^3 b}{3} + n_p A_{ps}(d_p - c)^2 + n_s A_s(d_s - c)^2 - \frac{(b - b_w)(c - h_f)^3}{3} \tag{7.10}$$

where c is the distance from the extreme compressive fiber to the neutral axis at the loading considered. It can be determined from equilibrium and compatibility equations (see Vol. 2). For rectangular sections use $b = b_w$ in Eq. (7.10).

If the cracked moment of inertia is to be determined with respect to the centroid of the cracked section, as theoretically it should be, the following expressions can be used and apply to prestressed and partially prestressed rectangular and T sections (Fig. 7.8):

$$\bar{y} = \frac{(b - b_w)h_f^2/2 + b_w c^2/2 + n_p A_{ps} d_p + n_s A_s d_s}{(b - b_w)h_f + cb_w + n_p A_{ps} + n_s A_s} \tag{7.11}$$

and

$$I_{cr} = \frac{b_w \bar{y}^3}{3} + \frac{b_w(c - \bar{y})^3}{3} + (b - b_w)h_f(\bar{y} - h_f/2)^2$$
$$+ \frac{(b - b_w)h_f^3}{12} + n_p A_{ps}(d_p - \bar{y})^2 + n_s A_s(d_s - \bar{y})^2 \tag{7.12}$$

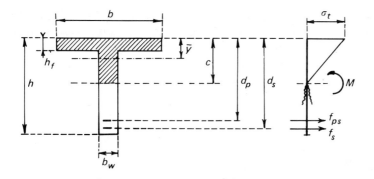

Figure 7.8 Typical stress diagram in a cracked prestressed section.

in which $\bar{y}$ = distance from extreme compressive fiber to the centroid of the cracked transformed section

c = distance from extreme compressive fiber to the neutral axis (zero stress) of the cracked section

b_w = b for rectangular sections

Equation (7.12) is valid for $\bar{y} \geq h_f$. Generally, c is larger than $\bar{y}$, except if tensile loading is used. Nevertheless, the expression remains valid.

Another more elegant method of calculating the cracked moment of inertia is from the moment versus curvature relationship, assuming the stress distribution along the section has already been established in previous calculations. That is:

$$\Phi_{cr} = \frac{\varepsilon_{ct}}{c} = \frac{M}{E_c I_{cr}} \tag{7.13}$$

where ε_{ct} is the strain on the extreme compressive fiber of concrete and the moment M includes the prestressing moment.

Equation (7.13) leads to:

$$I_{cr} = \frac{Mc}{E_c \varepsilon_{ct}} = \frac{Mc}{\sigma_t} \tag{7.14}$$

where σ_t is the stress on the extreme compressive fiber of concrete. Generally, c and σ_t are known at this stage of the design and M is the moment for which I_{cr} is being determined and includes the moment of the prestressing force about the centroid of the cracked section.

It is important to realize that I_{cr} as well as I_e are not constants but depend on the particular loading for which deflection is being calculated. Thus, one cannot superimpose values of deflections in the cracked state for different loadings. Instead, the loads have to be superimposed first and the corresponding deflection calculated.

Although I_{cr} values will differ depending on the formula used, often the related values of I_e and the corresponding deflections differ by a smaller proportion.

An example illustrating the above computations for a cracked and an uncracked member is given next.

7.5 EXAMPLE: DEFLECTION OF CRACKED OR UNCRACKED PRESTRESSED BEAM

Let us consider the pretensioned beam described in Example 4.9a. Let us also assume that the beam is designed as a partially prestressed beam according to the method developed in Vol. 2. This will allow us to study the case of a cracked or uncracked section, depending on the magnitude of the applied moment. The beam cross section is shown in Fig. 7.9.

The following information is given: span 70 ft, $A_c = 550$ in^2, $I_g = 82{,}065$ in^4, $f'_c = 5000$ psi, $E_c = 4.287 \times 10^6$ psi, $E_{ci} = 3.834 \times 10^6$ psi, $A_{ps} = 1.071$ in^2, $A_s = 1.80$ in^2, $F = 160.650$ kips, $f_{pe} = 150$ ksi, $F_i = F/0.83 = 193.554$ kips, $E_{ps} = 27 \times 10^6$ psi, $E_s = 29 \times 10^6$ psi, $n_p = 6.298$, $n_s = 6.765$, $w_G = 0.573$ klf, $w_D = 0.613$ klf, $w_L = 0.4$ klf; midspan moments: $M_G = 350.96$ kips-ft, $M_D = 375.463$ kips-ft, $M_L = 245$ kips-ft, $M_D + M_L = 620.463$ kips-ft, e_o at midspan $= 21.7$ in, e_o at support $= 7.9$ in, the prestressing steel profile has two draping points at 28 ft from the supports. The beam is designed not to crack under the effect of dead load and to crack under the full effect of dead and live loads. Let us compute the instantaneous deflections in each case.

(a) **Uncracked beam.** Two deflection values are of interest here: the initial instantaneous deflection at time of transfer Δ_i and the instantaneous deflection due to full effect of dead load and final prestressing force, Δ_D.

For Δ_i, the initial prestressing force and the initial elastic modulus of concrete are considered, assuming the self-weight of the beam acts as soon as the prestressing force is transferred. Using the equations of Fig. 7.4, we have:

$$\Delta_i = -\frac{F_i l^2}{8 E_{ci} I_g}\left[e_1 + (e_2 - e_1)\frac{4a^2}{3l^2}\right] + \frac{5 w_G l^4}{384 E_{ci} I_g}$$

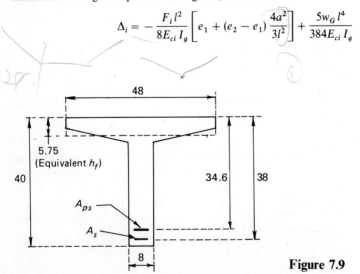

Figure 7.9

where e_1 and e_2 are the eccentricity of the prestressing force at midspan and at the supports, respectively. Thus:

$$\Delta_i = -\frac{193{,}554 \times (70 \times 12)^2}{8 \times 3.834 \times 10^6 \times 82{,}065}\left[21.7 + (7.9 - 21.7)\frac{4}{3}\left(\frac{28}{70}\right)^2\right]$$

$$+\frac{5 \times (70 \times 12)^4 \times 573/12}{384 \times 3.834 \times 10^6 \times 82{,}065}$$

$$\Delta_i = -1.0176 + 0.9838 = -0.0338 \text{ in}$$

Thus, an almost zero camber or deflection results at time of transfer. The elastic deflection due to dead load and prestressing force, assuming the concrete has acquired its specified strength, is estimated from the same formula as the one used for Δ_i but with E_c, F, and w_D, respectively, instead of E_{ci}, F_i, and w_G. Its value is found to be equal to:

$$\Delta_D = -0.75 + 0.94 = 0.19 \text{ in}$$

Note that Δ_D is not a real observable value because, by the time the prestressing force has reached its final value, the effect of time-dependent deflection would be prevalent. However, Δ_i is real.

(b) Cracked beam Let us compute the instantaneous deflection due to the effect of full external load, plus prestressing. It can be shown that the beam cracks under such loading. The corresponding depth of neutral axis (zero stress) at midspan is $c = 8.94$ in and the corresponding stress on the extreme compresive fiber of concrete is $\sigma_t = 1186$ psi. We need to compute the cracked moment of inertia I_{cr} and the effective moment of inertia I_e. The cracked moment of inertia will be computed from Eq. (7.14) in which the moment M includes $(M_D + M_L)$ and M_F, the moment of the prestressing force about the centroid of the cracked section. The location of the centroid of the cracked section $\bar{y}$ is given by Eq. (7.11). It can easily be shown that $\bar{y} \simeq 5.23$ in for the cross section considered. The moment due to the prestressing force is negative. Thus:

$$M_F = -F(d_p - \bar{y}) = -160.65(34.6 - 5.23)/12 = -393.191 \text{ kips-ft}$$

and

$$M = M_D + M_L + M_F = 620.463 - 393.191 = 227.272 \text{ kips-ft}$$

The cracked moment of inertia is then given by:

$$I_{cr} = \frac{Mc}{f_{ct}} = \frac{227.272 \times 12{,}000 \times 8.94}{1186} \simeq 20{,}558 \text{ in}^4$$

In order to determine the effective moment of inertia, we need the cracking moment of the midspan section, M_{cr}. Using Eq. (4.38), it can be shown that $M_{cr} \simeq 498.043$ kips-ft and $M_{dec} = 364.274$ kips-ft. Thus:

$$I_e = I_{cr} + \left(\frac{M_{cr} - M_{dec}}{M_a - M_{dec}}\right)^3 (I_g - I_{cr})$$

$$= 20{,}558 + \left(\frac{498.043 - 364.274}{620.463 - 364.274}\right)^3 (82{,}065 - 20{,}558) = 52{,}673 \text{ in}^4$$

and is less than I_g.

$w_D + w_L$

Replacing I_g, F_i, E_{ci}, and w_G, respectively, by I_e, F, E_c, and $w = 1.013$ klf in the expression used earlier for Δ_i, the instantaneous deflection at midspan of the cracked beam, assuming all the above loads are instantaneous, is determined as:

$$\Delta \simeq 1.25 \text{ in}$$

The reader may want to check that, using Eq. (7.10) as a first approximation for I_{cr}, a value of $I_{cr} \simeq 25{,}014$ in^4 would have been obtained with a corresponding I_e of 55,700 in^4. Both are not too different from the values calculated above. Note also that, for this particular example, the other approaches suggested for I_{cr} also lead to not too different values of I_e.

It is important to realize that this problem is fictitious in part, as in practice the live load and dead load are not applied simultaneously and the effects of time and the sustained part of the load have to be accounted for.

7.6 LONG-TERM DEFLECTION: SIMPLIFIED PREDICTION METHODS

It was pointed out repeatedly in the previous sections that the deflections calculated from the various derived expressions were "instantaneous elastic" deflections. We assumed that they were associated with a short-term loading. If the load is sustained, such as in the case of dead loads, the deflection increases with time, mainly because of the effects of creep and shrinkage of the concrete and relaxation of prestressing steel. In such a case, the deflection will be called the total deflection. The total deflection can be separated into two parts: an instantaneous elastic part and an additional long-term part. The first part is calculated as described in the previous sections and is assumed to remain constant for a given load. The additional part increases with time.

Figures 2.22 and 7.10 illustrate the effect of time on the deflection or prestressed and partially prestressed beams. It is interesting to note (Fig. 7.10 and Ref. 7.8) that the instantaneous deflection leads, for the beam and loading considered, to the development of one crack. However, as the time-dependent deflection increases with time, the number of cracks increases to 16 and their width increases significantly.

Continuous research efforts are being undertaken to predict time-dependent deflection in reinforced and prestressed concrete structures more reliably (Refs. 7.9 to 7.16). It has been customary in everyday design to estimate the additional long-term deflection by multiplying the instantaneous deflection Δ_i by an appropriate factor. The basic reason behind this approach comes from analyzing the effect of creep on the stress-strain response of concrete. The variation of creep strain under initial loading is illustrated in Fig. 7.11a. At any time t, the creep strain, assuming constant stress, can be obtained by multiplying the instantaneous strain under a given stress by the creep coefficient. Thus:

$$\varepsilon_{cC}(t) = \varepsilon_{ci} \times C_C(t) \tag{7.15}$$

Looking simultaneously at the stress-strain response of the material in its linear elastic range (Fig. 7.11), one can separate the instantaneous elastic strain from the

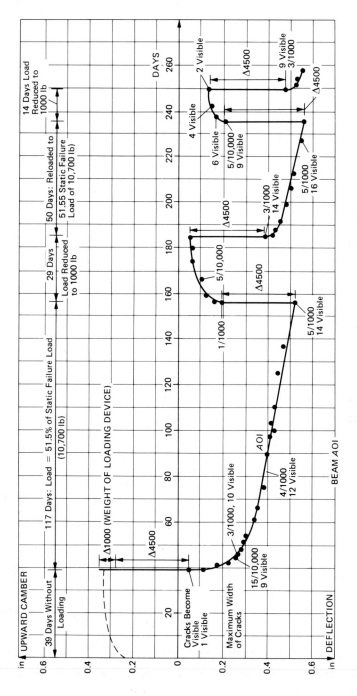

Figure 7.10 Deflection and cracking of a prestressed beam under sustained loadings versus time. (*Ref. 7.8. Courtesy Prestressed Concrete Institute.*)

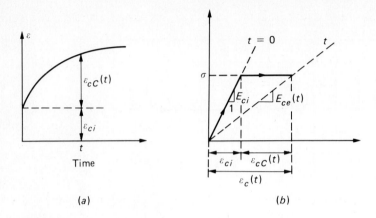

Figure 7.11 Relationship between (a) creep strain variation with time and (b) equivalent elastic modulus.

creep strain. By definition, the ratio σ/ε_{ci} is equal to the initial elastic modulus E_{ci}. With time, however, the ratio of stress to total strain decreases due to the creep strain. This effect can be simulated by using an equivalent modulus (Fig. 7.11b). Its value is obtained from:

$$E_{ce}(t) = \frac{\sigma}{\varepsilon_c(t)} = \frac{\sigma}{\varepsilon_{ci} + \varepsilon_{cc}(t)} = \frac{\sigma}{\varepsilon_{ci} + \varepsilon_{ci}\, C_c(t)}$$

$$= \frac{\sigma}{\varepsilon_{ci}(1 + C_c(t))} = \frac{E_{ci}}{1 + C_c(t)} \tag{7.16}$$

Any deflection formula is an inverse function of the modulus. In general, one can write:

$$\Delta = \frac{K}{E} \tag{7.17}$$

where K depends on dimensional and loading properties. Applied to the initial instantaneous deflection, Eq. (7.17) gives:

$$\Delta_i = \frac{K}{E_{ci}} \tag{7.18}$$

If the loading is sustained for a time t, the time-dependent deflection becomes:

$$\Delta(t) = \frac{K}{E_{ce}(t)} \tag{7.19}$$

Replacing K in Eq. (7.19) by its value from Eq. (7.18) and $E_{ce}(t)$ by its value from Eq. (7.16), leads to the total deflection at time t, assuming constant loading:

$$\Delta(t) = \Delta_i(1 + C_c(t)) \tag{7.20}$$

The above equation can be separated into two parts, a constant value Δ_i and a time-dependent value. Thus:

$$\Delta(t) = \Delta_i + C_c(t)\Delta_i \qquad (7.21)$$

At the end of the life of the member, it becomes

$$\Delta_U = \Delta_i + \Delta_i C_{CU} = \Delta_i(1 + C_{CU}) = \Delta_i + \Delta_{add} \qquad (7.22)$$

in which

$$\Delta_{add} = C_{CU}\,\Delta_i \qquad (7.23)$$

and Δ_U = life total deflection for the sustained loading considered
Δ_{add} = additional long-term deflection

Thus, the additional long-term deflection can be estimated by multiplying the instantaneous deflection by a factor. According to Eq. (7.23), the multiplier is equal to the ultimate creep coefficient. However, some other effects, such as those of shrinkage, relaxation, and the presence of nonprestressing steel, may be accounted for in the equation predicting Δ_{add}. Three such prediction equations are discussed next.

In Sec. 9.5.2.5, the ACI code suggests the following prediction equation for nonprestressed concrete members:

$$\Delta_{add} = \lambda\Delta_i \qquad (7.24)$$

where

$$\lambda = 2 - 1.2\,\frac{A_s'}{A_s} \geq 0.6 \qquad (7.25)$$

and Δ_i is the instantaneous deflection. It can be seen that for $A_s' = 0$, $\lambda = 2$, that is, about equal to an average value of creep coefficient.

Based on an extensive evaluation of parameters influencing Δ_{add} for typical precast prestressed members, Martin (Ref. 7.12) suggested the following equation:

$$\Delta_{add} = \lambda\Delta_i = \eta\,\frac{E_{ci}}{E_c}\,k_r\,C_{CU}\,\Delta_i \qquad (7.26)$$

where $\eta = F/F_i$ and F_i is the prestressing force immediately after transfer
$k_r = 1/(1 + A_s/A_{ps})$ when $A_s/A_{ps} \leq 2$
C_{CU} = ultimate creep coefficient
E_{ci} = elastic modulus of concrete at time of transfer

Different values of the multiplier λ were recommended for prestressed composite and noncomposite members depending on the effects of self-weight, prestressing, and superimposed dead load, if any. They are given in Table 3.4.1 of the PCI handbook (Ref. 7.6) and generally vary between 1.85 and 3.

Branson et al. (Refs. 7.4, 7.13, 7.14) suggested the following equation to predict the long-term additional deflection in prestressed noncomposite members:

$$\Delta_{add} = \left(\eta^{(\eta-1)} + \left(\frac{1+\eta}{2}\right)k_r\,C_{CU} \right)(\Delta_i)_{F_i} + k_r\,C_{CU}(\Delta_i)_G + K_{CA}\,k_r\,C_{CU}(\Delta_i)_{SD} \qquad (7.27)$$

where the subscripts G and SD refer to self-weight and superimposed dead load and K_{CA} is the age at loading factor for creep (Table 2.9). Other notations are the same as for Eq. (7.26).

In a recent paper, Tadros et al. (Refs. 7.1 and 7.15) rightly questioned the effectiveness of the k_r term in reducing downward deflection and pointed out that it can produce an opposite effect.

In an earlier publication, Tadros et al. (Ref. 7.16) studied a typical beam example where, depending on the method used, the predicted deflection varied from $+0.24$ in to -1.59 in. They also suggested a different approach to predict long-term deflection. A conclusion similar to theirs was reached in Ref. 7.17 where a partially prestressed concrete beam was analyzed. It is expected that the prediction of long-term deflection will see many additional refinements.

Because of the uncertainty associated with the prediction of long-term deflections, engineers should apply common sense in selecting and using any of the expressions predicting Δ_{add}. In most common design problems involving prestressed concrete a simple multiplier taken between 2 and 3 should be satisfactory in detecting if deflection-related problems are likely to occur.

7.7 LONG-TERM DEFLECTION BY INCREMENTAL TIME STEPS

1. Theoretical Approach

The additional long-term deflection is essentially caused by the simultaneous occurrence of creep and shrinkage of concrete and relaxation of the prestressing steel. As these effects are also associated with prestress losses, it is quite logical to combine the computations of deflections with those of prestress losses. A similar approach was suggested by Subcommittee 5 of ACI Committee 435 (Ref. 7.9). Theoretically, the computation procedure may include the following steps:

1. Divide the span into several segments (about 20 is more than adequate) to be each represented by its average or midsection.
2. Divide the design life of the structure into several time intervals. These are not equal intervals but have increasing lengths. Typical sets are used in Chap. 8 on prestress losses and in Table 7.1.
3. Select a time interval, starting in order by the first one. Determine the strain distributions, curvatures, and prestressing force at each section at the beginning of the time interval considered. For the first interval, these values correspond to the instantaneous effects and the initial prestressing force. Determine incremental creep and shrinkage strains and relaxation loss during the time interval. Compute new values of strains, curvatures, and prestressing force at each section at the end of the time interval. These will be used as reference values at the beginning of the next time interval. This procedure is then repeated for each time interval studied and at each section. Great care should be used to ensure

restoration of both equilibrium and strain compatibility at the end of each time interval.

4. By integrating or summing up along the beam the curvatures computed at the beginning of each time interval, the corresponding total time-dependent deflection can be determined. It will include the instantaneous deflection and the additional long-term deflection defined earlier.

The above procedure is time-consuming and the gain in accuracy over an approximate procedure is seldom justifiable. This is even more so because the prediction equations that govern the variations of creep and shrinkage with time, as well as the values of ultimate creep and shrinkage, are by no means accurate.

2. Simplified C Line Approach

The author has developed a technique to predict long-term deflections which combines theory with some simplified observations. It has the advantage of being relatively simple and sufficiently accurate, yet manageable in terms of computational effort. As the deflection is calculated at different time intervals, it allows for those special cases where values of deflections at intermediate times are needed. The procedure has the following characteristics and their justification:

1. A set of appropriate time intervals is selected, the total time-dependent loss of prestress and the percent prestress loss at the end of each time interval are estimated a priori. Note that, although an error may arise in estimating the total time-dependent prestress loss, little error will materialize in the estimates of the percent loss at the end of each time interval. This is because the laws governing creep, shrinkage, and relaxation, and their prediction equations, lead to very similar percent values at identical times (see Chap. 8).

   By estimating a priori the percent of prestress loss with time, the secondary effects of shrinkage and relaxation are indirectly accounted for and the major effect of creep on long-term deflection can be isolated and expanded.

2. Only one loading is considered during each time interval and comprises the combined effect of the prestressing force and the sustained external load. This is contrary to other methods where the loadings are separated into prestressing, self-weight, and superimposed dead load. Because of the combined loading, the beam is assumed equivalently prestressed by a force following the trajectory of the C line instead of the trajectory of the prestressing steel (Fig. 7.12). The C force is equal to F in magnitude. Its eccentricity at any section along the span is given by Eq. (4.15):

$$e_c = e_o - \frac{M}{F} \qquad (7.28)$$

where M is the externally applied moment and F varies with time, according to the assumed prestress losses. During any time interval, M and F are assumed constant, with the value of F taken at the beginning of the interval. As in most

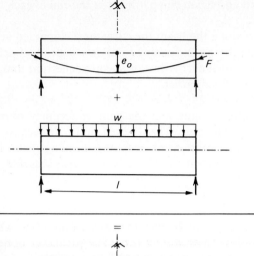

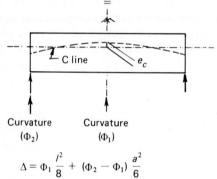

Curvature   Curvature
(Φ_2) (Φ_1)

$$\Delta = \Phi_1 \frac{l^2}{8} + (\Phi_2 - \Phi_1) \frac{a^2}{6}$$

Figure 7.12 C line approach used in computing deflection under combined prestressing and external load.

common cases the variation of e_o along the span is either linear or parabolic and M is either linear or parabolic, the profile of the trajectory of the C line will be either linear or parabolic.

3. Only two sections are considered in the computations of time-dependent variables, the midspan section and the support section. This is because the deflection of the beam for common linear or parabolic trajectories of the C line can be calculated in function of the curvatures of the midspan and support sections. Such expressions are given in Fig. 7.4 for the prestressing steel. They can be used directly here, noting that e_1 and e_2 become the eccentricities of the C force and are time dependent. In using these expressions, it is assumed that the C force, as well as its eccentricity, remain constant during the time interval considered.

Thus, according to the above described procedure, the deflection of the beam (Fig. 7.12) during each time interval is given by:

$$\Delta = \Phi_1 \frac{l^2}{8} + (\Phi_2 - \Phi_1) \frac{a^2}{6} \tag{7.4}$$

where a is given in Fig. 7.4 for several practical cases. The curvature at any section can be obtained either from Eq. (7.1) or from:

$$\Phi = \frac{\varepsilon_{ct} - \varepsilon_{cb}}{h} \tag{7.5}$$

At the end of each time interval, assuming uncracked section, ε_{ct} and ε_{cb} can be determined from:

$$\varepsilon_{ct} = \frac{F}{E_{ce}(t)A_c}\left(1 - \frac{e_c}{k_b}\right)$$

$$\varepsilon_{cb} = \frac{F}{E_{ce}(t)A_c}\left(1 - \frac{e_c}{k_t}\right) \tag{7.29}$$

where $E_{ce}(t)$ is the equivalent modulus of elasticity of the concrete as influenced by creep. (*Note:* k_t is negative.) Assuming that the short-term instantaneous modulus varies with time, the equivalent modulus of elasticity under sustained loading at any time t can be estimated, similarly to Eq. (7.16), from:

$$E_{ce}(t) = \frac{E_c(t)}{1 + C_C(\tau)} \tag{7.30}$$

where $E_c(t) =$ instantaneous elastic modulus of concrete at time t
  $t =$ age of concrete in days
  $C_C(\tau) =$ creep coefficient at time τ
  $\tau =$ time after loading $= t - t_A \geq 0$
  $t_A =$ age of concrete at loading

According to Tables 2.8 and 2.9 we have:

$$E_c(t) = 33\gamma_c^{1.5}\sqrt{f_c'(t)} \tag{7.31}$$

$$f_c'(t) = \frac{t}{b + ct}\, f_c'(28) \tag{7.32}$$

Thus:

$$E_c(t) = 33\gamma_c^{1.5}\sqrt{f_c'(28)}\,\sqrt{\frac{t}{b + ct}} = E_c\sqrt{\frac{t}{b + ct}} \tag{7.33}$$

where E_c is the design modulus of elasticity of concrete. Also using Table 2.9, we have

$$C_C(\tau) = \frac{\tau^{0.6}}{10 + \tau^{0.6}}\, C_{CU}\, K_{CH}\, K_{CA}\, K_{CS} \tag{7.34}$$

in which

$$\tau = t - t_A \tag{7.35}$$

Replacing $C_C(\tau)$ and $E_c(t)$ by their values from Eqs. (7.34) and (7.33) into Eq. (7.30) leads to:

$$E_{ce}(t) = E_c\left(\sqrt{\frac{t}{b+ct}}\right)\left[\frac{10 + (t - t_A)^{0.6}}{10 + (t - t_A)^{0.6}(1 + C_{CU}K_{CH}K_{CA}K_{CS})}\right] \quad (7.36)$$

in which the constants b and c and the factors K_{CH}, K_{CA}, and K_{CS} are given in Table 2.9.

For a given problem Eq. (7.36) is reduced to a time function multiplied by the design modulus of elasticity of concrete E_c.

Note, the above procedure remains valid when a superimposed dead load is applied. In such a case, an interval is selected to start at the date of application of the superimposed dead load and an adjusted value of the age at loading factor for creep is used. If the beam cracks, the curvatures can then be obtained from the M/EI relation Eq. (7.1) in which I is replaced by I_e and M includes the moment due to prestressing. Calculating the curvature from M/EI could also have been used directly in the above approach for the uncracked section.

The variation of concrete strain with time due to creep and age at loading is a dominant factor in any long-term deflection computation. Nondimensional graphs can be developed to predict such strains in common applications. A typical example (Ref. 7.18) is shown in Fig. 7.13. It is very convenient for use in preliminary estimates of long-term deflections, especially in segmental construction and composite structures, when excessive numerical computations are to be avoided.

3. Example

The same partially prestressed beam of Example 7.5 is considered. It is assumed that the beam is plant precast, steam cured, using type III cement and prestressed at one day of age after casting. As the superimposed dead load of 0.040 klf is small in relation to the weight ($w_G = 0.573$ klf), let us assume, for the sake of simplicity, that the whole dead load (0.613 klf) is applied at the time of prestress and represents the sustained load for which the long-term deflection is to be calculated. Thus, the sustained external moment is $M = M_D = 375.463$ kips-ft. The final prestressing force is $F = 160.165$ kips.

Let us also assume that the base value of the ultimate creep coefficient C_{CU} for the concrete material used equals 2.5 and the average relative humidity of the environment is 50 percent. Referring to Eq. (7.35) and Table 2.9, the following values of the various constants are obtained: $b = 0.70$, $c = 0.98$, $b_1 = 10$, $K_{CH} = 0.935$, $K_{CA} = 1.13$, $K_{CS} \simeq 0.86$. As the age of loading is one day, Eq. (7.36) becomes, for this example:

$$E_{ce}(t) = E_c\left(\sqrt{\frac{t}{0.7 + 0.98t}}\right)\left(\frac{10 + (t - 1)^{0.6}}{10 + (t - 1)^{0.6}(3.27)}\right)$$

where $E_c = 4.287 \times 10^6$ psi.

Several time intervals are selected (Table 7.1) and the percent prestress loss at the end of each interval is assumed. Note, the end of the first time interval is one day, the end of the second interval is seven days, and so on. The computations are summarized in Table 7.1 for each time interval. Let us follow them for the first interval:

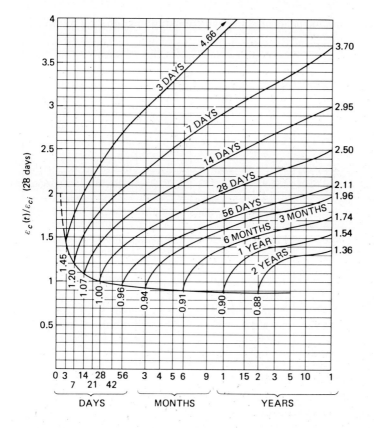

Figure 7.13 Typical design graph giving nondimensionalized concrete strains versus age and duration of loading. *(Ref. 7.18. Courtesy of the Prestressed Concrete Institute.)* *(Note:* ε_{ci} *(28 days) is the reference strain of a 28-day-old concrete subjected to short-term load.)*

The prestressing force is equal to its initial value immediately after transfer, that is:

$$F_i = \frac{F}{\eta} = \frac{160.65}{0.83} = 193.554 \text{ kips}$$

Values of the prestressing force at the end of the other intervals are given in the first line of Table 7.1 and correspond to: $F_i - (F_i - F) \times$ (percent loss). The equivalent modulus at $t = 1$ day is obtained from:

$$E_{ce}(1) = E_c \sqrt{\frac{1}{0.7 + 0.98}} = 0.772 E_c = 3.307 \times 10^6 \text{ psi}$$

Moduli values for the other time intervals are given in Table 7.1 and can be computed directly as the equation depends only on E_c and t.

Table 7.1 Summary of time-dependent deflection computations for the Example in Sec. 7.7

		End t_j of time interval (t_i, t_j), days					
		1	7	30	90	365	Life 50 × 365
Estimated % of time-dependent prestress loss		0	20	45	65	85	100
$F(t)$, kips		193.554	186.973	178.747	172.166	165.585	160.650
$E_{ce}(t)$, 10^6 psi		3.307	2.724	2.166	1.832	1.568	1.350
Midspan	ε_{ct}, 10^{-6}	120.9	150.7	195.5	236.9	283.4	335
	ε_{cb}, 10^{-6}	75.9	70.5	54.5	32.2	0.002	−32.8
	Φ_1, 10^{-6} in^{-1}	1.125	2.0	3.525	5.118	7.085	9.195
Support	ε_{ct}, 10^{-6}	33.7	39.6	47.6	54.2	60.9	68.6
	ε_{cb}, 10^{-6}	259	303.7	365.1	415.8	467.3	526.6
	Φ_2, 10^{-6} in^{-1}	−5.632	−6.602	−7.937	−9.04	−10.16	−11.45
Total Δ, in		$\simeq 0.00$	0.049	0.142	0.243	0.371	0.507

The eccentricity of the C force at midspan and the corresponding extreme fibers' strains and section curvature are given by Eqs. (7.28), (7.29), and (7.5):

$$e_c = 21.7 - \frac{375.463 \times 12}{193.554} = -1.58 \text{ in}$$

$$\varepsilon_{ct} = \frac{193,554}{3.307 \times 10^6 \times 550}\left(1 + \frac{1.58}{11.57}\right) = 120.9 \times 10^{-6}$$

$$\varepsilon_{cb} = \frac{193,554}{3.307 \times 10^6 \times 550}\left(1 - \frac{1.58}{5.51}\right) = 75.9 \times 10^{-6}$$

$$\Phi_1 = \frac{\varepsilon_{ct} - \varepsilon_{cb}}{40} = 1.125 \times 10^{-6} \text{ in}^{-1}$$

The eccentricity of the C force at the support section and the corresponding strains and curvature are given by:

$$e_c = e_o = 7.9 \text{ in (as } M = 0)$$

$$\varepsilon_{ct} = \frac{193,554}{3.307 \times 10^6 \times 550}\left(1 - \frac{7.9}{11.57}\right) = 33.7 \times 10^{-6}$$

$$\varepsilon_{cb} = \frac{193,554}{3.307 \times 10^6 \times 550}\left(1 + \frac{7.9}{5.51}\right) = 259 \times 10^{-6}$$

$$\Phi_2 = \frac{\varepsilon_{ct} - \varepsilon_{cb}}{40} = -5.63 \times 10^{-6}$$

The trajectory of the prestressing steel is bilinear and the moment varies parabolically. Thus, the trajectory of the C line (Eq. (7.28)) is biparabolic. In computing the deflection, we can assume as a first approximation that the expression given in Fig. 7.4 for a parabolic profile applies. Thus:

$$\Delta = \Phi_1 \frac{l^2}{8} + (\Phi_2 - \Phi_1) \frac{l^2}{48}$$

and the deflection at the end of the first time interval (one day) is:

$$\Delta = 1.125 \times 10^{-6} \times \frac{(70 \times 12)^2}{8} + (-5.63 - 1.125)10^{-6} \frac{(70 \times 12)^2}{48}$$

$$= 0.0992 - 0.0993 \simeq 0.$$

The computations for other intervals are given in Table 7.1. Note that the final value of long-term deflection is approximately 0.50 in. It is significantly different from the value otherwise obtained from Example 7.5a by multiplying Δ_i by C_{CU}.

Figure 7.14 illustrates the variation of curvatures at midspan and support sections with time for this example. It can be observed that the long-term effects were significant enough at midspan to change the stress and strain on the bottom fiber from compression to tension. If the sustained load had been higher, cracking would probably have developed.

The reader may want to check that the same numerical results would have been obtained if the curvature had been calculated from the M/EI relation, where M includes the moment due to prestressing.

It was assumed that the prestressing force is the same at midspan and at the support. In pretensioned members, it is smaller at the support due to the effect of bond development length. In posttensioned members, the loss of prestress due to anchor set may be significant.

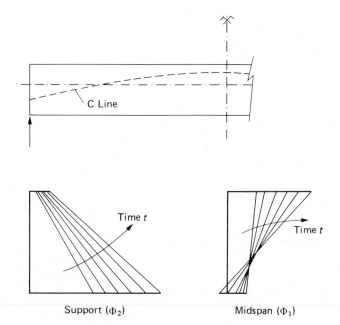

Support (Φ_2)   Midspan (Φ_1)

Figure 7.14 Typical variations of section curvatures with time *(Example in Sec. 7.7)*.

Because of smaller eccentricity, the flexural stress distribution at the support section is different from that at midspan. Hence, different creep, relaxation, and prestress loss effects will be observed at these two sections. Correction factors may be introduced to refine the deflection computations.

4. Age-Adjusted Effective Modulus

The effective modulus approach described in Eq. (7.30) is the simplest and most widespread approximate method to analyze the effect of creep on the deflection of a concrete structural element. However it can be shown that, for a given creep law, it may lead to a large error with respect to the theoretically exact solution, if aging of concrete is significant. The term "aging" refers to the change of the properties of concrete with the progress of its hydration. Hence, for a concrete member loaded at an early age, the effective modulus approach may be very inaccurate. Trost proposed a simple method to accommodate the effect of concrete aging by adding an aging coefficient to the effective modulus equation (Ref. 7.19). The method was later expanded and refined by Bazant (Ref. 7.20), following a rigorous formulation. The proposed age-adjusted effective modulus takes on the following most general form (Ref. 7.20):

$$E_{ce}(t, t_A) = \frac{E_c(t_A)}{1 + \chi(t, t_A)C_C(t, t_A)} \tag{7.37}$$

where $E_{ce}(t, t_A)$ = age-adjusted effective modulus at time t when loading occurs at time t_A

$E_c(t_A)$ = instantaneous elastic modulus at time t_A

$\chi(t, t_A)$ = aging coefficient at time t for a concrete member loaded at time t_A

$C_C(t, t_A)$ = creep coefficient at time t for a concrete member loaded at time t_A

Using a computerized analysis and a wide range of parameters, Bazant showed that the aging coefficient varies significantly with the age at loading t_A, the value of the creep coefficient, and whether a variable or a constant modulus is considered in the creep law. He also showed that, while other methods give exact solutions only when the applied stress is constant, the age-adjusted effective modulus applies for the case of constant stress, constant strain (as in a stress relaxation test) and the case of straining a member by differential creep.

The observed range of the aging coefficient was 0.71 to 1 when constant modulus was assumed in the creep law and 0.46 to 1 when variable modulus was considered. The lower values correspond to early ages at loading. As a first approximation, when loading occurs at normal ages and deflections are estimated at a later date, a value of $\chi \simeq 0.85$ is appropriate in common design applications. In such a case, the aging coefficient has essentially a delaying effect on the value of the effective modulus at a time t. A detailed discussion of the effect of the aging coefficient χ is given in Refs. 7.20 and 7.21.

7.8 DEFLECTION LIMITATIONS ✓

Although structural members can be properly designed for strength, they may develop excessive cambers or deflections over time. Hence, their behavior in service can be jeopardized. Fully prestressed bridge members, in which no tension or very little tension is allowed, often develop large cambers leading to an uneven road profile that seriously affects their riding properties. In building members, excessive deflection can cause serious damage to window frames, partitions, and other non-structural elements connected to them.

The ACI code provides a number of deflection limitations for building members. They are divided into two groups: one where deflection is likely to damage attached nonstructural members and the other where it is not. Limitations on the first group are more stringent. Two types of deflections are generally considered: the instantaneous deflection due to live load alone and its combination with the additional long-term deflection due to dead or sustained load. A summary of ACI maximum permissible deflections is given in Table 7.2. The British code CP-110 essentially follows an approach similar to ACI but the numerical values of permissible deflections are slightly different.

Permissible deflections prescribed in the AASHTO specifications for steel bridges are summarized in Table 7.3. No long-term effects are considered. Only the deflection due to live load plus impact is addressed. The permissible values are much more severe than those given in the ACI code for building members.

1. Example

Let us check if the deflections of the example beam described in Secs. 7.5 and 7.7 satisfy the ACI code limitations. Assume the beam is part of a floor construction and is attached to nonstructural elements likely to be damaged by large deflections.

It was found that the instantaneous deflection due to prestressing and self-weight at time of transfer was almost zero. However, in considering the final prestressing force and the superimposed dead load, a value of $\Delta_D \simeq 0.19$ in was found. If we assume an average of $\simeq 0.10$ in as a representative value of the instantaneous deflection due to total dead load and prestressing, a first estimate of Δ_{add} can be obtained by multiplying the above value by $C_{CU} = 2.5$. Thus $\Delta_{add} \simeq 2.5 \times 0.10 = 0.25$ in.

In the time-steps approach of the Example in Sec. 7, Δ_{add} was found to be equal to the total deflection (as the instantaneous deflection was zero). Thus, $\Delta_{add} \simeq 0.50$ in.

The total instantaneous deflection due to dead load and live load was calculated in Example 7.5b for the then-cracked beam and was found equal to 1.25 in. The instantaneous deflection of the uncracked beam under final prestress and full dead load was calculated in Example 7.5a and was found equal to 0.19 in. Thus, the instantaneous deflection due to live load only is obtained from:

$$\Delta_L = 1.25 - 0.19 = 1.06 \text{ in}$$

Let us check the deflection limitations given by ACI in Table 7.2 for live load plus additional long-term dead load:

$$\Delta_L + \Delta_{add} = 1.06 + 0.5 = 1.57 \leq \frac{l}{480} = 1.75 \text{ in} \qquad \text{OK}$$

Table 7.2 Maximum permissible computed deflections of the ACI code *(Courtesy American Concrete Institute)*

Type of member	Deflection to be considered	Deflection limitation
Flat roofs not supporting or attached to nonstructural elements likely to be damaged by large deflections	Immediate deflection due to the live load, L	$\dfrac{l\dagger}{180}$
Floors not supporting or attached to nonstructural elements likely to be damaged by large deflections	Immediate deflection due to the live load, L	$\dfrac{l}{360}$
Roof or floor construction supporting or attached to non-structural elements likely to be damaged by large deflections	That part of the total deflection which occurs after attachment of the nonstructural elements, the sum of the long-time deflection due to all sustained loads and the immediate deflection due to any additional live load‡	$\dfrac{l\S}{480}$
Roof or floor construction supporting or attached to non-structural elements not likely to be damaged by large deflections		$\dfrac{l\dagger\dagger}{240}$

† This limit is not intended to safeguard against ponding. Ponding should be checked by suitable calculations of deflection, including the added deflections due to ponded water, and considering long-time effects of all sustained loads, camber, construction tolerances, and reliability of provisions for drainage.

‡ The long-time deflection shall be determined in accordance with Section 9.5.2.3 or 9.5.4.2 but may be reduced by the amount of deflection which occurs before attachment of the nonstructural elements. This amount shall be determined on the basis of accepted engineering data relating to the time-deflection characteristics of members similar to those being considered.

§ This limit may be exceeded if adequate measures are taken to prevent damage to supported or attached elements.

†† But not greater than the tolerance provided for the nonstructural elements. This limit may be exceeded if camber is provided so that the total deflection minus the camber does not exceed the limitation.

7.9 DEFLECTION CONTROL

The deflection of prestressed and partially prestressed concrete members can be controlled to a great extent by properly selecting the magnitude and trajectory of the prestressing force. For commonly encountered reinforced concrete members,

Table 7.3 Maximum permissible deflections of the AASHTO specifications for steel bridges

Type of member	Deflection considered	Maximum permissible deflection	
		Vehicular traffic only	Vehicular and pedestrian traffic
Simple or continuous spans	Instantaneous due to service live load plus impact	$\dfrac{l}{800}$	$\dfrac{l}{1000}$
Cantilever arms		$\dfrac{l}{300}$	$\dfrac{l}{375}$

the ACI code waives deflection calculations, provided minimum thickness requirements are met. Minimum thicknesses for beams and slabs are given in Table 9.5a of the code in function of the span length. Similar limitations for prestressed concrete members may be developed in the future by ACI Committee 435 (Ref. 7.11).

Although it is difficult to establish rational limitations for prestressed members without referring to the prestressing force and its profile, two rules of thumb that were initially suggested by Lin (Ref. 1.11) may be of value:

1. Use a depth of prestressed member about 75 percent of the corresponding depth of a conventional reinforced concrete member.
2. Use $h \simeq 1.5$ to $2\sqrt{M_{\max}}$ where h = beam depth in inches and $M_{\max}$ = maximum moment in kips-feet. Note, as $M_{\max}$ includes the dead load moment, some iterations may be needed.

In practice, the depth of hollow core and solid slabs varies roughly between $l/45$ and $l/30$ and that of T or double T beams varies between $l/35$ and $l/25$.

For simple span highway bridges, constant beam depths often vary between $l/25$ and $l/15$.

In addition to staying within practical depth limitations, the designer can take some appropriate actions to reduce camber or deflection (Ref. 7.22). Some of these actions include:

1. Adding nonprestressed steel at appropriate locations to restrain the deformation of concrete due to creep and shrinkage. The time-dependent part of the camber or deflection can be significantly reduced.
2. Precambering at casting. This procedure attempts to balance the long-term deflection so as to achieve a final straight profile of the member.
3. Delaying or staging the application of prestress. The effect is to decrease time-dependent deflections as the concrete would have reached its full strength and maturity at time of prestressing. Stage stressing provides an in-between result.

REFERENCES

7.1. M. K. Tadros: "Designing for Deflection," reprint of a paper presented at the PCI Seminar on Advanced Design Concepts in Precast Prestressed Concrete, PCI Convention, Dallas, October 1979.

7.2. D. E. Branson: "Design Procedures for Computing Deflections," *ACI Journal*, vol. 75, no. 9, September 1968, pp. 730–742.

7.3. ———: *Deformation of Concrete Structures*, McGraw-Hill Book Company, New York, 1977.

7.4. ———: "The Deformation of Non-Composite and Composite Prestressed Concrete Members," *Deflection of Concrete Structures*, ACI-SP-43, American Concrete Institute, Detroit, 1974.

7.5. "Deflections of Continuous Concrete Beams," ACI Committee 435, *ACI Journal*, vol. 70, no. 12, December 1973, pp. 781–787.

7.6. *PCI Design Handbook: Precast Prestressed Concrete*, Prestressed Concrete Institute, Chicago, 1978.

7.7. A. E. Naaman and A. Siriaksorn: "Serviceability-Based Design of Partially Prestressed Beams, Part 1: Analytic Formulation," *PCI Journal*, vol. 24, no. 2, March/April 1979, pp. 64–89.

7.8. P. W. Abeles, E. I. Brown, II, and J. O. Woods, Jr.: "Preliminary Report on Static and Sustained Loading Tests," *PCI Journal*, vol. 13, no. 4, August 1968, pp. 12–32.

7.9. "Deflections of Prestressed Concrete Members," Subcommittee 5 of ACI Committee 435, *ACI Journal*, vol. 60, no. 12, December 1963, pp. 1697–1727.

7.10. *Deflections of Concrete Structures*, ACI Special Publication SP-43, American Concrete Institute, Detroit, 1974.

7.11. "Proposed Revisions to ACI Building Code and Commentary Provisions on Deflections," ACI Committee 435, *ACI Journal*, vol. 75, no. 6, June, 1978, pp. 229–238.

7.12. L. D. Martin: "A Rational Method for Estimating Camber and Deflections," *PCI Journal*, vol. 22, no. 1, January/February 1977, pp. 100–108.

7.13. D. E. Branson and K. M. Kripanarayanan: "Loss of Prestress, Camber, and Deflections of Non-Composite and Composite Prestressed Concrete Structures," *PCI Journal*, vol. 16, no. 5, September/October 1971, pp. 22–52.

7.14. A. F. Shaikh and D. E. Branson: "Non-Tensioned Steel in Prestressed Concrete Beams," *PCI Journal*, vol. 15, no. 1, February 1970, pp. 14–36.

7.15. M. K. Tadros, A. Ghali, and W. H. Dilger: "Effect of Non-Prestressed Steel on Prestress Loss and Deflection," *PCI Journal*, vol. 22, no. 2, March/April 1977, pp. 50–63.

7.16. ——— ——— and ———: "Time-Dependent Prestress Loss and Deflection in Prestressed Concrete Members," *PCI Journal*, vol. 20, no. 3, May/June 1975, pp. 86–98.

7.17. A. Siriaksorn and A. E. Naaman: "Serviceability-Based Design of Partially Prestressed Beams, Part 2: Computerized Design and Evaluation of Major Parameters," *PCI Journal*, vol. 24, no. 2, May/June 1979, pp. 40–60.

7.18. J. Muller: "Ten Years of Experience in Precast Segmental Construction," *PCI Journal*, vol. 20, no. 1, January/February 1975, pp. 28–61.

7.19. E. Trost: "Implications of the Superposition Principle in Creep and Relaxation Problems for Concrete and Prestressed Concrete" (in German), *Beton- und Stahlbetonbau* (Berlin-Wilmersdorf), no. 10, 1967, pp. 230–238, 261–269.

7.20. Z. P. Bazant: "Prediction of Creep Effects Using Age-Adjusted Effective Modulus Method," *ACI Journal*, vol. 69, no. 4, April 1972, pp. 212–217.

7.21. A. M. Neville and W. H. Dilger: *Creep of Concrete: Plain, Reinforced and Prestressed*, North Holland Publishing Co., chaps. 17–20, 1970.

7.22. A. R. Anderson: "Engineering for Camber," PCI Journal, vol. 16, no. 2, March/April 1971, pp. 7–9.

PROBLEMS

7.1 Determine the short- and long-term deflection (or camber) of the beam described in Prob. 5.1, assuming the beam is precast prestressed, has an ultimate creep coefficient $C_{CU} = 3$, and will be in an average environment having 70 percent relative humidity. The centroid of the steel has a linear profile with a single draping point at midspan and an eccentricity of 4 in at the supports. Make any other relevant assumptions if needed.

7.2 Check if the deflections of the beam described in Prob. 5.5 satisfy the ACI and AASHTO limitations on deflection. Assume $C_{CU} = 2.5$, $RH = 50$ percent, and make any other relevant assumption when necessary.

7.3 Going back to Prob. 6.4, determine the deflection at the end D of the cantilever for the design you have achieved. Make any relevant assumptions if necessary.

7.4 Going back to Prob. 4.6 compute the short- and long-term deflections assuming the beam is posttensioned seven days after casting and a parabolic tendon is used with the same eccentricities as arrived at in (g). Make any other relevant assumptions.

7.5 Consider the simple span precast prestressed concrete building member (Fig. P7.5) for which the following information is given: span center to center = 75 ft, live load = 40 psf, $f'_c = 6000$ psi, $f'_{ci} = 4500$ psi; allowable stresses: use ACI code; type of steel strands: $\frac{1}{2}$-in diameter, 270 ksi ultimate strength; area = 0.153 in^2 per strand, f_{pJ} at jacking = 190 ksi, f_{pi} at transfer = 175 ksi, $f_{pe} = 155$ ksi; section properties: $h = 32$ in, $A_c = 538$ in^2, $I = 49,329$ in^4, $y_b = 23.44$ in, $y_t = 8.56$ in, $Z_b = 2105$ in^3, $Z_t = 5763$ in^3, $w_G = 0.560$ klf.

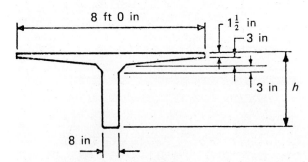

Figure P7.5

(a) Determine the minimum required prestressing force at midspan and the corresponding number of strands. Use the closest higher integer number. Select an acceptable strand layout and determine the actual eccentricities of the prestressing force at midspan and support. Assume a steel profile with a single draping point.

(b) Calculate the ultimate resisting moment of the midspan section according to the ACI procedure.

(c) Check shear requirements and determine the amount and spacing of stirrups along the span.

(d) Compute deflections and check deflection limitations according to the ACI code. (In estimating long-term deflection, assume $C_{CU} = 2$ and $RH = 40$ percent.)

(e) Compute the long-term deflection using the incremental time-step technique suggested in the text.

Make any other assumptions needed to complete the design.

Construction of a 20 million gallon (757,000 m³) prestressed concrete tank with precast walls pretensioned vertically and posttensioned circumferentially—Riverton Height's reservoir, Washington. (*Courtesy ABAM Engineers, Inc.*)

CHAPTER 8

PRESTRESS LOSSES

8.1 SOURCES OF LOSS OF PRESTRESS

The stress in the tendons (hence the prestressing force) of a prestressed concrete member continuously decreases with time. The total stress reduction during the lifespan of the member is called "total loss of prestress." As pointed out in Sec. 1.3, the total loss of prestress was the primary factor that hindered the early development of prestressed concrete. It is essential to estimate the magnitude of the total loss of prestress, since it leads to the value of the effective prestressing force needed for design.

The total loss of prestress is generally attributed to the cumulative contribution of some or all of the following sources:

1. Elastic shortening. Because the concrete shortens when the prestressing force (in full or in part) is applied to it, the tendons already bonded to the concrete shorten, simultaneously losing part of their stress.
2. Relaxation (or creep) of the stressed tendons. Relaxation is a property of the steel and is described under the heading "Relaxation" in Sec. 2.2.
3. Shrinkage of concrete. The gradual loss, with time, of free water from the concrete, called shrinkage, induces a shortening in the concrete which leads to a loss of stress in the attached tendons.
4. Creep of concrete. Creep is caused by the compressive stresses in the concrete. It induces a shortening strain in the concrete in excess of the elastic strain which increases with time and leads to a loss of stress in the attached tendons.

275

5. Friction. Friction loss occurs during tensioning of posttensioned tendons. It represents the difference in stress between the jacking end of the tendon and a section along the member.
6. Anchorage set. Many posttensioning anchorages of the wedge type require that the wedge "sets in" a certain distance in order to lock the tendon at end of jacking. This set (also called seating or slip) leads to a loss of stress in the tendon.
7. Other factors. These include restraining effects of adjoining elements and temperature effects, if any. As they depend on the type of structure, the corresponding stress loss cannot be covered here.

   Each of the above sources leads to a separate prestress loss in the tendons. Prestress losses occur either instantaneously or with time. Instantaneous losses in pretensioned members are generally reduced to the effect of elastic shortening of the members at transfer (or release) of prestress. In posttensioned members they include, in addition to the partial effect of elastic shortening, the effect of anchorage set and frictional losses between steel and concrete. Time-dependent losses include the effect of relaxation in the steel, as well as the effects of shrinkage and creep in the concrete. They affect all prestressed concrete elements.

   The main sources of prestress losses, the stage at which they occur, and the corresponding designation of stress loss in the tendon are summarized in Table 8.1. Each loss is attributed a literal value with a subscript containing a capital letter representing the particular source. For instance, Δf_{pR} represents the total stress loss in the *prestressing* tendons due to *Relaxation*. Two literal values are associated with time-dependent losses, one nondimensional and the other dimensional. The first value, such as Δf_{pR}, represents the total stress loss due to relaxation at the end

Table 8.1 Sources of prestress losses

| | Stage of occurrence | | Tendon stress loss | |
Source of prestress loss	Pretensioned members	Post-tensioned members	During time interval (t_i, t_j)	Total or during life
Elastic shortening of concrete (ES)	At transfer	At jacking	....	Δf_{pES}
Relaxation of prestressed tendons (R)	Before and after transfer	After transfer	$\Delta f_{pR}(t_i, t_j)$	Δf_{pR}
Shrinkage of concrete (S)	After transfer	After transfer	$\Delta f_{pS}(t_i, t_j)$	Δf_{pS}
Creep of concrete (C)	After transfer	After transfer	$\Delta f_{pC}(t_i, t_j)$	Δf_{pC}
Friction (F)	....	At jacking	....	Δf_{pF}
Anchorage set (A)	....	At transfer	....	Δf_{pA}
Total	Life	Life	$\Delta f_{pT}(t_i, t_j)$	Δf_{pT}

of the life of the structure, while the other, such as $\Delta f_{pR}(t_i, t_j)$, represents the loss during a particular time interval (t_i, t_j).

It is generally assumed that the tendons of a member are tensioned to the same stress and have identical properties. The total loss of stress in the tendons Δf_{pT} is the sum of the total losses due to each source. The magnitude of the total loss of prestress in a tendon varies along the member. In order not to add to an already congested notation, the variable x related to a particular section will not be introduced. It is thus assumed that prestress losses are being computed at a given section, most likely the critical section of interest. As a first approximation concrete gross-sectional properties will be used when needed. Transformed section properties could be used for better accuracy.

The object of this chapter is to show how prestress losses can be numerically determined.

8.2 TOTAL LOSSES IN PRETENSIONED MEMBERS

The total loss of prestress in pretensioned tendons comprises the following terms, each of which describes the total loss due to a particular source:

$$\Delta f_{pT} = \Delta f_{pES} + \Delta f_{pR} + \Delta f_{pS} + \Delta f_{pC} \tag{8.1}$$

The notation is explained in Table 8.1. Because the tendons are tensioned some time before the prestressing force is transferred to the concrete, the total stress loss due to relaxation can be divided into two parts, corresponding to two time intervals, namely:

$$\Delta f_{pR} = \Delta f_{pR}(t_o, t_t) + \Delta f_{pR}(t_t, t_l) \tag{8.2}$$

where t_o = time at jacking
t_t = time at transfer (or at release)
t_l = design life time of the member
$\Delta f_{pR}(t_o, t_t)$ = stress loss due to relaxation between t_o and t_t
$\Delta f_{pR}(t_t, t_l)$ = stress loss due to relaxation between t_t and t_l

Note that the elastic shortening occurs instantaneously at time of transfer t_t.

The total loss of prestress at a given section of a member can be related to the extreme stresses in the tendons, namely:

$$\Delta f_{pT} = f_{pJ} - f_{pe} \tag{8.3}$$

where f_{pJ} = stress in the prestressing tendons at end of jacking (this stress is assumed known as derived from the jacking force and should include the effect of the deflecting device, if any, in pretensioned members)
f_{pe} = effective stress in the tendons after all losses at section considered

The effective stress in the tendons can be derived from Eq. (8.3) as follows:

$$f_{pe} = f_{pJ} - \Delta f_{pT} \tag{8.4}$$

In pretensioned members the stress in the tendons immediately after transfer was defined in Sec. 3.10 and called f_{pi} for use in design. f_{pi} is the initial stress in the steel that the concrete experiences and is used to determine F_i, the initial value of the prestressing force. Although it is mostly used for the critical section (or sections) it can be defined at any section. The value of f_{pi} can be determined from:

$$f_{pi} = f_{pJ} - \Delta f_{pR}(t_o, t_t) - \Delta f_{pES} \qquad (8.5)$$

where Δf_{pES} and $\Delta f_{pR}(t_o, t_t)$ can be obtained from Eqs. (8.11), (8.14), and (8.19). Using Eqs. (8.1) to (8.5) leads to:

$$f_{pi} - f_{pe} = \Delta f_{pR}(t_t, t_l) + \Delta f_{pS} + \Delta f_{pC} \qquad (8.6)$$

Hence, the value of the ratio $\eta = f_{pe}/f_{pi}$ used in design and defined in Sec. 3.10 is directly related to prestress losses. Note that in Eq. (8.6) the reference times for shrinkage and creep may be different from t_o, the reference time for relaxation. For plant precast elements, they are generally taken equal to t_t.

A typical variation of stress in the prestressing tendons of a pretensioned member is schematically shown in Fig. 8.1. In estimating such stress variation a margin of error is generally accepted. Note that, because the sustained load on prestressed members may vary with time, losses can also be treated as limits rather than absolute values, hence leading to a band instead of a single curve.

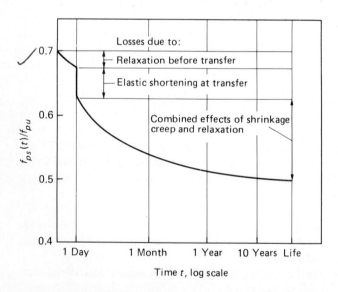

Figure 8.1 Typical variation of steel stress with time for pretensioned members. (precast, steam-cured).

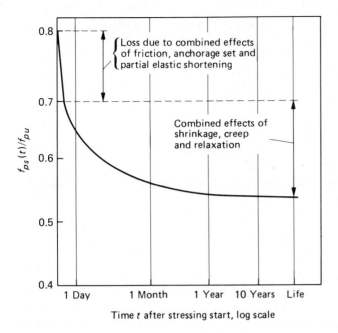

Figure 8.2 **Typical variation of steel stress with time for posttensioned members.**

8.3 TOTAL LOSSES IN POSTTENSIONED MEMBERS

The total loss of prestress in posttensioned tendons is given by:

$$\Delta f_{pT} = \Delta f_{pF} + \Delta f_{pA} + \Delta f_{pES} + \Delta f_{pR} + \Delta f_{pS} + \Delta f_{pC} \qquad (8.7)$$

Equation (8.7) differs from Eq. (8.1) by the two terms related to friction and anchorage set. If all tendons are tensioned simultaneously, no loss due to elastic shortening will occur. However, some elastic shortening loss takes place in an already anchored tendon when another tendon is being stressed. An overall value of Δf_{pES}, varying between zero and fifty percent of that obtained by pretensioning, is generally used throughout for all tendons. The initial design stress f_{pi}, leading to the initial prestressing force F_i, is then given by:

$$f_{pi} = f_{pJ} - \Delta f_{pES} - \Delta f_{pF} \qquad (8.8)$$

and the effective stress f_{pe} is obtained from Eq. (8.4).

A typical variation of stress with time in a posttensioned tendon is schematically shown in Fig. 8.2. Here, too, one has to allow for some margin of error in predicting such variation and a stress band instead of a single curve may be more representative.

8.4 METHODS FOR ESTIMATING PRESTRESS LOSSES

Many methods and design recommendations have been developed to predict prestress losses (Refs. 8.1 to 8.11). For design purposes, they can be classified according to three levels of difficulty and related computational accuracy. The corresponding three categories are identified as follows:

1. Lump sum estimate of the total loss of prestress Δf_{pT}
2. Lump sum estimates of the separate total loss due to each source such as creep and shrinkage, that is Δf_{pC}, Δf_{pS}, etc.
3. Accurate determination of losses by the time-step method

Each category is described in more detail in the following sections.

Note that the accurate determination of losses is more important for some structures than for others. The more accurate and detailed the technique of predicting losses, the more specific is the input information needed. While in the lump sum estimate of total loss (Δf_{pT}) one may only need to know if the structure is pretensioned or posttensioned, in the time-step method data on the basic materials properties, the shape of the structure, the age at loading, environmental conditions, and the like are necessary. A number of references are listed at the end of this chapter to provide a basic understanding of the various properties of concrete that are essential to the application of the time-step method in computing prestress losses (Refs. 8.12 to 8.31). Note that, although the ultimate strength of a prestressed member is insensitive to a misestimate in losses, an underestimate or an overestimate of losses can affect service behavior such as camber, deflection, and cracking.

1. Lump Sum Estimate of Total Losses

Lump sum estimates of the total prestress loss suggested in 1958 by the ACI-ASCE Joint Committee on Prestressed Concrete (Ref. 8.1) were discussed in Sec. 3.10. Table 3.12 gives some estimates found in the AASHTO specifications (Ref. 8.3) for routine design and can be used when applicable.

A study by Hernandez and Gamble (Ref. 8.28) of ten typical precast pretensioned bridge and building members where prestress losses were accurately predicted using the time-step method led to the following results: an average total loss of 45.48 ksi (313.6 MPa) with a standard deviation of 7.53 ksi (52 MPa) and a range of 33.91 to 59.98 ksi (234 to 413.6 MPa). The same beams were studied by Zia et al. (Ref. 8.11) who attempted to provide a simplified procedure giving lump sum estimates of separate losses. In addition, 12 standard double T beams selected from the PCI Design Handbook (Ref. 8.6) were analyzed. The average total loss predicted was around 42.4 ksi (292.3 MPa) with a range of the same order as that observed by Hernandez and Gamble. Based on their evaluation, Zia et al. (Ref. 8.11) recommended maximum limit values of total prestress loss applicable when the tendon stress immediately after anchoring does not exceed $0.83 f_{py}$. These limits are summarized in Table 8.2.

**Table 8.2 Recommended maximum stress loss not includ-
ing friction and anchorage set (Ref. 8.11)**

	Maximum loss† ksi (MPa)	
Type of strand	Normal weight concrete	Lightweight concrete
Stress-relieved strand	50 (345)	55 (380)
Low-relaxation strand	40 (276)	45 (311)

† For $f_{pi} \leq 0.83 f_{py}$.

The PCI Committee on Prestress Losses suggested a method of predicting losses based on the time-step technique (Ref. 8.5). However, simplified lump sum estimates were also given for common cases and are used in part in the PCI Handbook (Ref. 8.6).

It is this author's approach to use as a first approximation for normal weight concrete a lump sum total loss of 45 ksi (311 MPa) for pretensioned members and 35 ksi (242 MPa) for posttensioned members (excluding the effects of friction and anchorage set). An increment of 5 ksi (35 MPa) is used for lightweight concrete and a reduction of 5 ksi (35 MPa) is used for low-relaxation strands. Total loss values are reassessed and revised, if necessary, in the final design.

2. Lump Sum Estimate of Separate Losses

Several methods can be found in the technical literature to estimate the separate contribution of each source of loss (Refs. 8.3, 8.4, 8.8, and 8.11). The total loss of prestress is then obtained by summing up the separate contributions. Many of these methods have merit and can be used. However, because of the implied authority of the code, only the method described in the AASHTO specifications (Ref. 8.3) and recommended by ACI Committee 343 on Bridge Structures (Ref. 8.4) is reproduced in App. E. It requires some basic information on materials properties and environmental conditions, but it is self-explanatory and simple to implement. An example is developed in Sec. 8.9, case **b**.

3. Prestress Losses by the Time-Step Method

The time-step method is an accurate technique to predict losses and involves mostly time-dependent losses due to relaxation of the steel and shrinkage and creep of concrete. Instantaneous losses are computed as for the other methods.

In order to fully understand the time-step method, it is essential to realize that time-dependent losses are also interdependent. Figure 8.3 provides an overall diagram of how total losses are cumulated and illustrates the interfering causes and effects between the different sources of loss of prestress.

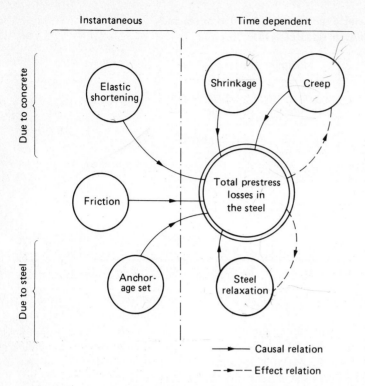

Figure 8.3 Interrelationships of causes and effects between prestress losses.

It is important to understand that the rate of stress loss due to one effect such as relaxation is being continuously altered by changes in the steel stress due to the other effects such as creep and shrinkage. The rate of creep, in its turn, is simultaneously altered by the changes in the steel stress due to relaxation and shrinkage. To account for these interdependent effects with time, a repetitive computation procedure using successive time intervals is used. Generally, the lengths of these time intervals are not equal and vary with the age of the concrete. The stress in the steel at the beginning of any time interval is taken equal to that at the end of the preceding interval. It is assumed constant during a time interval and used to compute incremental losses during that interval. The steel stress at the end of a time interval is obtained by subtracting the incremental stress losses from the stress at the beginning of the interval. More accuracy can be obtained by increasing the number of intervals. The procedure can be and has been computerized for efficiency (Refs. 8.28 to 8.30). The PCI Committee on Prestress Losses suggests a similar procedure based on four time intervals only (Ref. 8.5). A typical example of loss computations by the time-step method with a reasonable number of time intervals is covered in Sec. 8.9, case **a**.

The use of the accurate time-step method to compute prestress losses provides a unique case study to understand the important aspects of the structural behavior of concrete. In the following sections, the background information and the procedures needed to determine the loss of stress due to each source, whether instantaneous or time-dependent, is covered in detail.

8.5 LOSS DUE TO ELASTIC SHORTENING

The loss of stress in the steel due to elastic shortening of the concrete should be determined using the modulus of elasticity of the concrete at the time of prestressing. Its calculated value must be treated differently, however, depending on whether pretensioning or posttensioning is used.

1. Pretensioned Construction

It is assumed that the concrete at the time of transfer is instantaneously compressed and, therefore, experiences a sudden change in strain (Fig. 8.4). As the change in strain in the concrete is the same as that in the steel, the corresponding stress changes in each material are related. The following relation is generally used to estimate the stress loss due to elastic shortening:

$$\Delta f_{pES} = \left(\frac{E_{ps}}{E_{ci}}\right)(f_{cgs})_{F_i+G} = n_{pi}(f_{cgs})_{F_i+G} \tag{8.9}$$

where $(f_{cgs})_{F_i+G}$ represents the stress in the concrete at the centroid of the steel due to the prestressing force and the dead load of the member immediately after transfer. Its value is given by:

$$(f_{cgs})_{F_i+G} = (f_{cgs})_{F_i} + (f_{cgs})_G = \frac{F_i}{A_c} + \frac{F_i e_o^2}{I} - \frac{M_G e_o}{I} \tag{8.10}$$

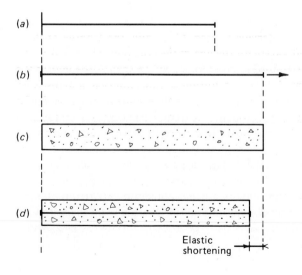

(a)

(b)

(c)

(d)

Elastic
shortening

Figure 8.4 Schematic representation of elastic shortening. (*a*) Free tendon. (*b*) Stressed tendon. (*c*) Unstressed concrete. (*d*) Shortening at transfer.

where $(f_{cgs})_{F_i}$ = stress in the concrete at the centroid of the prestressing steel due to the initial prestressing force F_i

$(f_{cgs})_G$ = stress in the concrete at the centroid of the prestressing steel due to self-weight of member

Note that in order to use Eq. (8.10), $F_i = A_{ps} f_{pi}$ must be known, hence f_{pi} must be estimated and, according to Eq. (8.5), the loss due to elastic shortening must be known. Therefore, Eq. (8.10) must be considered a first approximation in which F_i is estimated a priori and revised thereafter, if needed. The AASHTO specifications suggest an approximate value of $f_{pi} \simeq 0.63 f_{pu}$.

As f_{pi} is theoretically unknown, another way to estimate elastic shortening is to use f_{pJ}, the stress in the tendons immediately after jacking. It is a known quantity. The following relation can easily be derived:

$$\Delta f_{pES} = n_{pi} \left[\frac{f_{pi}}{f_{pJ}} (f_{cgs})_{F_J} + (f_{cgs})_G \right] \qquad (8.11)$$

in which:

$$F_J = A_{ps} f_{pJ} \qquad (8.12)$$

Replacing f_{pi} by its value from Eq. (8.5) in Eq. (8.11) gives:

$$\Delta f_{pES} = n_{pi} \left[\frac{f_{pJ} - \Delta f_{pR}(t_o, t_t) - \Delta f_{pES}}{f_{pJ}} (f_{cgs})_{F_J} + (f_{cgs})_G \right] \qquad (8.13)$$

in which Δf_{pES} is on both sides of the equation. Solving Eq. (8.13) for Δf_{pES} leads to:

$$\Delta f_{pES} = \frac{(f_{cgs})_{F_J}[f_{pJ} - \Delta f_{pR}(t_o, t_t)] + (f_{cgs})_G f_{pJ}}{f_{pJ}/n_{pi} + (f_{cgs})_{F_J}} \qquad (8.14)$$

where f_{pJ} and F_J are assumed given.

An example illustrating the use of Eqs. (8.9) and (8.14) is developed in Sec. 8.9, case **a**. Note that, once Δf_{pES} is obtained, the value of f_{pi} can be calculated from Eqs. (8.5) or (8.8).

2. Posttensioned Construction

The elastic shortening in posttensioned members varies from zero if all the tendons are tensioned simultaneously to half the value obtained for pretensioned members if an infinite number of sequential tensioning steps is used. In slabs where stretching of one tendon will have little effect on nonimmediately neighboring tendons, an elastic shortening value equal to one quarter that of equivalent pretensioned construction is often used.

Theoretically, the loss due to elastic shortening in a given tendon and the corresponding average loss for many tendons can be determined exactly. If N is the number of tendons sequentially tensioned in a beam, the elastic shortening loss in the jth tendon due to the tensioning of tendons $j + 1$ to N is given by the following expression:

$$(\Delta f_{pES})_j = n_{pi} \sum_{k=j+1}^{N} \frac{f_{pk} A_{pk}}{A_c} \left(1 + \frac{e_j e_k}{r^2} \right) \qquad (8.15)$$

where n_{pi} = moduli ratio at time of prestress = E_{ps}/E_{ci}
$\quad f_{pk}$ = stress in the kth tendon immediately after the last tendon has been
$\quad\quad$ tensioned
$\quad A_{pk}$ = area of kth tendon
$\quad e_j, e_k$ = eccentricities of jth and kth tendon with respect to the centroid of
$\quad\quad$ concrete section
$\quad A_c$ = area of concrete section (net area may also be used for better accuracy)
$\quad r = \sqrt{I/A_c}$ radius of gyration of section (net section properties may also be
$\quad\quad$ used for better accuracy)

Note that the last tendon N does not experience any elastic shortening. Its stress f_{pN} is known and equal to f_{pJ}. So, it is recommended to start computation with the last tendon proceeding backward. The average loss by elastic shortening to be applied throughout to the steel (ignoring the effects of friction) is then given by:

$$\Delta f_{pES} = \sum_{j=1}^{N} \frac{A_{pj}(\Delta f_{pES})_j}{\sum_{j=1}^{N} A_{pj}} \tag{8.16}$$

and if the tendons have all the same area

$$\Delta f_{pES} = \frac{1}{N} \sum_{j=1}^{N} (\Delta f_{pES})_j \tag{8.17}$$

8.6 LOSS DUE TO STEEL RELAXATION

At any time t and under normal temperature conditions, the stress in a stress-relieved prestressing steel tensioned initially to f_{pi} can be estimated as described in Sec. 2.2 from the following relation (Ref. 8.16):

$$f_{ps}(t) = f_{pi}\left[1 - \frac{\log(t)}{10}\left(\frac{f_{pi}}{f_{py}} - 0.55\right)\right] \tag{8.18}$$

where t is in hours and is not less than one hour, $\log t$ is to the base 10 and the ratio f_{pi}/f_{py} is not less than 0.55. Note that for $f_{pi} = 0.55 f_{py}$ practically no relaxation losses take place. The definition of f_{pi} in Eq. (8.18) applies to the initial stress in the free tendon and is different from that defined in Eqs. (8.5) and (8.8).

The PCI Committee on Prestress Losses (Ref. 8.5) suggests the use of the following equation to compute the relaxation loss of stress-relieved steel during an interval of time (t_i, t_j):

$$\Delta f_{pR}(t_i, t_j) = f_{ps}(t_i) - f_{ps}(t_j)$$

$$= \frac{f_{ps}(t_i)}{10}\left(\frac{f_{ps}(t_i)}{f_{py}} - 0.55\right)\log\left(\frac{t_j}{t_i}\right) \tag{8.19}$$

where the time is in hours and is not less than one hour and f_{pi}/f_{py} is not less than 0.55. A denominator of 45 instead of 10 is suggested for low-relaxation steel. Note

Table 8.3 Typical time-step computations of relaxation loss for a strand with two fixed ends not attached to a concrete member

	Time t_j at end of interval (t_i, t_j), days						Years	
	1	3	7	30	90	365	5	40
$f_{ps}(t_i)$, ksi	189.0	183.2	181.5	180.2	178.1	176.6	174.8	172.8
$f_{ps}(t_j)$, ksi	183.2	181.5	180.2	178.1	176.6	174.8	172.8	170.4
$\Delta f_{pR}(t_i, t_j)$, ksi	5.77	1.73	1.27	2.11	1.50	1.82	2.01	2.40
$\sum \Delta f_{pR}(t_i, t_j)$, ksi†	5.77	7.5	8.77	10.88	12.38	14.20	16.21	18.6
% Total relaxation	31.0	40.0	47.0	58.5	66.5	76.0	87.0	100.0
% Relaxation left	69.0	60.0	53.0	41.5	33.5	24.0	13.0	0

† Sum over preceding intervals.

that Eq. (8.19) cannot be exactly derived from (8.18) and is valid under normal temperature conditions only.

As mentioned in Sec. 8.4, the stress in the steel is being continuously altered with time and a number of reasonable time intervals are selected to compute relaxation loss during each interval, as well as its cumulative value over preceding intervals. An example of computation of relaxation loss is given in Table 8.3 for a free strand (not associated with a concrete element) at typical time intervals using Eq. (8.19) and assuming the following properties: $f_{pu} = 270$ ksi; $f_{py} = 245$ ksi; $f_{pi} = f_{ps}(1) = 189$ ksi; the starting time of an interval is taken the same as the end time of the immediately preceding interval.

A total relaxation loss of 18.6 ksi or about 10 percent of initial stress is observed. If, instead of using the time-step method over several time intervals, we use only a single interval between one hour and 40 years (350,400 hours), the total loss would be 23.20 ksi or 12.3 percent.

8.7 LOSS DUE TO SHRINKAGE ✓

The shrinkage strain of a given concrete material increases with time up to an ultimate value generally associated with the service life of the member. The ultimate shrinkage strain depends on many parameters, including composition of the concrete mix, characteristics of the aggregates, relative humidity of the environment, curing history, and the like. Some relevant background information on the shrinkage strain of concrete is covered in Sec. 2.3 and Table 2.9, as well as in Refs. 8.12, 8.14, 8.21, 8.24, 8.25, 8.26, and 8.31.

It is observed from Table 2.9 that the shrinkage strain at any time t for a concrete member can be predicted from the following general relation:

$$\varepsilon_S(t) = g(t)\varepsilon_{SU} K_{SH} K_{SS} \tag{8.20}$$

where $g(t)$ = time function

$\quad\quad t$ = time in days after the end of curing

$\quad\quad \varepsilon_{SU}$ = ultimate shrinkage strain of the concrete material

$\quad\quad K_{SH}$ = correction factor which depends on the average relative humidity of the environment where the structure is built

$\quad\quad K_{SS}$ = correction factor which depends on the shape and size of the member.

The time function $g(t)$ tends toward unity when t tends toward t_l, the service life of the structure. Hence, the shrinkage strain of a concrete member at the end of its service life is given by:

$$\varepsilon_S(t_l) \simeq \varepsilon_{SU} K_{SH} K_{SS} \tag{8.21}$$

Because the shrinkage strain is assumed uniform throughout the concrete member and because the prestressing steel undergoes the same change in strain as the concrete, the total stress loss in the prestressing steel due to shrinkage is obtained from:

$$\Delta f_{pS} = E_{ps}\, \varepsilon_S(t_l) = E_{ps} \varepsilon_{SU} K_{SH} K_{SS} \tag{8.22}$$

Typical values of ε_{SU}, K_{SH}, and K_{SS} are given in Sec. 2.3 and Table 2.9. Other recommended values can be found in various references (Refs. 8.5, 8.11, 8.23). In particular, a simple linear relation between water content in the concrete and ε_{SU} is often used, in practice, as an estimate of ultimate shrinkage strain. It is given by:

$$\varepsilon_{SU} = \left[2 + \frac{11}{230}(w - 220) \right] 10^{-4} \tag{8.23}$$

where w is the water content in pounds per cubic yard. Using the SI system, Eq. (8.23) becomes:

$$\varepsilon_{SU} = \left[2 + \frac{11}{1337}(w - 1279) \right] 10^{-4} \tag{8.24}$$

where w is in newtons per cubic meter.

In using the time-step method to compute losses, it is necessary to estimate the shrinkage strain at any time t and the incremental strain during a time interval (t_i, t_j). For this a time function is needed. Two such functions which apply to tests where constant environmental conditions prevail can be used when applicable. The time function suggested by ACI Committee 209 (Ref. 8.12) is described in Table 2.9 and given by:

$$g(t) = \frac{t}{b + t} \tag{8.25}$$

where t is in days and b is a parameter taken equal to 35 for moist-cured concrete and 55 for steam-cured concrete.

Another time function is often used to predict the variation of shrinkage and creep strain of concrete with time. It is given by:

$$g(t) = 0.157 \ln(t) - 0.115 \tag{8.26}$$

where t is the time in days after curing. Equation (8.26) is valid for $t \geq 2.08$ days and $t \leq 1214$ days. For $t = 2.08$, $g(t) = 0$ and for $t = 1214$, $g(t) = 1$. Note that, contrary to Eq. (8.25), this equation is not sensitive to the conditions of curing.

The magnitude of the stress loss in the prestressing steel due to shrinkage over a time interval (t_i, t_j) can be obtained from:

$$\Delta f_{pS}(t_i, t_j) = |f_{ps}(t_j) - f_{ps}(t_i)|$$
$$= E_{ps}\,\varepsilon_{SU}\,K_{SH}\,K_{SS}[g(t_j) - g(t_i)] \tag{8.27}$$

where $f_{ps}(t_i)$ = stress in the prestressing steel at time t_i
$f_{ps}(t_j)$ = stress in the prestressing steel at time t_j

The starting time for the first time interval is assumed to represent the end of the curing period of the concrete.

Using the time function given by Eq. (8.25) in Eq. (8.27) leads to:

$$\Delta f_{pS}(t_i, t_j) = E_{ps}\,\varepsilon_{SU}\,K_{SH}\,K_{SS}\,\frac{b(t_j - t_i)}{(b + t_i)(b + t_j)} \tag{8.28}$$

where b is taken equal to 35 for moist-cured concrete and 55 for steam-cured concrete.

If the time function given by Eq. (8.26) is used in Eq. (8.27), the stress loss due to shrinkage over the time interval (t_i, t_j) takes the following form:

$$\Delta f_{pS}(t_i, t_j) = 0.157 E_{ps}\,\varepsilon_{SU}\,K_{SH}\,K_{SS}\,\ln\left(\frac{t_j}{t_i}\right) \tag{8.29}$$

where the times are in days, are not more than 1214 days, and the starting time of the first interval must be set equal to 2.08 days. The stress loss at 1214 days is assumed the same as the loss at the end of service life.

Table 8.4 Typical time-step computations of shrinkage losses assuming only shrinkage is considered (Eq. 8.28)

	Time t_j at end of interval (t_i, t_j); days						Years	
	1	3	7	30	90	365	5	40 (life)
$f_{ps}(t_i)$, ksi	172	171.69	171.12	170.14	166.84	163.95	161.79	161.02
$\Delta f_{ps}(t_i, t_j)$, ksi	0.31	0.57	0.98	3.30	2.89	2.16	0.77	0.18
$f_{ps}(t_j)$, ksi	171.69	171.12	170.14	166.84	163.95	161.79	161.02	160.84
$\sum \Delta f_{ps}(t_i, t_j)$, ksi†	0.31	0.88	1.86	5.16	8.05	10.21	10.98	11.16
% Total shrinkage	3	8	17	46	72	91	98	100
% Shrinkage left	97	92	83	54	28	9	2	0

† Sum over preceding intervals

Table 8.5 Typical time-step computations of shrinkage losses assuming only shrinkage is considered (Eq. (8.29))

	Time t_j at end of interval (t_i, t_j), days				
	7	30	90	365	1214 max
$f_{ps}(t_i)$, ksi	172	169.87	167.31	165.38	162.92
$\Delta f_{ps}(t_i, t_j)$, ksi	2.13	2.56	1.93	2.46	2.11
$f_{ps}(t_j)$, ksi	169.87	167.31	165.38	162.92	160.81
$\sum \Delta f_{ps}(t_i, t_j)$, ksi†	2.13	4.69	6.62	9.08	11.19
% Total shrinkage	19	42	59	81	100
% Shrinkage left	81	58	41	19	0

† Sum over preceding intervals.

An example is given next, assuming only shrinkage losses take place. In Sec. 8.9, an example where all prestress losses are simultaneously considered is developed. Note that Eqs. (8.25) to (8.29) apply only if constant environmental conditions are assumed.

1. Example

To illustrate the use of Eq. (8.28) or Eq. (8.29), let us compute the shrinkage loss in the tendons of a prestressed concrete member, assuming no other loss is taking place. The following input data are given: $\varepsilon_{SU} = 5 \times 10^{-4}$, $f_{pi} = 172$ ksi, $E_{ps} = 28{,}000$ ksi, relative humidity $H = 60$ percent, volume-to-surface ratio of member $= 1.5$, moist-cured concrete. Using Tables 2.9 and 2.10 leads to: $K_{SH} = 0.80$, $K_{SS} = 1$. Hence, the ultimate shrinkage strain for the member is given by: $\varepsilon_{SU} K_{SH} K_{SS} \simeq 4 \times 10^{-4}$.

Computed shrinkage losses during typical time intervals using either Eq. (8.28) or Eq. (8.29) are summarized in Tables 8.4 and 8.5. The total stress loss due to shrinkage at the end of the service life is the sum of all losses over the various time intervals. It can also be computed from Eq. (8.22).

Note that the rate of loss, using Eq. (8.28), is smaller at early age and higher thereafter than that using Eq. (8.29). Note also that more than 50 percent of the stress loss due to shrinkage generally occurs within a month and more than 80 percent within a year of the end of curing.

8.8 LOSS DUE TO CREEP

The creep strain of concrete depends on many factors, such as time, age at loading, relative humidity, type of aggregates, and the like. Some relevant background information is given in Sec. 2.3 and Table 2.9. Equation (2.7) can be used to predict the creep strain of a concrete material at any time t, assuming constant stress situation:

$$\varepsilon_C(t) = C_C(t)\varepsilon_{ci} \tag{8.30}$$

where ε_{ci} = instantaneous (initial) elastic strain

$C_C(t)$ = creep coefficient at time t

At the end of the service life of the structure assumed subjected to a constant stress, Eq. (8.30) takes the following form:

$$\varepsilon_{CU} = C_{CU}\varepsilon_{ci} \tag{8.31}$$

where ε_{CU} = ultimate creep strain of concrete material

C_{CU} = ultimate creep coefficient

Typical values of ultimate creep coefficients are given in Table 2.11. An average value of 2.35 is suggested by ACI Committee 209 for routine design (Ref. 8.12).

The determination of the loss of prestress due to the creep of concrete is very similar to the procedure followed for shrinkage. The ultimate creep strain ε_{CU} of a given material at the end of its service life is first estimated and correction factors are used to account for member size and shape, age at loading, and relative humidity. Then a relation $g(t)$ is used to assess the creep strain $\varepsilon_C(t)$ at a certain time t. Once the strain is known, the strain change over a time interval is determined and the corresponding stress loss is computed.

Using Eq. (8.31) and Table 2.9 leads to the creep strain at any time t for a concrete structural member:

$$\varepsilon_C(t) = g(t)C_{CU}K_{CH}K_{CA}K_{CS}\varepsilon_{ci} \tag{8.32}$$

where $g(t)$ = time function

K_{CH} = correction factor depending on the average relative humidity of the environment where the structure is built

K_{CA} = age at loading factor

K_{CS} = shape and size factor

The time function suggested by ACI Committee 209 is given by:

$$g(t) = \frac{t^{0.60}}{10 + t^{0.60}} \tag{8.33}$$

where t is in days. A time function identical to that given by Eq. (8.26) for shrinkage is also used for creep.

The creep strain change in the concrete member, assuming constant stress or constant ε_{ci} over an interval of time (t_i, t_j), is given by:

$$\Delta\varepsilon_C(t_i, t_j) = [g(t_j) - g(t_i)]C_{CU}K_{CH}K_{CA}K_{CS}\varepsilon_{ci} \tag{8.34}$$

As steel and concrete are assumed bonded and, hence, undergo the same change in strain, the change in stress in the steel over a time interval (t_i, t_j) can be readily computed (assuming elastic response) from:

$$\Delta f_{pC}(t_i, t_j) = E_{ps}\Delta\varepsilon_C(t_i, t_j)$$

$$= E_{ps}[g(t_j) - g(t_i)]C_{CU}K_{CH}K_{CA}K_{CS}\varepsilon_{ci} \tag{8.35}$$

For a prestressed concrete member, ε_{ci} is the elastic strain in the concrete at the centroid of the prestressing steel due to the applied sustained loading, that is, the combination of prestressing force and dead load. It can be easily computed from the corresponding stress. However, as the prestressing force varies with time due to creep and other losses, the stress also varies and should be computed preferably at the beginning of any time interval considered. Hence, we have:

$$\varepsilon_{ci} = \frac{f_{cgs}(t_i)}{E_c} \tag{8.36}$$

where $f_{cgs}(t_i)$ = stress in the concrete at the centroid of the steel at time t_i due to the prestressing force and dead load

E_c = modulus of elasticity of concrete

Replacing ε_{ci} by its value from Eq. (8.36) in Eq. (8.35) leads to the stress loss in the prestressing steel due to creep over a time interval (t_i, t_j):

$$\Delta f_{pC}(t_i, t_j) = n_p C_{CU} K_{CH} K_{CA} K_{CS} f_{cgs}(t_i)[g(t_j) - g(t_i)] \tag{8.37}$$

in which $n_p = E_{ps}/E_c$, n_{pi} can replace n_p at early ages, and $f_{cgs}(t_i)$ can be computed from:

$$f_{cgs}(t_i) = \frac{f_{ps}(t_i)A_{ps}}{A_c}\left(1 + \frac{e_o^2}{r^2}\right) - \frac{M_D e_o}{I} \tag{8.38}$$

Note that in using Eq. (8.38) in the time-step method to compute prestress losses, the value of the stress in the prestressing steel at time t_i, $f_{ps}(t_i)$, depends on the total losses, including creep, over all time intervals preceding the interval considered. Hence, it can be computed from:

$$f_{ps}(t_i) = f_{pi} - \sum_{\substack{preceding \\ intervals}} \Delta f_{pT}(t_i, t_j) \tag{8.39}$$

An example is given next, assuming only creep losses take place. In Sec. 8.9 an example is developed where all prestress losses are simultaneously considered.

1. Example

In order to illustrate the stress loss computations due to creep, let us consider a pretensioned prestressed concrete tie (tensile member) of section 10×10 in, reinforced with four half-inch diameter stress-relieved strands placed symmetrically at each corner. The following information is provided: $f_{pu} = 270$ ksi, $f_{pi} = 189$ ksi, $A_{ps} = 4 \times 0.153 = 0.612$ in^2, $E_{ps} = 28{,}000$ ksi, $f_c' = 5000$ psi, $f_{ci}' = 4000$ psi, $E_c = 4.287 \times 10^3$ ksi, $E_{ci} = 3.834 \times 10^3$ ksi, $n_{pi} = E_{ps}/E_{ci} = 7.3$, $n_p = E_{ps}/E_c = 6.53$, $C_{CU} = 2$, $e_o = 0$. The concrete is steam-cured, the age at loading t_A is one day, and the relative humidity of the environment is 57 percent. Referring to Tables 2.9 and 2.10, the following values of the various correction factors can be easily derived: $K_{CA} = 1.13$, $K_{CH} = 0.885$, and $K_{CS} = 0.915$ for a volume-to-surface ratio of 2.5 in. The corresponding ultimate creep coefficient for the member is given by:

$$C_{CU} K_{CA} K_{CH} K_{CS} \simeq 1.83$$

Let us assume the fictitious situation where only creep losses are considered, that is, no other effect but creep is taking place. It is also assumed that the prestressed tie is subjected to the effect of the prestressing force only, that is, no external force or moment are present.

For instance, let us use the time function given by Eq. (8.26). The stress loss due to creep over a time interval (t_i, t_j) is obtained from Eq. (8.37) in which $g(t)$ is replaced by its value from Eq. (8.26). Hence:

$$\Delta f_{pC}(t_i, t_j) = n_p \, C_{CU} \, K_{CH} \, K_{CA} \, K_{CS} \, f_{cgs}(t_i) \times 0.157 \ln\left(\frac{t_j}{t_i}\right)$$

that is:

$$\Delta f_{pC}(t_i, t_j) = 1.83 \times 0.157 \times n_p \, f_{cgs}(t_i) \times \ln\left(\frac{t_j}{t_i}\right)$$

A number of time intervals are selected and the corresponding computations of stress loss due to creep are shown in Table 8.6. For the correct use of the above equation, the starting time of the first time interval is set equal to 2.08 days, and the starting time of any other interval is set equal to the end time of the immediately preceding interval. The stress in the steel at the beginning of the first time interval is taken equal to f_{pi}. It is assumed that the initial moduli ratio prevails during the first time interval while n_p prevails during the following intervals. Because there is no bending moment and the prestressing force is concentric, Eq. (8.38) leads to a uniform compressive stress given by:

$$f_{cgs}(t_i) = \frac{f_{ps}(t_i) A_{ps}}{A_c}$$

For the first time interval, it leads to:

$$f_{cgs}(t_i) = \frac{189{,}000 \times 0.612}{100} = 1156.7 \text{ psi}$$

All other computations are summarized in Table 8.6. It can be observed that, similarly to shrinkage, more than 80 percent of creep losses occur within a year of age.

Table 8.6 Typical time-step computations of creep losses assuming only creep is considered (Eq. (8.37))

	Time t_j at end of interval (t_i, t_j), days				
	7	30	90	365	1214 max
Moduli ratio n_{pi} or n_p	7.30	6.53	6.53	6.53	6.53
$f_{ps}(t_i)$, ksi	189	186.06	182.95	180.64	177.73
$f_{cgs}(t_i)$, psi	1156.7	1138.6	1119.6	1105.5	1087.7
$\Delta f_{pC}(t_i, t_j)$, ksi	2.944	3.109	2.307	2.904	2.452
$f_{ps}(t_j)$, ksi	186.06	182.95	180.64	177.73	175.28
$\sum \Delta f_{sC}(t_i, t_j)$, ksi†	2.944	6.053	8.360	11.264	13.72
% Total creep	21	44	61	83	100
% Creep left	79	56	39	17	0

† Sum over preceding intervals.

8.9 EXAMPLE: COMPUTATION OF PRESTRESS LOSSES FOR A PRETENSIONED MEMBER

Compute the prestress losses at midspan at various time intervals for the simply supported pretensioned beam shown in Fig. 8.5, given the following information: $A_{ps} = 2.142$ in^2, $e_o = 14.56$ in, $f_{pJ} = 189$ ksi, $F_J = 404.8$ kips, $f_{py} = 245$ ksi, $f_{pu} = 270$ ksi, $E_{ps} = 28,000$ ksi, $f'_c = 5000$ psi, $f'_{ci} = 4000$ psi, $E_c = 33\gamma_c^{1.5}\sqrt{f'_c}$, $n_{pi} = 7.3$, $n_p = 6.53$, $\gamma_c = 150$ pcf, $\varepsilon_{SU} = 5 \times 10^{-4}$, $C_{CU} = 2$, relative humidity $H = 70$ percent, span $l = 70$ ft, live load $= 320$ plf, dead weight $= 494$ plf, $A_c = 474$ in^2, $I_g = 21,540$ in^4, $r = 6.74$ in, $y_t = 5.94$ in, $y_b = 18.06$ in, $Z_t = 3626$ in^3, $Z_b = 1193$ in^3, $k_t = 2.52$ in, $k_b = 7.65$ in.

(margin notes) $n_f = \dfrac{E_{ps}}{E_c}$ $n_{pi} = \dfrac{E_{ps}}{E_{ci}}$ $r = \sqrt{\dfrac{I}{A_c}}$

Assume that the beam is steam-cured, that release of the steel (stress transfer) is achieved 24 hours after tensioning, and that curing ends at time of transfer.

(a) Compute prestress losses by the time-step method. For the value of the volume-to-surface ratio of the beam of approximately 1.91 in, Table 2.10 leads to the following shape and size correction factors for shrinkage and creep: $K_{SS} \simeq 0.97$ and $K_{CS} \simeq 0.98$.

Table 2.9 is used to determine the humidity correction factors for shrinkage and creep, assuming a relative humidity $H = 70$ percent:

$$K_{SH} = 1.40 - 0.01H = 1.40 - 0.70 = 0.70$$

and

$$K_{CH} = 1.27 - 0.0067H = 1.27 - 0.0067 \times 70 \simeq 0.80$$

Table 2.9 is also used to estimate the age at loading factor for creep, assuming $t_A = 1$ day:

$$K_{CA} = 1.13 t_A^{-0.095} = 1.13$$

A number of appropriate time intervals are selected for the computations, as shown in Table 8.7. A typical example of computation is developed next for each type of prestress loss and the overall results are summarized in Table 8.7.

For instance, the relaxation loss at 24 hours, just before transfer can be determined from Eq. (8.19), using a value of $t_i = t_o = 1$ hour and a corresponding initial stress in the steel equal to f_{pJ}. That is:

$f_{pJ} = f_{ps}(t_i)$

(margin note: A_5 for low relaxation steel:)

$$\Delta f_{pR}(1, 24) = \frac{189}{10}\left(\frac{189}{245} - 0.55\right)\log\left(\frac{24}{1}\right) = 5.776 \text{ ksi}$$

This result can be seen in the second column of Table 8.7.

(margin note)
$$\Delta f_{pR}(t_i, t_j) = \frac{f_{ps}(t_i)}{10}\left(\frac{f_{ps}(t_i)}{f_{py}} - 0.55\right)\log\left(\frac{t_j}{t_i}\right)$$

time in hrs.
and $\neq 1$ hr.

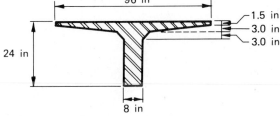

24 in

96 in

1.5 in
3.0 in
3.0 in

8 in

Figure 8.5

Table 8.7 Time-step computations of prestress losses for example beam

	Time t_j at end of time interval (t_i, t_j), days					Years		Cumulative total loss over 40 years
Moduli ratio n_{pi} or n_p	1	7	30	90	365	5	40 (life)	
Moduli ratio n_{pi} or n_p	7.3	7.3	6.53	6.53	6.53	6.53	6.53	....
$f_{ps}(t_i)$, ksi	189.00	169.47	162.82	155.54	149.47	143.74	140.62	....
$f_{cgs}(t_i)$, ksi	....	1.886	1.716	1.530	1.374	1.227	1.147	....
$\Delta f_{pC}(t_i, t_j)$, ksi	....	3.72	3.813	2.89	2.82	1.783	0.914	15.94
$\Delta f_{pS}(t_i, t_j)$, ksi	....	0.904	2.282	2.545	2.36	0.967	0.242	9.30
$\Delta f_{pR}(t_i, t_j)$, ksi	5.776	2.03	1.18	0.629	0.546	0.368	0.304	10.833
Δf_{pES}, ksi	13.754	....	....	....	....	....	....	13.754
$\Delta f_{pT}(t_i, t_j)$, ksi	19.53	6.648	7.27	6.06	5.726	3.119	1.461	49.83
$f_{ps}(t_j)$, ksi	169.47	162.82	155.54	149.47	143.74	140.62	139.16	
$\sum \Delta f_{pT}(t_i, t_j)$, ksi†	19.53	26.19	33.46	39.53	45.26	48.37	49.83	
$\dfrac{\sum \Delta f_{pT}(t_i, t_j)}{\Delta f_{pT}} \times 100\%$†	39%	52%	67%	79%	91%	97%	100%	
$\dfrac{f_{ps}(t_j)}{f_{pu}}$	0.63	0.60	0.58	0.55	0.53	0.52	0.515	

† Sum over preceding intervals.

294

The stress loss due to elastic shortening which takes place at transfer can be determined from Eq. (8.14). For this, the following values are needed:

$$(f_{cgs})_G = -\frac{M_G e_o}{I} = -\frac{3{,}630{,}900 \times 14.56}{21{,}540} \simeq -2454 \text{ psi} = -2.454 \text{ ksi}$$

[handwritten: $F_J = A_{ps} \, f_{pJ}$]

$$(f_{cgs})_{F_J} = \frac{F_J}{A_c} + \frac{F_J e_o^2}{I} = \frac{404{,}800}{474} + \frac{404{,}800 \times 14.56^2}{21{,}540} \simeq 4838 \text{ psi} = 4.838 \text{ ksi}$$

Assuming that $n_{pi} = 7.3$ at time of transfer, Eq. (8.14) gives:

[handwritten: eqn 8.14 page 284]

[handwritten: Δf_{pES} elastic shortening]
$$\Delta f_{pES} = \frac{4.838(189 - 5.776) - 2.454 \times 189}{189/7.3 + 4.838} \simeq 13.754 \text{ ksi}$$

[handwritten right margin:
$$\Delta f_{pES} = \frac{(f_{cgs})_{FJ}\left[f_{pJ} - \Delta f_{pR}(t_o, t_i)\right]}{f_{pJ}/n_{pi} + (f_{cgs})_{FJ}} + (t_{cgs})_G f_{pJ}$$
]

The value of f_{pES} is shown in the second column of Table 8.7. It is used to determine the stress in the prestressing steel immediately after transfer. That stress is taken as the initial stress at the beginning of the second time interval and is given by $f_{ps}(t_i) = f_{ps}(1) = 189 - 5.776 - 13.754 = 169.47$ ksi. Note that the use of the approximate Eq. (8.9) to estimate the loss due to elastic shortening is illustrated in case **b** of this Example.

In order to estimate the loss of stress due to shrinkage over a time interval (t_i, t_j), Eq. (8.28) is used, assuming $b = 55$. It gives:

[handwritten: shrinkage]
$$\Delta f_{ps}(t_i, t_j) = 28{,}000 \times 5 \times 10^{-4} \times 0.70 \times 0.97 \frac{55(t_j - t_i)}{(55 + t_i)(55 + t_j)}$$

[handwritten: eqn 8.28 page 288]
$$= 9.506 \frac{55(t_j - t_i)}{(55 + t_i)(55 + t_j)}$$

Although the reference time for creep and shrinkage is different from the reference time for relaxation, the difference of one day will be neglected in this example to simplify the presentation. Applying the above equation to the second time interval between 1 and 7 days leads to:

$$\Delta f_{ps}(1, 7) = 9.506 \frac{55(7 - 1)}{(55 + 1)(55 + 7)} \simeq 0.904 \text{ ksi}$$

This value is shown in the third column of Table 8.7.

In order to estimate the loss of stress due to creep over a time interval (t_i, t_j), Eq. (8.37) is used. That is: *[handwritten: creep]*

[handwritten: eqn 8.37 page 291]
$$\Delta f_{pc}(t_i, t_j) = n_p \times 2 \times 0.80 \times 1.13 \times 0.98 f_{cgs}(t_i)[g(t_j) - g(t_i)]$$

$$= 1.772 n_p \, f_{cgs}(t_i)[g(t_j) - g(t_i)]$$

where the time function $g(t)$ can be obtained from Eq. (8.33). Hence: *[handwritten: page 290]*

$$\Delta f_{pc}(t_i, t_j) = 1.772 n_p \, f_{cgs}(t_i)\left[\frac{t_j^{0.60}}{10 + t_j^{0.60}} - \frac{t_i^{0.60}}{10 + t_i^{0.60}}\right]$$

Applying the above equation to the second time interval between one and seven days and using n_{pi} instead of n_p, leads to:

$$\Delta f_{pc}(1, 7) = 1.772 \times 7.3 \times f_{cgs}(t_i)\left[\frac{7^{0.60}}{10 + 7^{0.60}} - \frac{1^{0.60}}{10 + 1^{0.60}}\right]$$

$$= 1.971 f_{cgs}(t_i)$$

page 291

$$f_{cgs}(t_i) = \frac{f_{ps}(t_i) \, A_{ps}}{A_c}\left(1 + \frac{e_o^2}{r^2}\right) - \frac{M_D \, e_o}{I}$$

The value of $f_{cgs}(t_i)$ is given by Eq. (8.38):

$$f_{cgs}(1) = \frac{169.47 \times 2.142}{474}\left(1 + \frac{14.56^2}{6.74^2}\right) - \frac{3630.9 \times 14.56}{21,540} = 1.886 \text{ ksi}$$

Hence:

$$\Delta f_{pC}(1, 7) = 1.9704 \times 1.8854 = 3.72 \text{ ksi}$$

This value is shown in the third column of Table 8.7. The procedure is continued for all time intervals in which the stress in the steel at the beginning of an interval is taken equal to that at the end of the preceding interval. The results are summarized in Table 8.7. The total losses obtained after a service life of 40 years amount to 49.83 ksi, that is about 26 percent of the initial stress. Among these losses, the relaxation loss amounts to 10.833 ksi and the creep loss to 15.94 ksi. The above values should be compared with either the lump sum estimate of total losses (say about 45 ksi) or the lump sum estimates of separate losses, such as described next.

Note that, if we did not use the time-step method, the relaxation loss over 40 years (350,400 hours) would have been $\Delta f_{pR}(1,350,400) \simeq 23.20$ ksi, which is more than twice the value obtained.

(b) Compute prestress losses using the AASHTO specifications (App. E). Using the notation of this text, we have:

Elastic shortening:

$$\Delta f_{pES} = n_{pi}(f_{cgs})_{F_i+G}$$

where $(f_{cgs})_{F_i+G}$ is given by Eq. (8.10) in which F_i is replaced by $A_{ps}f_{pi}$ and f_{pi} is estimated equal to $0.63f_{pu}$ or 170.1 ksi. Hence:

$$(f_{cgs})_{F_i+G} = \frac{170.1 \times 2.142}{474} + \frac{170.1 \times 2.142 \times 14.56^2}{21,540} - \frac{3630.9 \times 14.56}{21,540} = 1.90 \text{ ksi}$$

and the loss due to elastic shortening becomes:

$$\Delta f_{pES} = 7.3 \times 1.90 = 13.87 \text{ ksi}$$

Shrinkage:

$$\Delta f_{pS} = 17,000 - 150H = 17,000 - 150 \times 70 = 6500 \text{ psi} = 6.5 \text{ ksi}$$

Creep of concrete:

$$\Delta f_{pC} = 12(f_{cgs})_{F_i+G} - 7(f_{cgs})_{SD}$$

As no superimposed dead load exists in this problem, $(f_{cgs})_{SD}$ is nil and

$$\Delta f_{pC} = 12 \times 1.9 = 22.80 \text{ ksi}$$

Relaxation of prestressing steel:

$$\Delta f_{pR} = 20 - 0.4\Delta f_{pES} - 0.2(\Delta f_{pS} + \Delta f_{pC})$$

$$\Delta f_{pR} = 20 - 0.4 \times 13.87 - 0.2(6.5 + 22.8) = 8.59 \text{ ksi}$$

Total losses:

$$\Delta f_{pT} = 13.87 + 6.5 + 22.80 + 8.59 = 51.76 \text{ ksi}$$

It can be observed that, at least for this particular problem, the total loss of prestress using the approximate code approach is not too different from that obtained by the time-step method.

(c) Determine the value of η for case a. The ratio η is defined by Eq. (3.3) as:

$$\eta = \frac{F}{F_i} = \frac{f_{pe}}{f_{pi}}$$

The stress in the steel immediately after transfer, defined as f_{pi}, is shown in Table 8.7 as $f_{ps}(t_j)$ where $t_j = 1$ day, that is $f_{pi} = 169.47$ ksi. The stress in the steel after all losses or effective prestress f_{pe} is given in Table 8.7 by $f_{ps}(t_j)$ where $t_j = 40$ years or life, that is $f_{pe} = 139.295$ ksi. Hence, $\eta = 139.295/169.47 \simeq 0.822$.

8.10 LOSS DUE TO FRICTION

If in a posttensioned member a tendon is tensioned to a certain stress at the anchorage, a section of this tendon located a distance x from the anchorage will experience a smaller stress. This is due to the stress losses generated by friction between the tendon and its duct between the two sections.

Because the loss due to friction is instantaneous and in order not to encumber the notation, the variable "time" will be dropped. Instead, the variable x, representing distance, will be used. Hence, $f_{ps}(x)$ will represent the stress in the prestressing tendon at distance x from the jacking end at the particular time considered.

1. Analytical Formulation

Frictional losses are generally treated differently if a linear effect or an angular effect is considered. The linear effect is explained by the law of dry friction which leads to a frictional force between two surfaces in contact, directly proportional to the vertical force between them. The angular friction effect is illustrated by the analysis of a belt around a fixed drum (Fig. 8.6). It can be shown that, if a load T_2 is

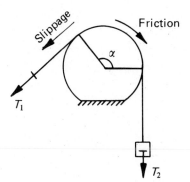

Figure 8.6 Belt friction.

Duct

Tendon

Figure 8.7 Wobbling effect.

applied at one end of the belt, the load T_1, necessary at the other end to just initiate slippage in the direction of T_1, is given by:

$$T_1 = T_2\, e^{\mu\alpha} \qquad (8.40)$$

where μ is the coefficient of static angular friction and α is the angle between T_1 and T_2 in radians.

For posttensioned tendons, the linear effect is called the wobbling effect. It is due to the fact that, for a theoretically linear tendon, the tendon or its duct are never exactly linear (Fig. 8.7), thus leading to some form of friction. It is a general practice to treat the wobbling effect of friction as an additional angular friction instead of a linear one. This leads to the widely accepted formula giving the stress in the steel at a section s (Fig. 8.8) along the tendon as a function of the stress at the anchorage:

$$f_{ps}(s) = f_{pJ}\, e^{-(\mu\alpha + Ks)} \qquad (8.41)$$

where f_{pJ} = stress in the prestressing steel at the jacking end or at the anchorage
μ = coefficient of angular friction
K = wobble coefficient, per unit length
α = change in angle between the force at the anchorage and the force at s, in radians
s = curvilinear length of tendon from anchorage to section s

In general, the curvilinear abscissa s is well approximated for beams by its horizontal projection x which leads to:

$$f_{ps}(x) = f_{pJ}\, e^{-(\mu\alpha + Kx)} \qquad (8.42)$$

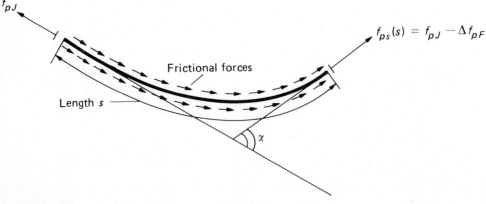

Figure 8.8

The ACI building code accepts the approximation of Eq. (8.42) by the following equation:

$$f_{ps}(x) = \frac{f_{pJ}}{1 + \mu\alpha + Kx} \tag{8.43}$$

provided $\mu\alpha + Kx \leq 0.30$. Note that Eq. (8.43) is derived from Eq. (8.42) by keeping the first term of the Taylor expansion of the exponential $e^{-(\mu\alpha + Kx)}$ and neglecting higher-order terms.

ACI Committee 343 on Bridge Structures (Ref. 8.4) proposes an extension of Eq. (8.42) to account for the horizontal angular change α_h, if any, of the prestressing steel profile, in addition to the vertical angular change α. Although the presence of α_h is treated similarly to the presence of α, it will be assumed in the following analytical developments that only α is present.

The frictional coefficients α and K depend on many factors which include the type of prestressing steel (wires, strands, or bars), the type of duct, and the surface conditions of both (rusted or galvanized).

A range of values recommended by ACI Committee 343 (Ref. 8.4) is given in Table 8.8. Recommended average design values can also be found in the AASHTO specifications (reproduced in App. E) when experimental data for the materials used are not available. Note, however, that in some particular applications more accurate test data may be necessary.

Table 8.8 Friction coefficients for post-tensioned tendons (*Adapted from Ref. 8.4 with permission of the American Concrete Institute*)

Type of tendon	Wobble coefficient K		Curvature coefficient μ
	Per foot	Per meter	Per radian
Tendons in flexible metal sheathing			
Wire tendons	0.0010–0.0015	0.0033–0.0049	0.15–0.25
7-wire strand	0.0005–0.0020	0.0016–0.0066	0.15–0.25
High-strength bars	0.0001–0.0006	0.0003–0.0020	0.08–0.30
Tendons in Rigid Metal Duct			
7-wire strand	0.0002	0.00066	0.15–0.25
Pregreased tendons Wire tendons and			
7-wire strand	0.0003–0.0020	0.0010–0.0066	0.05–0.15
Mastic-Coated Tendons Wire tendons and			
7-wire strand	0.0010–0.0020	0.0033–0.0066	0.05–0.15

2. Graphical Representation

As the stress in the prestressing steel may be needed at several sections along the member, a graphical representation of Eq. (8.42) is often used. As this equation leads to an exponential curve on an arithmetic graph, a semilog graph paper in which a straight line is obtained is generally preferred. This can be observed by taking the natural logarithm of Eq. (8.42), which leads to:

$$\ln (f_{ps}(x)) = \ln (f_{pJ}) - (\mu\alpha + Kx) \qquad (8.44)$$

Note that, if the value of α is zero, constant, or directly proportional to x, Eq. (8.44) leads to a straight line on a semilog graph, where $f_{ps}(x)$ is represented on the log scale and x on the arithmetic scale.

It can be shown that α is proportional to x, as a first approximation, when the profile of the tendon is parabolic, which is generally the case for beams. Also, it can be shown that α is proportional to s in circular tendons. For the cases where α is proportional to x, Eq. (8.42) can be written in the following form (Ref. 8.32):

$$f_{ps}(x) = f_{pJ} e^{-\lambda x} \qquad (8.45)$$

where

$$\lambda x = \mu\alpha + Kx \qquad (8.46)$$

Equation (8.45) leads to a straight line on a semilog graph with a slope equal to $-\lambda$. The value of λ can be obtained from Eq. (8.46) as:

$$\lambda = \mu \frac{\alpha}{x} + K \qquad (8.47)$$

It is important to understand that the value of λ is constant only along a tendon segment whose profile is represented by the same analytic equation. For a profile made out of different parabolas, several values of λ will be obtained for each parabolic segment.

Graphical representations of the variation of steel stress along the tendon are often used in combination with charts which lead to the value of $(\mu\alpha + Kx)$ or $e^{-(\mu\alpha + Kx)}$. However, the ready availability of electronic calculators eliminates the need for such charts. The following example illustrates the graphical representation and the corresponding computations.

3. Example: Loss Computations Due to Friction

A three-span continuous prestressed concrete beam has a tendon profile (Fig. 8.9) made out of a series of parabolic segments. Assuming $\mu = 0.15, K = 7.5 \times 10^{-4}/\text{ft}, f_{pJ} = 189$ ksi, $f_{pu} = 270$ ksi, and $E_{ps} = 28{,}000$ ksi, compute $f_{ps}(x)$ at several sections along the span and plot $f_{ps}(x)$ versus x on an arithmetic scale and on a semilog scale, respectively. Because of symmetry, only half the structure is first considered.

The value of α between the jacking end A and any point x can be obtained from the slopes of the parabolic profile at these points. The equation of a parabola can easily be derived knowing its vertex and one of its points. For instance, the first parabola of the end

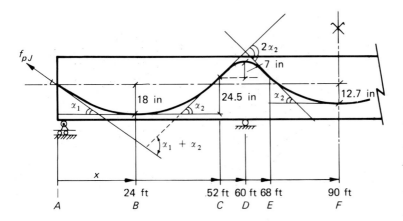

Figure 8.9

span (ABC) leads to the following equation:

$$e_o(x) = 1.5 - \frac{1.5}{24^2}(x-24)^2$$

where $e_o(x)$ and x are in feet and x is taken with respect to the jacking end. The slope at any point with respect to the horizontal axis (x axis) is given by:

$$\frac{\partial e_o(x)}{\partial x} = -\frac{3}{24^2}(x-24)$$

As a first approximation, the slope at any point x is equal to the angle in radians between the tangent at x and the horizontal axis or the tangent at the vertex B. For instance, referring to Fig. 8.9, the angle between the jacking end A and the vertex B is given by the above equation for $x = 0$, that is:

$$\alpha_1 \simeq \tan \alpha_1 = \frac{3 \times 24}{24^2} = 0.125 \text{ rad}$$

Similarly, the angle between the vertex B and point C (Fig. 8.9) is given by:

$$\alpha_2 = \left| -\frac{3}{24^2}(52-24) \right| = 0.1458 \text{ rad}$$

Hence, the angle α between the tangents at two points, such as A and C, is the sum of angles of their tangents with respect to the horizontal axis. That is $\alpha = \alpha_1 + \alpha_2$.

Table 8.9 summarizes the values of α obtained at different sections, chosen to correspond to every end section and vertex of each parabola on the profile, as well as a section equally distant between them. Corresponding stresses in the prestressing steel are determined from Eq. (8.42). It can be seen that, at a section 90 feet from the support, the stress loss due to friction amounts to about 30 ksi or 16 percent of the stress at the jacking end. If only one end of the tendon is jacked, the stress loss at the other end would be about twice that at 90 ft. As such loss is too excessive, both ends of the tendons should be jacked. This

Table 8.9 Variation of steel stress due to friction for the tendon of the Example in Sec. 8.10

Section	A		B		C		D		E		F
x, ft	0	12	24	38	52	56	60	64	68	79	90
α, rad	0	0.0625	0.125	0.1978	0.2706	0.3434	0.4162	0.4890	0.5618	0.6346	0.7074
$\mu\alpha + Kx$	0	0.01838	0.03675	0.05817	0.07959	0.09351	0.1074	0.1213	0.1353	0.1544	0.1736
$e^{-(\mu\alpha + Kx)}$	1	0.9818	0.9639	0.9435	0.9235	0.9107	0.8981	0.8857	0.8735	0.8569	0.8406
$f_{ps}(x)$, ksi	189	185.56	182.17	178.32	174.54	172.12	169.74	167.4	165.1	161.96	158.9
$\Delta f_{ps}(x)$, ksi	0	3.44	6.83	10.68	14.46	16.88	19.26	21.6	23.9	27.5	30.10

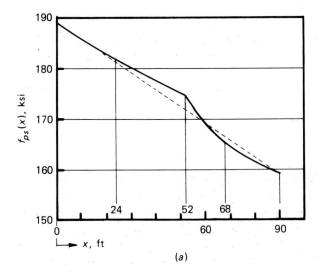

(a)

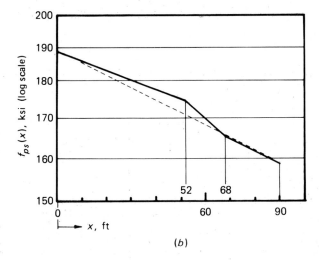

(b)

Figure 8.10 Typical stress reduction in the steel due to friction. (*a*) Arithmetic scale. (*b*) Semilog scale.

would lead to a stress profile symmetrical with respect to the midspan section of the interior span.

Figure 8.10*a* shows the variation of $f_{ps}(x)$ versus x plotted on an arithmetic scale. It can be seen that the curve is made out of three slightly nonlinear segments. Perfectly linear segments are obtained when $f_{ps}(x)$ is plotted versus x on a semilog graph, such as seen in Fig. 8.10*b*.

If we compare Figs. 8.10*a* and *b*, we notice that they are almost identical. This is why an arithmetic scale is often used in practice and straight-line segments are shown for each parabolic portion of the tendon. They are assumed to well approximate the exponential segments. This result greatly simplifies the evaluation of loss due to the anchorage set, as explained next.

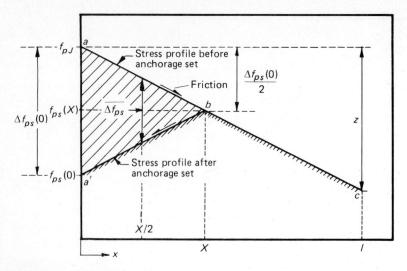

Figure 8.11 Schematic stress variation in the tendons from jacking end before and after anchorage set.

8.11 LOSS DUE TO ANCHORAGE SET

Let us assume that the stress in the tendon just before anchoring is represented by a straight line ac on a semilog paper (Fig. 8.11). Note that the stress at the jacking end is highest and the frictional force along the tendon is pointed toward the positive direction of x. If the anchorage sets a certain distance δ, there is a sudden drop in stress at the jacking end and the frictional force changes direction over a certain distance. The loss of stress at the jacking end is felt with gradually decreasing magnitude in the various sections of the tendons up to a distance X. X is the abscissa of the section of tendon after which the loss due to the anchorage set is not felt and the stress in the steel is the same as just before the anchoring operation. Note also that, because the reversed frictional force depends on the same frictional coefficients as the initial frictional force, the slopes of the two lines ab and $a'b$ (Fig. 8.11), representing the stress variations in the tendon just before and after anchorage set are equal and opposite. The stress profile after anchoring is now represented by $a'bc$. In order to quantitatively define this profile, two unknowns must be determined, namely: the loss of stress at the jacking end $\Delta f_{ps}(0)$ and the distance X.

Due to the symmetry between ab and $a'b$, the loss of stress at the jacking end (Fig. 8.11) can be written as:

$$\Delta f_{ps}(0) = f_{pJ} - f_{ps}(0) = 2[f_{pJ} - f_{ps}(X)] \tag{8.48}$$

where $f_{ps}(X)$ is the stress in the tendon at section X. Notice that the section at $x = 0$ experiences a loss in stress equal to $\Delta f_{ps}(0)$ while the section at X does not experience any loss in stress.

The average loss in stress $\overline{\Delta f_{ps}}$ in the tendon between the jacking end and section X can be obtained from the following equation:

$$\overline{\Delta f_{ps}} \, X = \int_0^X \Delta f_{ps}(x) \, dx = 2 \int_0^X (f_{ps}(x) - f_{ps}(X)) \, dx \qquad (8.49)$$

Furthermore, the total shortening of the tendon over a distance X is equal to the anchorage set δ. This corresponds to an average loss in strain equal δ/X. Using Hooke's law, the average loss of stress in the tendon is related to the average loss of strain by:

$$\overline{\Delta f_{ps}} = E_{ps} \frac{\delta}{X} \qquad (8.50)$$

Using Eq. (8.45) in Eq. (8.49) and neglecting higher-order terms in the Taylor expansion of $e^{-\lambda x}$, it can be shown that:

$$\overline{\Delta f_{ps}} \simeq f_{pJ} \lambda X \simeq f_{pJ} - f_{ps}(X) = \frac{1}{2}(f_{pJ} - f_{ps}(0)) = \frac{\Delta f_{ps}(0)}{2} \qquad (8.51)$$

and from Eqs. (8.50) and (8.51) the following relation is obtained:

$$\Delta f_{ps}(0) = 2E_{ps} \frac{\delta}{X} \qquad (8.52)$$

If in Eq. (8.52) we replace the first term by its value from Eq. (8.51), the value of X can be determined by (Ref. 8.32):

$$X = \sqrt{\frac{E_{ps}\,\delta}{f_{pJ}\,\lambda}} \qquad (8.53)$$

Note that the distance X depends both on f_{pJ} and the frictional characteristics of the tendon represented by λ. Note also that Eq. (8.53) is valid, provided λ is constant over the tendon profile considered. Table 8.10 shows typical values of λ for different tendon profiles. Generally different values of λ can be observed for different segments of the tendon and X may not necessarily involve only one segment. It is probably with this in mind that, in design practice (Ref. 8.5), the following formula is used to determine X (Fig. 8.11):

$$X = \sqrt{\frac{E_{ps}\,\delta}{(z/l)}} \qquad (8.54)$$

where l = span or a known distance along the cable (assuming jacking from one
    end only)

z = loss in stress along $l = f_{pJ} - f_{ps}(l)$ which can be computed from
    Eq. (8.42).

Table 8.10 Recommended values of λ and X for typical tendon profiles

Tendon profile	Description	$\lambda = (\mu\alpha + Kx)/x$	X if less than l
Linear	f_{pJ} ... x	K	$X = \sqrt{\dfrac{E_{ps}\delta}{Kf_{pJ}}}$
Parabolic	f_{pJ} ... a ... b	$(2\mu a/b^2 + K)$	$X = \sqrt{\dfrac{E_{ps}\delta}{(2\mu a/b^2 + K)f_{pJ}}}$
Circular	f_{pJ} ... R	$\dfrac{\mu}{R} + K$	$X = \sqrt{\dfrac{E_{ps}\delta}{(\mu/R + K)f_{pJ}}}$
Any shape or combination of shapes (approximate approach)	$f_{ps}(x)$ ... f_{pJ} ... z ... x l	$\lambda = \left(\dfrac{z}{l}\right)\dfrac{1}{f_{pJ}}$	$X = \sqrt{\dfrac{E_{ps}\delta}{(z/l)}}$

It can be shown that Eqs. (8.53) and (8.54) are essentially identical. Note that (z/l) is the slope of line abc (Fig. 8.11) in the semilog scale and is considered a first approximation of the average slope representing stress loss in the arithmetic scale. In practice, one additional difficulty may be encountered. That is, the calculated value of X may be larger than the span length l or may include more than one curvilinear segment of the tendon profile. In order to answer this question, it is convenient to first define the area lost under the stress versus abcissa diagram of the tendon, due to slippage generated by anchorage set.

1. Area Lost

Referring to Eq. (8.52), we can write:

$$\frac{\Delta f_{ps}(0)}{2} X = E_{ps}\delta \qquad (8.55)$$

The first term of Eq. (8.55) is equal to the area of the triangle aba' in Fig. 8.11 (shaded area) and is the area between the stress diagram prior to anchorage set and the stress diagram after anchorage set. It can be described as the area lost due to anchorage set. More generally, this result can be stated as:

> the total area between the stress diagrams in the steel just before and after anchorage set is equal to $E_{ps} \delta$.

It is important to note that the exponential curve representing $f_{ps}(x)$ is generally flat in the portion of interest and is approximated by a straight line. This is why, in practice, the semilog graph representation of $f_{ps}(x)$ is assumed without much attention to the log values and the areas are calculated as if arithmetic scales are present. The following two examples illustrate the use of this result.

2. Example: Loss Due to Anchorage Set

Going back to Sec. 8.10 and Fig. 8.9, let us assume that the following questions are to be answered:

1. Using $f_{pJ} = 189$ ksi, what are the values of X and $\Delta f_{ps}(0)$, assuming $\delta = 0.2$ in?
2. If the code specifications allow a temporary stress at the jacking end of up to $0.80 f_{pu}$ (provided the stress is less than or equal to $0.70 f_{pu}$ after anchorage set), what is the jacking stress you would recommend to the jacking crew? Draw the stress profile before and after anchorage set.
3. Using $f_{pJ} = 189$ ksi, what are the values of X and $\Delta f_{ps}(0)$ if $\delta = 0.5$ in?

The first question can be answered by referring to Eq. (8.53), that is:

$$X = \sqrt{\frac{E_{ps} \delta}{\lambda f_{pJ}}} = \sqrt{\frac{28{,}000 \times 0.2/12}{0.001531 \times 189}} = 40.15 \text{ ft}$$

where λ is obtained from Table 8.10 and is equal to:

$$\lambda = 2\mu \frac{a}{b^2} + K = 2 \times 0.15 \times \frac{18/12}{24^2} + 7.5 \times 10^{-4} = 0.001531$$

Note that the approximate code approach would have led to an almost equal value in this case. That is:

$$X = \sqrt{\frac{E_{ps} \delta}{z/l}} = \sqrt{\frac{28{,}000 \times 0.2/12}{(189 - 174.54)/52}} = 40.96 \text{ ft}$$

The corresponding loss of stress at the jacking end is given by Eq. (8.52):

$$\Delta f_{ps}(0) = 2 \times 28{,}000 \times \frac{0.2/12}{40.15} = 23.25 \text{ ksi}$$

The stress profile in the steel before release is represented in Fig. 8.12 by the segmented line *abcd*. After anchorage set, the new stress profile is given by *febcd*. Note that the highest stress is at point *e* and corresponds to a value of 177.4 ksi. As the code allows for a value of 189 ksi and in order to achieve such a value, one can temporarily overstress at the jacking end. The difference between 189 and 177.4 is 11.6 ksi or about 12 ksi. Therefore, in order to answer question 2, it is proposed to recommend a stress at the jacking end of $(189 + 12) = 201$ ksi, which is less than $0.80 f_{pu} = 216$ ksi. The new stress profile (Fig. 8.12) follows the segmented

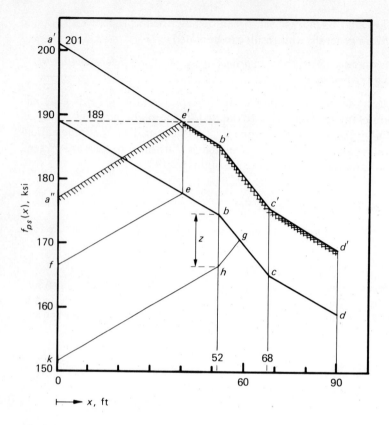

Figure 8.12

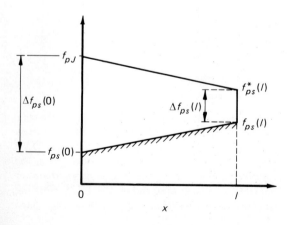

Figure 8.13 Typical tendon stress variation in short members before and after anchorage set.

line $a'b'c'd'$ before release and $a''e'b'c'd'$ after anchorage set. Note that the lines $a'b'c'd'$ and $abcd$ are almost parallel. This observation prompted the use of $aa' \simeq ee' \simeq 12$ ksi in determining the overstress at the jacking end.

Let us answer question 3 where the prestressing system used has an anchorage set $\delta = 0.5$ in. If we apply Eq. (8.53), using the same value of λ over the tendon profile, we would get $X = 63.49$ ft. This means that X is somewhere along the second parabolic portion say between b and c on Fig. 8.12 and Eq. (8.53) is not valid throughout as λ changes from segment ab to bc. The solution can be achieved by using the area lost result described in Sec. 8.11. Referring to Fig. 8.12, the steel stress profile before and after anchorage set is represented by the segmented lines $abgcd$ and $khgcd$, respectively, where two quantities are unknowns, namely X, the abscissa of point g, and $z = \overline{bh}$, the stress difference between b and h. We can write four equations (Fig. 8.12):

$$\text{Area lost} = \frac{(\Delta f_{ps}(0) + z) \times 52}{2} + (X - 52)\frac{z}{2} = E_{ps}\delta$$

$$\frac{z}{2} = f_{ps}(52) - f_{ps}(X) = 174.54 - f_{ps}(X)$$

$$f_{ps}(X) = f_{ps}(52) - (f_{ps}(52) - f_{ps}(68))\frac{(X - 52)}{(68 - 52)}$$

$$= 174.54 - (174.54 - 165.1)\frac{X - 52}{16}$$

$$\Delta f_{ps}(0) = 2(f_{pJ} - f_{ps}(52)) + z = 2(189 - 174.54) + z$$

Their solution leads to: $X = 58.4$ ft, $f_{ps}(0) = 36.6$ ksi, and $z = 7.68$ ksi. The above procedure is exact but cumbersome. If one accepts the practical code approximation of Eq. (8.54), where z is the loss over 90 feet length, the value of X would be:

$$X = \sqrt{\frac{28,000 \times 0.5/12}{(189 - 158.9)/90}} = 59.06 \text{ ft}$$

which is almost the same as the more exact value derived above.

3. Example: Loss Due to Anchorage Set in Short Beams

In the previous example it was assumed that the anchorage set δ and the span length l were such that X was less than l. It may happen that, in short beams, the X calculated from Eq. (8.53) is larger than l. Hence, Eq. (8.53) cannot be used as it becomes obvious that X must equal l. In such a case, the area lost approach explained in Section 8.11 (1. Area Lost) is used to determine the loss of stress $\Delta f_{ps}(0)$ at the jacking end (Fig. 8.13). Equation (8.55) leads to:

$$\text{Area lost} = \left(\frac{\Delta f_{ps}(0) + \Delta f_{ps}(l)}{2}\right)l = E_{ps}\delta$$

and, referring to Fig. 8.13, we have:

$$\Delta f_{ps}(0) = \Delta f_{ps}(l) + 2(f_{pJ} - f^*_{ps}(l))$$

where $f^*_{ps}(l)$ can be computed from Eq. (8.45) in function of f_{pJ}. If a single parabola is involved in the profile such as for a simply supported beam, the solution of the two above equations leads to:

$$\Delta f_{ps}(0) = E_{ps}\frac{\delta}{l} + f_{pJ}\lambda l$$

$$\Delta f_{ps}(l) = E_{ps}\frac{\delta}{l} - f_{pJ}\lambda l$$

Referring to Fig. 8.13, it can be seen from the stress profile after anchorage set, that jacking must be applied only from one end, as there is no benefit in jacking the other end. This may be true in some other cases also, even when X is less than l. In general, a tradeoff must be made between the benefit of additional jacking and that of labor and material cost involved.

8.12 CONCLUDING REMARKS

This chapter has presented a range of methods for calculating prestress losses and their variation with time. If accuracy is not important and only the total losses are needed, the lump sum methods should be used as a first approximation. However, there are cases where the time-step method or similar approaches are necessary. Corresponding computations can be efficiently run, using computers or programmable electronic calculators. A large number of time intervals, as well as time intervals reflecting the real life of the member, can be handled in the repetitive process to improve accuracy. Accuracy in estimating total losses, as well as losses at a certain point in time, may be important. It is generally not desirable to be conservative as service conditions such as camber, deflection, and cracking are directly affected.

REFERENCES

8.1. ACI-ASCE Committee 423 (formerly 323): "Tentative Recommendations for Prestressed Concrete," *ACI Journal*, vol. 54, no. 7, January 1958, pp. 545–578.

8.2. *Building Code Requirements for Reinforced Concrete and Commentary on Building Code Requirements for Reinforced Concrete*, American Concrete Institute (ACI 318–77), Detroit, Michigan, 1977.

8.3. *Standard Specifications for Highway Bridges*, 12th ed., American Association of State Highway and Transportation Officials (AASHTO), Washington, D.C., 1977, 496 pp.

8.4. ACI Committee 343: "Analysis and Design of Reinforced Concrete Bridge Structures," Report 343-77 (also reprinted in ACI Manual of Concrete Practice), chap. 9, American Concrete Institute, Detroit, Michigan, 1977, pp. 73–80.

8.5. PCI Committee on Prestress Losses: "Recommendations for Estimating Prestress Losses," *PCI Journal*, vol. 20, no. 4, July/August 1975, pp. 43–75. Also comments in vol. 21, no. 2, March/April 1976.

8.6. *PCI Design Handbook—Precast and Prestressed Concrete*, 2d ed., Prestressed Concrete Institute, Chicago, Illinois, 1977, pp. 3-28, 3-31.

8.7. D. E. Branson and K. M. Kripanarayanan: "Loss of Prestress, Camber, and Deflection of Non-Composite and Composite Prestressed Concrete Structures," *PCI Journal*, vol. 16, no. 5, September/October 1971, pp. 22–52.

8.8. R. J. Glodowski and J. J. Lorenzetti: "A Method for Predicting Prestress Losses in a Prestressed Concrete Structure," *PCI Journal*, vol. 17, no. 2, March/April 1972, pp. 17–31.

8.9. H. N. Grouni: "Prestressed Concrete: a Simplified Method for Loss Computation," *ACI Journal*, vol. 70, no. 2, February 1973.

8.10. M. K. Tadros, A. Ghali, and W. H. Dilger: "Time-Dependent Prestress Loss and Deflection in Prestressed Concrete Members," *PCI Journal*, vol. 20, no. 3, May/June 1975, pp. 86–98.

8.11. P. Zia, H. K. Preston, N. L. Scott, and E. B. Workman: "Estimating Prestress Losses," *Concrete International*, vol. 1, no. 6, June 1979, pp. 32–38.

8.12. ACI Committee 209: "Prediction of Creep, Shrinkage, and Temperature Effects in Concrete Structures," Special Publication SP-27, Designing for Effects of Creep, Shrinkage, and Temperature, American Concrete Institute, Detroit, Michigan, 1971, pp. 51–93.

8.13. ACI Committee 435: "Deflections of Reinforced Concrete Flexural Members," *ACI Journal*, vol. 63, no. 6, June 1966, pp. 637–674.

8.14. T. C. Hansen and A. H. Mattock: "Influence of Size and Shape of Member on the Shrinkage and Creep of Concrete," *ACI Journal*, vol. 63, no. 2, February 1966, pp. 267–290.

8.15. J. A. Hanson: "Prestress Loss as Affected by Type of Curing," *PCI Journal*, vol. 9, no. 2, April 1967, pp. 69–93.

8.16. D. D. Magura; M. A. Sozen, and C. P. Siess: "A Study of Stress Relaxation in Prestressing Reinforcement," *PCI Journal*, vol. 9, no. 2, April 1964, pp. 13–57.

8.17. W. Podolny, Jr. and T. Melville: "Understanding the Relaxation in Prestressing," *PCI Journal*, vol. 14, no. 4, August 1969, pp. 43–54.

8.18. H. N. Grouni: "Loss of Prestress Due to Relaxation After Transfer," *ACI Journal*, vol. 75, no. 2, February 1978, pp. 64–66.

8.19. E. Schultchen and T. Huang: "Relaxation Losses in 7/16 in. Diameter Special Grade Prestressing Strands," Fritz Engineering Laboratory Report, No. 339.4, Lehigh University, Bethlehem, Pennsylvania, July 1969.

8.20. A. D. Ross: "Creep of Concrete Under Variable Stress," *ACI Journal*, vol. 29, no. 9, March 1958, pp. 739–758.

8.21. I. Lyse: "Shrinkage and Creep of Concrete," *ACI Journal*, vol. 31, no. 8, February 1960, pp. 775–782.

8.22. D. E. Branson and M. L. Christiason: "Time-Dependent Concrete Properties Related to Design Strength and Elastic Properties, Creep, and Shrinkage," ACI-SP 27, Designing for Effects of Creep, Shrinkage, and Temperature, American Concrete Institute, Detroit, 1971, pp. 257–277.

8.23. J. J. Brooks and A. M. Neville: "Estimating Long-Term Creep and Shrinkage From Short-Term Tests," *Magazine of Concrete Research*, vol. 27, no. 90, March 1975.

8.24. B. L. Meyers and D. E. Branson: "Design Aid for Predicting Creep and Shrinkage Properties of Concrete," *ACI Journal*, vol. 69, no. 9, September 1972, pp. 551–555.

8.25. PCI Committee on Design Handbook: "Volume Changes in Precast Prestressed Concrete Structures," *PCI Journal*, vol. 22, no. 5, September/October 1977, pp. 38–53.

8.26. Z. P. Bazant and L. Panula: "Creep and Shrinkage Characterization for Analyzing Prestressed Concrete Structures," *PCI Journal*, vol. 25, no. 3, May/June 1980, pp. 86–122.

8.27. T. Huang: "Prestress Losses in Pretensioned Concrete Structural Members," Report no. 339.9, Fritz Engineering Laboratory, Lehigh University, Bethlehem, Pennsylvania, August, 1973, 100 pp.

8.28. H. D. Hernandez and W. L. Gamble: "Time-Dependent Prestress Losses in Pretensioned Concrete Construction," Structural Research Series no. 417, Civil Engineering Studies, University of Illinois, Urbana, May 1975, 171 pp.

8.29. R. Sinno and H. L. Furr: "Hyperbolic Functions of Prestress Loss and Camber," *Journal of the Structural Division, ASCE*, vol. 96, no. ST4, April 1970, pp. 803–821.

8.30. ———: "Computer Program for Computing Prestress Loss and Camber," *PCI Journal*, vol. 17, no. 5, September/October 1972, pp. 27–38.

8.31. G. E. Troxell, H. E. Davis, and J. W. Kelley: *Composition and Properties of Concrete*, 2d ed., McGraw-Hill Book Company, Inc., New York, 1968.

8.32. T. Huang: "Anchorage Take-up Loss in Post-Tensioned Members," *PCI Journal*, vol. 14, no. 4, August 1969, pp. 30–35.

PROBLEMS

8.1 A pretensioned prestressed concrete tie (tensile member) of cross section shown in Fig. P8.1 is tensioned by four half-inch diameter stress-relieved prestressing strands. Stress transfer is achieved 24 hours after pouring. Steam-curing is used during this initial period. The following information is provided: $f'_c = 5000$ psi; $f'_{ci} = 4000$ psi at transfer; $f_{pJ} = 189$ ksi; $f_{pu} = 270$ ksi; $f_{py} = 230$ ksi; $A_{ps} = 4 \times 0.153 = 0.612$ in^2; $\varepsilon_{SU} = 3.4 \times 10^{-4}$; $C_{CU} = 1.8$; $E_{ps} = 28{,}000$ ksi; relative humidity $H = 50$ percent; normal weight concrete. It is assumed that the tie is not subjected to any bending moment.

Calculate the loss of prestress in the steel after 1, 3, 7, 30, 60, 365 days, 5 years, 40 years. Order your calculations in a table.

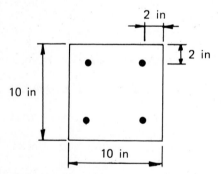

Figure P8.1

8.2 Calculate the total loss of stress in the steel at midspan of the pretensioned beam shown in Fig. P8.2 at different periods of time, given the following information. The beam is steam-cured and the steel is released after 24 hours of tensioning. Use: $\varepsilon_{SU} = 4 \times 10^{-4}$; $C_{CU} = 1.8$; $f_{pu} = 270$ ksi; $f_{pJ} = 190$ ksi; $f_{py} = 243$ ksi; $f'_c = 6000$ psi; $f'_{ci} = 4500$ psi at transfer; $A_{ps} = 0.459$ in^2; $e_o = 10.64$ in; span $l = 40$ ft; $E_{ps} = 27{,}000$ ksi; $\gamma_c = 150$ pcf; $H = 80$ percent.

(a) Calculate the loss of stress in the steel after 1, 7, 14, 30, 60, 365, and 1214 days, using the time function given by Eq. (8.26) in determining shrinkage and creep losses. (b) Determine prestress losses using the procedure recommended in the AASHTO specifications and compare total losses with those obtained in (a).

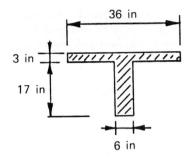

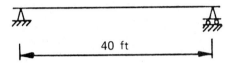

Figure P8.2

8.3 A precast concrete column (Fig. P8.3) is pretensioned in plant with four strands (0.5-in diameter; 270-ksi strength). Three months later it is transported to site, erected, and fixed to its base footing by posttensioning with four unbonded rods (diameter 5/8 in; strength 160 ksi). It is assumed that just before tensioning the rods, the stress in the pretensioned strands is 165 ksi. Posttensioning the rods is achieved in two successive steps, two opposite rods at a time. Assume that the stress in the rods (being tensioned) immediately after anchoring is 112 ksi.

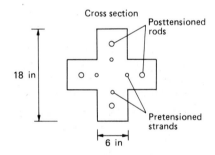

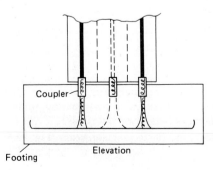

Figure P8.3

Determine the value of the prestressing force just after completion of the posttensioning operation. The following information is given: $f'_c = 6000$ psi; $E_c = 4.70 \times 10^6$ psi; $E_{ps} = 28 \times 10^6$ psi; $n_p = 5.96$; $A_c = 180$ in^2; area of four strands $= 4 \times 0.153 = 0.612$ in^2; area of one posttensioning rod $= 0.307$ in^2; force of one posttensioning rod $= 34.38$ kips at 112 ksi. Use gross area of concrete instead of net area wherever the area is used.

8.4 Show that for a parabolic cable profile, the angle α_x (Fig. P8.4) between the tangents at two points of abscissa 0 and x, respectively, is as a first approximation proportional to x. Derive the corresponding value of λ given in Table 8.10 for a parabolic profile.

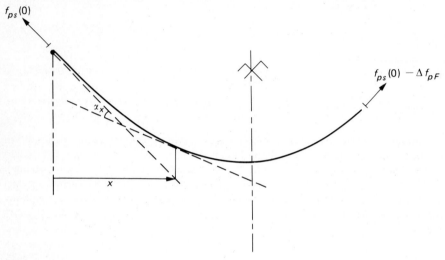

Figure P8.4

8.5 In a nuclear power vessel hoop tendons are posttensioned between buttresses which are 240° apart as shown in Fig. P8.5. The following information is given: $\mu = 0.20$; $K = 5 \times 10^{-4}$/ft; $E_{ps} = 27,000$ ksi; $f_{pJ} =$ stress in the prestressing steel at jacking end just prior to transfer $= 200$ ksi; anchorage set $\delta = 0.5$ in; 1 rad $= 57.29°$.

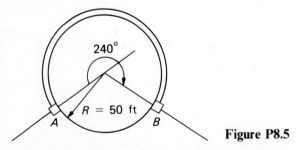

Figure P8.5

(a) Calculate the stress in the steel at the anchorage immediately after transfer (i.e., after anchorage set) assuming circular tendons. (b) If jacking is applied at only one end, what is the stress at the other end? (c) Would you recommend stressing from both ends?

8.6 A simply supported posttensioned prestressed concrete rectangular beam (Fig. P8.6) requires a final prestressing force F at midspan (after all losses) equal to 265 kips. The

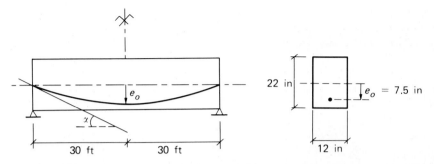

Figure P8.6

following information is given: $f'_c = 5000$ psi; $f'_{ci} = 4000$ psi; $\bar{\sigma}_{ti} = -189$ psi; $\bar{\sigma}_{ci} = 24{,}000$ psi; $\bar{\sigma}_{ts} = -213$ psi; $\bar{\sigma}_{cs} = 2000$ psi; $\gamma_c = 150$ pcf; span $l = 60$ ft; live load $= 200$ plf; friction coefficient $\mu = 0.25$; wobble coefficient $K = 10 \times 10^{-4}$/ft; anchorage set $\delta = 0.2$ in; $E_{ps} = 27{,}000$ ksi; $f_{py} = 245{,}000$ psi; $f_{pu} = 270{,}000$ psi. The steel profile follows a single parabola having its vertex at midspan.

(*a*) Assuming a stress in the prestressing steel at the jack $f_{pJ} = 190$ ksi, determine the stress profile in the steel along the beam before and after anchorage set (assuming stressing from one end only). Should stressing be considered from both ends? (*b*) Assuming overstress is temporarily allowed, determine the new stress profile after anchorage set for which the stress at any point along the beam would be less than or equal to 190 ksi. (*c*) Based on the results derived in (*b*), determine total prestress losses at midspan using the AASHTO specifications and assuming the following: $\varepsilon_{SU} = 4 \times 10^{-4}$ in/in; $C_{CU} = 2$; $H = 70$ percent; moist-cured concrete. (*d*) Determine the effective stress in the steel at midspan after all losses and the required number of strands if the area of one strand is 0.153 in^2.

A composite bridge using precast prestressed concrete beams provides a functional solution to an overcrossing in Walla Walla, Washington. (*Courtesy Prestressed Concrete Institute.*)

CHAPTER 9

ANALYSIS AND DESIGN
OF COMPOSITE BEAMS

9.1 TYPES OF PRESTRESSED CONCRETE COMPOSITE BEAMS

Composite construction implies the use, in a single structure acting as a unit, of different structural elements made with similar or different structural materials. Common examples include steel-concrete beams or columns, orthotropic steel decks, and sandwich panel construction. The two structural materials are utilized in the best possible manner: in a steel-concrete composite beam the steel is designed to carry tension and the concrete to carry compression.

In a composite member where only concrete is used as a material, the concrete is placed in at least two separate stages generally leading to two different unit weights and properties. This is the case for composites made with a precast reinforced or prestressed concrete element combined with a concrete element cast in place at a different time. When the precast element is also prestressed, it essentially simulates the function of the steel in a steel-concrete system; it will carry the tension without cracking.

The extensive application of composite construction to reinforced and prestressed concrete is relatively recent (Refs. 9.1 to 9.3). The use of prestressed concrete composite beams in the United States seriously took off in the early fifties mostly in highway bridges where precast prestressed I beams of the AASHTO or BPR (Bureau of Public Roads) types were used in combination with cast-in-place reinforced concrete slabs. Prestressed composite construction in buildings materialized only a decade later (Ref. 9.4). It has achieved significant growth since. The

317

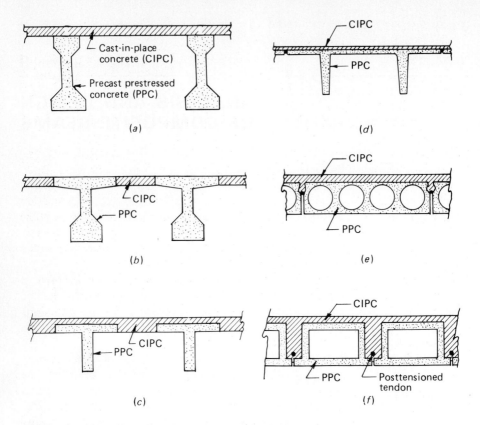

Figure 9.1 Typical cross sections of composite beams.

cause of such development is made clearer in the next section where the advantages of composite construction are stressed.

Typical cross sections of prestressed concrete composite beams are shown in Fig. 9.1. Several alternatives can be identified whereas the cast-in-place (CIP) slab is poured either on top of the precast prestressed concrete (PPC) beams, or in between them, or both. The cast-in-place slab can act as a structural element allowing lateral transfer of loading such as in Figs. 9.1a,b and c or as a topping having a continuous interface with the precast elements (Figs. 9.1d and e). The former case is typical of bridge applications while the latter is typical of building applications. An interesting application is also shown in Fig. 9.1f where the cast-in-place element is itself a beam that may be subsequently reinforced or posttensioned.

This chapter deals mostly with composite beams where a cast-in-place slab is poured on top (or on top and in between) of precast-prestressed girders. The reader should find no conceptual difficulty in extending the approach to any other case.

9.2 ADVANTAGES OF COMPOSITE CONSTRUCTION

Composite construction can result in appreciable savings in construction cost and is likely to remain a major alternative in many applications (Ref. 9.5). Some of its often cited advantages are clarified below:

1. Total construction time is substantially reduced when precast concrete elements are used.
2. Pretensioning in plant is more cost effective than posttensioning on site. Because the precast-prestressed concrete element is factory produced and contains the bulk of the reinforcement, rigorous quality control and higher mechanical properties can be achieved at relatively low cost. The cast-in-place concrete slab does not need to have high mechanical characteristics and thus is very suitable to field conditions. It is not uncommon to have a precast beam with a concrete compressive strength of 7 to 9 ksi (48 to 62 MPa) and a cast-in-place slab with a specified concrete compressive strength of 4 ksi (28 MPa) only.
3. The precast-prestressed concrete units are erected first and can be used to support with no additional scaffolding (or shoring) the forms needed for the cast-in-place slab. In order to save removal time the forms themselves can be replaced by stay-in-place precast (reinforced or prestressed) concrete panels as shown in Fig. 9.2.
4. In addition to its contribution to the strength and stiffness of the composite member, the cast-in-place slab provides an effective means to distribute the loads in the lateral direction. When the slab is used as a topping (Figs. 9.1d and e) and is essentially not reinforced it eliminates the need for costly longitudinal watertight joints between the precast units. However, to improve watertightness, a light steel mesh may be added to the slab, particularly along the joints.
5. The cast-in-place slab can be poured continuously over the supports of precast units placed in series, thus providing continuity to a simple span system (Fig. 9.3). The continuity after slab hardening is assumed to act only for superimposed dead loads and live loads (Ref. 9.6).
6. In principle composite construction offers a very effective means of properly

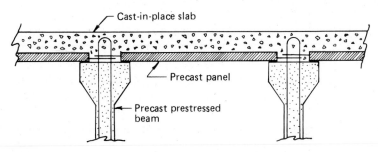

Figure 9.2

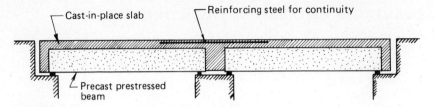

Figure 9.3 Precast prestressed beams made continuous for live loads by a cast-in-place slab.

controlling deflections by properly sizing and designing the precast and cast-in-place elements. However, in practice, it is difficult to accurately predict deflections. This is particularly true for composite construction where two different concretes of different properties and ages at loading are used and where composite action is achieved only after a certain time. In addition, the restraint provided by the cast-in-place slab can cause increases in the prestress losses of the precast element and thus influences time-dependent deflections (Refs. 9.7 and 9.8).

One of the few limitations of composite construction is the size (overall dimensions and weight) of precast prestressed units that can be transported and erected. In the United States precast beams spanning up to 50 m (160 ft) are not uncommon.

9.3 PARTICULAR DESIGN ASPECTS OF PRESTRESSED COMPOSITE BEAMS

The design of prestressed concrete composite beams can be essentially reduced to that of noncomposite beams provided their differences are understood and accounted for. Particular design differences comprise:

1. The loading stages and their relation to whether the beam responds as a composite or noncomposite beam.
2. The transformed effective flange width and corresponding transformed section properties.
3. The horizontal shear at the interface between the precast beam and the cast-in-place slab.

Other design considerations such as flexure by allowable stresses or ultimate strength, shear, torsion, and deflection are to a great extent very similar to those of noncomposite monolithic beams.

Each of the above-mentioned differences and similarities is addressed respectively in the next sections followed by a complete example illustrating how to accommodate them.

Table 9.1 Loading stages in prestressed composite beams during service

| Resisting section | Extreme loading stage | Loading combination | | Incremental maximum external moment |
		Unshored construction	Shored construction	
Precast section	1	$F_i + GP$	$F_i + GP$	....
	2	$(F_i, F) + GP + S + A$	$(F_i, F) + GP$	M_p
Composite section	3	....	$S + A$	....
	4	$SD + L + I$	$S + A + SD + L + I$	M_c

Notation: (F_i, F) = initial prestressing force, effective prestressing force or any value in between; GP = precast girder self-weight; S = cast-in-place slab weight; A = additional weight acting with slab (diaphragms); SD = superimposed dead load; L = live load; I = impact load if any.

9.4 LOADING STAGES, SHORED VERSUS UNSHORED BEAMS

The various loadings affecting composite beams can be separated into two groups, one involving the precast section alone and the other involving the composite section (Table 9.1). Within each group two extreme loadings (a minimum and a maximum loading) can be identified for design. Loadings affecting the precast-prestressed beam are the prestressing force, the self-weight of the beam, and, if the beam is *unshored*, the weight of the cast-in-place slab and other elements such as diaphragms poured at the same time as the slab.

Shoring (Fig. 9.4) implies the use, during slab pouring and curing, of temporary supports or shores under the precast beam and/or the slab, to relieve the beam from supporting the weight of the slab by itself. After hardening the shores are removed, the weight of the slab is released and the beam acting now as a composite beam sustains the additional weight of the slab. Further loadings by superimposed dead loads and live loads are resisted by the composite beam. Differences in loading between the shored and unshored cases are identified in Table 9.1 where the symbols used are explained. As, depending on the loadings, either the precast or the composite section properties apply, the maximum external moment is given for each case. The maximum moment for the precast section is defined by M_p and the additional maximum moment on the composite section by M_c. Final stresses under combined loadings are obtained by superimposing the stresses induced by M_p on the precast section to those induced by M_c on the composite section. Their determination is covered in Sec. 9.7.

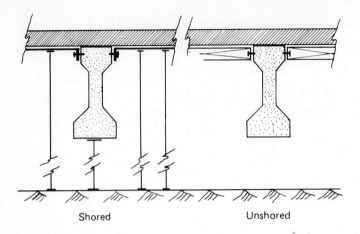

Shored   Unshored

Figure 9.4

9.5 EFFECTIVE AND TRANSFORMED FLANGE WIDTH AND SECTION PROPERTIES

1. Effective Flange Width

Theoretically, when a monolithic beam with an infinitely wide flange is subjected to flexural loading, the compressive stress on the top fiber is not constant but varies transversely across the flange due to what is known as "shear lag" (Fig. 9.5a) (Refs. 9.5 and 9.9). The stress is maximum at the level of the web and decreases away from it. It would be too complex to utilize the theoretical solution to calculate the compressive stresses in the flange. Instead, a simplified approach is adopted in design whereas the flange is assumed to have an equivalent effective width b_e over which the flexural stress is assumed uniform transversely. In principle the

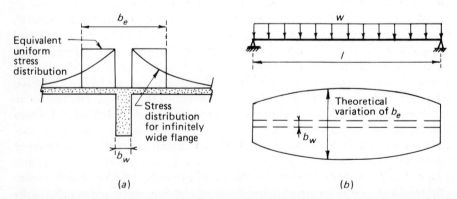

Figure 9.5 Distribution of compressive stresses in flanged beams and effective width b_e.

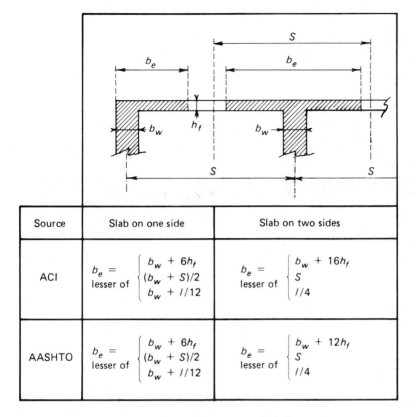

Source	Slab on one side	Slab on two sides
ACI	$b_e =$ lesser of $\begin{cases} b_w + 6h_f \\ (b_w + S)/2 \\ b_w + l/12 \end{cases}$	$b_e =$ lesser of $\begin{cases} b_w + 16h_f \\ S \\ l/4 \end{cases}$
AASHTO	$b_e =$ lesser of $\begin{cases} b_w + 6h_f \\ (b_w + S)/2 \\ b_w + l/12 \end{cases}$	$b_e =$ lesser of $\begin{cases} b_w + 12h_f \\ S \\ l/4 \end{cases}$

Figure 9.6 Design values of effective flange width b_e.

effective flange at a given section is determined to carry the same load as the real one. The effective width itself is shown to vary along the span length (Fig. 9.5b) and depends on the type of loading (concentrated versus uniform loads), the dimensional properties of the section, and even the time-dependent parameters of the concrete material such as creep and deflection.

For practical purposes, the ACI code and AASHTO specifications prescribe design values of the effective flange width b_e for reinforced and prestressed concrete noncomposite and composite flanged beams. They are summarized in Fig. 9.6 for beams with either a slab on one side only, such as end beams, or a slab on two sides such as interior beams. They are to be used in determining the part of the section that resists the loads. Values given by AASHTO apply to nonisolated beams only (that are part of a slab system); if the beam is isolated AASHTO imposes more stringent values of b_e.

In using I type precast sections, the value of b_w in Fig. 9.6 can be replaced by b_v, the width of the section at the interface with the slab. If a precast T section is used, it is safer to use b_w of the T, otherwise engineering judgment should be exercised in determining b_e. Note that some of the single T and double T standard precast-prestressed sections given in the PCI handbook have flange widths larger than the b_e values recommended in Fig. 9.6. However, as they are very efficiently designed and manufactured under plant conditions, such excess is considered acceptable.

The British code CP-110 recommends an effective flange width not to exceed the lesser of (1) the width of the web plus one-fifth the distance between points of zero moments, or (2) the actual width of the flange. For a continuous beam the distance between points of zero moments can be taken as 70 percent of the clear span.

2. Transformed Flange Width

A prestressed composite beam is assumed to behave elastically under service load. The strain distribution along the entire section due to a bending moment is assumed linear and the stresses are calculated from the strains using Hooke's law. As two different materials are involved, the precast beam and the cast-in-place slab, and as each has a different modulus of elasticity, different stresses are generated in each material for the same strain. To account for this difference in design one of the two elements is transformed into a fictitious element having the same modulus of elasticity as the other one. This is done using the concept of transformed section generally applied to the cast-in-place slab. The slab section of depth h_f and width b_e (Fig. 9.7) is transformed into an equivalent section having same depth h_f and a transformed width b_{tr}. It can be easily shown that b_{tr} is given by:

$$b_{tr} = b_e \frac{(E_c)_{\text{CIPC}}}{(E_c)_{\text{PPC}}} = b_e n_c \qquad (9.1)$$

where $(E_c)_{\text{CIPC}}$ = modulus of elasticity of the cast-in-place concrete slab
$\quad (E_c)_{\text{PPC}}$ = modulus of elasticity of the precast-prestressed concrete beam
$\quad\quad n_c$ = ratio of the above moduli

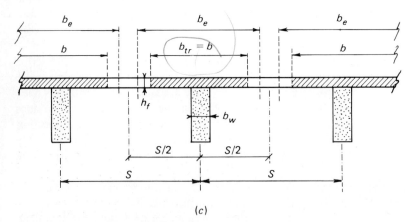

(c)

Figure 9.7 Transformed versus effective flange widths.

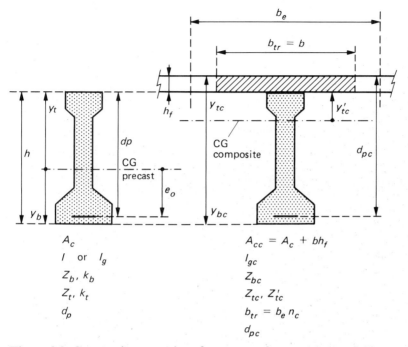

Figure 9.8 Geometric properties of precast and composite sections.

Equation (9.1) ensures that, under bending (linear strain diagram), the total compressive force in the actual slab of width b_e is the same as the force in the transformed slab of width b_{tr} and having same modulus of elasticity as the precast beam.

3. Cross Section Properties

The use of a transformed width b_{tr} leads to a fictitious slab having the same strength and modulus as those of the precast beam. Consequently, the composite section can also be considered transformed into an equivalent monolithic (noncomposite) section having the same strength and modulus as the precast beam. Its geometric properties are determined and directly used in flexural design in the same manner as monolithic noncomposite sections (Fig. 9.8). In order to correlate with the notation used earlier to describe the flange width in general let us call:

$$b = b_{tr} \tag{9.2}$$

thus the area of the composite section is given by:

$$A_{cc} = A_c + h_f b_{tr} = A_c + h_f b = A_c + h_f b_e n_c \tag{9.3}$$

where A_c is the cross-sectional area of the precast concrete beam.

Other section properties listed in Fig. 9.8 are determined according to established procedures. The subscript c is used to differentiate them from those of the precast beam.

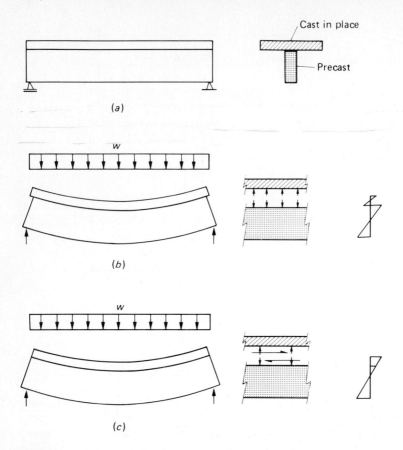

Figure 9.9 (*a*) **Typical precast beam and cast-in-place slab.** (*b*) **Noncomposite action (zero interface shear).** (*c*) **Composite action (full interface shear).**

9.6 INTERFACE SHEAR OR HORIZONTAL SHEAR

1. Evaluation of Horizontal Shear

The success of composite action depends on the shear resistance at the interface between the precast and the cast-in-place element to allow full transfer of stresses. If no shear resistance existed and a load is applied to the composite beam shown in Fig. 9.9a, the slab would slide with respect to the beam (Fig. 9.9b) and the system would act as if two separate elements were used. However, if sufficient shear resistance is provided, the slip between the two elements can be prevented and composite action can be counted on (Fig. 9.9c). Thus a good connection between the two components of the composite system is essential. This can be done by artificially roughening the interface surface, providing a bonding agent, and/or using shear connectors mostly in the form of extended stirrups. The problem of shear transfer in composite beams has generated substantial research work and is relatively well documented (Refs. 9.10 to 9.15).

The horizontal shear stress at the interface between the precast beam and the cast-in-place slab is generated by the loads acting on the composite section only. They are the additional dead load and live load if the precast beam is unshored to which the slab weight is added when the precast beam is shored. The horizontal shear stress due to bending is equal in magnitude to the vertical shear stress (Fig. 9.10) and can be derived for the interface surface as shown in Sec. 6.5 assuming elastic uncracked section:

$$v_h = \frac{Q_c}{b_v I_{gc}} \Delta V$$ (9.4)

where Q_c = first static moment about the centroid of the composite section of the portion of section above the shear plane considered (that is essentially the flange)

b_v = width of shear plane considered (width of interface surface)

I_{gc} = gross moment of inertia of composite section

ΔV = shear force acting on composite section only

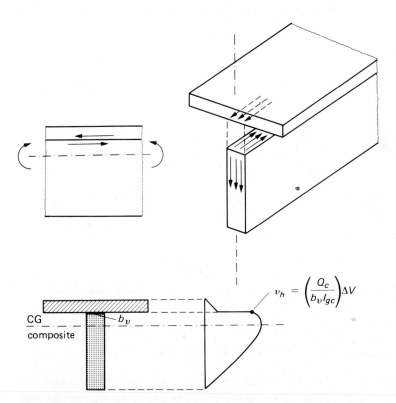

Figure 9.10 Interface horizontal shear stress.

For the same reasons cited in Sec. 6.5 and for design purposes the horizontal shear stress can be taken as:

$$v_h = \frac{\Delta V}{b_v d_{pc}} \qquad (9.5)$$

where d_{pc} is the distance from the extreme compressive fiber of the composite section to the centroid of the prestressing steel. It is also equal to the vertical shear stress generated by the same loading.

If ultimate strength design is considered the use of ΔV in Eq. (9.5) becomes doubtful as flexural cracking (or the neutral axis of bending) can cross the interface and the total shear force will be developed to maintain force equilibrium in the section. Thus at ultimate Eq. (9.5) is replaced by:

$$v_{uh} = \frac{V_u}{b_v d_{pc}} \qquad (9.6)$$

where V_u is the factored total shear force and v_{uh} is the required design horizontal shear strength at the interface. Another way of analyzing the problem is through the shear friction approach described in Sec. 9.6. The ACI code recommends the use in Eq. (9.6) of d_{pc} or $0.8h_c$ whichever is larger. If a partially prestressed precast beam is used d_{pc} should be replaced by the distance to the centroid of the tensile force at ultimate.

2. ACI Code Provisions for Horizontal Shear

In Secs. 17.5.2 and 17.5.3 of the 1977 ACI code, provisions are given to assure proper transfer of shear forces across the interface surface of a composite member. However, in view of possibly inconsistent results, these sections are deleted in the proposed revisions of the code (Ref. 9.16). Hence, an investigation of horizontal shear strength is required in all composite members and is treated similarly to other effects, such as vertical shear, bending, and the like.

Shear transfer resistance. As for vertical shear, torsion, and bending, the design for horizontal shear transfer is based on ultimate strength requirements. Shear forces are used by the ACI code, but shear stresses are preferably used in this text.

The design is based on satisfying the following relation:

$$v_{uh} \le \phi v_{nh} \qquad (9.7)$$

where v_{uh} = factored design shear stress (Eq. (9.6))
v_{nh} = nominal horizontal shear strength
ϕ = strength reduction factor = 0.85

The following limiting values of v_{nh} are recommended by the ACI code:

1. $v_{nh} = 80$ psi (0.55 MPa) when contact surfaces are clean, free of laitance, intentionally roughened to a full amplitude of $\frac{1}{4}$ in (6.3 mm), and ties are not provided.

2. $v_{nh} = 80$ psi (0.55 MPa) when minimum ties are provided in accordance with Eqs. (6.39) and (6.40) and the contact surfaces are not intentionally roughened but are clean and free of laitance. This means that the minimum shear reinforcement in the precast beam is extended to the cast-in-place slab.
3. $v_{nh} = 350$ psi (2.4 MPa) when minimum ties are provided and the contact surfaces are intentionally roughened, clean, and free of laitance.

   Tie spacing shall not exceed four times the least dimension of supported element (mostly slab thickness) or 24 in (60 mm). Ties for horizontal shear may consist of simple bars or wires, multiple leg stirrups, and vertical legs of smooth or deformed welded wire fabric. Ties must be fully anchored into all interconnected elements.

   If v_{uh} exceeds 350 psi (2.4 MPa) at the section considered, the design shall proceed in accordance with Sec. 11.7 of the code which is based on evaluating the shear friction at the interface and is described next. In any case, the value of v_{uh} shall not exceed $0.2f'_c$ or 800 psi (5.52 MPa), whichever is smaller. Otherwise, section dimensions have to be changed.

Tie Reinforcement Using Shear Friction. According to the commentary of the ACI code, the design approach for shear friction assumes that a horizontal crack will form at the interface between the precast and the cast-in-place elements. Shear or tie reinforcement of area A_{vf} is provided to prevent this crack from causing failure. The slip of one crack face with respect to the other is accompanied by crack opening due to the roughness of the crack surfaces. Such opening is sufficient to cause yielding of the reinforcement crossing the crack, hence providing a clamping force $A_{vf}f_y$ across the crack faces. The applied shear is then resisted by a combination of (1) friction between the crack faces, (2) resistance of the protrusions on the crack faces to shearing, and (3) dowel action of the reinforcement. However, in the shear friction method it is assumed that all the shear resistance is due to friction between the crack faces. Hence, artificially high values of the friction coefficient μ are used in order to correlate predicted shear strengths with test results (Refs. 9.10 to 9.15).

   The required area of shear-friction reinforcement is given by:

$$A_{vf} = \frac{H_u}{\phi f_y \mu} \tag{9.8}$$

where H_u = total horizontal shear force at the interface (see explanation below)
   f_y = design yield strength of shear reinforcement not to exceed 60 ksi (414 MPa)
   $\phi = 0.85$
   μ = frictional coefficient to be taken equal to 1.4 when concrete is placed monolithically, 1 when concrete is placed against hardened concrete with intentionally roughened surface, and 0.7 when concrete is placed against steel (a value of 0.4 is recommended in the PCI handbook for concrete placed against smooth concrete)

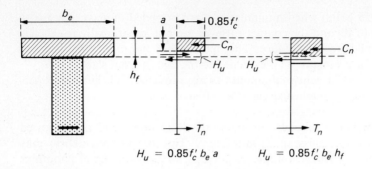

$$H_u = 0.85 f'_c b_e a \qquad H_u = 0.85 f'_c b_e h_f$$

Figure 9.11 Interface shear at ultimate flexural capacity.

The value of H_u can be estimated by summing up the horizontal factored shear stresses over half the span for a simply supported beam and, as a first approximation, over $1/4$ the span for a continuous beam (assuming the points of zero moments are at the $1/4$ span). The corresponding value of A_{vf} obtained is to be distributed over the distance considered.

Alternatively H_u may be determined as the horizontal force needed at the interface to allow the nominal moment resistance of the midspan section or section of maximum moment to develop (Fig. 9.11). Its value depends on whether the neutral axis at ultimate is in the flange or in the web (precast beam). Using the ACI assumptions for the stress block at ultimate leads to:

$$H_u = \text{the smaller of} \begin{cases} 0.85 f'_c b_e a \\ \text{and} \\ 0.85 f'_c b_e h_f \end{cases} \qquad (9.9)$$

where f'_c is for the cast-in-place slab, and a is the depth of the stress block at ultimate.

The author would like to suggest, as another alternative, a slightly different design approach which correlates with the approach generally used for shear and torsion in which the requirements are checked at a given section. Using the same principle of shear friction, the nominal horizontal shear resistance at the interface can be written as:

$$v_{nh} = \frac{A^*_{vf} f_y \mu}{b_v s} \qquad (9.10)$$

where A^*_{vf} = area of shear-friction reinforcement extending across the interface over a distance s

s = spacing of shear-friction reinforcement, assumed vertical and other notations are as defined earlier.

For the purpose of design we want:

$$v_{uh} = \phi v_{nh} \qquad (9.11)$$

Replacing v_{uh} by its value from Eq. (9.6) and v_{nh} by its value from Eq. (9.10) in

Eq. (9.11) leads to:

$$A_{vf}^* = \frac{V_u}{\phi f_y \mu} \frac{s}{d_{pc}}$$

(9.12)

Similarly to shear reinforcement the value of A_{vf}^* is to be determined at each section.

Seldom a horizontal shear stress of more than 350 psi is encountered. Many composite members used in buildings such as T's, double T's and hollow-cored slabs where the cast-in-place "topping" is in contact with the precast unit throughout its width, do not require any tie reinforcement (i.e., the shear stress is less than 80 psi (0.55 MPa)).

An example illustrating some of the computations for horizontal shear transfer is given in Sec. 9.11e.

9.7 FLEXURE: WORKING STRESS ANALYSIS AND DESIGN

Most of the concepts developed in Chap. 4 for the analysis and design of non-composite beams apply to composite beams. The section properties of both the precast beam and the transformed composite beam (Fig. 9.8) are used when needed. The adapted notation and terminology are to be correlated with those of Chap. 4.

1. Extreme Loadings

Whether the beam is shored or not, two extreme loading conditions for the composite system can be identified and essentially bound all others in terms of extreme fiber stresses on the precast beam (Fig. 9.12). The first extreme loading is the initial loading under initial prestressing force and self-weight of the precast beam, that is $(F_i + M_{GP}$ or $M_{min})$; the second extreme loading corresponds to the cumulative effects of the final prestressing force, self-weight of beam, weight of cast-in-place slab, additional dead load, and live load, that is $(F + M_{max})$. In between these two loadings the beam changes from noncomposite to composite and the corresponding bending moments generate different types of stresses. In order to follow the variations of stresses the bending moment M_{max} is broken down into two parts, M_p and M_c, representing the maximum bending moments on the precast and composite sections respectively (see also Table 9.1).

2. Stress Inequality Conditions

Five stress inequality conditions can be written for the composite beam and are of the form:

$$(\text{Actual stress}) \left\{ \begin{matrix} \geq \\ \text{or} \\ \leq \end{matrix} \right\} (\text{allowable stress})$$

(9.13)

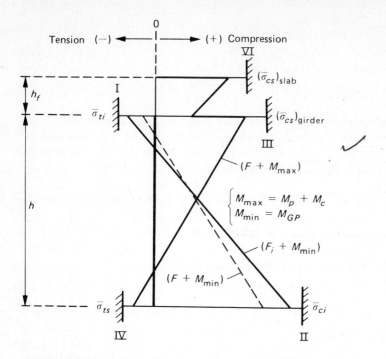

Figure 9.12 Extreme loading's stresses in composite sections.

Four allowable stresses are binding for the precast beam and one for the cast-in-place slab. The two stress inequality conditions for the first extreme loading are identical to the noncomposite beam case (Table 4.2). They are given in Table 9.2 and are numbered I and II. Two other stress inequality conditions (numbered III and IV) can be written for the precast beam under the second extreme loading and one condition can be written for the cast-in-place slab (numbered VI). In addition, the practicality condition which states that the prestressing steel must be placed inside the section also applies here and is given the same number (V) as in Tables 4.2 and 4.3.

Let us develop stress conditions III, IV, and VI for the composite beam assuming unshored construction first. Referring to the stress-versus-loading diagrams of Fig. 9.13 and summing up the stresses on the top fiber of the precast beam we have:

$$\frac{\eta F_i}{A_c}\left(1 - \frac{e_o}{k_b}\right) + (M_{GP} + M_S)/Z_t + (M_{SD} + M_L)\frac{y'_{tc}}{I_{gc}} \leq \bar{\sigma}_{cs} \qquad (9.14)$$

where M_{GP} = bending moment due to self-weight of precast prestressed beam or girder

M_S = bending moment due to cast-in-place slab

M_{SD} = bending moment due to superimposed dead load

M_L = bending moment due to live load

y'_{tc} = distance from centroid of composite beam to top fiber of precast beam = $h - y_{bc}$

In the above equation we have used the final value of the prestressing force. However, the effect of time, when needed, should not be ignored; it is described as "time lapse" in Fig. 9.13 and the notation (F_i, F) is used to remind the reader that any value between F_i and F may also apply.

If the precast beam is shored during casting and curing of the cast-in-place slab, the slab weight will apply only when the section acts as a composite section. Referring to Fig. 9.14, the third stress inequality condition leads to:

$$\frac{\eta F_i}{A_c}\left(1 - \frac{e_o}{k_b}\right) + M_{GP}/Z_t + (M_S + M_{SD} + M_L)\frac{y'_{tc}}{I_{gc}} \leq \bar{\sigma}_{cs} \qquad (9.15)$$

Equations (9.14) and (9.15) can be rewritten in a single form emphasizing the separate effects of the moments acting on the precast and composite beam, respectively, that is:

$$\frac{\eta F_i}{A_c}\left(1 - \frac{e_o}{k_b}\right) + \frac{M_p}{Z_t} + \frac{M_c}{Z'_{tc}} \leq \bar{\sigma}_{cs} \qquad \frac{y'_{tc}}{I_{gc}} = z'_{tc} \quad (9.16)$$

Table 9.2 Useful ways of writing the stress inequality conditions for composite sections

Way	Stress condition	Inequality equation
1	I	$(F_i/A_c)[1 - (e_o/k_b)] + M_{GP}/Z_t \geq \bar{\sigma}_{ti}$
	II	$(F_i/A_c)[1 - (e_o/k_t)] - M_{GP}/Z_b \leq \bar{\sigma}_{ci}$
	III	$(\eta F_i/A_c)[1 - (e_o/k_b)] + M_p/Z_t + M_c/Z'_{tc} \leq \bar{\sigma}_{cs}$
	IV	$(\eta F_i/A_c)[1 - (e_o/k_t)] - M_p/Z_b - M_c/Z_{bc} \geq \bar{\sigma}_{ts}$
2	I	$e_o \leq k_b + (1/F_i)(M_{GP} - \bar{\sigma}_{ti} Z_t)$
	II	$e_o \leq k_t + (1/F_i)(M_{GP} + \bar{\sigma}_{ci} Z_b)$
	III	$e_o \geq k_b + (1/F_i)[(M_p + M_c Z_t/Z'_{tc} - \bar{\sigma}_{cs} Z_t)/\eta]$
	IV	$e_o \geq k_t + (1/F_i)[(M_p + M_c Z_b/Z_{bc} + \bar{\sigma}_{ts} Z_b)/\eta]$
3	I	$F_i \leq (M_{GP} - \bar{\sigma}_{ti} Z_t)/(e_o - k_b)$
	II	$F_i \leq (M_{GP} + \bar{\sigma}_{ci} Z_b)/(e_o - k_t)$
	III	$F = \eta F_i \geq (M_p + M_c Z_t/Z'_{tc} - \bar{\sigma}_{cs} Z_t)/(e_o - k_b)$
	IV	$F = \eta F_i \geq (M_p + M_c Z_b/Z_{bc} + \bar{\sigma}_{ts} Z_b)/(e_o - k_t)$
All	V	$e_o \leq (e_o)_{mp} = y_b - (d_c)_{min}$ = maximum practical eccentricity
	VI	$(M_c/Z_{tc}) \leq (\bar{\sigma}_{cs})_{slab}$

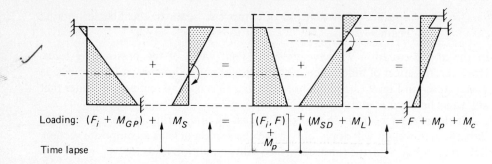

Loading: $(F_i + M_{GP})$ + M_S = $\left[(F_i, F)\right]$ + $(M_{SD} + M_L)$ = $F + M_p + M_c$

$+$

M_p

Time lapse

Figure 9.13 Stress vs. loading diagrams: unshored composite beam.

where M_p = sum of external bending moments acting on the precast beam
$\quad\quad M_c$ = sum of external bending moments acting only on the composite beam
$\quad\quad Z'_{tc} = I_{gc}/y'_{tc}$

The values of M_p and M_c are given for the unshored case by:

$$M_p = M_{GP} + M_S \tag{9.17}$$

$$M_c = M_{SD} + M_L \tag{9.18}$$

and for the shored case by:

$$M_p = M_{GP} \tag{9.19}$$

$$M_c = M_S + M_{SD} + M_L \tag{9.20}$$

Any additional load such as weight of diaphragms or weight of attached elements not accounted for in Eqs. (9.17) to (9.20) can be added to the values of M_p or M_c where appropriate (Table 9.1).

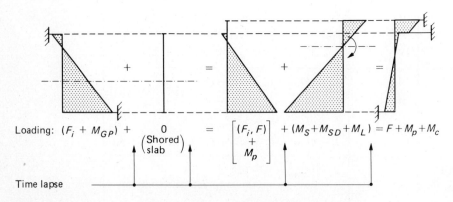

Loading: $(F_i + M_{GP})$ + $\begin{array}{c}0\\ \left(\text{Shored}\atop\text{slab}\right)\end{array}$ = $\left[(F_i, F)\right]$ + $(M_S + M_{SD} + M_L) = F + M_p + M_c$

$+$

M_p

Time lapse

Figure 9.14 Stress vs. loading diagrams: shored composite beam.

Following exactly the same approach, the fourth stress inequality condition related to the bottom fiber of the composite section at maximum load leads to:

$$\frac{\eta F_i}{A_c}\left(1 - \frac{e_o}{k_t}\right) - \frac{M_p}{Z_b} - \frac{M_c}{Z_{bc}} \geq \bar{\sigma}_{ts} \tag{9.21}$$

in which the notation is defined earlier.

The last stress inequality condition states that the maximum compressive stress in the cast-in-place slab at full service load is less than or equal to the allowable compressive stress in the slab, thus:

$$\frac{M_c}{Z_{tc}} \leq (\bar{\sigma}_{cs})_{\text{slab}} \tag{9.22}$$

All the above stress inequality conditions are summarized in Table 9.2 and can be directly correlated with Tables 4.2 and 4.3 given for the noncomposite case. Three different ways of writing these equations are proposed, each emphasizing a particular variable, namely, the stress, the eccentricity e_o, and the prestressing force.

The sixth condition corresponding to Eq. (9.22) is largely satisfied in the great majority of cases. The five others can be treated in a manner exactly similar to their handling in Chap. 4. Thus the working stress analysis and design of composite beams are reduced to those of noncomposite beams and the concepts developed in Chap. 4 apply with the equations given in Table 9.2. Some particular points are addressed next. An example is covered in Sec. 9.11b.

3. Feasible Domain, Limit Kern, Steel Envelopes

The following design similarities exist between a composite and a noncomposite beam:

1. The geometric feasibility domain can be built for the precast prestressed beam using the four stress inequality conditions (Table 9.2, way 2). Thus all the feasible combinations of F_i and e_o can be identified.
2. If a precast beam is properly selected a feasible domain will generally exist. It is then very likely that the prestressing force can be determined from stress condition IV (Table 9.2, way 3) in which e_o is replaced by $(e_o)_{mp}$.
3. Equations giving the limits of the upper and lower limits of the limit kern k_t' and k_b' for the precast beam are the same as Eqs. (4.22) and (4.23) used for the noncomposite beam.
4. It can be easily shown that the envelopes of the prestressing steel in the precast beam are given at any section along the span by:

$$e_{ou} \geq \begin{cases} k_b\left(1 - \frac{\bar{\sigma}_{cs}}{\sigma_g}\right) + \frac{M_p + M_c\, Z_t/Z_{tc}'}{F} & (9.23) \\[2mm] \text{and} \\[2mm] k_t\left(1 - \frac{\bar{\sigma}_{ts}}{\sigma_g}\right) + \frac{M_p + M_c\, Z_b/Z_{bc}}{F} & (9.24) \end{cases}$$

$$e_{o1} \leq \begin{cases} k_b \left(1 - \dfrac{\bar{\sigma}_{ti}}{\sigma_{gi}}\right) + \dfrac{M_{GP}}{F_i} & (9.25) \\[2mm] \text{and} \\[2mm] k_t \left(1 - \dfrac{\bar{\sigma}_{ci}}{\sigma_{gi}}\right) + \dfrac{M_{GP}}{F_i} & (9.26) \end{cases}$$

5. Assuming the cracking moment of the composite beam is larger than M_p, its value (in excess of M_p) for a simply supported beam may be obtained from:

$$\Delta M_{cr} = \frac{Z_{bc}}{Z_b} \left[\eta F_i(e_o - k_t) - M_p\right] - f_r Z_{bc} \qquad (9.27)$$

where f_r is the modulus of rupture of the concrete of the precast beam. Thus the total moment leading to cracking in the composite beam is given by:

$$M_{cr} = \Delta M_{cr} + M_p \qquad (9.28)$$

9.8 FLEXURE: ULTIMATE STRENGTH ANALYSIS AND DESIGN

Failure in composite beams often occurs at the interface between the precast beam and the cast-in-place slab where excessive slip may develop under increased loading. If however an adequate connection is provided to assure shear transfer, prestressed concrete composite beams behave at ultimate much in the same way as noncomposite beams.

The simplest and fastest approach to analyze a composite beam at ultimate is to assume a monolithic section (with transformed slab width) similarly to what was used in the working stress design approach of the previous section. In such a case the analysis and design at ultimate are identical to those of noncomposite beams and the provisions developed in Chap. 5 as well as the design flowchart (Fig. 5.13) apply, provided d_{pc} is used instead of d_p. The following remarks are in order:

1. Let us consider a composite beam made with a precast beam and a cast-in-place slab of effective width b_e larger than b_w or b_v. If at ultimate the neutral axis falls within the slab, the ultimate strength will depend on the compressive strength of the slab and $(f'_c)_{slab}$ can be used in determining the ultimate moment. If the transformed width b_{tr} and the compressive strength of the precast beam are used instead of b_e and $(f'_c)_{slab}$, a slight error is introduced as b_{tr}/b_e is equal to the moduli ratio instead of the strength ratio. However, in general this error introduces in the value of ultimate moment a very insignificant error, as ultimate resistance mostly depends on the strength of the steel and the location of neutral axis, which varies very little.

   If the neutral axis falls in the precast beam and if a linear strain distribution is assumed at ultimate, a stress discontinuity occurs at the interface between the precast beam and the cast-in-place slab (of different properties). This discontinuity is present even when the ACI equivalent stress block at ultimate is used.

Again in such a case it is much simpler (with little loss in accuracy) to assume a monolithic section, equal to the section with transformed slab width, and proceed with the design.

2. When the composite beam is made of a precast T element with a topping such as in Figs. 9.1*d* and *e* (i.e., slab width equal to width of interface) it is recommended to use an equivalent monolithic T section having a flange width equal to the average between b_{tr} of the topping and b of the precast beam, and a flange depth equal to the sum of thicknesses of the topping and the flange of the precast T beam. Such approximation will speed up the design and the accuracy in the prediction of ultimate resistance is not jeopardized.

The above remarks apply to common design situations. A more exact analysis could be applied for special cases. An example illustrating ultimate strength analysis of a composite beam is given in Sec. 9.11**c.**

9.9 DESIGNING FOR SHEAR AND TORSION

The analysis and design for shear and torsion of composite beams are similar to those of monolithic noncomposite beams provided adequate shear transfer is ensured at the interface. The design is based on ultimate strength resistance to factored loads and the provisions given in Chap. 6 for noncomposite beams as well as the design flow charts (Figs. 6.13 and 6.25) apply here. The following design hints are suggested.

1. Use the geometric properties of the composite beam with the transformed cast-in-place slab and the mechanical properties of the precast beam. Thus the values of f'_c and b_w of the precast beam (b_w for T section or smallest width of I section), and d_{pc} of the composite beam (instead of d_p) are used in the equations. If a tapered web is present an average value of b_w can be taken.
2. Consider in the design the total factored shear forces independently of whether they are applied to the precast or composite section.
3. Extend stirrups, when needed, into the cast-in-place slab and anchor them with a 90° bend or equivalent (such as closed hooks). This will generally allow the design to satisfy minimum tie requirements for horizontal shear transfer at the interface and improve shear transfer resistance under repeated loads (Ref. 9.11).

An example of shear design in a composite beam is given in Sec. 9.11.

9.10 DEFLECTIONS

The determination of deflections in prestressed concrete composite beams presents a number of conceptual difficulties that are significantly more complex than those encountered in Chap. 7 for noncomposite beams. The following are the main reasons:

1. The beam acts as a noncomposite beam in its early life and as a composite beam after hardening of the slab.
2. Different time-dependent materials properties for the precast beam and the cast-in-place slab, and different ages at loading are present.
3. Because of 1 and 2 above the time-dependent deflections are substantially influenced by the time and sequence of construction operations. At least two additional time-dependent deflections must be evaluated, one for the precast and one for the composite beam.
4. Once the slab is added, subsequent prestress losses in the precast beam lead to additional long-term deflection whose effect may be magnified by the restraint provided by the slab.
5. Differential shrinkage and creep between the precast beam and the cast-in-place slab influence long-term deflection (Ref. 9.17).

There is no simple answer to such a problem and although some solutions are proposed in the technical literature they still carry a number of simplifying assumptions and inherent uncertainties. References 9.7, 9.8 and 9.18 may be consulted for specific problems. In general, however, the simpler the approach the better it is, provided engineering judgment is exercised.

For common design problems the author recommends the following sequence:

1. Estimate instantaneous elastic deflection due to prestressing force and self-weight of precast beam. For this use an average value of force between F and F_i and an average value of modulus between E_{ci} and E_c, call it $(\Delta_i)_1$.
2. Estimate long-term additional deflection of precast beam up to the time at which the slab is added. Call it $(\Delta_{add})_2 = \lambda_1(\Delta_i)_1$ where λ_1 is a coefficient to be estimated (same as in Chap. 7).
3. Determine instantaneous deflection in precast beam due to slab weight. Call it $(\Delta_i)_3$. Note $(\Delta_i)_3$ is zero if the beam is shored.
4. Compute resulting deflection at time of addition of the slab, that is $\Delta_4 = (\Delta_i)_1 + (\Delta_{add})_2 + (\Delta_i)_3$.
5. Determine instantaneous deflection due to additional dead load acting on composite section, call it $(\Delta_i)_5$.
6. Compute resulting deflection in the composite beam, call it $\Delta_6 = \Delta_4 + (\Delta_i)_5$.
7. Estimate additional long-term deflection in the composite beam, call it $(\Delta_{add})_7 = \lambda_2 \Delta_6$ where λ_2 is a coefficient to be estimated.
8. Compute the instantaneous deflection due to live load using composite section properties.
9. Check if deflection requirements of Table 7.2 are satisfied.

Of course the coefficients λ_1 and λ_2 will depend on the material properties and the time lapse between operations. Some common average values may be derived from Ref. 9.18 and the PCI handbook. Note that the additional long-term deflection in the composite beam that can affect elements attached or connected to the beam is given by Δ_7 only. A typical example of deflection computations for a composite beam is given in Sec. 9.11f.

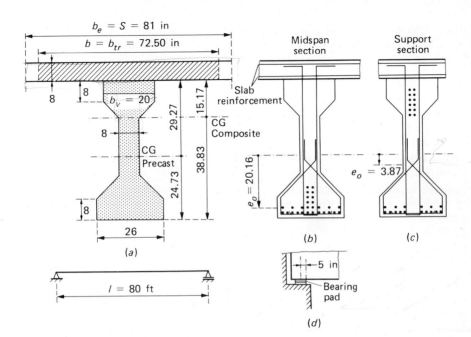

Figure 9.15 Example Beam. (*a*) Cross section. (*b*) Strand layout at midspan. (*c*) Strand layout at supports. (*d*) Bearing pad position at supports.

9.11 EXAMPLE: PRESTRESSED COMPOSITE BRIDGE BEAM

A simply supported typical interior bridge beam with a span of 80 ft (24.4 m) is considered. It consists of a precast pretensioned AASHTO type IV girder and an 8-in (20-cm) thick cast-in-place slab. Unshored construction is assumed. The spacing S between beams is 6.75 ft (2.06 m). Cross-sectional dimensions are given in Fig. 9.15a. The precast beam and the slab are made of normal weight concrete with $\gamma_c = 150$ pcf (23.6 kN/m³). The mechanical properties of the precast beam are: $f'_c = 5000$ psi; $f'_{ci} = 4000$ psi; $E_c = 4.3 \times 10^6$ psi; $E_{ci} = 3.85 \times 10^6$ psi.

For the slab we have $f'_c = 4000$ psi and $E_c = 3.85 \times 10^6$ psi. Thus the modular ratio between slab and beam is $n_c = 3.85/4.3 = 0.895$. The prestressing tendons consist of half-inch-diameter strands with area per strand equal to 0.153 in² and strength $f_{pu} = 270$ ksi. Total prestress losses of 45 ksi are assumed, leading to an effective prestress $f_{pe} = 144$ ksi. The initial stress after transfer, f_{pi}, is estimated at 174 ksi giving $\eta = 144/174 \simeq 0.83$.

Allowable stresses for the precast beam are as follows: $\bar{\sigma}_{ti} = -6\sqrt{f'_{ci}} = -379.5$ psi; $\bar{\sigma}_{ci} = 0.6 f'_{ci} = 2400$ psi; $\bar{\sigma}_{ts} = -3\sqrt{f'_c} = -212.1$ psi; $\bar{\sigma}_{cs} = 0.4 f'_c = 2000$ psi. For the cast-in-place slab we have $\bar{\sigma}_{cs} = 0.4 f'_c = 1600$ psi. It is assumed that the interface surface between the precast beam and the cast-in-place slab is intentionally roughened for proper shear transfer.

The weight of the precast beam is 0.822 klf and the weight of the slab is 0.675 klf. In addition, the slab supports an asphalt topping equivalent to a superimposed dead load of

Table 9.3 Bending moments and shear forces for example beam

Loading	Moments at midspan, kips-ft	Shear forces at first critical section, kips	Moments at first critical shear section, kips-ft	Resisting section
Precast beam	657.5	30.40	94.941	Precast
Cast-on-place slab	540.0	24.98	77.963	section
	$M_p = 1197.5$	55.38	172.904	
Asphalt	199.72	9.25	28.875	Composite
Live load + impact	889.19	46.49	182.7	section
	$M_c = 1088.97$	55.74	211.575	

(handwritten note at left: Calculation page 575)

(handwritten note in table: 182.7 → page 575)

0.25 klf on the composite beam. The live load consists of AASHTO HS20 truck loading augmented by an impact coefficient $I = 0.244$. A wheel distribution factor of $S/5.5$ is used. A summary of corresponding bending moments at midspan and shear forces and moments at $h_c/2$ from the face of the support is given in Table 9.3. As $h_c = 62$ in and a 10-in-wide bearing pad is used at the support (Fig. 9.15d), the first critical section is at $(62 + 10)/2 = 36$ in or 3 ft from the center of the support. The reader may want to refer to Sec. 14.9 (Composite Bridge) to find how the live load values given in Table 9.3 were obtained. If a detailed design is to be pursued, Table 9.3 can be extended to cover every twentieth of the span.

In a preliminary design a smaller precast girder (AASHTO type III, Fig. 14.6) was tried but led to an almost inexistent feasibility domain for F_i and e_o. Thus a type IV girder was selected and should be largely satisfactory. In the following rapid design steps the approach prescribed by ACI and described in this chapter will be followed. It is essentially the same as the AASHTO approach except for some limiting values such as load factors which will be pointed out.

a. Composite section properties. The effective width has to be determined first, that is:

(handwritten: page 323, eqn 9.6, Table Fig)

$$b_e \leq \begin{cases} b_v + 12h_f = 116 \text{ in} \rightarrow \text{(by AASHTO)} \\ S = 81 \text{ in} \\ l/4 = 240 \text{ in} \end{cases}$$

Thus:

$$b_e = 81 \text{ in}$$

and

(handwritten: eqn 9.1)

$$b_{tr} = b_e n_c = 81 \times 0.895 = 72.50 \text{ in}$$

The area of the composite section with transformed slab is given by:

(handwritten: eqn 9.3)

$$A_{cc} = A_c + b_{tr} h_f = 789 + 72.5 \times 8 = 1369 \text{ in}^2$$

The other geometric properties of the precast beam and the composite beam with transformed slab are summarized in Table 9.4.

Table 9.4 Properties of precast and composite sections for example beam

Precast beam	Composite beam (with transformed slab)
$A_c = 789$ in^2	$A_{cc} = 1369$ in
$y_t = 29.27$ in	$y_{tc} = 23.17$ in
$y_b = 24.73$ in	$y_{bc} = 38.83$ in
$h = 54$ in	$h_c = 62$ in
$I_g = 260,700$ in^4	$I_{gc} = 634,229$ in^4
$Z_t = 8907$ in^3	$Z_{tc} = 27,373$ in^3
$Z_b = 10,540$ in^3	$Z_{bc} = 16,334$ in^3
$k_t = -13.36$ in	$Z'_{tc} = 41,808$ in^3
$k_b = 11.29$ in	$h_f = 8$ in
$b_v = 20$ in	$b_e = 81$ in
$b_w = 8$ in	$b = b_{tr} = 72.5$ in

$$\frac{I_{gc}}{y_{tc}} - h_f$$

$$2 + c$$

b. Prestressing force. Let us determine the required prestressing force and check stresses at midspan. As it is very likely that a feasible domain exists, stress condition IV at equality will be used to determine the minimum value of F assuming an eccentricity e_o equal to the maximum practical eccentricity $(e_o)_{mp} = 20.16$ in. Note that the value of $(e_o)_{mp}$ was arrived at after a first computational trial. Using Table 9.2, way 3, we have:

$$F = \eta F_i = \frac{(M_p + M_c Z_b/Z_{bc} + \bar{\sigma}_{ts} Z_b)}{e_o - k_t}$$

$$= \frac{1197.5 \times 12,000 + 1088.97 \times 12,000 \times 10,540/16,334 - 212.1 \times 10,540}{20.16 + 13.36}$$

$$F = 613,567 \text{ lb} = 613.567 \text{ kips}$$

Each strand can carry a final force of $0.153 \times 144 = 22.032$ kips. The minimum required number of strands is:

$$N = \frac{613.567}{22.032} = 27.85$$

It is rounded off to 28 strands thus leading to:

$$\begin{cases} F = 616.90 \text{ kips} \\ N = 28 \text{ strands} \\ F_i = F/\eta = F/0.83 = 743.25 \text{ kips} \\ e_o = 20.16 \text{ in at midspan} \end{cases}$$

Details of strand layout for the midspan and support sections are given in Figs. 9.15b and c. The steel profile is selected to have two draping points each at 30 ft from the support. Twelve strands are draped bringing the eccentricity of the prestressing force at the supports to 3.87 in.

The reader may want to check that the extreme fiber stresses at midspan for the above values of prestressing forces and eccentricity are as follows (use Table 9.2, way 1):

For the precast beam:

$$\sigma_{ti} = 145 \text{ psi} > \bar{\sigma}_{ti} = -379.5 \text{ psi}$$

$$\sigma_{ci} = 1615 \text{ psi} < \bar{\sigma}_{ci} = 2400 \text{ psi}$$

$$\sigma_{cs} = 1312 \text{ psi} < \bar{\sigma}_{cs} = 2000 \text{ psi}$$

$$\sigma_{ts} = -202 \text{ psi} > \bar{\sigma}_{ts} = -212.1 \text{ psi}$$

For the cast-in-place slab:

$$\sigma_{cs} = 478 \text{ psi} < (\bar{\sigma}_{cs})_{\text{slab}} = 1600 \text{ psi}$$

It can be seen from the stresses other than σ_{ts} that the beam has still a lot of capacity and can be used for longer spans.

c. Ultimate moment requirements. Let us determine the ultimate moment resistance ϕM_n. It applies to the composite section. Following the flow chart (Fig. 5.13) we have:

$$A_{ps} = 28 \times 0.153 = 4.284 \text{ in}^2$$

$$\rho_p = \frac{A_{ps}}{bd_{pc}} = \frac{4.284}{72.5 \times 57.43} = 0.00103$$

$$f_{ps} = f_{pu}\left(1 - 0.5\rho_p \frac{f_{pu}}{f'_c}\right) = 270\left(1 - 0.5\rho_p \frac{270}{5}\right) = 262.5 \text{ ksi}$$

$$q = \rho_p \frac{f_{ps}}{f'_c} = 0.05408$$

Check depth of neutral axis at ultimate:

$$c = \frac{1.18qd_{pc}}{\beta_1} = \frac{1.18 \times 0.05408 \times 57.43}{0.80} = 4.58 \text{ in} < h_f = 8 \text{ in}$$

The neutral axis is in the flange. Analyze as a rectangular section with $q < 0.30$:

$$\phi M_n = \phi[A_{ps} f_{ps} d_{pc}(1 - 0.59q)] = 56{,}270 \text{ kips-in} = 4689 \text{ kips-ft}$$

The moment resistance is to be compared with the required strength design moment obtained from factored loads. Referring to Table 9.3 and using the AASHTO specifications for load factors, we have:

$$M_u = 1.3[M_D + \tfrac{5}{3}M_{L+I}] \quad 199.72$$

$$M_u = 1.3[657.5 + 540 + 199.8 + \tfrac{5}{3}889.17] = 3743 \text{ kips-ft}$$

The corresponding value of M_u using ACI is 3468 kips-ft. Thus in any case:

$$M_u < \phi M_n \qquad\qquad\qquad\qquad \text{OK}$$

The cracking moment can be computed from Eqs. (9.27) and (9.28) and leads to:

$$M_{cr} = M_p + \Delta M_{cr} = 1197.5 + 1536.5 = 2734 \text{ kips-ft}$$

Hence

$$1.2M_{cr} = 3281 \text{ kips-ft}$$

and the ultimate resisting moment ϕM_n also satisfies the requirement of being at least 20 percent larger than M_{cr}.

d. Vertical shear. Let us check vertical shear at $h/2$ from face of support.

Referring to Figs. 9.15b and c and the assumed steel profile, it can be shown that e_o at the section under investigation is approximately 5.5 in and that the corresponding value of d_{pc} is equal to either 42.77 in or $0.8h_c = 49.6$ in depending on the formula used. Using AASHTO and Table 9.3, we have:

$$V_u = 1.3[V_D + \tfrac{5}{3}V_{L+I}]$$

$$= 1.3[30.4 + 24.98 + 9.25 + \tfrac{5}{3}46.49]$$

$$= 184.75 \text{ kips}$$

Thus

$$\frac{v_u}{\phi} = \frac{V_u}{\phi b_w d_{pc}} = \frac{184,750}{0.85 \times 8 \times 49.6} = 548 \text{ psi}$$

The properties of the transformed composite section will be used in computing v_c, the shear resistance of concrete. Referring to the flow chart (Fig. 6.13) v_c is to be taken as the smaller of v_{ci} or v_{cw}. v_{ci} is given by Eq. (6.28):

$$v_{ci} = 0.6\sqrt{f_c'} + \frac{V_G}{b_w d_{pc}} + \frac{1}{b_w d_{ps}}\left(\frac{\Delta V_u \times \Delta M_{cr}}{\Delta M_u}\right)$$

The value of V_G applies to the own weight of the composite section, hence (Table 9.3):

$$V_G = V_{GP} + V_S = 30.40 + 24.98 \doteq 55.38 \text{ kips}$$

The values of ΔV_u and ΔM_u apply to the factored superimposed dead load and live load. Hence (Table 9.3):

$$\Delta V_u = 1.3[9.25 + \tfrac{5}{3}(46.49)] = 112.753 \text{ kips}$$

$$\Delta M_u = 1.3[28.875 + \tfrac{5}{3}(182.7)] = 433.39 \text{ kips-ft}$$

The value of ΔM_{cr} is obtained from Eq. (9.27) in which, assuming unshored construction, $M_p = M_{GP} + M_S$. Hence:

$$\Delta M_{cr} = \frac{Z_{bc}}{Z_b}[F(e_o - k_t) - M_p] - f_r Z_{bc}$$

$$= \left[\frac{16,334}{10,540}[616,900(5.5 + 13.36) - 2,074,842] + 424.2 \times 16,334\right]\bigg/12,000$$

$$= 1812 \text{ kips-ft}$$

Thus, the shear-flexure cracking resistance v_{ci} is given by:

$$v_{ci} = 0.6\sqrt{5000} + \frac{55,380}{8 \times 49.6} + \frac{1}{8 \times 49.6}\left(\frac{1812 \times 112,753}{433.39}\right)$$

$$= 42.44 + 139.56 + 1188 \simeq 1370 \text{ psi}$$

The web-shear cracking resistance is given by Eq. (6.30):

$$v_{cw} = 3.5\sqrt{f'_c} + 0.3\sigma_g + \frac{V_p}{b_w d_{pc}}$$

Referring to Figs. 9.15b and c and noting that the draping point of the strands is taken at 30 ft from the support, it can be shown that:

$$V_p = F \sin \alpha = 616.90 \left(\frac{20.16 - 3.87}{30 \times 12} \right) = 27.915 \text{ kips}$$

If the average prestress σ_g is computed for the composite section (safer design), the value of v_{cw} becomes:

$$v_{cw} = 3.5\sqrt{5000} + 0.3\frac{616,900}{1369} + \frac{27,914}{8 \times 49.6}$$

$$= 247.48 + 135.19 + 70.35 \simeq 453 \text{ psi}$$

The shear resistance of concrete is taken as the lesser of v_{ci} or v_{cw}. Hence $v_c = v_{cw} = 453$ psi.

As $v_u > \phi v_c$ stirrups are required. Assuming #3 U stirrups with area $A_v = 0.22$ in^2 and $f_y = 60,000$ psi, the required spacing is given by Eq. (6.37):

$$s = \frac{A_v f_y}{(v_u/\phi - v_c)b_w} = \frac{0.22 \times 60,000}{(548 - 453)8} = 17.37 \text{ in}$$

Round it off to $s = 16$ in. It is less than $0.75h_c$. It can also be shown that A_v is more than $(A_v)_{min}$.

The reader may want to check that, if the alternative conservative method (Fig. 6.13) was used, the computed value of v_c would have been equal to 742 psi for a value of $V_u d_{pc}/M_u = 1$. As v_c is limited by $5\sqrt{f'_c}$, it is taken equal to $v_c = 5\sqrt{f'_c} = 354$ psi. The corresponding stirrup spacing is 8.5 in or about half the spacing required when using the more accurate method.

Note that all stirrups are extended and anchored in the slab, providing ties for shear transfer.

e. Horizontal shear. The total shear force calculated in the previous section is used at ultimate:

$$V_u = 184.75 \text{ kips}$$

The interface shear is given by:

$$v_{uh} = \frac{V_u}{b_v d_{pc}} = \frac{184,750}{20 \times 49.6} \simeq 186 \text{ psi}$$

As the interface surface is intentionally roughened and all shear reinforcement (more than minimum) is extended and anchored in the slab the nominal horizontal shear resistance can be taken as $v_{nh} = 350$ psi. As $v_{uh} < \phi v_{nh}$ no additional shear transfer reinforcement is required. Note that the condition for tie spacings to be not more than $4h_f$ is also satisfied.

Let us assume to illustrate the procedure described in Section 9.6 (Tie Reinforcement Using Shear Friction) that shear-friction reinforcement is to be determined for the same above calculated shear force.

Using Eq. (9.12) and assuming same spacing s as used in the design of vertical stirrups, we have:

$$A_{vf}^* = \frac{V_u}{\phi f_y \, \mu \, d_{pc}} \, s = \frac{184.75}{0.85 \times 60 \times 1} \, \frac{16}{49.6} = 1.17 \text{ in}^2$$

That is, 1.17 in^2 of the tie reinforcement is needed at a spacing of 16 in at the section considered.

If we use Eq. (9.8) as recommended by ACI, the value of H_u must be determined first. It can be taken equal to the compressive force in the slab at nominal moment capacity of the midspan section. Using the depth of the neutral axis obtained in Example 9.11c and Eq. (9.9), we have:

$$H_u = 0.85 f'_c \, b_e \, a = 0.85 \times 4 \times 81 \times (4.58 \times 0.80) \simeq 1009 \text{ kips}$$

and from Eq. (9.8):

$$A_{vf} = \frac{H_u}{\phi f_y \, \mu} = \frac{1009}{0.85 \times 60 \times 1} = 19.78 \text{ in}^2$$

The above value is to be distributed along the interface surface of half the span, i.e., 40 ft. It leads on the average to about 0.5 in^2 per linear foot or 0.66 in^2 per 16-in spacing. In comparing with the value of A_{vf}^* obtained above at the first critical section, one should keep in mind that A_{vf}^* will vary along the span and is larger near the supports.

f. Deflections. The steps suggested in Sec. 9.10 will be followed. The member is not cracked under service loads and thus I_g or I_{gc} will be used throughout. Only the answers are given next. The reader is also referred to Fig. 7.4 which gives the various analytic formulas for deflections. Unshored construction is assumed.

1. Using an average force between F and F_i and an average modulus between E_c and E_{ci}, the deflection due to the prestressing force and the self-weight of the precast beam is:

$$(\Delta_i)_1 = -1.26 + 0.71 \simeq -0.55 \text{ in}$$

It is a camber.

2. Additional long-term deflection until addition of slab (using $\lambda_1 \simeq 1$):

see page 259 $\longrightarrow$ $$(\Delta_{add})_2 = \lambda_1 (\Delta_i)_1 = -0.55 \text{ in}$$

3. Instantaneous deflection in precast beam due to weight of cast-in-place slab (use E_c):

$$(\Delta_i)_3 = 0.55 \text{ in}$$

4. Resulting deflection:

$$\Delta_4 = (\Delta_i)_1 + (\Delta_{add})_2 + (\Delta_i)_3 = -0.55 \text{ in}$$

5. Instantaneous deflection in composite beam due to asphalt weight:

$$(\Delta_i)_5 = 0.085 \text{ in}$$

6. Resulting deflection in composite:

$$\Delta_6 = \Delta_4 + (\Delta_i)_5 = -0.465 \text{ in}$$

7. Additional long-term deflection in the composite beam (use $\lambda_2 \simeq 1.2$)

$$(\Delta_{add})_7 = \lambda_2 \Delta_6 = -0.56 \text{ in}$$

Thus the total long-term deflection under sustained loads will be a camber of $(-0.465 - 0.56) \simeq -1.02$ in.

It can be shown that the instantaneous deflection in the composite beam due to live load plus impact is of the order of 0.38 in and that deflection limitations are satisfied whether the ACI code or the AASHTO specifications are considered.

REFERENCES

9.1. C. P. Siess: "Composite Construction for I-Beam Bridges," *Transactions ASCE*, vol. 114, 1949, pp. 1023–1045.

9.2. F. J. Samuely: "Some Recent Experience in Composite Precast and In Situ Concrete Construction with Particular Reference to Prestressing," *Proceedings of the Institute of Civil Engineers*, vol. 1, 1952, pt. 1, no. 30, pp. 222–259.

9.3. R. H. Evans and A. S. Parker: "Behavior of Prestressed Concrete Composite Beams," *ACI Journal*, vol. 52, no. 6, 1955, pp. 861–881.

9.4. ACI-ASCE Committee 333: "Tentative Recommendations for Design of Composite Beams and Girders for Buildings," *ACI Journal*, vol. 32 (Proceedings vol. 57), no. 6, December 1960, pp. 609–628.

9.5. G. M. Sabnis: "Handbook of Composite Construction Engineering" Van Nostrand-Reinhold Co., New York, 1979.

9.6. C. L. Freyermuth: "Design of Continuous Highway Bridges with Precast Prestressed Concrete Girders," *PCI Journal*, vol. 14, no. 2, April 1969, pp. 14–36.

9.7. D. E. Branson: "The Deformation of Non-Composite and Composite Prestressed Concrete Members," ACI Special Publication SP-43-4, Deflections of Concrete Structures, 1974, pp. 83–127.

9.8. M. K. Tadros, A. Ghali, and W. H. Dilger: "Time-Dependent Prestress Loss and Deflection in Prestressed Concrete Members," *PCI Journal*, vol. 20, no. 3, May/June 1975, pp. 86–98.

9.9. S. P. Timoshenko and J. P. Goodier: *Theory of Elasticity*, 3d ed. McGraw-Hill Book Company, New York, 1970, pp. 262–268.

9.10. N. W. Hanson: "Precast-Prestressed Concrete Bridges: (2) Horizontal Shear Connections," *PCA Journal*, vol. 2, no. 2, May 1960, pp. 38–58.

9.11. J. C. Saemann and G. W. Washa: "Horizontal Shear Connections Between Precast Beams and Cast-In-Place Slabs," *ACI Journal*, vol. 61, no. 11, November 1964, pp. 1383–1409.

9.12. J. C. Badoux and C. L. Hulsbos: "Horizontal Shear Connection in Composite Concrete Beams Under Repeated Loads," *ACI Journal*, vol. 64, no. 12, December 1967, pp. 811–819.

9.13. A. H. Mattock and N. M. Hawkins: "Research on Shear Transfer in Reinforced Concrete," *PCI Journal*, vol. 17, no. 2, March/April 1972, pp. 55–75.

9.14. A. H. Mattock, W. K. Li, and T. C. Wang: "Shear Transfer in Lightweight Reinforced Concrete," *PCI Journal*, vol. 21, no. 1, January/February 1976, pp. 20–39.

9.15. A. F. Shaikh: "Proposed Revisions to Shear-Friction Provisions," *PCI Journal*, vol. 23, no. 2, March/April 1978, pp. 12–21.

9.16. ACI Committee 318: "Proposed Revisions to Building Code Requirements for Reinforced Concrete (ACI 318–77) and Commentary on Building Code Requirements for

Reinforced Concrete (ACI 318–77)," *Concrete International*, vol. 2, no. 1, January 1980, pp. 56–60.

9.17. A. H. Mattock: "Precast-Prestressed Concrete Bridges: (5) Creep and Shrinkage Studies," *PCA Journal*, vol. 3, no. 2, May 1961, pp. 32–66.

9.18. L. D. Martin: "A Rational Method for Estimating Camber and Deflection of Precast Prestressed Members," *PCI Journal*, vol. 22, no. 1, January/February 1977, pp. 100–108.

PROBLEMS

9.1 Repeat the example problem of Sec. 9.11 assuming $f'_c = 6000$ psi and a span of 100 ft, everything else being the same. Assume that live load plus impact lead to a midspan moment $M_{L+I} = 1143$ kips-ft and a shear force of 46.5 kips at 3 ft from the center of supports. Analyze two possible alternatives: (*a*) unshored and (*b*) shored construction. In each case build the feasibility domain for F_i and e_o and provide an acceptable layout of the prestressing tendons at midspan and at the supports. Compare the two solutions with respect to all design aspects (shear, deflections, etc.) and draw related conclusions.

9.2 A composite floor system is made out of simply supported precast prestressed rectangular beams and a cast-in-place reinforced concrete slab. Relevant dimensions and properties are given next.

Precast beams
$f'_c = 5000$ psi; $f'_{ci} = 4000$ psi; $\bar{\sigma}_{ti} = -190$ psi; $\bar{\sigma}_{ci} = 2400$ psi; $\bar{\sigma}_{ts} = -424$ psi; $\bar{\sigma}_{cs} = 2250$ psi.

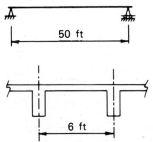

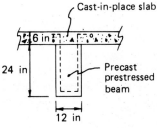

Figure P9.2

Cast-in-place slab
$f'_c = 4000$ psi; $(\bar{\sigma}_{cs})_{slab} = 1800$ psi.

Other information
Live load = 50 psf on slab; Lightweight concrete slab with $\gamma_c = 110$ pcf; Normal weight concrete beam with $\gamma_c = 150$ pcf; Prestressing strands having $\frac{1}{2}$-in diameter and 270 ksi strength; $f_{pe} = 150$ ksi; $f_{py} = 240$ ksi; $\eta = F/F_i = 0.85$.

Provide a complete design for the prestressed beams, that is:
(*a*) Determine geometric properties of precast beam and composite section.
(*b*) Determine F and e_o at midspan. Find limit kern.
(*c*) Determine limit zone. Select steel profile.
(*d*) Check ultimate moment at midspan.
(*e*) Find stirrups—check ties for horizontal shear.
(*f*) Check deflections.
Take any corrective action when necessary.

Parking structure with one-way posttensioned slabs supported on posttensioned continuous beams. (*University of Illinois at Chicago Circle.*)

CHAPTER 10

CONTINUOUS BEAMS AND INDETERMINATE STRUCTURES

10.1 ADVANTAGES AND FORMS

There are inherent advantages in using continuous (or statically indeterminate) beams instead of simply supported (or statically determinate) beams. Everything else being equal, continuity leads to smaller design moments, smaller deflections, and higher rigidity against lateral loads. It allows redistribution of stresses under overload conditions and ensures a higher margin of safety against collapse. Most reinforced concrete structures are monolithically cast in place and most steel structures are made continuous.

Because they have to resist smaller design moments, continuous beams are generally shallower than simple span beams and need lesser amounts of materials. Additional savings can be expected in the cost of posttensioning continuous beams as several spans can be prestressed with the same continuous tendon (saving labor costs) and only two anchorages are needed per tendon. Furthermore, by giving the tendon an undulating profile, the same tendon can be used to resist both the positive and negative moments. However, the cost effectiveness of continuity in prestressed concrete members depends on many other factors, such as span length, design criteria, construction conditions, and the like. It is often observed that the economy associated with the use of prestressed concrete instead of other structural materials would be nonexistent in many cases if the elements were not precast.

The benefits of continuity are often offset by drawbacks that are particular to the use of prestressed concrete. These drawbacks include: (1) practical difficulties in laying out and grouting undulating tendons, (2) tendon frictional losses which may become excessive, (3) a more complex design procedure (as noted later in this chapter) where parasitic secondary moments must be accounted for, (4) the effects on connected members of shortening due to the horizontal compression, as well as creep and shrinkage, and, most importantly, (5) the difficulty in handling moment reversals. Contrary to that in reinforced concrete where the reinforcement is tailored to resist external moments by cutting or bending up bars, the prestressing force in prestressed continuous beams is generally dimensioned to resist the maximum moment along several spans and the corresponding tendons are essentially run throughout the various spans. Hence little flexibility is left at sections where moment reversals may occur and where the same reinforcement must resist both positive and negative moments. This is particularly serious when ultimate strength, hence cracked section analysis, is considered. However, the increased acceptance of partial prestressing, where both prestressed and conventional nonprestressed reinforcement are used, may eliminate this disadvantage.

The above arguments suggest that the benefits of continuity in prestressed concrete may be often offset by its drawbacks. There are, however, many applications where continuity is clearly preferable. Examples include continuous slabs (one- and two-way slabs), medium- and long-span bridges, and applications where precast prestressed elements are made continuous by posttensioning on site.

Continuity can be achieved in several ways. Typical examples are shown in Fig. 10.1. In cases (a) and (b) the structure is monolithically cast in place and posttensioned thereafter. In cases (c) and (d) precast pretensioned elements are used to create the basic configuration, then they are jointed and posttensioned in situ to achieve continuity. The elements are generally designed to resist, by pretensioning, their own weight as well as handling stresses while posttensioning provides the additional resistance to counteract superimposed dead loads and live loads. Case (e) of Fig. 10.1 shows a typical segmental construction where precast elements are first designed as successive cantilevers to resist construction loads, then as part of a continuous system when the structure is ready for service. In all the above cases a typical tendon profile is shown. In general, the profile follows the deflected shape of the structure or the moment diagram due to a uniform load (such as dead load) plotted positive downward.

10.2 NECESSARY ANALYTIC BACKGROUND

The detailed treatment of continuous prestressed concrete beams and frames in their various forms requires space and efforts that go much beyond the scope of this text. Here only the case of a cast-in-place monolithic structure with the same continuous tendon running throughout will be considered. However, the background information and the design approach suggested in this chapter should provide a sufficient basis to allow the user to extend the analysis to different cases. The procedure presented can be followed throughout using hand computations or, even better, an electronic calculator. The following analytic background is needed:

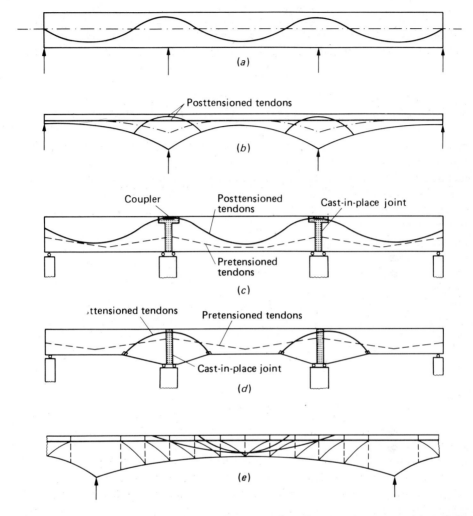

Figure 10.1 Examples of continuous beams and tendon arrangements. (*a*) and (*b*) Cast-in-place construction. (*c*) and (*d*) Precast elements made continuous. (*e*) Cantilever construction.

1. Analysis of statically indeterminate structures subjected to external loads. Any method such as matrix structural analysis or moment distribution is acceptable. Moment distribution will be used in this chapter to determine the total moment due to prestressing. The total prestressing moments allow the determination of the secondary moments.
2. Analysis and design of simply supported prestressed concrete beams under working stresses, as described in Chap. 4.
3. Background given in Secs. 10.3 to 10.10.

An example illustrating the most important aspects of design is covered in sufficient detail in Sec. 10.15. A number of properties, theorems, and corollaries are stated without proof. The reader may want to review the material in Refs. 10.1 to 10.6 for additional information.

10.3 SIGN CONVENTION AND SPECIAL NOTATION

In dealing with indeterminate structures it is very important to follow a consistent sign convention as one does not know a priori the sign of the secondary moments (defined in Sec. 10.4). Their value and sign are derived from the analysis. The sign convention set in Sec. 4.5 remains valid here. In particular, the eccentricity of the prestressing force $e_o(x)$ at any section x is assumed positive when F is below the neutral axis and negative when it is above it (Fig. 10.2). For vertical members, positive is to the right and negative is to the left. The same sign convention holds for the C line.

The prestressing force F produces a compression in the concrete and is assumed positive. The primary moment generated by F at any section x and the corresponding eccentricity $e_o(x)$ have opposite signs.

The following notations (some of which are explained later in this chapter) will be used:

e = eccentricity in general (used mostly in Figures to reduce the burden of subscripts)

$e_o(x)$ = eccentricity of the centroid of the prestressing steel at section x

$e_c(x)$ = eccentricity of the C line at section x

$e_{oc}(x)$ = eccentricity of the Zero-Load-C line (ZLC line) at section x. The ZLC line in a statically indeterminate structure is due to the sole effect of prestressing (no external loads).

$M(x)$ = external moment, in general, at section x

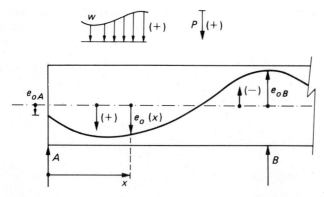

Figure 10.2 Sign convention for tendon eccentricity.

$M_1(x) =$ primary moment due to the prestressing force at section x.
$$M_1(x) = -Fe_o(x)$$
$M_2(x) =$ secondary moment at section x due to prestressing
$M_F(x) =$ total moment at section x due to prestressing. $M_F(x) = M_1(x) + M_2(x)$

M_{1A}, $M_{1B} =$ primary moments at particular sections A and B
M_{2A}, $M_{2B} =$ secondary moments at particular sections A and B

The reference to section x is ignored in the above notation if it is obvious that no confusion will result in the mathematical treatment. This is especially true when referring to equations from Chap. 4, in which the critical section is generally implied.

10.4 SECONDARY MOMENTS AND ZERO-LOAD-C LINE

It was shown in Secs. 4.2 and 4.12 that the prestressing force F and an external moment M acting at any section of a simply supported beam can be resolved into a force $C = F$ acting at a distance $\delta = M/F$ from the line of action of F. The geometric lieu of the C force along the various sections of a member was defined as the C line or pressure line (also called thrust line). For a simply supported beam the eccentricity of the C line with respect to the centroid of the section was defined (Sec. 4.12) as:

$$e_c(x) = e_o(x) - \frac{M(x)}{F} \tag{10.1}$$

where $e_o(x)$ is the eccentricity of the prestressing force and $M(x)$ the external moment at section x. Equation (10.1) suggests that if no external moment is applied, the eccentricity of the C force at any section is identical to that of the prestressing steel, hence, the C line coincides with the trajectory of the steel. Thus, in a simply supported beam (or statically determinate structure), the effect of prestressing is reduced to that of the tendons alone at each section. The supports do not provide any restraint to the deformation of the structure and the prestressing moment is given by $M_F(x) = -Fe_o(x)$. This is not the case, however, for prestressed continuous beams where intermediate supports restrict the free deformation of the structure, hence leading to support reactions called secondary reactions. Secondary reactions act like concentrated loads on a simply supported beam. They generate at each section a moment called secondary moment (also called parasitic moment or hyperstatic moment). Therefore, under the sole effect of prestressing (i.e., in presence of no external loads) two types of moments are generated at each section of a continuous beam: the primary moment defined as for a simply supported beam by $M_1(x) = -Fe_o(x)$ and the secondary moment $M_2(x)$ generated by the secondary reactions. The moment due to prestressing at each section x becomes:

$$M_F(x) = M_1(x) + M_2(x) = -Fe_o(x) + M_2(x) \tag{10.2}$$

Since, in general, $M_F(x)$ is determined from the analysis of the structure, say by moment distribution, the secondary moment is derived from Eq. (10.2) as:

$$M_2(x) = M_F(x) - M_1(x) = M_F(x) + Fe_o(x) \qquad (10.3)$$

Secondary moments are secondary in nature but not in magnitude. They can represent a significant portion of the prestressing moment and, hence, must be accounted for in design. Advantage can be taken of their presence and may lead to savings in the prestressing force.

Because of the existence of secondary moments, the C line under the sole effect of prestressing, defined here as the Zero-Load-C line (ZLC line), because no external load is applied, does not coincide with the prestressing steel. Its eccentricity at any section x is given by:

$$e_{oc}(x) = -\frac{M_F(x)}{F} = e_o(x) - \frac{M_2(x)}{F} \qquad (10.4)$$

Equation (10.4) suggests that the prestressing force acts as if it had an eccentricity $e_{oc}(x)$. Thus, $e_{oc}(x)$ can also be described as the effective eccentricity of the prestressing force. In presence of an external moment $M(x)$ the eccentricity of the C line becomes:

$$e_c(x) = e_{oc}(x) - \frac{M(x)}{F} = e_o(x) - \frac{M(x) + M_2(x)}{F} \qquad (10.5)$$

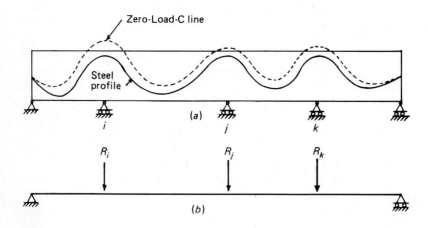

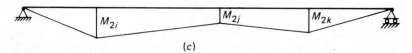

Figure 10.3 (*a*) **Tendon profile and ZLC line.** (*b*) **Secondary reactions.** (*c*) **Secondary moment diagram.**

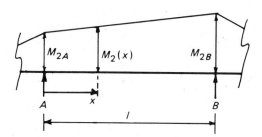

Figure 10.4

which suggests that $M_2(x)$ acts like an additional external moment on the section. When the secondary moments are equal to zero, Eq. (10.5) is reduced to Eq. (10.1) used in simply supported beams. The tendon profile in the continuous beam is then said to be concordant.

If both the ZLC line and the steel profile are known, Eq. (10.4) leads to:

$$\frac{M_2(x)}{F} = e_o(x) - e_{oc}(x) = \Delta e_{oc}(x) \tag{10.6}$$

$\Delta e_{oc}(x)$ can be considered as a fictitious increase or decrease in the eccentricity of the prestressing steel at section x, due to the presence of secondary moments.

A typical tendon profile and corresponding ZLC line are plotted in Fig. 10.3a. The secondary reactions and corresponding secondary moments are schematically shown in Figs. 10.3b and c. As the secondary reactions are generated by prestressing, they form a system of forces with a null resultant. Since they act as concentrated forces at the supports, the variation of secondary moment between consecutive supports is linear. Therefore, for a typical span with left support A and right support B (Fig. 10.4), the secondary moment at any section x can be computed from the secondary moments at the supports as:

$$M_2(x) = M_{2A}\left(1 - \frac{x}{l}\right) + M_{2B}\left(\frac{x}{l}\right) \tag{10.7}$$

The existence of secondary moments and other properties of interest in continuous beams are illustrated in the next section.

10.5 EXAMPLE: SECONDARY MOMENTS AND CONCORDANCY PROPERTY

The existence of secondary moments can be illustrated by the simple example of a two-span continuous beam prestressed by a straight tendon with eccentricity e throughout (Fig. 10.5a). Assume that only the effect of prestressing is considered, that is no external loads are acting. If the intermediate support B was nonexistent, the beam would be simply

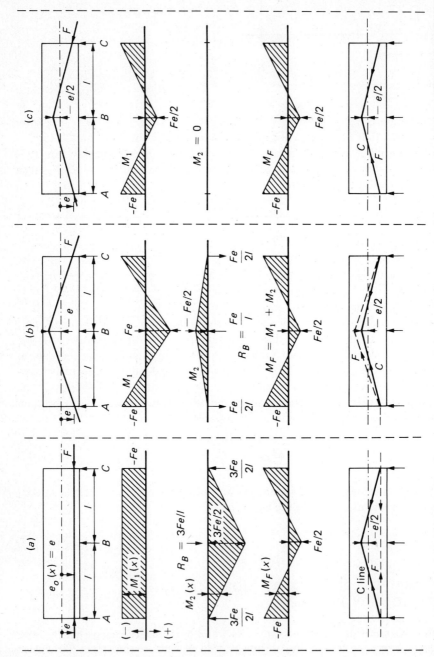

Figure 10.5 Examples illustrating the variations of primary moments, secondary moments, prestressing moments, and ZLC line in continuous beams.

supported at A and C and would camber under the effect of prestressing. The presence of support B restrains the movement of the beam and hence generates a reaction R_B. The magnitude and direction of R_B are such that R_B should create a deflection at B equal and opposite to the camber created by prestressing. Referring to Fig. 7.4 and assuming a simply supported beam with span $2l$, the deflection due to R_B is given by:

$$\Delta_1 = \frac{R_B(2l)^3}{48EI}$$

The camber due to prestressing is given by:

$$\Delta_2 = -\frac{Fe(2l)^2}{8EI}$$

Setting the sum $(\Delta_1 + \Delta_2)$ equal to zero leads to:

$$R_B = \frac{3Fe}{l}$$

Prestressing the continuous beam ABC of Fig. 10.5a generates at each section x a primary moment $M_1(x) = -Fe$ and a secondary moment $M_2(x)$ induced by the secondary reaction R_b. $M_2(x)$ is obtained by treating the beam AC as simply supported with a concentrated force R_B at B. The primary and secondary moment diagrams for this case are plotted in Fig. 10.5a, as well as their sum, the prestressing moment $M_F(x)$. Two observations can be made: (1) the secondary moment can be significant in magnitude and (2) the prestressing moment in a continuous beam can be substantially different from the primary moment otherwise obtained if the beam was simply supported. Given the prestressing moment, the ZLC line can be determined from Eq. (10.4) and is also shown in Fig. 10.5a. Note that it deviates substantially from the steel profile. It has the same eccentricity e at the end supports and an eccentricity $-e/2$ at B.

Two other cases are covered in Fig. 10.5. In case b the eccentricity of the prestressing force at B is changed from $+e$ (case a) to $-e$, while the eccentricities at the supports are kept the same. Using the deflection equations given in Fig. 7.4, the reaction R_B is calculated in a manner similar to case a. The primary, secondary, and total moments are determined and plotted in Fig. 10.5b. It can be observed that, although both the primary and secondary moment diagrams are different from case a, the resulting prestressing moment diagram is the same. Hence, the corresponding ZLC line is also the same.

Finally, in case c of Fig. 10.5 the steel profile is modified to show an eccentricity $-e/2$ at support B. This profile is the same as the ZLC line found in cases a and b. Following the same analytic steps, it is found that the secondary reaction R_B vanishes. Hence, the secondary moment is zero at any section and the prestressing moment becomes equal to the primary moment (Fig. 10.5c). It is observed that the prestressing moment and the ZLC line are the same as for cases a and b. Moreover, in case c the ZLC line coincides with the trajectory of the steel. When this occurs, the tendon profile is said to be "concordant." Therefore, for case c we have a concordant steel profile while for cases a and b we have nonconcordant steel profiles.

The above example suggests an additional important result: the three different steel profiles of cases a, b, and c in which only the eccentricity at the intermediate support B was varied, led to the same ZLC line. This result is due to a property of the "linear transformation" explained in the next section.

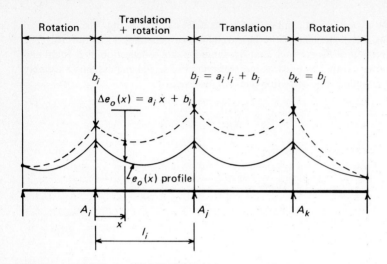

Figure 10.6 Example of linear transformation illustrating rotational effect, translational effect, or their combination.

10.6 LINEAR TRANSFORMATION

Let us consider a continuous beam with a given tendon profile (Fig. 10.6). Let us modify the profile by subjecting it to some finite displacements at the intermediate supports, without changing its intrinsic shape between supports and keeping the same eccentricities at the end supports. Such a transformation is called a linear transformation. Along any span the linear transformation is equivalent to a rotation, a translation, or both. Calling $\Delta e_o(x)$ the displacement between the reference profile and the transformed profile at any section x of a span, its value can be obtained from

$$\Delta e_o(x) = ax + b \qquad (10.8)$$

where x is the distance from the left support and a and b are constants. The term ax essentially leads to a rotation with respect to the left support while b represents a translation. The following theorem stated without proof is originally due to Guyon (Refs. 10.1 and 10.2):

A linear transformation of a tendon profile does not modify the ZLC line of the tendon.

In other words, the linearly transformed profile produces the same ZLC line as the reference profile. Hence moments, shear forces, and stresses due to prestressing remain the same. However, the primary and secondary moments will change.

The property of the linear transformation applies not only to beams but also to frames with rigid joints. If the continuous beam or frame is fixed at one or two end supports, it would apply even if the eccentricities at the fixed ends are modified.

Referring back to Fig. 10.5, it is observed that the tendon profiles for cases b and c are in effect linearly transformed from the profile of case a and, therefore, they all lead to the same ZLC line.

10.7 CONCORDANT TENDONS

A concordant tendon profile produces a ZLC line coinciding with it at any section, that is $e_{oc}(x) = e_o(x)$. Thus, the secondary reactions are null and the secondary moments vanish everywhere. A number of properties related to concordant tendons are stated next without proof. Additional information can be found in Refs. 10.5 and 10.6.

1. *The principle of superposition is valid for concordant tendons, namely, superposing one or more concordant profiles leads to a concordant profile and superposing a concordant profile with a nonconcordant profile leads to a nonconcordant profile.*
2. *Any real moment diagram in a continuous beam on nonsettling supports produced by any combination of external loadings (whether transverse loads or moments) plotted to any scale along the centroidal axis of the beam, defines the location of a concordant tendon. Thus, if a tendon is placed along a trajectory proportional to a moment diagram, it will be concordant.*
3. *The reciprocal of the property 2 above is also true: the profile of any concordant tendon measured from the centroid of the section is a moment diagram for some real system of loadings.*
4. *Any ZLC-line trajectory is a concordant tendon profile.*

Property 2 can be used to determine a concordant tendon profile. For instance, given a moment diagram defined by $M(x)$ at any section x, a tendon profile can be defined by:

$$e_o(x) = \frac{M(x)}{K} \tag{10.9}$$

where K is an arbitrary constant. The value of K can be determined so that the tendon profile obtained remains within the section, that is $|e_o(x)| \le (e_o)_{mp}$. For this all the critical sections (supports and midspan or near midspan sections) are considered and at each critical section i a value K_i is determined as follows:

$$K_i = \left[\frac{M(x)}{(e_o)_{mp}} \right]_i \tag{10.10}$$

K is taken as the largest value of all $|K_i|$ obtained and is used throughout the member in Eq. (10.9). Of course, this implies that the same prestressing force is used throughout.

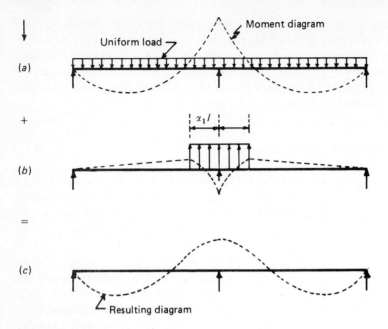

Figure 10.7 Example of superposition of moment diagrams to obtain a smooth concordant profile.

Note that the moment diagram $M(x)$ is not necessarily due to a single loading. If, for instance, a moment due to a uniform load such as dead load is used, it will have sharp peaks (negative moments) at the supports. Hence, sharp, unacceptable peaks will also be present in the corresponding profile. In order to smooth the profile another loading is appropriately selected to generate positive moments at the support and is superimposed to the first loading. A schematic representation of this procedure is shown in Fig. 10.7 and an example is developed in Sec. 10.15, case *a*.

It is important to point out that, although concordant tendons can be easily determined, they are not necessarily more desirable than nonconcordant tendons. They simplify the design procedure by eliminating secondary moments. However, a nonconcordant tendon profile can be selected in such a way as to magnify the effect of prestressing and hence can lead to significant savings in the prestressing steel.

10.8 CONCEPT OF EQUIVALENT LOADS

The effects of a prestressing tendon acting on a beam can be replaced by the effects of equivalent loads. These loads can easily be identified if the free body diagram of the concrete without the steel is drawn. Equivalent loads include concentrated horizontal and vertical forces, moments at the external supports, and transverse forces along the tendon profile. Transverse forces are generated by the curvature or the change in profile of the tendon. They can be in the form of a concentrated force (due to an abrupt change in the slope of the tendon profile), a uniform load, or a distributed variable load. For a curved tendon, the magnitude of the equivalent transverse load over a unit length is equal F/ρ_x, where ρ_x is the radius of curvature of the tendon at the section considered.

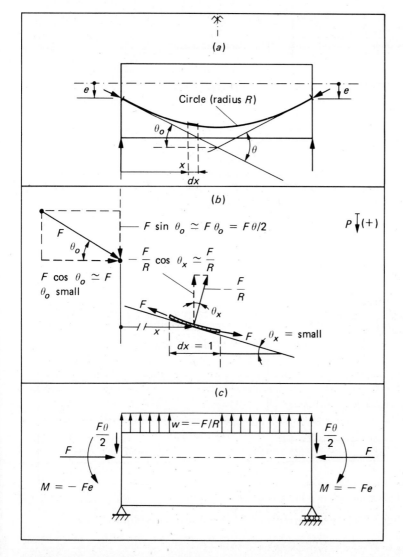

Figure 10.8 Concept of equivalent load. (*a*) Tendon profile. (*b*) Transverse load due to prestressing. (*c*) Free body diagram of concrete beam.

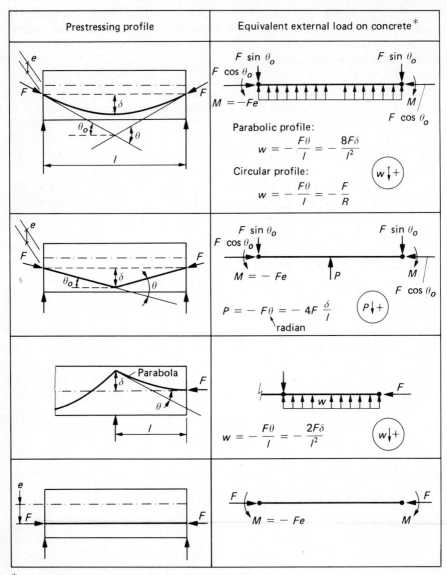

*Assuming shallow members for δ in function of θ.

Figure 10.9 Equivalent load formulas for typical tendon profiles in beams.

To illustrate the equivalent-load concept, let us consider a simply supported beam prestressed by a tendon carrying a force F and having a circular profile with radius R (Fig. 10.8a). Let us determine the equivalent transverse load on the concrete. The change in curvature between section x and $x + dx$ produces a transverse negative reaction $-(F/R)dx$ (Fig. 10.8b). For a unit length, the reaction becomes $-F/R$. Its projection on the vertical axis is $-F/R \cos \theta_x$. However, as θ_x is very small, $\cos \theta_x \simeq 1$, and the vertical reaction on the concrete beam is equal $-F/R$ per unit length. This is essentially equivalent to a uniform external load on the beam of value $w = -F/R$. The corresponding free body diagram of the beam, including the end loads, is shown in Fig. 10.8c. The beam can be analyzed for these loads and should show exactly the same state of stress as if the prestressing force of Fig. 10.8a was present. In particular, external equilibrium equations must be satisfied. For the beam of Fig. 10.8c, they lead to:

$$\text{Horizontal forces:} \qquad F + (-F) = 0$$

$$\text{Moments:} \qquad -Fe + Fe = 0$$

$$\text{Vertical forces:} \qquad F\frac{\theta}{2} + F\frac{\theta}{2} + wl = 0$$

The last equation is verified because $wl = -(F/R)l$ and $l/R = \theta$.

Examples of equivalent loads for typical steel profiles are shown in Fig. 10.9 and should cover the majority of practical cases. The given formulas generally assume that the eccentricity is small with respect to the span, that is, relatively shallow members are used. Note that for a straight tendon the transverse load is zero. Only the axial force and the end moments are present. In order for the moments generated by the equivalent loads to have the correct sign, the loads are assumed positive downward and negative upward. This explains the negative sign associated with the loads of Fig. 10.9.

When a continuous undulating tendon is used, the length over which the same profile exists is not the same as the span length. This is illustrated in Fig. 10.10, where a typical example of equivalent loads for a continuous beam is shown. For instance, the load w_4, assuming a parabolic portion of tendon, is given by $-F\theta_4/l'_4$ or $-8F\delta_4/(l'_4)^2$. Several such loads can be present along the same span.

Note that for any other nonstandard steel profile, the equivalent transverse loading can be found from the theoretical curvature of the profile or from the following steps:

1. Plot the primary moment diagram $M_1(x)$ throughout the various spans as if there were no supports.
2. From the moment diagram, determine the corresponding shear diagram either graphically or algebraically.
   (*Note:* $V(x) = dM_1(x)/dx$.)
3. From the shear diagram, determine the loading diagram.
   (*Note:* $w(x) = -dV(x)/dx$.)

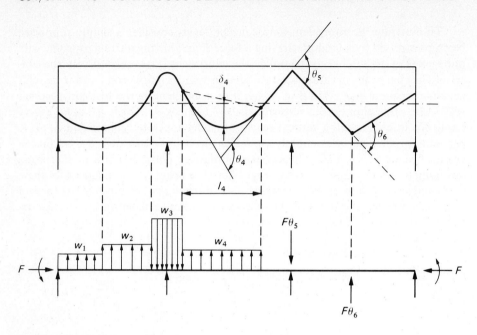

Figure 10.10 Equivalent loads in a continuous beam.

If an analytic equation for the steel profile $e_o(x)$ is given, the above procedure should not present any difficulty, as $M_1(x) = -Fe_o(x)$.

The concept of equivalent load is very convenient. It is particularly useful in determining the total moment due to prestressing, $M_F(x)$, and in the load-balancing approach described in Sec. 10.13. An example illustrating the use of the equivalent load concept to determine the prestressing moment is developed in Sec. 10.15, case d.

10.9 PRESTRESSING MOMENT AND ELASTIC STRESSES

To determine the prestressing moment in a continuous beam with a given tendon profile, the following steps can be followed:

1. Determine the equivalent loads as suggested in Sec. 10.8.
2. Determine the fixed-end moments generated by the equivalent loads in each span.
3. Determine by moment distribution the end moments in each span. These are the prestressing moments at the supports. (An example is given in Sec. 10.15, case **d**.)
4. Determine the secondary moments at the supports, using Eq. (10.3).

5. Determine the secondary moments at any section x between supports using Eq. (10.7).
6. Determine the prestressing moment at any section x using Eq. (10.2).

Steps 2 and 3 assume that the moment distribution method is used. However, any other analytic procedure for indeterminate structures would be acceptable. As the moment-distribution method utilizes fixed-end moments, a summary of fixed-end moments for common loading cases is given in Fig. 10.11. Steps 4 to 6 can be condensed in a single equation applied to a typical span with end supports A and B. It is given by:

$$M_F(x) = -Fe_o(x) + (M_{FA} + Fe_{oA})\left(1 - \frac{x}{l}\right) + (M_{FB} + Fe_{oB})\frac{x}{l} \qquad (10.11)$$

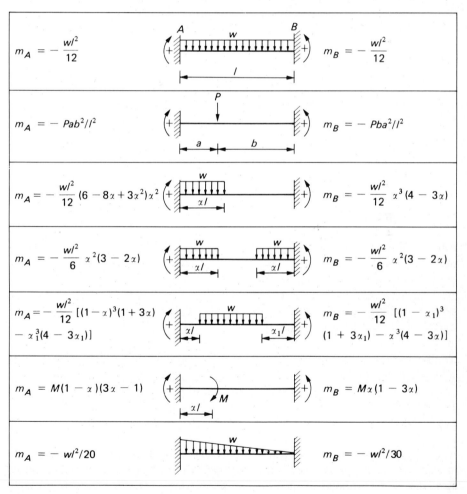

Figure 10.11 Fixed-end moments for members with constant stiffness.

where M_{FA} and M_{FB} are the prestressing moments and e_{oA} and e_{oB} the tendon eccentricities at supports A and B of a typical span. In Eq. (10.11) the first term represents the primary moment at section x and the remaining terms represent the secondary moment. Given the secondary moments, the ZLC line can be determined by its eccentricity from Eq. (10.4). It should be pointed out that the eccentricity of the ZLC line depends on the profile of the tendons but is independent of the magnitude of the prestressing force. Hence it could be determined assuming a unit force.

In order to determine the stresses in the concrete at any section, the equations developed in Chap. 4 and the stress inequality conditions given in Table 4.2 can be used directly provided the eccentricity of the prestressing steel is replaced by that of the ZLC line. Another way of using these equations is to keep the eccentricity of the steel as is, but add the secondary moment to the external moments. For instance, Eq. IV of Table 4.2 applied to a section of a continuous beam can be written in one of two ways:

$$\frac{F}{A_c}\left(1 - \frac{e_{oc}}{k_t}\right) - \frac{M_{max}}{Z_b} \geq \bar{\sigma}_{ts} \tag{10.12}$$

where e_{oc} has replaced e_o, or

$$\frac{F}{A_c}\left(1 - \frac{e_o}{k_t}\right) - \frac{M_{max} + M_2}{Z_b} \geq \bar{\sigma}_{ts} \tag{10.13}$$

where M_2 is added to the external moment. The above equations and those given in Table 4.2 apply to any section x.

10.10 DESIGN AIDS

It has become increasingly clear from the preceding sections that the profile of the steel along a continuous beam represents an important design parameter. Although graphical solutions can be used and particular analytic solutions for a given problem can be devised, a general formulation of the profile for the most common practical cases and its effects on the primary and secondary moments will lead to substantial time savings in design.

The steel profile can be made of segmented straight lines, parabolic curves, circular curves, or a combination thereof. Most likely an undulating shape is achieved in a continuous beam. It is generally convenient to assume parabolic parts because of the simple analytic representation of a parabola. Khachaturian and Gurfinkel (Ref. 10.7) have developed equations for the steel profile of continuous beams, assuming either parabolic parts or parts represented by a parabola and a fourth-degree curve. Their assumed configurations for a typical end span and a typical intermediate span, for the case where parabolic parts were used, are shown in Figs. 10.12a and b. Note that the steel profile of the end span is made out of three different parabolas, while two parabolas (three parts) are used for a symmetrical intermediate span. The corresponding equations for $e_o(x)$ at any section

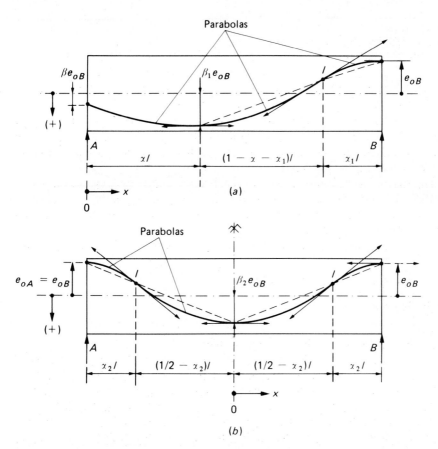

Figure 10.12 Typical tendon profiles suggested in Ref. 10.7. (*a*) Profile for a typical end-span made out of three parabolas. (*b*) Profile for a typical interior span.

are given in Figs. 10.13 and 10.14. They are expressed in terms of the support eccentricity e_{oB} and other nondimensional coefficients explained in the figures. The sign convention adopted in this text is still valid. Hence, e_{oB} is negative and the factors β, β_1, and β_2 as shown, for instance, in Fig. 10.13 are negative. In general, they can be negative or positive. In developing the above equations, the following assumptions were made: (1) the members are shallow, hence, the horizontal component of F is assumed equal to F and e_o is small with respect to l, (2) a horizontal tangent exists at the common point between the first two parabolas of the end span, (3) horizontal tangents exist at all supports and at midspans of the inter-mediate spans, and (4) the prestressing force was assumed constant throughout. In addition, a special condition was derived to ensure a common tangent at the inflection point I between two successive parabolas. It is given in Fig. 10.13 for a typical end span and in Fig. 10.14 for a typical interior span. It can be shown in

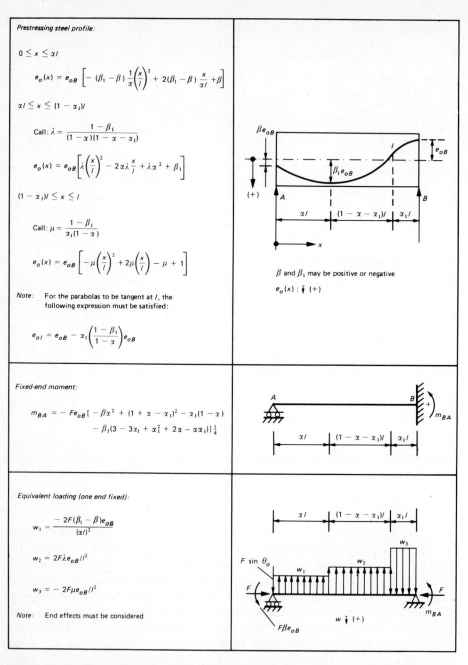

Prestressing steel profile:

$0 \le x \le \alpha l$

$$e_o(x) = e_{oB}\left[-(\beta_1 - \beta)\frac{1}{\alpha}\left(\frac{x}{l}\right)^2 + 2(\beta_1 - \beta)\frac{x}{\alpha l} + \beta\right]$$

$\alpha l \le x \le (1 - \alpha_1)l$

Call: $\lambda = \dfrac{1 - \beta_1}{(1 - \alpha)(1 - \alpha - \alpha_1)}$

$$e_o(x) = e_{oB}\left[\lambda\left(\frac{x}{l}\right)^2 - 2\alpha\lambda\frac{x}{l} + \lambda\alpha^2 + \beta_1\right]$$

$(1 - \alpha_1)l \le x \le l$

Call: $\mu = \dfrac{1 - \beta_1}{\alpha_1(1 - \alpha)}$

$$e_o(x) = e_{oB}\left[-\mu\left(\frac{x}{l}\right)^2 + 2\mu\left(\frac{x}{l}\right) - \mu + 1\right]$$

Note: For the parabolas to be tangent at *l*, the following expression must be satisfied:

$$e_{ol} = e_{oB} - \alpha_1\left(\frac{1 - \beta_1}{1 - \alpha}\right)e_{oB}$$

β and β_1 may be positive or negative

$e_o(x):$ ↓ $(+)$

Fixed-end moment:

$$m_{BA} = -Fe_{oB}\left[-\beta\alpha^2 + (1 + \alpha - \alpha_1)^2 - \alpha_1(1 - \alpha) - \beta_1(3 - 3\alpha_1 + \alpha_1^2 + 2\alpha - \alpha\alpha_1)\right]\frac{1}{4}$$

Equivalent loading (one end fixed):

$$w_1 = \frac{-2F(\beta_1 - \beta)e_{oB}}{(\alpha l)^2}$$

$$w_2 = 2F\lambda e_{oB}/l^2$$

$$w_3 = -2F\mu e_{oB}/l^2$$

Note: End effects must be considered

Figure 10.13 Expressions for the eccentricities, fixed-end moment, and equivalent loads for a typical end span. *(Adapted from Ref. 10.7.)*

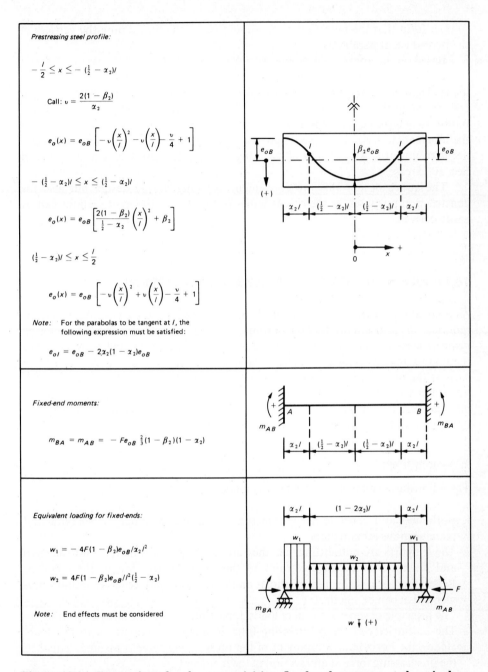

Prestressing steel profile:

$$-\frac{l}{2} \le x \le -(\tfrac{1}{2} - \alpha_2)l$$

Call: $\upsilon = \dfrac{2(1 - \beta_2)}{\alpha_2}$

$$e_o(x) = e_{oB}\left[-\upsilon\left(\frac{x}{l}\right)^2 - \upsilon\left(\frac{x}{l}\right) - \frac{\upsilon}{4} + 1\right]$$

$$-(\tfrac{1}{2} - \alpha_2)l \le x \le (\tfrac{1}{2} - \alpha_2)l$$

$$e_o(x) = e_{oB}\left[\frac{2(1 - \beta_2)}{\tfrac{1}{2} - \alpha_2}\left(\frac{x}{l}\right)^2 + \beta_2\right]$$

$$(\tfrac{1}{2} - \alpha_2)l \le x \le \frac{l}{2}$$

$$e_o(x) = e_{oB}\left[-\upsilon\left(\frac{x}{l}\right)^2 + \upsilon\left(\frac{x}{l}\right) - \frac{\upsilon}{4} + 1\right]$$

Note: For the parabolas to be tangent at l, the following expression must be satisfied:

$$e_{ol} = e_{oB} - 2\alpha_2(1 - \alpha_2)e_{oB}$$

Fixed-end moments:

$$m_{BA} = m_{AB} = -Fe_{oB}\,\tfrac{2}{3}(1 - \beta_2)(1 - \alpha_2)$$

Equivalent loading for fixed-ends:

$$w_1 = -4F(1 - \beta_2)e_{oB}/\alpha_2 l^2$$

$$w_2 = 4F(1 - \beta_2)e_{oB}/l^2(\tfrac{1}{2} - \alpha_2)$$

Note: End effects must be considered

Figure 10.14 Expressions for the eccentricities, fixed-end moments and equivalent loads for a typical interior span. *(Adapted from Ref. 10.7.)*

simpler form that the inflection point should be on the line joining the vertices of the two adjacent parabolas.

Based on the above steel profile, Khachaturian and Gurfinkel developed direct expressions for the equivalent loads and the fixed-end moments generated by such loads. These are summarized in Figs. 10.13 and 10.14 with the notation and sign convention of this text. They allow the quick use of the method of moment distribution to determine the total moments due to prestressing. Note that, in order for the moments to have the correct sign, the load is assumed positive downward and negative upward. An example illustrating the use of these formulas is given in Sec. 10.15, case *c*.

The above design aids are convenient not only during design but also for construction purposes, as the eccentricities of the steel at different sections can be easily derived.

10.11 WORKING STRESS ANALYSIS AND DESIGN

As pointed out in Sec. 10.2, the analysis of prestressed concrete continuous beams requires, in addition to the background needed to analyze simply supported beams, the background developed in the preceding sections and the knowledge of a method to analyze statically indeterminate structures subjected to external loads, such as the moment distribution method. Analysis and design steps are proposed below but are not unique. Many paths can be followed to arrive at the same answer and the user may want to develop his own approach or introduce appropriate modifications to fit a particular type of problem.

1. Assumptions

The following assumptions are generally made:

1. Both steel and concrete act within their linear elastic range of behavior and plane sections remain plane.
2. The supports are nonsettling (i.e., they are fixed against transverse movements) and do not provide any restraint to the axial deformation of the member. Restricting axial deformation (short- and long-term) may generate tertiary moments that can be significant.
3. The principle of superposition of loads, forces, moments, and stresses is valid.
4. The eccentricity of the prestressing force is small in comparison to the span. Hence, the horizontal component of the prestressing force is assumed equal to the prestressing force at any section of the member.
5. Friction is negligible, that is, the prestressing force is assumed constant throughout the length of the member. In practice, the design is revised after a first iteration to account for frictional effects.
6. Gross-sectional properties are used at first.

2. Analysis or Investigation

Given materials and sectional properties, tendon profile, prestressing force, and loading conditions, the following steps are proposed:

1. For the given prestressing force and steel profile, determine the equivalent loading on the member. (Use information provided in Secs. 10.8 and 10.10.)
2. For the equivalent loads obtained in 1, determine (say by the method of moment distribution) the prestressing moments $M_F(x)$ at the supports. (Use information provided in Secs. 10.9 and 10.10.)
3. Determine the secondary moments at the supports and at any section x, using Eqs. (10.3) and (10.7).
4. Determine the eccentricity of the zero load C line (ZLC line) from Eq. (10.4).
5. Determine the maximum and minimum moments at each section x due to external loads.
6. The rest of the analysis essentially follows the steps used in Chap. 4 for simply supported beams, except that $e_{oc}(x)$ replaces $e_o(x)$ in the computation of stresses, as pointed out in Sec. 10.9.

It was shown in Sec. 10.9 that, instead of using $e_{oc}(x)$ to compute the stresses, $e_o(x)$ could have been used, provided the secondary moment is added to the external moments. This approach may also seem attractive when the steel envelopes are built and deflections are computed.

3. Design

Two major approaches can be followed to design continuous beams. In the first approach the mechanical and practical constraints of the problem are directly used to arrive at an appropriate tendon profile and prestressing force. In the second approach, a tendon profile, which satisfies the practical constraints of the problem, is selected a priori, then the prestressing force is determined (and the profile adjusted if needed) to satisfy the mechanical constraints of the problem. The first approach is very similar to an optimization problem. It can be substantially more involved than the second and would require the use of optimization techniques. Nevertheless, there have been some attempts to provide a solution within reach of the designer (Ref. 10.8). Only the second approach will be followed here. It leads, according to the author's experience, to a sure and fast solution, even when revisions and iterations are considered.

In order further to simplify the problem, it will be assumed that the concrete cross section is provided. If not, a section may be dimensioned as if the beam was simply supported; then its depth is reduced by about 20 percent. Some revision may be needed. Assuming the material properties and the loading conditions are known the problem is reduced to finding an acceptable prestressing force and tendon profile.

Two methods are described next: one assumes that a nonconcordant steel profile will be used and the other assumes that a concordant steel profile will be arrived at.

Nonconcordant steel profile. The following steps are proposed:

1. Determine minimum and maximum moments at each section due to external loads.
2. Select a practical tendon profile that has a maximum practical eccentricity at each critical section. Critical sections are support sections with maximum negative moments and midspan or near midspan sections with maximum positive moments. A steel profile made out of parabolas can be used as suggested in Sec. 10.10.
3. Determine the equivalent loads for the above profile (see Sec. 10.8). As the prestressing force F is unknown, the equivalent loads are determined in function of F. (A unit value can also be assumed for F, provided the moments obtained in steps 3 and 4 are, in the final design, multiplied by the design value of F.)
4. Determine (say by the method of moment distribution) the moments due to prestressing at the supports. Use Sec. 10.9 and Fig. 10.11 if necessary for the fixed-end moments.
5. Determine the secondary moments at the supports and at other critical sections (Eqs. (10.3) and (10.7)).
6. Determine the eccentricity of the ZLC line at each of the critical sections (Eq. (10.4)).
7. Determine the needed prestressing force at each of the critical sections using the eccentricity of the ZLC line and the same stress inequality conditions described in Chap. 4 for simply supported beams. Select the highest value obtained, call it F, and assume it is used throughout. Check if stresses at all critical sections are satisfactory. If not, the same remedies as used in Chap. 4 for simply supported beams can be followed, provided e_o is replaced by e_{oc}.

Note that, by adjusting the steel profile, the secondary moments can be beneficially used to magnify or reduce the effect of the prestressing force. They can also be eliminated, leading to a concordant profile. The steel profile can be linearly transformed without changing the ZLC line, thus providing some flexibility in design. Enough tolerance should be allowed in the steel profile selected to accommodate some variations in the actual prestressing force due to friction. In finalizing the design, the limit kern and the limit zone can be built to check if the steel is within the limit zone at each section, hence ensuring that none of the allowable stresses are violated.

A typical design example with a nonconcordant steel profile is given in Sec. 10.15, case **c**.

Concordant Steel Profile. A concordant steel profile can be obtained as described in Sec. 10.7. For such a profile, the secondary moments are zero and the design approach is similar to that of simply supported beams (Chap. 4). The following steps are suggested:

1. Determine the maximum and minimum moments at each section due to external loads.
2. Using the eccentricities of the concordant profile, determine the required prestressing force at each critical section. Select the highest value obtained, call it F, and use it throughout the member.
3. Check if F is acceptable at all critical sections, that is, check if the stress inequality conditions (Tables 4.2 and 4.3) are satisfied. If they are, build the limit zone and check if the steel profile is within the limit zone at each critical section. If it is, stresses are satisfied everywhere. Otherwise, a different profile must be sought, as described in Sec. 10.7, and steps 2 and 3 must be repeated.

Of course, it is assumed in step 2 that a required prestressing force can be computed for each critical section. If this is not possible (no feasible domain), the assumed section properties or the steel profile must be appropriately modified.

A typical design example with a concordant steel profile is developed in Sec. 10.15, case **b.**

10.12 LIMIT KERN AND LIMIT ZONE

The concepts of limit kern and limit zone described in Sec. 4.12 for statically determinate beams remain the same for continuous beams. The eccentricity of the C line at any section should remain within the upper and lower limits of the kern to satisfy all allowable stresses. The only difference with the case of simply supported beams is that the eccentricity of the C line takes into account not only the external moments but also the secondary moments due to prestressing. Note that the secondary moments, from the C line viewpoint, can be considered as external moments due to secondary reactions and can be added to the external moments due to external loads. In a reciprocal manner, given the limit kern, the limit zone or steel envelopes can be obtained by plotting from the upper and lower limits of the kern two vectors of magnitude $(M_{max} + M_2)/F$ and $(M_G + M_2)/F_i$ or $(M_{min} + M_2)/F$ whichever is smaller. This last procedure takes into consideration the fact that the minimum moment which, in a continuous beam, is influenced by the live load, is different from the dead load moment. If the secondary moments are not added to the external moments, a limit zone for the Zero-Load-C line is obtained. This limit zone for the ZLC line is the same as if the beam is statically determinate or if a concordant cable is used leading to zero secondary moments. In such a case, the limit zones for the ZLC line and for the steel become the same. The above procedure is graphically illustrated in Fig. 10.15 for a typical span section where the secondary moment is assumed positive. For a span section where the maximum and minimum moments are positive, Eqs. (4.22) and (4.23) for the limit kern are

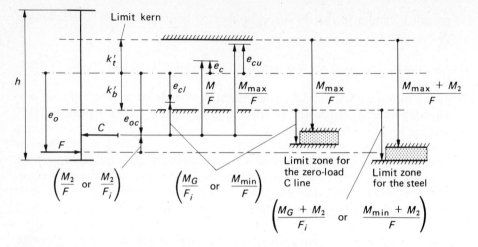

Figure 10.15 Graphical illustration of the relation between the limit kern and the limit zone for a typical span section with positive secondary moment.

valid and the envelopes of the steel defining the limit zone become:

$$e_{ou}(x) = k'_t + \frac{M_{max}(x) + M_2(x)}{F} \tag{10.14}$$

$$e_{ol}(x) = \text{the smaller of} \begin{cases} k'_b + \dfrac{M_G + M_2}{F_i} \\ \text{and} \\ k'_b + \dfrac{M_{min} + M_2}{F} \end{cases} \tag{10.15}$$

Note that M/F, in general, is equivalent to an eccentricity and is positive downward. When the moments are negative, such as at the supports, the values of k'_t and k'_b have to be redefined and the comments made at the end of Sec. 4.12 can be used for a speedy solution. A typical example of limit zone is shown in Fig. 10.28.

10.13 LOAD-BALANCING METHOD

1. General Approach

The analysis of prestressed concrete statically indeterminate structures is hindered by the existence of secondary moments due to prestressing. The load-balancing method proposed by T. Y. Lin reduces the analysis to that of a nonprestressed structure in which consideration of secondary moments is essentially bypassed (Refs. 10.5 and 10.9). It offers a particularly simple and elegant means to design continuous beams, slabs, shells, and frames.

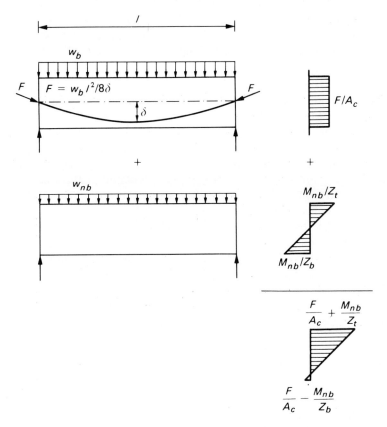

Figure 10.16 Typical stresses under balanced and unbalanced loads.

Balancing the external load consists of selecting a prestressing force and steel profile which create a transverse load exactly equal and opposite to the external load. Section 10.8 and Fig. 10.9 provide the necessary background for such procedure. For instance, to balance a uniform load w_b in a simply supported beam, a parabolic steel profile with zero end eccentricities can be selected (Fig. 10.16). The prestressing force F needed will be a function of the load to be balanced, w_b, and the acceptable sag δ of the tendon. As the transverse load created by the tendons balances exactly the external load, a uniform state of stress F/A_c develops throughout the beam. The beam remains essentially level and no deflection or camber is observed. Note that, in order to balance the load, the end eccentricities were taken equal to zero, otherwise an end moment which disturbs the uniform state of stress is generated. The same approach for load balancing can be applied to a continuous beam, noting that eccentricities at intermediate supports are not necessarily equal to zero because in such a case the moments on each side of the support balance each other. If due to load balancing the continuous beam remains level, no secondary moments are generated (under F and the balanced load w_b) and the beam can be analyzed for the unbalanced load w_{nb} as if it was a continuous nonprestressed beam in which a uniform compression F/A_c is present. The moments $M_{nb}(x)$ induced by the unbalanced loads are then calculated by elastic analysis and the corresponding stresses at the extreme fibers $M_{nb}(x)/Z_t$ and $-M_{nb}(x)/Z_b$ determined. Resulting stresses due to the uniform compression and the unbalanced moments are computed and compared with the allowable stresses. This is illustrated in Fig. 10.16 for the case of a simply supported beam.

The balanced load need not necessarily be uniform. It can be a concentrated load, a uniform load, or a combination thereof (Fig. 10.17). For a uniform load, a shallow parabolic tendon profile is generally selected, while a linear profile with a sharp directional change is used for a concentrated load. Assuming shallow members, the principle of superposition holds, hence, the combined profile of Fig. 10.17 would balance both the applied uniform and concentrated loads.

In an approach similar to that followed in Sec. 10.11, it will here also be assumed that the concrete cross section is given and that the design is reduced to finding the prestressing force and its profile. Referring to a continuous beam (Fig. 10.18) where a typical end span, intermediate span, and cantilever span are shown, the following design steps are suggested, assuming uniform loadings and relatively shallow members:

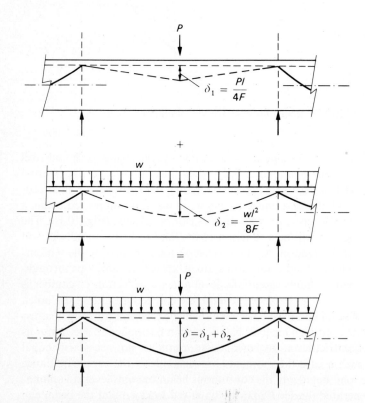

Figure 10.17 Superposition in load balancing.

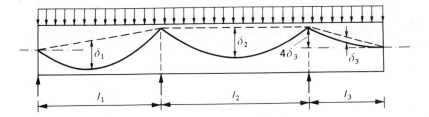

Figure 10.18 Typical load balancing in a continuous beam.

1. Select the balanced load w_b. It is generally taken equal to the dead load plus the sustained part of the live load, if any. This will ensure a level structure even under long-term effects. Note that for each span a different value of the uniform balanced load may be used.
2. Select a steel profile made out of parabolas, having maximum practical eccentricities at the intermediate supports and maximum feasible sags in span. Zero eccentricities must be present at the end supports. At cantilever ends, if any, the slope of the tendon has also to be zero.
3. Determine the prestressing force required in each span to balance the balanced load of that span. (Use the relations given in Fig. 10.9.) Select the highest value obtained, call it F and assume it is adopted throughout the member. The sags are then adjusted (reduced) in those spans where the required force was less than F. Their final values are obtained from the relationship between F and the balanced load.

    Note that, in cantilever spans, because of the implied relationship between the sag and the eccentricity at the near-support section (Fig. 10.18), the near-support eccentricity may have to be adjusted to achieve the required sag and this will influence the sag in the adjacent span.
4. Compute the unbalanced moment $M_{nb}(x)$ due to the unbalanced load w_{nb} by elastic analysis as if the beam was continuous nonprestressed.
5. Check if the stresses at critical sections and other key sections are within allowable limits. Stresses are given by:

$$\frac{F}{A_c} + \frac{M_{nb}(x)}{Z_t} \tag{10.16}$$

and

$$\frac{F}{A_c} - \frac{M_{nb}(x)}{Z_b} \tag{10.17}$$

If stresses are acceptable, the design can be pursued. If they are not, the design should be revised. Generally, either the prestressing force has to be increased, or the concrete cross section has to be modified.
6. Modify the theoretical tendon profile shown in Fig. 10.18 by providing smooth transitions over the supports and check the effects of such modification. It is likely that secondary moment will be generated. These moments are often neglected in slabs (Chap. 11), but may be significant in beams.

The above steps have generally assumed that no major problem occurs. With some experience, the user should be able to resolve difficulties if they occur and avoid them by proper dimensioning in the preliminary phase of the design. An example of load balancing applied to a two-span continuous beam is given in Sec. 10.15, case **e.**

2. Load-Balancing of Edge-Supported Slabs

The principle of load balancing is very appropriate for the design of slabs, especially two-way slabs, as these structures are highly indeterminate and their analysis or design by any other technique necessitates a number of approximations. Slabs are covered in detail in Chap. 11. Here, only edge supported slabs are addressed to illustrate the use of load balancing (Fig. 10.19). Assuming a uniform load, w_b, to be balanced and assuming parabolic tendons are used, the load-balancing relationship applied to edge-supported slabs becomes:

$$\frac{8F_x \delta_x}{l_x^2} + \frac{8F_y \delta_y}{l_y^2} = w_b \qquad (10.18)$$

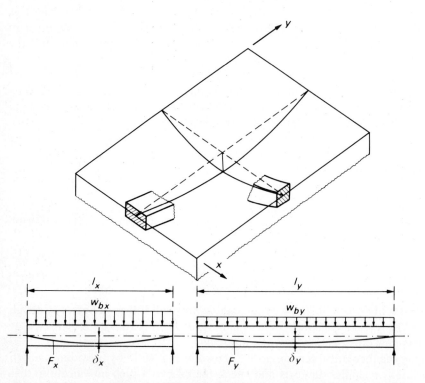

Figure 10.19 Two-dimensional load balancing in prestressed concrete slabs.

where F_x and F_y are the forces in the x and y directions of span l_x and l_y and δ_x and δ_y are the corresponding tendon sags (Fig. 10.19). The values of δ_x and δ_y are selected as maximum practical. However, an infinite set of values F_x and F_y may satisfy Eq. (10.18). It is generally more economical to balance a larger portion of w_b in the shorter direction. That portion is selected in such a way that the resulting prestressing force in the longer direction generates a uniform compression of the order of 150 psi ($\simeq 1$ MPa). In practice, such compression is believed necessary to avoid cracking. An average compression of 125 psi (0.86 MPa) is considered a minimum (Sec. 11.8).

Example: Load-Balancing of an Edge-Supported Slab

A 30 × 50 ft rectangular slab is simply supported on walls on four sides. It is to be designed to carry, in addition to its own weight, a live load of 100 psf. The following information is given: $f'_c = 5000$ psi, $f_{pe} = 168$ ksi. The prestressing steel consists of single-strand unbonded tendons with 0.6-in diameter covered by a plastic tubing which gives them an outside diameter of 0.75 in. The strand cross-sectional area is 0.216 in², leading to an effective force per strand of 36.288 kips. It is estimated that 20 percent of the live load can be considered a sustained load.

Let us provide a quick dimensioning of the slab thickness and the required prestressing steel.

Referring to Fig. 10.19 where the notation is explained, let us estimate the slab thickness assuming a span-to-depth ratio of 40 for the shorter span, that is:

$$h = \frac{l_x}{40} = \frac{30 \times 12}{40} = 9 \text{ in}$$

The tendons are to be set in two layers normal to each other and forming a grid with zero eccentricities along the edges of the slab. The lowest layer is chosen for the short direction. Assuming a $\frac{3}{4}$-in net cover to the lowest layer and assuming that the two layers touch each other, leads to the following sags: $\delta_x = 3.375$ in and $\delta_y = 2.625$ in.

Assuming lightweight sand is used in the concrete, the dead weight of the slab is estimated at 100 psf. The balanced load is chosen as the sum of the dead weight and the sustained part of the live load, that is $w_b = 100 + 0.2 \times 100 = 120$ psf. The corresponding unbalanced load is $w_{nb} = 80$ psf.

Let us determine the prestressing force needed in each direction assuming a slab strip with unit width. The effective prestressing force in the longitudinal direction is assumed to achieve a uniform compression of 150 psi, hence:

$$F_y = 150 \times 9 \times 12 = 16,200 \text{ lb/ft}$$

The effective force in the short direction is then obtained from Eq. (10.18):

$$\frac{8F_x \delta_x}{l_x^2} + \frac{8F_y \delta_y}{l_y^2} = w_b$$

that is:

$$\frac{8F_x \times 3.375/12}{(30)^2} + \frac{8 \times 16,200 \times 2.625/12}{(50)^2} = 120$$

or
$$0.0025F_x + 11.34 = 120$$

which leads to $F_x = 43,464$ lb/ft.

As one strand develops an effective force of 36,288 lb, the strand spacing will be:

Short direction $\quad s_x = \dfrac{36,288 \times 12}{43,464} \simeq 10$ in

Long direction $\quad s_y = \dfrac{36,288 \times 12}{16,200} \simeq 26.9$ in

These are acceptable spacings in comparison to the slab thickness. In practice, a value of 27 in will be used for s_y.

The above prestressing combined with the effect of balanced load produces uniform compressive stresses in both the short and the long directions of the slab. They are given by:

$$\sigma_{gx} = \frac{F_x}{A_c} = \frac{43,464}{12 \times 9} = 402.4 \text{ psi}$$

$$\sigma_{gy} = \frac{F_y}{A_c} = \frac{16,200}{12 \times 9} = 150 \text{ psi}$$

Let us determine the moments and corresponding stresses due to the unbalanced load $w_{nb} = 80$ psf. Using the 1963 edition of the ACI building code, where coefficients are provided to calculate moments in slabs, we have for the case of a slab simply supported on its edges and having $l_x/l_y = 0.60$:

$$M_x = 0.081 \times 80 \times 30^2 = 5832 \text{ lb-ft} = 69,984 \text{ lb-in/ft}$$

$$M_y = 0.010 \times 80 \times 50^2 = 2000 \text{ lb-ft} = 24,000 \text{ lb-in/ft}$$

Given section moduli $Z_t = Z_b = 162$ in^3/ft in each direction, the above moments produce flexural stresses on the concrete top and bottom fibers given by:

Short direction $\quad \sigma = \pm \dfrac{69,984}{162} = \pm 432$ psi

Long direction $\quad \sigma = \pm \dfrac{24,000}{162} = \pm 148.15$ psi

The resulting stresses will be:

Short direction $\quad \sigma_t = 402.4 + 432 = 834.4$ psi

$$\sigma_b = 402.4 - 432 = -29.6 \text{ psi}$$

Long direction $\quad \sigma_t = 150 + 148.15 = 298.15$ psi

$$\sigma_b = 150 - 148.15 = 1.85 \text{ psi}$$

where the subscripts t and b are for top and bottom fibers. It can be observed that the above stresses are acceptable. The design, however, is not completed. Ultimate strength requirements and cracking conditions must be checked. It is very likely that additional nonprestressed reinforcement will be added in span and near the end anchorages to resist stress concentrations and avoid concrete splitting.

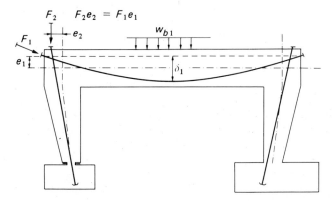

Figure 10.20 Typical load balancing in prestressed concrete frames.

3. Load Balancing of Frames

The load-balancing approach can be easily applied to the various members of a frame. Under balanced load uniform compression is present in all members and the frame can be analyzed for the unbalanced load as a nonprestressed frame. However, shortening of the various members due to the uniform compression must be accounted for in the design. It leads to moments and reactions that can be determined by elastic analysis and should be added to the effects of unbalanced loads.

In selecting the steel profile for load balancing, the forces and eccentricities at the joints between two members must be selected in such a way that the primary moments are equal (Fig. 10.20). The profile in each member depends on the balanced load. The balanced load can be the effect of an end moment such as in the legs of the frame shown in Fig. 10.20 where no direct transverse load is applied. In such a case a linear tendon is used.

The load-balancing method, applied to frames, can be very convenient for a fast first approximation design. However, because of the effects of shortening of the various members due to prestressing, elastic indeterminate analysis of the frame is needed and is to be added to that of the unbalanced load. As computerized techniques such as matrix structural analysis are available and could be efficiently used, it seems that one would gain in directly using them in accordance with the method described in Sec. 10.11 for beams. Nevertheless, the load-balancing method will help in providing a first estimate of the prestressing force in each member which reduces the number of iterations incurred in the design.

4. Limitations of Load Balancing

It is shown in the previous sections and in the example (Sec. 10.15) that load balancing can be a very efficient method for preliminary determination of the prestressing force and its profile. However, it has some serious drawbacks listed below:

1. The steel profile obtained by load balancing in continuous beams is characterized by abrupt changes in direction over the supports. This is, of course, not acceptable. If the profile is smoothed along a transition region over the supports, load balancing would not be valid any more and an analysis similar to that described in Sec. 10.11 would be required. Therefore, in some cases it is better to follow right away the procedure described in Sec. 10.11. However, in shallow continuous slabs, the change in angle at the supports is so small that it can be accommodated over a very small transition region of the same order as the width of the support. Practically, load balancing can then still be considered valid (see Chap. 11).
2. Load balancing does not directly account for moments induced by tendons anchored eccentrically at the end supports. The eccentricity of the steel centroid at the end supports must be zero.
3. Load balancing does not consider the effects of friction, if any, along the member. However, this limitation is also present in the other methods, at least in the preliminary design phase.

10.14 ULTIMATE STRENGTH ANALYSIS

The analysis of a statically indeterminate prestressed concrete structure differs from that of a statically determinate structure by two main aspects: (1) secondary moments must be accounted for in the analysis and (2) the attainment of ultimate moment resistance at one critical section does not necessarily lead to the collapse of the structure; this is because plastic hinges may form at several critical sections and redistribution of moments occurs.

1. Secondary Moments

It was pointed out earlier that secondary moments can be considered as additional external moments at each section. In Sec. 18.10.3 the ACI code specifies that the moments used in computing the required strength shall be the sum of the moments due to factored loads (including redistribution, if permitted, as explained in Sec. 10.14) and secondary moments with a load factor of 1. The secondary moments are to be determined on the basis of effective prestress, not the force in the tendons at ultimate behavior (Ref. 10.10). Hence, referring to the required strength for one of the most common load combinations, we have:

$$M_u(x) = 1.4M_D(x) + 1.7M_L(x) + M_2(x) \qquad (10.19)$$

where $M_2(x)$ is the secondary moment at section x, obtained from elastic analysis. The ultimate moment resistance $\phi M_n(x)$ at a given section, can be computed as described in Chap. 5.

If the most critical section in a continuous beam has sufficient rotational capacity at ultimate, it behaves as a plastic hinge. Failure of the member will not follow the formation of the first plastic hinge. Instead, redistribution of moments occurs and, if the loading is increased, another section reaches its ultimate resistance, leading to another plastic hinge. A collapse mechanism develops in which each hingeing section provides a resistance equal to its own ultimate strength resistance. Failure of the member can then be predicted by limit analysis as illustrated next.

2. Limit Analysis

Limit analysis is based on the formation of plastic hinges in a structure, leading to a collapse mechanism. Theoretically, a plastic hinge shows an elastoplastic moment-rotation response and is quite representative of the behavior of steel structures. In reinforced and prestressed concrete the cracking of members at one critical section reduces substantially the corresponding stiffness and a pseudo-plastic response may develop. Such a response may simulate a plastic hinge if sufficient ductility exists at the critical section (Ref. 10.11). Adequate ductility can be achieved if a low reinforcing index q is used, if nonprestressed bonded reinforcement is added to the section, and if sufficient resistance to shear cracking is provided.

To illustrate the application and benefits of limit analysis, let us consider a continuous beam with two equal spans, subjected to an increasing uniform load w. Let us assume that the flexural resistance of the intermediate support section B is equal in magnitude to that of the near-midspan section. These sections are assumed to have sufficiently ductile behavior to allow the formation of plastic hinges.

A uniform load w, under which the beam behaves elastically, leads to a maximum moment at the intermediate support B, and a moment diagram shown in Fig. 10.21a. If the load is increased, it will eventually reach a value called w_{uB} for which the ultimate resistance M_n at section B is attained. (*Note:* ϕM_n is used instead of the nominal resistance M_n to incorporate right away the ACI design approach.) Section B is assumed to have an elastoplastic moment-curvature response as described in Fig. 10.21b. The value ϕM_n can be defined as the plastic moment of section B. The magnitude of the plastic moment at B is then given by:

$$\phi M_n = \frac{w_{uB} l^2}{8}$$

(10.20)

from which w_{uB} can be determined as:

$$w_{uB} = \frac{8\phi M_n}{l^2}$$

(10.21)

Because of the rotational capacity at section B, the beam does not collapse as the moment resistance is not yet attained at the near-midspan section. For all practical purposes, if the applied load exceeds w_{uB}, the support section B can be replaced by a plastic hinge offering a known resistance or plastic moment ϕM_n. The structure is essentially reduced to a statically determinate structure because a redundant moment has been replaced by a known moment. If the load is increased further, it will reach a value w_u at which a second plastic hinge forms at the near-midspan

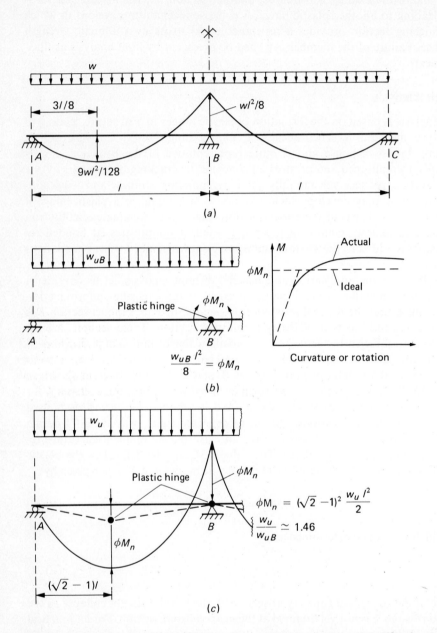

Figure 10.21 Moment redistribution in continuous beams. (a) Elastic moment diagram. (b) First plastic hinge. (c) Collapse mechanism and plastic moment diagram.

section (Fig. 10.21c). This section is assumed to have the same elastoplastic behavior as section B with an equal magnitude of plastic moment ϕM_n. The two plastic hinges added to the end support hinge constitute a mechanism leading to the collapse of the beam.

A mechanism can be analyzed by statics as known values of plastic moments ϕM_n are given at each hinge. For the present case (Fig. 10.21c) it is necessary to identify first the near-midspan section at which the plastic hinge develops. This is done by differentiating the equilibrium equation of moments at that section and setting the result equal to zero. The corresponding abcissa for this example is found equal to $(\sqrt{2} - 1)l$. Using moment equilibrium equations at the near midspan section and at the support B leads to the value of w_u for which collapse of the structure occurs, namely:

$$w_u = \frac{2\phi M_n}{l^2(\sqrt{2} - 1)^2}$$

The value of w_u is substantially higher than w_{uB}. Their ratio is:

$$\frac{w_u}{w_{uB}} = \frac{2}{(\sqrt{2} - 1)^2 8} \simeq 1.46$$

The above result suggests that the load that led to failure of the continuous beam is 46 percent higher than that which would have produced failure if the beam was simply supported. The formation of a plastic hinge at B has allowed, in effect, a redistribution of moments from support B (at which the moment did not increase beyond ϕM_n) to the near-midspan section. Redistribution of moments is recognized by researchers (Ref. 10.10) and permitted by the ACI code as described next.

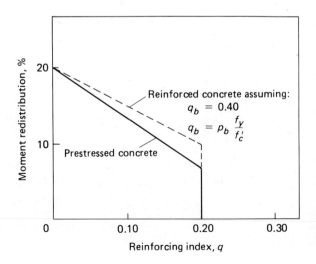

Figure caption: Reinforced concrete assuming:
$q_b = 0.40$
$q_b = p_b \dfrac{f_y}{f_c'}$

Prestressed concrete

Axis labels: Moment redistribution, % — Reinforcing index, q

Figure 10.22 ACI code allowable moment redistribution.

3. Redistribution of Moments

To account for the potential rotational capacity of reinforced and prestressed concrete structures, the ACI code (Sec. 18.10.4) allows a limited redistribution of moments in continuous beams containing a minimum amount of bonded reinforcement. It essentially states that, when the reinforcing index q given by Eq. (5.25) is less than 0.20, negative moments at the support can be increased or decreased by not more than 0.20 $(1 - q/0.30)$. This last relation is graphically shown in Fig. 10.22. The modified negative moments shall be used for calculating moments at sections within span for the same loading arrangement. If a T section is used, the reinforcing index q_w associated with the web, and described in Fig. 5.12, replaces q in the above relations.

10.15 EXAMPLE: DESIGN OF A PRESTRESSED CONTINUOUS BEAM

Let us determine the prestressing steel needed for the posttensioned two-equal-span continuous bridge beam of cross section described in Fig. 10.23. (See also Sec. 14.9.) The following design information is provided: $l = 120$ ft; section properties: $A_g = 30$ ft^2, $y_t = y_b = 2.5$ ft $= 30$ in, $Z_t = Z_b = 43$ ft^3, $I_g = 107.5$ ft^4, $k_b = 1.433$ ft $= 17.2$ in, $k_t = -17.2$ in; concrete properties and allowable stresses: $f'_c = 6000$ psi, $f'_{ci} = 4800$ psi, $\bar{\sigma}_{ti} = -208$ psi, $\bar{\sigma}_{ts} = -232$ psi, $\bar{\sigma}_{ci} = 2880$ psi, $\bar{\sigma}_{cs} = 2400$ psi; steel stresses: $f_{pi} = 181$ ksi, $f_{pe} = 145$ ksi,

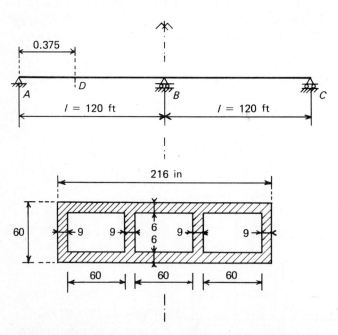

Figure 10.23 Span and cross section of example beam.

$\eta = f_{pe}/f_{pi} = 0.80$; loadings: $w_G = 4.5$ klf, $w_L = 3.2$ klf. Assume a maximum practical eccentricity $(e_o)_{mp} = 26$ in which applies to both the span and support sections.

Note that the section is completely symmetrical and is considered in its entirety as a single section. The steel stress along ghe beam is assumed constant, that is, frictional losses are neglected at least in the initial design phase.

Because of symmetry only two critical sections will be analyzed: the intermediate support section B and the near-midspan section D taken at $\frac{3}{8}l$ from the left support. The moments generated at each of these sections are summarized in Table 10.1 and can easily be verified. Several design cases are considered next.

(a) Determine a concordant tendon profile. For this, the following property will be used: any moment diagram or combination of moment diagrams due to any external loading plotted to any scale is a location for a concordant tendon.

To determine the profile, a combination of two loadings will be used: a uniform downward load throughout the length (here the dead load w_G is used) and a uniform upward load w over a fifth of the span adjacent to the support B (Fig. 10.7). The second load is intended to smoothe the profile over the support, since it produces a moment opposite to the dead load moment.

Because of symmetry, only one span AB is considered. Noting that the reaction at A is $3w_G l/8$, the moment at any section x from end support A due to the dead load w_G can be written as:

$$M_G(x) = w_G l^2 \frac{1}{2} \frac{x}{l} \left(\frac{3}{4} - \frac{x}{l} \right) = w_G l^2 \times \text{(factor)}$$

It is generally convenient first to compute separately the factor in function of x/l and then to compute the corresponding moment for any uniform load such as w_G. The numerical values for this case are given in Table 10.2.

Table 10.1 Moments for example beam

	Near-midspan section D	Intermediate support section B
x from left support	$\frac{3}{8}l = 45$ ft	$l = 120$ ft
Dead load moment, kips-ft	4556.2	$-8,100$
Live load moment, kips-ft	4374	$-5,832$
Minimum moment, kips-ft	3463	$-8,100$
Maximum moment, kips-ft	8930	$-13,932$

Table 10.2 Summary of computations to determine a concordant tendon profile

x/l		0.1	0.2	0.3	(3/8)
Dead load $w_G = 4.5$ klf	Factor	0.0325	0.055	0.0675	0.0703
	Moment, kips-ft (factor) $w_G l^2$	2106	3564	4374	4556
Additional load $w = 10$ klf over $0.2l$	Factor $\times 10^{+4}$	-3.8	-7.6	-11.4	-14.25
	Moment, kips-ft (factor) wl^2	-54.7	-109.4	-164.2	-205.2
Resulting moment,† kips-ft		2051	3455	4210	4351
$e_o(x) = \dfrac{M(x)}{K}$, in $K = 221.81$		9.25	15.58	18.98	19.62

† Rounded off values

The moment at any section x, due to the second loading w (uniform over $0.2l$) can be calculated as follows: determine the fixed-end moments at A and B (use Fig. 10.11); determine the end moment at B by moment distribution (because of symmetry only one span is used); determine the vertical reaction at A by satisfying the following equation leading to the end moment at B:

$$R_A l - w(0.2l)\left(\frac{0.2l}{2}\right) + 0.0162 wl^2 = 0$$

which leads to $R_A = -0.0038 wl$. Hence, the moment at any section of the span is given by:

For $x/l \le 0.8$

$$M(x) = R_A x = -0.0038\left(\frac{x}{l}\right) wl^2$$

For $0.8 \le x/l \le 1$

$$M(x) = R_A x + \frac{w}{2}(x - 0.8l)^2 = wl^2\left[-0.0038\frac{x}{l} + \frac{1}{2}\left(\frac{x}{l} - 0.8\right)^2\right]$$

The two above expressions can be written each as wl^2 miltiplied by a factor. The choice of w should be such that it would lead to a reduction in moments at and near the support B to smoothe the profile sufficiently. Some iteration may be needed. A value of $w = 10$ klf is used for this example and leads to the moments shown in Table 10.2.

The resulting moments are then calculated and the steel profile is determined using Eq. (10.9):

$$e_o(x) = \frac{M(x)}{K}$$

0.4	0.5	0.6	0.7	0.8	0.9	1
0.070	0.0625	0.045	0.0175	−0.02	−0.0675	−0.125
4536	4050	2916	1134	−1296	−4374	−8100
−15.2	−19	−22.8	−26.6	−30.4	15.8	162
−218.9	−273.6	−328.3	−383	−437.8	227.52	2332.8
4317	3776	2588	751	−1734	−4146	−5767
19.46	17.02	11.67	3.39	−7.82	−18.69	−26.0

where K is determined from the critical section that is B for which e_o is taken equal to the maximum practical eccentricity. Hence:

$$K = \frac{M_B}{(e_o)_{mp}} = \frac{-5767}{-26} = 221.81$$

The corresponding eccentricities at various sections are given in Table 10.2 and the steel profile is plotted in Fig. 10.24. The reader may want to check if the profile is within the feasible domain of the critical sections, as described in the next case.

(b) Determine the prestressing force, assuming that a concordant steel profile will be arrived at in the design. The problem is essentially reduced to the case of a simply supported beam where no secondary moments due to prestressing exist. Two critical sections will be analyzed to find the prestressing force, the support, and the near midspan sections B and D in a way similar to a simply supported cantilever beam. As the eccentricity of the ZLC line is the

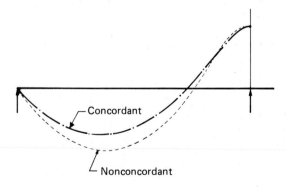

Concordant

Nonconcordant

Figure 10.24

same as that of the steel, the equations of Table 4.2, way 2, can be used to build the feasibility domain at each of these sections. These are given below, where e_o is expressed in inches and F_i in kips:

At support B:

Condition I $\qquad e_o = 17.2 + \dfrac{112{,}656}{F_i}$

Condition II $\qquad e_o = -17.2 + \dfrac{311{,}196}{F_i}$

Condition III $\quad e_o = 17.2 - \dfrac{13{,}932}{F_i}$

Condition IV $\quad e_o = -17.2 + \dfrac{187{,}432}{F_i}$

At near midspan section D:

Condition I $\qquad e_o = 17.2 + \dfrac{70{,}128}{F_i}$

Condition II $\qquad e_o = -17.2 + \dfrac{268{,}668}{F_i}$

Condition III $\quad e_o = 17.2 - \dfrac{88{,}962}{F_i}$

Condition IV $\quad e_o = -17.2 + \dfrac{112{,}402}{F_i}$

Note that in order to use Table 4.2 for the support section B, the section is assumed temporarily inverted with the negative moments and the eccentricity both becoming positive. The feasibility domains for the above two sections are plotted in Fig. 10.25. For clarity the domain for B is plotted with the e_o axis positive upward, while the domain for D is plotted with the axis positive downward. It can be observed that there is a feasibility domain for each section. The minimum prestressing force in each case is given by the intersection of line IV with the maximum practical eccentricity line, $|(e_o)_{mp}| = 26$ in. As the force at B is larger than at D, it controls the design. Hence, the prestressing force F_i is given by the abcissa of point A of Fig. 10.25. That is:

$$\frac{1}{F_i} \simeq 23 \times 10^{-5}$$

or

$$F_i \simeq 4339 \text{ kips}$$

and

$$F = 0.8F_i = 3471 \text{ kips}$$

Note that the above result can be also obtained from stress condition IV in which e_o is

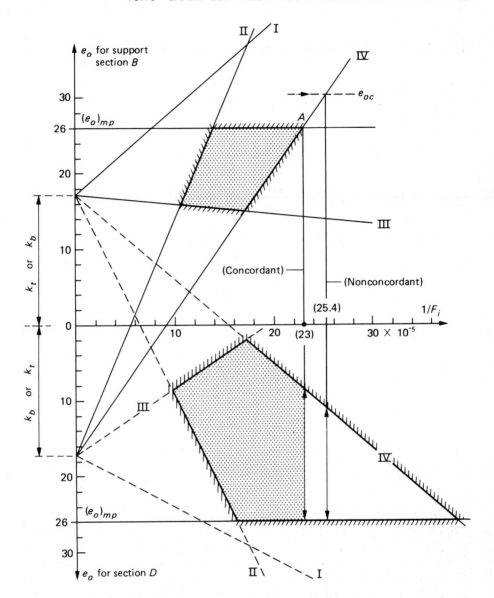

Figure 10.25 Feasibility domains for the support- and near-midspan sections.

replaced by $(e_o)_{mp} = 26$ in. The same prestressing force is used throughout the beam. Its eccentricity at section D, as shown in the domain of Fig. 10.25, can vary from about 8.6 to 26 in without violating any allowable stress. This flexibility will be found very convenient in selecting the tendon profile. For instance, the concordant tendon profile determined in case **a** with an eccentricity $e_o = 19.62$ in at section D (Table 10.2) and $e_o = -26$ in at section B is an acceptable profile and can be used here.

(c) Determine the prestressing force assuming a nonconcordant profile. Using the design aids of Sec. 10.10, let us consider a tendon profile made out of three parabolas and passing by the maximum eccentricities at sections B and D, that is $e_{oD} = 26$ in and $e_{oB} = -26$ in. Assume the following coefficients for the profile: $\beta = 0$; $\beta_1 = -1$; $\alpha = 0.375$; $\alpha_1 = 0.15$. The fixed-end moment at support B (Fig. 10.13) is given by:

$$m_B = -Fe_{oB} \tfrac{1}{4}[-\beta\alpha^2 + (1 + \alpha - \alpha_1)^2 - \alpha_1(1 - \alpha) - \beta_1(3 - 3\alpha_1 + \alpha_1^2 + 2\alpha - \alpha\alpha_1)]$$

$$m_B = -Fe_{oB} \tfrac{1}{4}[0 + 1.5006 - 0.0938 + 3.2663]$$

$$m_B = -Fe_{oB} \frac{4.673}{4} = -1.168Fe_{oB}$$

Because of symmetry, the fixed-end moment at B is also the end moment at B. Hence, the moment at B due to prestressing is given by:

$$M_{FB} = -1.168Fe_{oB}$$

As the primary moment is given by:

$$M_{1B} = -Fe_{oB}$$

the secondary moment is:

$$M_{2B} = M_{FB} - M_{1B} = -0.168Fe_{oB}$$

As e_{oB} is negative, it can be seen that the secondary moment is positive and, therefore, beneficial. The corresponding eccentricity of the Zero-Load-C line at B is given by:

$$(e_{oc})_B = e_{oB} - \frac{M_{2B}}{F} = e_{oB} + 0.168e_{oB}$$

$$(e_{oc})_B = 1.168e_{oB} = -1.168 \times 26 = -30.37 \text{ in}$$

Hence, the prestressing steel placed at an eccentricity of -26 in acts as if it was placed at -30.37 in. Going back to the domain of Fig. 10.25, we can plot a horizontal line for $(e_{oc})_B$ with magnitude 30.37 in and read the value of F_i from its intersection with line IV. It leads to:

$$\frac{1}{F_i} = 25.4 \times 10^{-5}$$

hence

$$F_i = 3937 \text{ kips}$$

and

$$F = 0.8F_i = 3150 \text{ kips}$$

The value of F can also be derived analytically from stress condition IV of Table 4.2, in which e_o is replaced by $(e_{oc})_B$, that is:

$$F = \frac{|M_{max}| + \bar{\sigma}_{ts} Z_b}{|(e_{oc})_B| - k_t}$$

$$F = \left(\frac{13,932 \times 12,000 - 232 \times 43 \times 12^3}{30.37 + 17.2}\right)10^{-3} = 3152 \text{ kips}$$

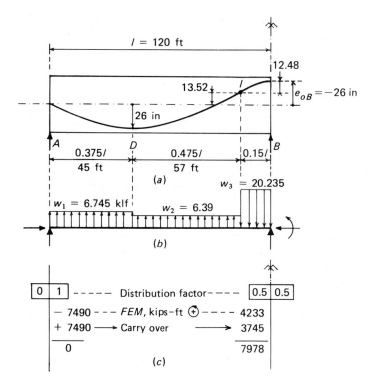

Figure 10.26 Determination of prestressing moment at support section. (a) Tendon profile. (b) Equivalent load. (c) Moment distribution.

This value of F is about 10 percent smaller than the value obtained in case **a** for the concordant profile and illustrates the beneficial effect of a nonconcordant profile.

For the above value of F, the secondary moment at B can be calculated as:

$$M_{2B} = -0.168Fe_{oB}/12 = 1147 \text{ kips-ft}$$

The secondary moment at section D can be determined from Eq. (10.7) and leads to $M_{2D} = 430$ kips-ft. The ZLC line at D will have an eccentricity of:

$$e_{oc} = e_{oD} - \frac{M_{2D}}{F} = 26 - \frac{430 \times 12}{3152} = 24.36 \text{ in}$$

It is inside the domain at D, hence, all stresses are satisfied. The steel profile for this case is also plotted in Fig. 10.24 for comparison with the concordant profile.

The reader may also want to check that a concordant steel profile would have been obtained if the following values were used: $e_{oB} = -26$ in; $e_{oD} = 22$ in; $\beta = 0$; $\alpha = 0.375$; $\alpha_1 = 0.17$. It can be arrived at by trial and error, using the design aids of Sec. 10.10 directly. This new concordant profile is different from the one obtained in case **a**.

(d) For the nonconcordant steel profile given in case **c** and a value of $F = 3152$ kips, determine the prestressing moment at B using the equivalent load approach. The steel profile is made out of three parabolas with characteristics shown in Fig. 10.26a.

The equivalent uniform loads over the three beam segments are given by:

$$w_1 = -\frac{8F\delta}{l_1^2} = \frac{8 \times 3152 \times 26/12}{(2 \times 45)^2} = -6.745 \text{ klf}$$

$$w_2 = -\frac{8F\delta}{l_2^2} = \frac{8 \times 3152 \times (26 + 13.52)/12}{(2 \times 57)^2} = -6.39 \text{ klf}$$

$$w_3 = \frac{8F\delta}{l_3^2} = \frac{8 \times 3152 \times 12.48/12}{(2 \times 18)^2} = 20.235 \text{ klf}$$

The load is assumed positive downward. Using the information given in Fig. 10.11, the fixed-end moments due to w_1 are determined as follows:

$$(m_A)_1 = -\frac{w_1 l^2}{12}(6 - 8\alpha + 3\alpha^2)\alpha^2$$

$$= \frac{6.745 \times 120^2}{12}(6 - 8 \times 0.375 + 3 \times 0.375^2)0.375^2$$

$$= 3894.84 \text{ kips-ft}$$

$$(m_B)_1 = -\frac{w_1 l^2}{12}(4 - 3\alpha)\alpha^3$$

$$= \frac{6.745 \times 120^2}{12}(4 - 3 \times 0.375)0.375^3$$

$$= 1227.14 \text{ kips-ft}$$

Similarly, it can easily be shown that the fixed-end moments due to w_2 and w_3 are:

$$(m_A)_2 = 3886.3 \text{ kips-ft}$$

$$(m_B)_2 = 5665.65 \text{ kips-ft}$$

$$(m_A)_3 = -290.93 \text{ kips-ft}$$

$$(m_B)_3 = -2659.33 \text{ kips-ft}$$

Summing up the effects of the above three loads leads to the resulting fixed-end moments:

$$m_A = 3894.84 + 3886.3 - 290.93 \simeq 7490 \text{ kips-ft}$$

$$m_B = 1227.14 + 5665.65 - 2659.33 \simeq 4233 \text{ kips-ft}$$

To obtain the end moments, the moment distribution method is applied to one span, as shown in Fig. 10.26c. It leads to the moment due to prestressing at support B:

$$M_{FB} = 7978 \text{ kips-ft}$$

The prestressing moment at B is the sum of the primary and secondary moments, namely

$$M_{FB} = M_{1B} + M_{2B} = -Fe_{oB} + M_{2B}$$

from which the secondary moment is determined:

$$M_{2B} = M_{FB} + Fe_{oB} = 7978 + 3152(-26/12) \simeq 1149 \text{ kips-ft}$$

The above value of M_{2B} is essentially the same as that obtained in case **c**.

(e) Determine the prestressing force, using load balancing. Let us attempt to balance the dead load plus 10 percent of the live load, that is:

$$w_b = 4.5 + 0.10 \times 3.24 = 4.824 \text{ klf}$$

and

$$w_{nb} = 0.9 w_L = 2.916 \text{ klf}$$

As the balanced load is a uniform load, a parabolic steel profile is selected. It should have zero eccentricity at the end supports. Because of symmetry, only one span is considered; thus $e_{oA} = 0$. In order to obtain the maximum sag (or the smallest prestressing force), the maximum practical eccentricities at the intermediate support B and in span at the vertex of the parabola are used. Hence, the eccentricity at the vertex is $e_{ov} = 26$ in (Fig. 10.27). The abcissa x_v of the vertex with respect to the left support A must be such that the chosen parabola passes by a point of zero eccentricity at A and a point of eccentricity $e_{oB} = -26$ in at B. The following relation can easily be derived:

$$x_v^2 = \left(\frac{e_{ov} - e_{oA}}{e_{ov} - e_{oB}} \right) (l - x_v)^2$$

from which the value of x_v is computed. For this example $x_v = 49.7$ ft.

The steel profile is shown in Fig. 10.27. The sag of the parabola at the vertex is determined from:

$$\delta = e_{ov} - (e_{oA} + e_{oB}) \frac{x_v}{l} = 26 + 26 \frac{49.7}{120} \simeq 36.8 \text{ in}$$

Given w_b and δ the prestressing force can be obtained from (Fig. 10.9):

$$w_b = \frac{8 F \delta}{l^2}$$

or

$$F = \frac{w_b l^2}{8 \delta} = \frac{4.824 \times 120^2}{8 \times 36.8/12} = 2831.5 \text{ kips}$$

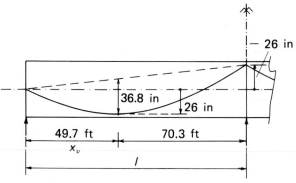

Figure 10.27 Tendon profile for load balancing.

A uniform compressive stress is generated in the section under the effects of w_b and F. Therefore the beam is level. Let us determine the maximum moments induced by the nonbalanced load:

At the support section B:

$$M_{nb} = -\frac{w_{nb}\,l^2}{8} = -\frac{2.916 \times 120^2}{8} = -5248.8 \text{ kips-ft}$$

At the section of maximum positive moment assumed at $x = \frac{3}{8}l$:

$$M_{nb} = \frac{3w_{nb}\,l^2}{32} = \frac{3 \times 2.916 \times 120^2}{32} = 3936.6 \text{ kips-ft}$$

The extreme fiber stresses in the concrete resulting from the uniform compression due to F and the effects of unbalanced moments are given by Eqs. (10.16) and (10.17):

At the support section B:

$$\sigma_t = \frac{F}{A_c} + \frac{M_{nb}}{Z_t} = \frac{2831.5 \times 10^3}{30 \times 144} - \frac{5248.8 \times 12,000}{43 \times 1728}$$

$$= 655.4 - 847.7 \simeq -192 \text{ psi}$$

$$\sigma_b = \frac{F}{A_c} - \frac{M_{nb}}{Z_b} = 655.4 + 847.7 \simeq 1503 \text{ psi}$$

At the section of maximum positive moment:

$$\sigma_t = \frac{F}{A_c} + \frac{M_{nb}}{Z_t} = \frac{2831.5 \times 10^3}{30 \times 144} + \frac{3936.6 \times 12,000}{43 \times 1728}$$

$$= 655.4 + 635.7 \simeq 1291 \text{ psi}$$

$$\sigma_b = \frac{F}{A_c} - \frac{M_{nb}}{Z_b} = 655.4 - 635.7 \simeq 20 \text{ psi}$$

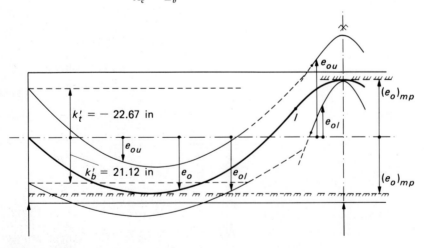

Figure 10.28

The above stresses are acceptable. A similar conclusion would be reached if F_i was used instead of F. Note that, in practice, the steel profile must show a smooth transition over the support B and, therefore, an analysis as described in Sec. 10.11 would be ultimately required. Nevertheless, the load-balancing method gave a very good first approximation value of the prestressing force comparable to those obtained in cases **b** and **c** above.

(f) The reader may want to pursue this example to determine the limit zone and the ultimate moment resistance of the beam. The limit zone is shown in Fig. 10.28 for the nonconcordant steel profile used in case **c** and offers a relatively large margin of flexibility. It will be found that ultimate strength requirements are also largely satisfied.

REFERENCES

10.1. Y. Guyon: "A Study of Continuous Beams and of Statically Redundant Systems in Prestressed Concrete," *Annales de l'Institut Technique du Bâtiment et des Travaux Publics*, vol. 8, September 1945, Translation no. 33 by Cement and Concrete Association, London, 1951.

10.2. ———: *Prestressed Concrete*, vol. 2, John Wiley & Sons, New York, 1960.

10.3. F. Leonhardt: "Continuous Prestressed Concrete Beams," *ACI Journal*, vol. 49, March 1953, pp. 617–636.

10.4. A. L. Parme and G. H. Paris: "Designing for Continuity in Prestressed Concrete Structures," *ACI Journal*, vol. 47, September 1951, pp. 54–64.

10.5. T. Y. Lin: *Design of Prestressed Concrete Structures*, 2d ed., John Wiley & Sons, New York, 1963, Chaps. 10 and 11.

10.6. J. R. Libby: *Modern Prestressed Concrete: Design Principles and Construction Methods*, 2d ed., Van Nostrand-Reinhold, New York, 1977, Chap. 8.

10.7. N. Khachaturian and G. Gurfinkel: *Prestressed Concrete*, McGraw-Hill Book Company, New York, 1969, Chap. 10.

10.8. J. Fauchart: "Prestressing of Continuous Beams" (in French), *Annales des Ponts et Chausées*, 2d term, 1978, pp. 7–25.

10.9. T. Y. Lin: "Load Balancing Method for Design and Analysis of Prestressed Concrete Structures," *ACI Journal*, vol. 60, no. 6, June 1963, pp. 719–742.

10.10. ——— and K. Thornton: "Secondary Moment and Moment Redistribution in Continuous Prestressed Concrete Beams," *PCI Journal*, vol. 17, no. 1, January/February 1972, pp. 1–20. See also discussion of above paper by A. H. Mattock and closure by the authors in *PCI Journal*, vol. 17, no. 4, July/August 1972, pp. 86–88.

10.11. K. J. Thompson and R. Park: "Ductility of Prestressed and Partially Prestressed Concrete Sections," *PCI Journal*, vol. 25, no. 2, March/April 1980, pp. 46–69.

PROBLEMS

10.1 Determine the trajectory of the Zero-Load-C line (ZLC line) for a rectangular beam with two fixed ends, prestressed by a straight tendon parallel to the concrete centroid.

10.2 Several prestressed concrete rectangular beams with different steel profiles are shown in Figs. P10.2a to P10.2d. They are assumed fixed at their left support and roller supported at their right support. They all have the same cross section $b \times h$, an available maximum practical eccentricity $(e_o)_{mp} = \pm 0.4h$, and a prestressing force F. For each case shown:

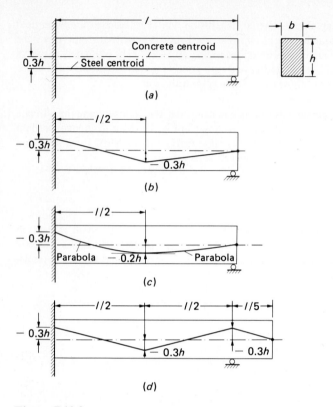

Figure P10.2

(a) determine the equivalent loading and the corresponding prestressing moment through-out the span; (b) plot the primary, secondary, and prestressing moment diagrams; (c) determine the Zero-Load-C line (ZLC line); and (d) suggest minor adjustments leading to a concordant profile. Note that in this problem, units are irrelevant and solutions can be obtained in function of l, F, and h. One can assume $l = 1$, $F = 1$, $h = 1$ and proceed. However, if numerical values are preferred, the following values can be used: $l = 50$ ft, $F = 100$ kips, $h = 20$ in.

10.3 Consider the prestressed posttensioned concrete beam with cross section shown in Fig. P10.3a. It is to be evaluated in three different design cases: two, three, and four equal continuous spans (Fig. P10.3b). The following information is provided: $I = 41{,}300$ in^4, $A_c = 576$ in^2, $Z_t = 3990$ in^3, $Z_b = 3020$ in^3, $y_t = 10.33$ in, $y_b = 13.67$ in, $k_t = -5.24$ in, $k_b = 6.94$ in, $\bar{\sigma}_{ti} = \bar{\sigma}_{ts} = -3\sqrt{f'_c}$, $\bar{\sigma}_{ci} = \bar{\sigma}_{cs} = 0.4f'_c$, $f'_c = 5000$ psi, $(d_c)_{min} = 3$ in, $LL = 0.6$ klf, $w_G = 0.6$ klf. Assume an average initial stress in the prestressing steel immediately after release of 183 ksi and an average final stress after all losses of 148 ksi. For each case A, B, and C: (a) determine the required prestressing force and its profile by elastic analysis (assume a cable profile made out of parabolas having zero eccentricities at the end supports, and assume, at first, that the inflection points between successive parabolas are at about $L/6$ from the intermediate supports); (b) determine the prestressing force by load balancing; (c) determine a concordant cable profile.

Note: Make any necessary assumptions in your design.

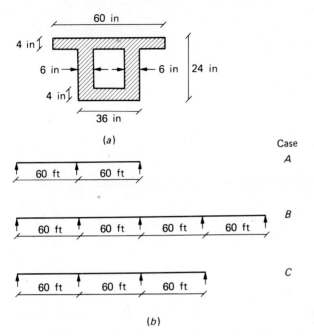

Figure P10.3

10.4 A three-span continuous prestressed slab is 24 in high and spans 60 ft between supports. Assume $k_t = k_b = k'_t = k'_b = 4$ in, that is, no tension is allowed, and $\eta = 0.80$. Let us assume that after a preliminary design we found a concordant steel profile with a final prestressing force of 192 kips. Eccentricities of the concordant profile at A, B, C, and D are shown in Fig. P10.4. In order to correctly place the tendons in the concrete with a sufficient cover at C, we linearly transform the tendon profile by lowering the eccentricity at point C by three inches. Assume that the problem is then solved and everything else is satisfied. (a) Show qualitatively the secondary moment diagram along the slab and compute the secondary moments at B, C, and D. Make sure their sign is right. (b) What are the values of external moments M_{max} and M_{min} which could be applied at sections B and C without violating any allowable stress?

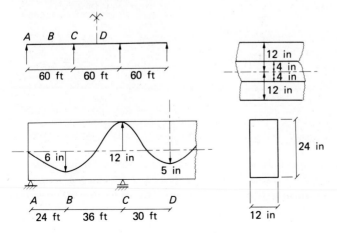

Figure P10.4

(a) Test of a $\frac{1}{3}$ scale model of a two-way flat plate posttensioned with un-bonded tendons. *(Courtesy Ned Burns, University of Texas at Austin.)*

(b) Typical tendon layout in a two-way flat plate. *(Courtesy Post-Tensioning Institute.)*

CHAPTER 11

PRESTRESSED CONCRETE SLABS

11.1 SLAB SYSTEMS

Slabs are structural elements that provide mostly flat, horizontal surfaces used as floors, roofs, and decks. They can be constructed in any of many ways, using various structural materials and composite systems. Depending on the selected configuration, they can be made to resist bending either primarily in one direction (one-way action) or in two directions (two-way action). The deformed surface of a slab under load has a single curvature, as in beams, for one-way slabs and a double curvature for two-way slabs.

One-way slabs are supported (or assumed supported) along two parallel edges. One-way prestressed concrete slabs do not necessarily imply a monolithically cast rectangular plate, as shown in Fig. 11.1. They can be made of precast beams placed adjacent to each other between parallel lines of support, posttensioned T beams as for joist-type slabs, or of a composite system made out of precast beams and a cast-in-place slab. In these cases the slab is mostly called a "deck" or a "slab

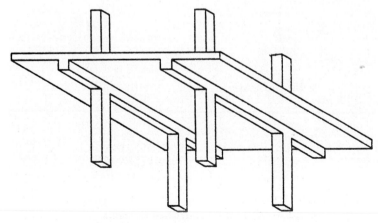

Figure 11.1 Typical one-way slab.

(*a*) Composite deck with hollow-cored slabs.

(*b*) Adjacent T beams.

Figure 11.2 One-way slab decks using precast prestressed elements.
(*Courtesy Material Service Corporation.*)

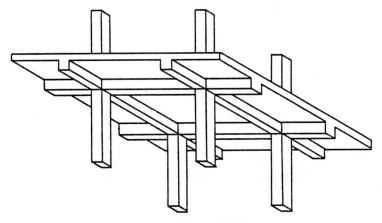

Figure 11.3 Two-way beam-supported slab.

deck." Two examples of one-way slab systems using precast prestressed elements are shown in Fig. 11.2. The analysis and design of slab decks are essentially reduced to the analysis and design of beams as covered in the preceding chapters and in Chap. 14. Some particular design aspects of monolithically cast one-way rectangular slabs are covered in Sec. 11.2.

Two-way prestressed concrete slabs belong mostly to the following groups: edge-supported slabs, flat slabs, and flat plates. Slabs supported along their four edges on beams or walls, with and without restraint, are called *edge-supported slabs*. A typical example is shown in Fig. 11.3 and a design example is covered in Sec. 10.13. In order to induce sufficient two-way action in edge-supported slabs and for design purposes, the ratio of long-to-short span must be less than about 2. Otherwise, they can be essentially analyzed as one-way slabs. *Flat slabs* are slabs supported on columns without beams in between (Fig. 11.4). They are generally strengthened around the columns using a drop panel or a column capital or both. A particular variety is the waffle (or grid slab) in which voids are created from the lower surface by cubicle-like reusable forms (Fig. 11.4b). A solid slab having same thickness everywhere, including around the columns, is called a *flat plate* (Fig. 11.4c).

The flat plate slab construction is the most common form of prestressed concrete slab construction. Flat plates are extensively used in residential and commercial buildings of all types, including hotels, parking structures, and the like. They offer numerous advantages including a minimal slab depth with an unobstructed bottom surface which can be formed and finished easily (Ref. 11.1). Construction time is reduced and substantial cost savings may be achieved. In the United States, most prestressed concrete flat plates are cast in place and posttensioned using unbonded tendons. The tendons are generally placed from one edge to the opposite edge of the slab, spanning several panels. They are stressed from and anchored at the periphery of the slab. The connections with the supporting columns are generally cast monolithically with the slab. However, in the particular case of lift slabs,

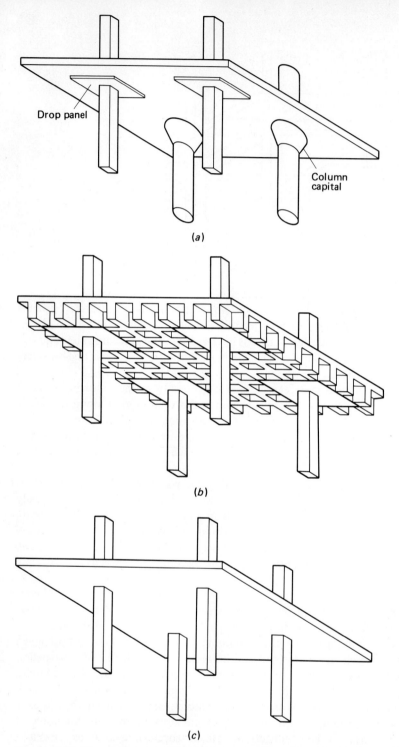

Figure 11.4 Types of two-way slabs. (*a*) **Flat slabs.** (*b*) **Waffle slabs.** (*c*) **Flat plate.**

the flat plate is cast at ground level and then lifted to its final position along preerected steel or concrete columns. Connections using steel collars or corbels are then attached between the slabs and the columns.

There are many analysis and design methods that are applicable to two-way slabs; some are simpler than others. Only one approximate method, the equivalent frame method, which is generally applicable to all types of two-way slabs and is recommended by the ACI code, will be covered here. A great deal can be learned by referring to related chapters on slabs in other books and textbooks on reinforced and prestressed concrete (Refs. 11.2 to 11.6). After a brief coverage of the design of one-way slabs, this chapter will mostly focus on the design of prestressed concrete two-way flat plates. Related requirements of the current edition of the ACI code and salient recommendations found in two reports by ACI-ASCE Committee 423 (Refs. 11.7, 11.8) are incorporated in the following sections.

11.2 ONE-WAY SLABS

A typical example of a simple span and continuous one-way slab is shown in Figs. 11.5a and b. The design of slabs starts generally by assuming a slab depth using the span as a guide. For common applications where a relatively light live load is used, the span-to-depth ratio ranges between 25 and 35 for single spans and

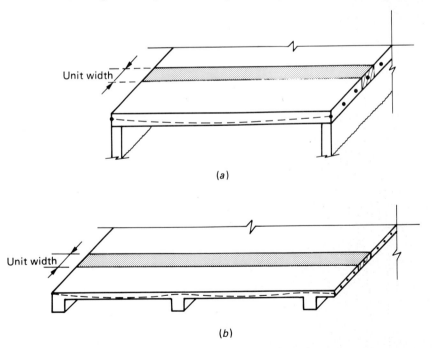

(a)

(b)

Figure 11.5   One-way slabs. (a) Simple span supported on walls. (b) Continuous spans supported on beams.

between 30 and 45 for continuous spans. Values obtained are rounded off to the next quarter- or half-inch. Other minimum requirements related to fire endurance, as described in Sec. 11.8, should also be considered in selecting an appropriate depth. Once the depth is obtained, the designer proceeds by considering a unit wide strip of slab as a beam and determining the prestressing force according to the procedures described in Chaps. 4 and 10. To determine the prestressing force, the load-balancing approach is generally preferred for continuous slabs. The balanced load is mostly taken equal to the dead load and a tendon profile made out of parabolic segments with zero eccentricities at the end supports and maximum practical eccentricities at intermediate supports and other critical sections is selected (Fig. 11.6a). The required prestressing force per unit width of slab is then

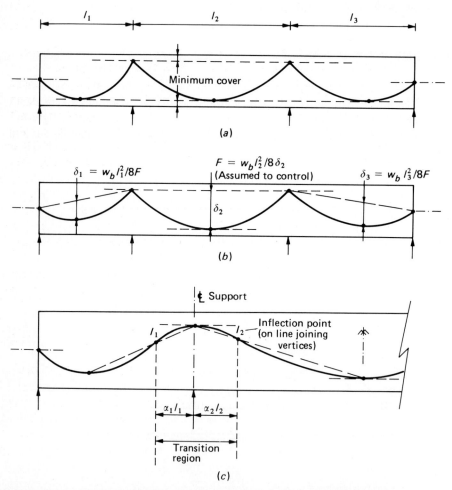

Figure 11.6 Design evolution of tendon profile. (*a*) Assumed initial profile. (*b*) Profile for balanced load. (*c*) Final profile smoothed over supports.

determined and the steel profile adjusted accordingly (Fig. 11.6b). The force is then translated into tendons at a given spacing. For the final design, the tendon profile is smoothed over the intermediate supports (Fig. 11.6c). The distance over which the transition is made is taken equal to about $0.1l$ on each side of the support. The effect of smoothing the tendon on the possible generation of secondary moments is often neglected in practice for slabs.

When the main flexural reinforcement in one-way slabs consists of unbonded tendons, the ACI code (Sec. 18.9) recommends the addition of a minimum amount of ordinary bonded reinforcement. The purpose of such reinforcement is to provide an alternative load-carrying system in the event of a catastrophic failure or abnormal loading of one span (such as due to fire) of a continuous slab that leads to distress in the other spans. The minimum area of bonded reinforcement shall be computed from:

$$A_s = 0.004A \tag{11.1}$$

where A is the area of that part of the cross section between the flexural tension face and the center of gravity of the gross section. The above reinforcement shall be distributed over the precompressed tensile zone as close as possible to the extreme tensile fiber. Its length shall be at least one-third of the clear span spacing in positive moment regions and one-sixth of the clear span on each side of the support in negative moment regions. The above minimum reinforcement may be considered to contribute to the ultimate resistance of the member. Moreover, ACI-ASCE Committee 423 (Ref. 11.8) recommends that, as a minimum bonded ordinary reinforcement is provided, a fictitious allowable tensile stress of $-9\sqrt{f'_c}$ to $-12\sqrt{f'_c}$ may be used in design, thus leading to a smaller amount of prestressing steel. However, the use of a lower magnitude of allowable stress may be preferable when durability is a controlling criterion, such as for parking structures in severe climates. The above provisions suggest that slabs using unbonded tendons are essentially treated as partially prestressed members.

In computing the nominal moment resistance of slabs with unbonded tendons and until a generally acceptable formula is developed, the following value of the stress in the prestressing steel is recommended (Ref. 11.2):

$$\begin{cases} f_{ps} = f_{pe} + 15 & \text{in ksi} \\ \text{or } f_{ps} = f_{pe} + 103.5 & \text{in MPa} \end{cases} \tag{11.2}$$

Equation (11.2) is considered to satisfactorily represent the stress in unbonded tendons at nominal resistance of one-way slabs with span-to-depth ratios up to 35.

According to the ACI code Sec. 7.12, a minimum amount of ordinary reinforcement normal to the main flexural reinforcement shall be provided for shrinkage and temperature stresses in one-way slabs. The following minimum ratio of reinforcement area to gross concrete area but not less than 0.0014 is specified:

Slabs where grade 40 or 50 deformed bars are used	0.0020
Slabs where grade 60 deformed bars or welded wire fabric (smooth or deformed) are used	0.0018

Slabs where reinforcement with yield strength exceeding
60 ksi measured at a yield strain of 0.35 percent is used $\dfrac{0.0018 \times 60}{f_y}$

For f_y given in megapascals, this last expression becomes $0.0018 \times 414/f_y$.

The above reinforcement shall not be spaced farther apart than five times the slab thickness or 18 in (46 cm).

Once the main reinforcement has been determined in one-way slabs, the design proceeds by checking shear, deflection, and other requirements exactly as for beams. The ACI requirement that the design nominal moment M_n be at least 1.2 times the cracking moment should theoretically apply to one-way slabs. However, based on recent tests where the behavior of slabs beyond cracking and up to failure was investigated, ACI-ASCE Committee 423 recommends that the above requirement be waived for one-way and two-way slabs prestressed with unbonded tendons (Ref. 11.8). This is because the load-deflection response of such slabs did not show any abrupt change at the onset of cracking and the minimum amount or reinforcement in slabs prestressed with unbonded tendons exceeds the minimum reinforcement of ordinary reinforced concrete slabs anyway.

In order to avoid excessive shortening of the slab and reduce its effects on attached columns and walls, it is recommended to limit the length of slab between construction joints to less than 150 ft (46 m). For longer lengths, the effect of slab shortening on columns and walls should be reviewed (Ref. 11.8).

11.3 EXAMPLE: ONE-WAY SLAB DESIGN

A typical bay slab of a parking structure with several bays consists of a 60-ft-wide one-way slab system with five consecutive spans, 25 ft center-to-center. The slab is supported on columns through 2-ft-wide beams placed along column lines (Fig. 11.7a). Provide a preliminary design for the slab, assuming unbonded tendons are specified and given the following information: normal weight concrete, $\gamma_c = 150$ pcf, $f'_c = 5000$ psi, $f'_{ci} = 3500$ psi, $E_c = 4.287 \times 10^6$ psi, minimum concrete cover to the centroid of the prestressed reinforcement $= 1\frac{1}{4}$ in, the tendons are strands with $f_{pu} = 270$ ksi, and assumed $f_{pe} \simeq 160$ ksi. A uniform live load of 100 psf is prescribed.

Let us select a slab depth $h = 8$ in which corresponds to a span-to-depth ratio $l/h = 37.5$ (well within the range recommended in Sec. 11.2) and let us provide a design for a 1-ft-wide strip of slab, assuming it acts as a one-way continuous beam. The corresponding dead weight of the strip per linear foot is $w_G = 100$ plf.

Prestressing steel. Load balancing will be used to determine the prestressing steel. A theoretical tendon profile made out of parabolas with maximum practical eccentricities or minimum cover at support sections and at midspan or near midspan sections is selected. The vertex of each parabola falls at midspan for a typical interior span and at $0.414l$ (near midspan) from the exterior span. The corresponding sags are shown in Fig. 11.7b, and the prestressing force to balance a uniform load $w_b = w_G = 0.1$ klf is given by:

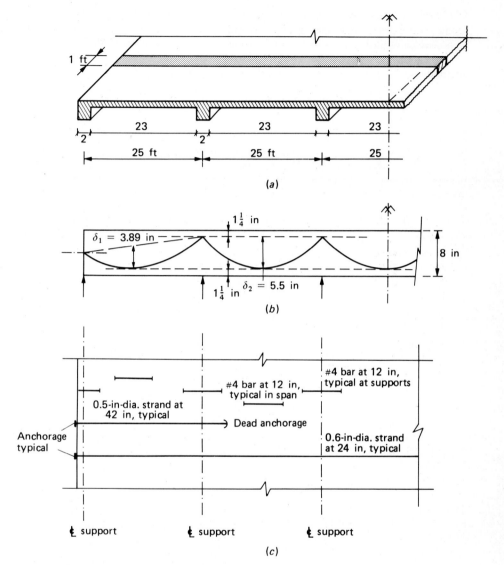

Figure 11.7 Slab of example problem.

For the exterior span:

$$F_1 = \frac{w_b l_1^2}{8\delta_1} = \frac{0.1 \times \overline{25}^2}{8 \times 3.89/12} = 24.1 \text{ kips}$$

For a typical interior span:

$$F_2 = \frac{w_b l_2^2}{8\delta_2} = \frac{0.1 \times \overline{25}^2}{8 \times 5.5/12} = 17.05 \text{ kips}$$

One of two different alternatives can be followed, namely, either (1) keep the same steel profile with different prestressing forces at interior and exterior spans, or (2) select the highest value of the prestressing force F_1 throughout the slab and adjust the tendon sag in the interior spans to achieve the same balanced load as the exterior spans. For the second alternative, the sag at an interior span would be given by:

$$\delta_2 = \frac{w_b l_2^2}{8F_1} = \frac{0.1 \times 25^2}{8 \times 24.1} \times 12 = 3.89 \text{ in}$$

This means that the tendon eccentricity at midspan is reduced from 2.75 to 1.14 in. Hence, the corresponding value of d_p is reduced by about 24 percent and the nominal resistance of the cracked section at ultimate is at least equally reduced. Therefore, it is decided to follow the first alternative which leads to a lesser overall amount of steel and probably some cost savings. Thus, two series of tendons are proposed. The tendons of the first series should provide a force $F_2 = 17.05$ kips per foot width of slab and should run continuously throughout the length of the slab with a theoretical profile as shown in Fig. 11.7b. The tendons of the second series should provide a force $F_1 - F_2 = 7.05$ kips per foot width of slab and will be placed in the exterior spans only. They will be anchored in the first interior span with dead anchors placed at about a quarter span from the first interior support and they will have the same profile as the first series of tendons in the exterior spans. In practice, the above requirements can be achieved using, for instance, the following tendon layout. The first series of tendons consists of 0.6-in-diameter prestressing strands spaced at 24 in. Each has a cross-sectional area of 0.217 in^2 and carries a final prestressing force of $0.217 \times 160 = 34.72$ kips that is equivalent to 17.36 kips per foot width of slab. The second series of tendons consists of 0.5-in-diameter prestressing strands spaced at 42 in. They each have a cross-sectional area of 0.153 in^2 and carry a final prestressing force of 24.48 kips that is equivalent to 7 kips per foot width of slab.

Bonded reinforcement. Because unbonded tendons are used, the ACI code prescribes the addition of a minimum amount of bonded bar reinforcement at critical sections. The following required values per foot width of slab can easily be obtained (Eq. (11.1)), assuming $f_y = 60$ ksi:

$$A_s = 0.004A = 0.004 \times 12 \times 4 = 0.192 \text{ in}^2$$

This can be achieved using a #4 bar every 12 inches that is equivalent to $A_s = 0.20$ in^2. The length of these bars should be at least 7 ft 8 in in both positive and negative moment regions and they should have a minimum clear concrete cover of $\frac{3}{4}$ in.

Reinforcement layout. The final reinforcement layout in the main direction is schematically described in Fig. 11.7c. It leads to the following summary of results per foot width of slab:

Exterior span: at $0.414l_1$ from exterior support.

$$A_{ps} = 0.1085 + 0.044 = 0.1525 \text{ in}^2$$
$$A_s = 0.20 \text{ in}^2$$
$$e_o = 2.75 \text{ in}$$
$$d_p = 6.75 \text{ in}$$
$$d_s = 7 \text{ in}$$

Similar values are obtained at the first interior support except that the tendon eccentricity is negative. Note that the average prestress for the exterior span is given by:

$$\sigma_g = \frac{F}{A_c} = \frac{0.1525 \times 160 \times 10^3}{12 \times 8} \simeq 254 \text{ psi}$$

Interior span: at midspan

$$A_{ps} = 0.1085 \text{ in}^2$$

$$A_s = 0.20 \text{ in}^2$$

$$e_o = 2.75 \text{ in}$$

$$d_p = 6.75 \text{ in}$$

$$d_s = 7 \text{ in}$$

$$\sigma_g = \frac{0.1085 \times 160 \times 10^3}{12 \times 8} \simeq 181 \text{ psi}$$

Similar values are obtained at interior supports except that the tendon eccentricity is negative.

Of course, the actual tendon profile is smoothed over the supports as described in Fig. 11.6c and potential secondary moments so generated are assumed negligible.

Unbalanced load. The unbalanced load is equal to the live load, that is $w_{nb} = 0.1$ klf. Using available tables for moments in continuous beams and assuming alternative span loadings when needed, leads to the approximate values of moments shown in the second line of Table 11.1. Note that the negative moments are given both at the centerline of the first interior support and at the left face of the support. The latter values are used in design. The moments due to the unbalanced load generate stresses which are added to the average stress in the section under balanced condition and compared to allowable stresses. Stresses are shown in Table 11.1, lines 3 to 5. Considering that the moment at the face of the support is used in design, all stresses are acceptable. In particular, no tensile stress exceeds in magnitude the recommended code limit of $-6\sqrt{f_c'}$ or -424 psi.

Ultimate moment. The strength design moments or factored moments required by the ACI code are given in Table 11.1, line 6; the nominal moment resistance of the critical sections are given in line 7. They were obtained, as for beams, according to the procedure described in Chap. 5, using the value of f_{ps} obtained from Eq. (11.2) for unbonded tendons (considered safer than that given by Eq. (5.13)). An example of calculation is shown next for the near-midspan section of the exterior span.

Using Eq. (11.2), we have:

$$f_{ps} = f_{pe} + 15 = 160 + 15 = 175 \text{ ksi} = 175,000 \text{ psi}$$

Referring to the flow chart, Fig. 5.13, and taking into account the nonprestressed steel, we have:

$$d = \frac{A_{ps} f_{ps} d_p + A_s f_y d_s}{A_{ps} f_{ps} + A_s f_y} \simeq 6.85 \text{ in}$$

$$q = \frac{A_{ps} f_{ps}}{bdf_c'} + \frac{A_s f_y}{bdf_c'} = 0.09413$$

$$\phi M_n = \phi(A_{ps} f_{ps} + A_s f_y)d(1 - 0.59q) = 225,262 \text{ lb-in}$$

Table 11.1 Numerical results for Example 11.4

Line	Item	Exterior span	First interior support		Typical interior span
		Near midspan	Center of support	Left face of support	Midspan
1	Moment due to balanced load or dead load, $w_b = w_G = 0.1$ klf	$0.078w_b\,l^2$	$-0.1071w_b\,l^2$	$\dfrac{w_b(l-1)}{2}(1-0.21l)$ kips-ft	$0.0364w_b\,l^2$
2	Moment due to unbalanced load or live load $w_{nb} = w_L = 0.1$ klf	$0.10w_{nb}\,l^2$	$-0.1205w_{nb}\,l^2$	$\dfrac{w_{nb}(l-1)}{2}(1-0.24l)$ kips-ft	$0.0805w_{nb}\,l^2$
3	Uniform stress due to balanced loading condition, psi	254	254	254	181
4	Maximum stresses due to unbalanced load, psi	±586	∓706	∓563	±472
5	Maximum stresses under service loads, top and bottom fiber, psi	840 -332	-452 960	-309 817	653 -291
6	Strength design moment, $\lvert M_u = 1.4M_D + 1.7M_L\rvert$, lb-in	209,400	266,092	208,080	140,858
7	Design nominal moment resistance ϕM_n, lb-in	225,262	225,262	225,262	182,539
8	A_{ps}, in^2 per foot width of slab	0.1525	0.1525	0.1525	0.1085
9	A_s, in^2 per foot width of slab	0.20	0.20	0.20	0.20

The above moment is shown in Table 11.1, line 7. Note that if the strength design moment was found higher than the nominal moment resistance at the support, moment redistribution could have been used. In this case it would allow up to about 14 percent of the negative moment to be redistributed (see Sec. 10.14). Such redistribution may need to be considered at a typical interior support section where the nominal moment resistance is the same as that of the midspan section of a typical interior span and is smaller than that of the first interior support.

Shear. The maximum shear force occurs at the exterior face of the first interior support. It can be taken conservatively equal to $(1.15/2)wl_n$ where l_n is the clear span. For this problem, the factored shear force per unit width of slab is obtained as:

$$V_u = \frac{1.15}{2}(1.4w_b + 1.7w_{nb})l_n$$

$$= \frac{1.15}{2} \times 0.31 \times 23 = 4.10 \text{ kips}$$

It leads to a factored shear stress:

$$v_u = \frac{V_u}{bd} \simeq \frac{4100}{12 \times 6.85} \simeq 50 \text{ psi}$$

The concrete shear resistance v_c is at least equal $2\sqrt{f'_c} = 141.4$ psi. As v_u is less than ϕv_c the shear resistance of the slab is acceptable.

Deflection. The slab is assumed level under the effect of the balanced load (dead load) and the prestressing force. Hence, only the unbalanced load (live load) will create deflections. Using tables of coefficients for continuous beams and assuming alternative span loadings, the maximum deflection is shown to occur in the exterior span with a value:

$$\Delta_i \simeq 0.01 \frac{w_{nb} l^4}{E_c I_g}$$

$$\simeq 0.01 \frac{100/12 \times (25 \times 12)^4}{4.287 \times 10^6 \times 8^3} \simeq 0.31 \text{ in}$$

The above deflection is smaller than the limiting value $l/360 = 0.83$ in allowed by the code.

Temperature reinforcement. A minimum amount of ordinary reinforcement normal to the main reinforcement is provided to resist shrinkage and temperature stresses. Assuming a reinforcement ratio of 0.0018 leads to the following area of steel:

$$A_s = 0.0018 \times 12 \times 8 = 0.17 \text{ in}^2 \text{ per foot of slab}$$

It is achieved by placing #4 bars at 14-in spacings transverse to the main reinforcement and along the tensile face of the slab.

11.4 CHARACTERISTICS OF TWO-WAY FLAT SLABS

1. Load Path

In order to understand the behavior of flat slabs and their reinforcing arrangement, it is informative to follow the path along which the load is transferred from the slab to the supporting columns. Let us consider a typical interior panel of a flat plate subjected to a uniform load w per unit area. The panel can conceptually be modeled into a central slab portion surrounded by broad strips of slabs acting as beams and spanning between columns in each direction (Fig. 11.8). The strip-beams are called column strips and are bounded in each direction by middle strips of slab. This panel representation simulates a slab (the central portion) supported on beams (the column strips) along its edges. Let us consider the load w on a unit area at the center of the slab. It can be assumed that such a load is transferred in each direction from the central portion of the slab to the edge beams (column strips) which in turn transfer the load to the columns. It can be observed that the fraction of load transferred from the central portion along the long direction is delivered to the column strip spanning in the short direction. This fraction of the load added to that carried directly in the short direction sums up to 100 percent of the load

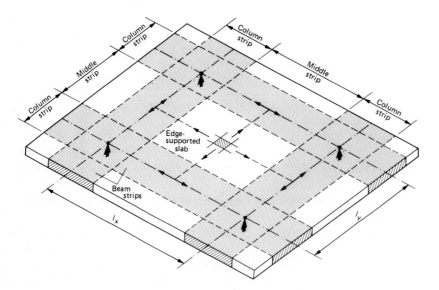

Figure 11.8 Conceptual representation of a flat plate into equivalent slab and beam.

applied to the panel. A similar reasoning can be made by following the path of the load along the short direction first. Hence, for two-way slab systems supported on columns, the applied load must be considered in its entirety in each direction analyzed. This conclusion may seem to contradict the procedure used for edge- or beam-supported slabs in which only a fraction of the load is assumed carried in each direction. However, for such slabs the support beams must transfer the load to the columns and are in effect part of the slab system.

2. Reinforcement Layout

Let us consider the same typical interior panel described in Fig. 11.8 and its representation by a central slab portion surrounded by column strips. Using load balancing, we can conceptually arrive at a tendon layout made out of the superposition of two systems of tendons, namely, (1) for the central portion of the slab, a tendon grid with relatively large spacing as for edge-supported slabs (Fig. 11.9a) and (2) concentrated bands of tendons along the column strips as for beams (Fig. 11.9b). Superposition of the two systems of tendons leads to the final tendon layout shown in Fig. 11.9c. It can be shown that a more rigorous analysis will lead to a very similar tendon arrangement, such as obtained from the equivalent frame method described in Sec. 11.6. Note that even a reinforced concrete slab would essentially have a reinforcing bar layout similar (in plan) to that of the tendons of Fig. 11.9.

3. Theoretical Distribution of Moments

The theoretical distribution of moments in a typical panel of a flat plate subjected to vertical loading can be obtained using, for instance, Wertergaard's analysis. The resulting distribution is schematically illustrated in Fig. 11.10, assuming bending in the long direction. A similar distribution can be obtained when the analysis is undertaken for the short direction. Note that a section of the moment surface parallel to the long direction leads to a moment diagram very similar to that obtained for a typical span of a continuous beam. Moreover, the transverse distribution of moments in a slab is not uniform. This implies that portions of the slab, across its width, will be subjected to higher moments than others and the corresponding reinforcement distribution will be uneven. However, for the purpose of design, the slab is generally divided into a column strip and a middle strip within which the transverse distribution of moments is considered constant (Fig. 11.11). The distribution of moments currently suggested in the ACI code for reinforced concrete slabs and by ACI-ASCE Committee 423 (Ref. 11.7) for prestressed concrete flat plates reflects the theoretical distribution and is explained for prestressed flat plates in Sec. 11.7.

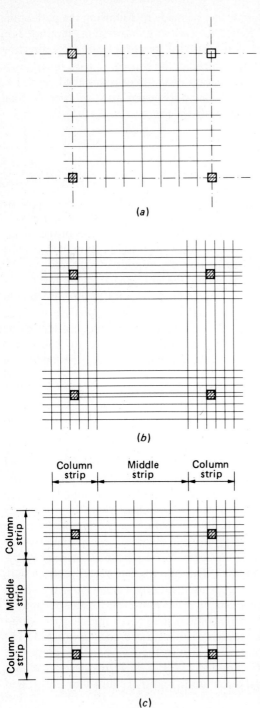

(a)

(b)

Column strip | Middle strip | Column strip

Column strip

Middle strip

Column strip

(c)

Figure 11.9   Conceptual arrangement of tendons in flat plates. (a) Tendons for central portion of slab. (b) Beam tendons along the column strips. (c) Superposition of the two systems.

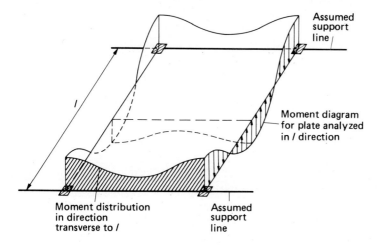

Figure 11.10 Theoretical distribution of moments in a flat plate analyzed in one direction.

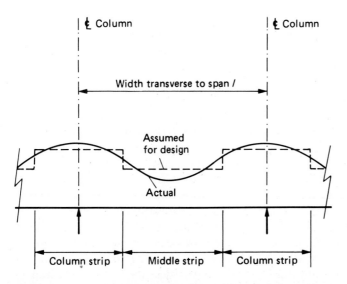

Figure 11.11 Assumed design distribution of moments across the width of a flat plate.

4. Special Notations

Most of the standard notations used throughout this text prevail in this chapter. However, in order to accommodate some of the differences with the ACI code notations, the following terms will be used for flat plates in the next sections:

l = span, in general, in the main direction where analysis for flexure is undertaken

l_a, l_b = longer and shorter span of a given panel measured center-to-center of supports

l_x, l_y = dimensions of panel center-to-center of columns in the x and y directions respectively (they are used when a particular panel is considered)

l_1 = span in the direction where the analysis for flexure is undertaken (it is mostly used when the equivalent frame is considered)

l_2 = width of equivalent frame (it is sometimes used as the span transverse to l_1)

d = distance from extreme compression fiber to centroid of tensile force in the steel (both prestressed and nonprestressed)

Note that the terms l_1 and l_2, which were used in continuous one-way slabs and beams to characterize consecutive spans, are different from the above defined values. The x and y directions are used as the principal directions when a slab involving several panels in each direction is considered.

11.5 ANALYSIS AND DESIGN METHODS

Two-way flat slabs are highly indeterminate structures. Their analysis should always satisfy the basic principles of statics and should, theoretically, take into account the restraints against rotation and translation offered by the supporting system. Available tools for the analysis and design of concrete flat slabs range from ready-to-use tables and charts to complex finite element programs. Approximate methods are also available for their analysis and include yield line analysis, the direct design (analysis) method of the ACI code, and the equivalent frame method.

Although yield line analysis applies to reinforced concrete slabs, there is so far little evidence to prove its validity for prestressed concrete slabs as their rotational capacity after cracking and the resulting moment redistribution may not be sufficient to develop yielding lines. Similarly, the direct design method described in the ACI code has been derived for reinforced concrete slabs on the basis of their rotational and moment redistribution capacity after cracking, as evidenced by tests. Hence, it is not generally applicable to prestressed concrete slabs. The equivalent frame method of analysis has been shown to apply both to reinforced and to prestressed concrete two-way slab systems and is described in more detail in Sec. 11.6.

Numerous studies have led to particular design methods for prestressed concrete flat slabs. Saether proposed a method based on structural membrane theory

(Ref. 11.9), while an elasticity based approach was developed by Rozvany and Hampson (Ref. 11.10). A direct design method was proposed by Wang (Ref. 11.11) and design charts were developed by Parme (Ref. 11.12). However, the most widely used design method for prestressed concrete flat slabs is the load-balancing method developed by T. Y. Lin (Ref. 11.13) and described in Sec. 10.13. It is often used in combination with the equivalent frame method of analysis where the unbalanced load is considered.

1. Load Balancing

The load-balancing approach is the fastest and most convenient method to treat prestressed concrete flat slabs. More than any other method, it allows a clear visualization of the relationship between the distribution of moments and the tendon arrangement to balance these moments. This can be simply illustrated by considering a slab with an irregular pattern of columns (Fig. 11.12) for which other analytic procedures which are generally derived for ideal situations become tedious. Even the arbitrary concept of dividing the slab into column strips and middle strips gets confusing for the slab of Fig. 11.12, as a column strip in one panel may become a middle strip in another panel. However, going back to the basic idea of load balancing, the slab can be divided by a grid made out of lines passing by every row of columns in both the longitudinal and transverse directions. The primary system of tendons (say, along the lettered grid lines) is given a profile, whereas high

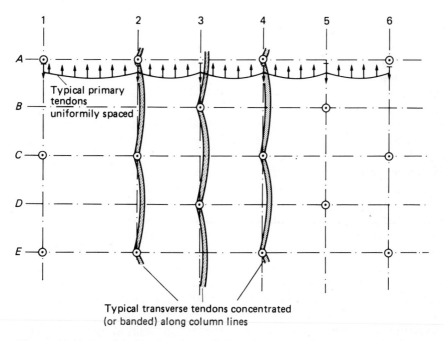

Figure 11.12 Load balancing for a slab with an irregular pattern of columns.

points (where reverse curvatures and downward loads occur) are placed at the numbered grid lines. They are given a uniform spacing. The primary system of tendons creates a downward reaction *at* the numbered grid lines which is reacted by the secondary system of tendons placed in bands *along* the numbered grid lines. The secondary tendons which are grouped together form like-beams that carry the load to the columns in the transverse direction. It can be seen that, following a very simple reasoning, a tendon arrangement can be arrived at in an otherwise rather complex problem. Since the balanced load generally represents a large portion of the total service load, it is believed that, where slabs are concerned, differences between the balanced-load approach and a more rigorous analysis will be relatively small.

11.6 ANALYSIS BY THE EQUIVALENT-FRAME METHOD

The equivalent-frame method is the most widely used method in the United States to analyze concrete slabs. It applies to *all* types of two-way slab systems, including beam-supported slabs, waffle slabs, flat slabs with drop panels, and flat plates (Ref. 11.14). It essentially reduces the analysis of a two-way slab system to two independent analyses of a one-way slab or planar frame system. Theoretically it also applies to one-way slabs; however, in such a case it may not be the most efficient procedure.

Recent tests of large structural models have shown that the equivalent-frame method satisfactorily predicts factored moments and shears in prestressed concrete slab systems (Refs. 11.15 to 11.17). Although other analytic methods, such as the elastic theory of thin plates and the finite-element method, are acceptable, the equivalent frame method is recommended by ACI-ASCE Committee 423 for the analysis of prestressed concrete flat plates (Refs. 11.7 and 11.8). It is described in sufficient detail in the ACI code and its commentary, Sec. 13.7. Note, however, that Secs. 13.7.6.3, 13.7.7.4, and 13.7.7.5 are not applicable to prestressed concrete slabs, as they are essentially derived on the basis of moment redistribution in reinforced concrete slabs.

In the equivalent frame method, the three-dimensional structural system is represented by a series of two-dimensional frames which are then analyzed for loads acting in the plane of the frames. The moments determined at the critical sections of the frame are distributed to the slab sections. The two-dimensional equivalent frames are centered on column lines longitudinally and transversely through the structure (Fig. 11.13*a*). Each frame consists of a row of columns and a continuous slab-beam strip bounded laterally by the centerline of the panel on each side of the columns. For vertical loading, each floor may be analyzed separately, assuming the far ends of the attached columns are fixed at the floors above and below (Fig. 11.13*b*). The equivalent frame in each direction is assumed to carry 100 percent of the applied load. Alternative loadings of spans must be considered as for continuous beams to determine maximum effects.

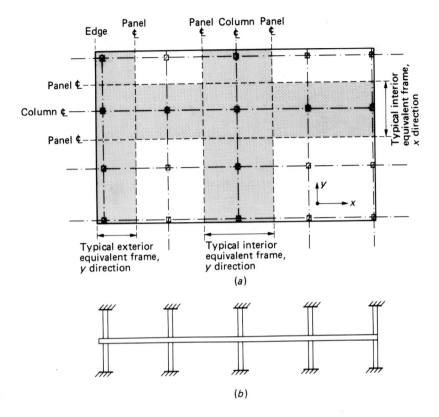

**Figure 11.13 Typical equivalent frames as used in the equivalent-frame method.
(a) Plan. (b) Elevation in the x direction.**

To clarify the analysis, the equivalent frame can be separated into three parts
(Fig. 11.14): (1) the horizontal slab strip, including any beams spanning in the
direction of the frame, (2) the columns or other supporting members extending
above and below the slab, and (3) the elements of the structure that provide
moment transfer between the horizontal and vertical members. Moment transfer
between horizontal and vertical components depends on the connection between
them and their relative flexibility or relative stiffness. For slender columns or for
soft column-slab connections such as in the case of lift slabs, the stiffness of the
columns may be neglected. For the general case, however, the rotational resistance
of the columns must be accounted for. To do this, an equivalent column is assumed. It consists of the actual columns above and below the slab plus an attached
torsional member transverse to the direction of the span for which moments are
being determined and bound on each side by the panel centerlines (Fig. 11.14). The
attached torsional member is assumed to have a constant cross section consisting
of the larger of (1) a portion of slab having a width equal to the column, bracket, or
capital in the direction of the span for which moments are being determined, or

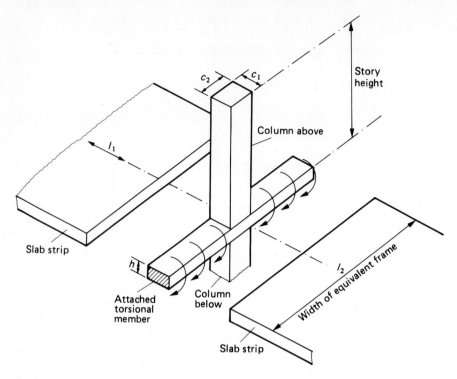

Figure 11.14 Main elements of the equivalent frame.

(2) for monolithic or fully composite construction, the portion of slab specified in (1) plus that part of the transverse beam, if any, above and below the slab taken not greater than four times the slab thickness. According to Sec. 13.7.4 of the ACI code, the flexibility (inverse of stiffness) of the equivalent column shall be taken as the sum of the flexibilities of the actual columns above and below the slab and the flexibility of the attached torsional member. That is:

$$\frac{1}{K_{ec}} = \frac{1}{\sum K_c} + \frac{1}{K_t} \tag{11.3}$$

where K_{ec} = flexural stiffness of equivalent column (above and below)

$\quad K_c$ = flexural stiffness of actual column (for members with constant cross section the flexural stiffness is equal to $4E_c I/l$.)

$\quad K_t$ = torsional stiffness of attached torsional member (see Eq. (11.4))

The summation applies to the column above and below the slab. Stiffnesses are expressed in terms of moment per unit rotation. In computing K_c, the moment of inertia outside of joints or column capitals may be based on the gross area of concrete. The torsional stiffness of the attached torsional member can be calculated from the following expression:

$$K_t = \sum \frac{9E_{cs}\, C}{l_2(1 - c_2/l_2)^3} \tag{11.4}$$

where E_{cs} = modulus of elasticity of slab concrete

l_2 = width of equivalent slab or length of span transverse to l_1

l_1 = length of span in direction that moments are being determined measured face-to-face of supports

c_2 = size of rectangular or equivalent rectangular column, capital or bracket measured transverse to l_1

C = cross-sectional constant for the attached torsional member

The summation refers to the torsional elements on either side of a column. Hence, for an interior column or an edge column of an interior frame, $\Sigma = 2$ while $\Sigma = 1$ for a corner column. The constant C may be evaluated from the cross section by dividing it into separate rectangular parts and carrying out the following summation:

$$C = \sum \left(1 - 0.63 \frac{x}{y}\right) \frac{x^3 y}{3} \tag{11.5}$$

where x and y are the shorter and longer overall dimensions of each rectangular part considered, respectively. For the slab of Fig. 11.14 only one rectangular part exists for which $x = h$ and $y = c_1$.

Given the stiffnesses of the equivalent column and the slab-beam, the frame can be analyzed for vertical loads by any acceptable method such as the moment distribution method. Then the reinforcement is provided and a similar procedure is carried out for the other longitudinal and transverse frames.

11.7 DESIGN DISTRIBUTION OF MOMENTS AND TENDONS

The equivalent frame method leads to the moments at the critical sections of the continuous slab-beam or equivalent frame considered. The negative moments so obtained apply at the centerline of the supports. Since the actual support is not a line support but is made of the columns and a relatively wide band of slab spanning transversely between columns, the critical section for the negative moment may be taken at the face of the supporting column but in no case at a distance greater than $0.175l_1$ from the center of the column (ACI code, Sec. 13.7). The critical section is assumed the same across the entire slab width. Note that actual moments are not uniformly distributed but vary across the width of the slab, as shown earlier in Fig. 11.10. To distribute the moments obtained from the equivalent-frame method, the total width of the slab is divided, for the purpose of design, into a column strip and a middle strip (Fig. 11.15). The column strip is assumed to have a width, on each side of the column, equal to the smaller of $l_1/4$ or

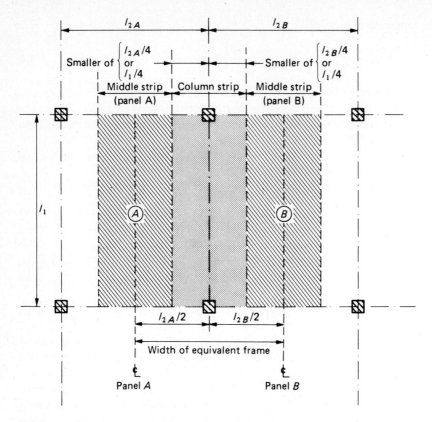

Figure 11.15 Design definitions of slab strips.

$l_2/4$, where l_1 and l_2 are the dimensions of the panel center-to-center of supports (ACI, Sec. 13.2). The middle strip comprises, on each side of the column, the part of slab bounded by the column strip and the centerline of the panel. The moments are assumed constants (transversely) within the boundaries of a column strip or a middle strip. For panels with length-to-width ratio not exceeding 1.33, the following approximate distribution of moments and corresponding tendons is recommended (Refs. 11.7 and 11.8):

Simple spans: 55 to 60 percent to the column strip with the remainder to the middle strip

Continuous spans: 65 to 75 percent to the column strip with the remainder to the middle strip

As a general guide, the spacing of tendons or bundles of tendons should not exceed four times the slab thickness in column strips and six times the slab thickness in middle strips. Note that the ultimate strength of flat plates is primarily

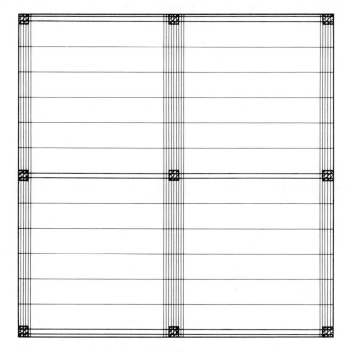

Figure 11.16 Example of banded tendon distribution.

controlled by the total amount of tendons in each direction. However, as indicated by tests, tendons passing through the columns or directly around their edges contribute more to the load-carrying capacity than other tendons. Hence, it is recommended to place some tendons through the columns or around their edges.

Although the tendon distribution described above follows essentially the distribution of moments, it has evolved due to construction difficulties and related costs. At this time, the preferred method of placing tendons in two-way flat plates in the United States is the banded tendon distribution illustrated in Fig. 11.16. In this method the number of tendons required across the entire design strip (width of equivalent frame) are banded close to the column lines in one direction and uniformly distributed in the other direction. This creates in effect a two-directional prestressed one-way slab system in which the portions of slab where tendons are banded act as the support beams. ACI-ASCE Committee 423 recommends placing at the slab perimeter at least two tendons through the columns in the distributed direction (Ref. 11.8). The spacing of the tendons which are distributed should not exceed six times the slab thickness. The banded tendon distribution method offers the advantage of a simplified reinforcing system, hence speed of construction and related savings in cost. As the ultimate load capacity of two-way slabs is mostly dependent on the total amount of steel, it is believed that banding the tendons should not lead to a significant loss in ultimate capacity when compared to the other tendon distribution described earlier in Fig. 11.9.

11.8 PRELIMINARY DESIGN INFORMATION

The preliminary design process for slabs generally consists of first selecting a slab depth and then providing the necessary amount of reinforcement to satisfy service load and ultimate strength criteria. The following remarks will help in reducing the design iterations and arriving quickly at an acceptable solution.

1. Slab Thickness and Reinforcement Cover

The span-to-depth ratio of two-way slabs designed for live loads of the order of 50 psf (2.4 kN/m^2) commonly ranges from 40 to 55. Higher values are associated with the presence of edge beams or drop panels. The depth of two-way slabs may also be controlled by fire-rating requirements. For a given fire endurance, the thickness requirements of concrete slabs, whether plain, reinforced, or prestressed, are essentially the same. Recommended minimum thicknesses vary from about three to seven inches (7.5 to 17.5 cm) for fire endurances of one to four hours and for various types of aggregates. For the same range of fire endurance, the recommended clear concrete cover over the reinforcement ranges from about $\frac{3}{4}$ in (19 mm) to $1\frac{1}{4}$ in (32 mm) for restrained slabs and from about $\frac{3}{4}$ to 2 in (19 to 50 mm) for unrestrained slabs. Such fire safety requirements must be considered when the design would otherwise call for a very thin slab.

Similarly to one-way slabs, it is generally recommended to limit the total length of two-way slabs to about 150 ft (46 m) between construction joints in each direction. For longer lengths the effects of slab shortening on the attached columns and walls should be considered in design (Ref. 11.8).

2. Average Prestress

The average prestress is defined as the final prestressing force after losses divided by the total area of concrete. For slabs prestressed with unbonded tendons, a minimum average prestress of 125 psi (0.86 MPa) and a maximum average prestress of 500 psi (3.5 MPa) are recommended by ACI-ASCE Committee 423. The minimum value is meant to limit excessive tension and cracking, while the maximum value is meant to limit excessive elastic shortening and creep. For applications such as parking structures, where durability against application of deicer chemicals is an important factor, a minimum average prestress level of 175 to 200 psi (1.2 to 1.4 MPa) is recommended. Actual values of average prestress for currently built flat plates seem to range mostly between 175 and 300 psi (1.2 and 2.1 MPa). Note that, in addition to the prestressed unbonded tendons, a minimum amount of ordinary bonded reinforcement is required by the ACI code, as described in Sec. 11.9.

11.9 FLAT PLATES: DESIGN FOR FLEXURE

1. Working Stress Design

Similarly to the design of prestressed concrete beams, the determination of the prestressing force in flat plates is essentially based on service load analysis. The load-balancing approach is generally preferred at this stage and as, according to the equivalent frame method, the flat plate is analyzed as two independent one-way slabs, the design procedure is essentially reduced to that of continuous beams and one-way slabs (Chap. 10). In summary, the balanced load is commonly taken equal to the dead load and the prestressing force is determined by load balancing, assuming a tendon profile made out of parabolas with zero eccentricities at the end supports and maximum practical eccentricities at intermediate supports. No secondary moments are generated for the theoretical tendon profile and a uniform compressive stress exists throughout the slab. The equivalent-frame method is then used to determine the moments due to the unbalanced load. Corresponding stresses are superimposed to the uniform stress induced by prestressing and the resulting stresses are compared to code-allowable stresses. The procedure is repeated and refined as needed.

When using the equivalent-frame method of analysis, the following allowable stresses at service loads are recommended for the design of solid posttensioned flat plates with bonded or unbonded tendons (Ref. 11.7):

a. Compression in concrete
   Negative moment areas around columns $\qquad$ $0.30f'_c$
b. Tension in concrete in slabs with an average final prestress of
   125 psi (0.86 MPa) or higher
   For positive moments without addition of nonprestressed
   reinforcement $\qquad$ $-2\sqrt{f'_c}$
   For positive moments with addition of nonprestressed
   reinforcement $\qquad$ $-6\sqrt{f'_c}$
   For negative moments without addition of nonprestressed
   reinforcement $\qquad$ 0
   For negative moments with addition of nonprestressed
   reinforcement $\qquad$ $-6\sqrt{f'_c}$

All other allowable stresses given in the ACI code apply. For stresses under initial conditions use f'_{ci} instead of f'_c. Deviations from the above values are permissible when more rigorous methods of analysis are used and when it is shown that the slab will perform satisfactorily under all design conditions.

A note is in order here. In applying the load-balancing approach to flat plates, the effect of reverse tendon curvature which provides a smooth transition over the supports (Fig. 11.6c), leading generally to a nonconcordant tendon profile, is commonly neglected in design. This is done on the basis that secondary moments are small and the ultimate moment capacity is not influenced. However, such an approximation is not always warranted, as secondary moments may substantially influence the elastic analysis and are to be included in the strength design moment which dictates the required nominal resistance. To determine the secondary moments, a loading equivalent to the effects of prestressing can be applied (see Sec. 10.8) and the equivalent frame method used to determine the total moment due to prestress. The secondary moment is then obtained from the difference between the total prestressing moment and the primary moment (Eq. (10.3)).

2. Ultimate Strength Design

Posttensioned flat plates must meet the ultimate strength requirements of the ACI code. This is achieved by comparing the nominal moment resistance at all critical sections to the factored moments (or strength design moment) at these sections. The factored moment is calculated as for continuous beams, noting that it should generally include the secondary moment, if any, with a factor of 1 (Eq. (10.19)).

The nominal moment resistance at each critical section is calculated as for beams according to the procedures described in Chap. 5. Nonprestressed reinforcement, when used, should be considered; moment redistribution, as described in Sec. 10.14, is allowed. Note that reaching the nominal resistance at a critical section does not necessarily imply failure of the structure. A collapse mechanism may develop for which a substantially higher load may be applied before failure occurs.

When unbonded tendons are used, some additional provisions given in the ACI code must be considered. They are summarized next.

Unbonded Tendons. A number of recent tests have shown that Eq. (5.13) overestimates the amount of stress increase in unbonded tendons in two-way flat plates and flat slabs. Until a generally acceptable formula is developed, the following expression which was recommended in the 1963 ACI code can be used (Ref. 11.2):

$$\begin{cases} \quad f_{ps} = f_{pe} + 15 & \text{in ksi} \\ \text{or} \quad f_{ps} = f_{pe} + 103.5 & \text{in MPa} \end{cases} \tag{11.6}$$

Equation (11.6) is considered satisfactory for flat plates and flat slabs with span-to-depth ratios up to 35. The Committee also recommends that the ACI requirement for the factored nominal resistance (ϕM_n) to exceed 1.2 times the cracking moment be waived for two-way posttensioned systems with unbonded tendons.

According to the ACI code, Sec. 18.9, two-way flat plates prestressed with unbonded tendons should contain a minimum area of bonded ordinary reinforcement, as follows:

1. In positive moment areas where the magnitude of tensile stress in concrete at service load does not exceed $| -2\sqrt{f'_c} |$, bonded reinforcement is not required.

2. In positive moment areas where the magnitude of tensile stress in concrete at service load exceeds $|-2\sqrt{f_c'}|$, the minimum area of bonded reinforcement shall be computed from:

$$A_s = \frac{|N_c|}{0.5f_y} \tag{11.7}$$

where N_c is the tensile force in the concrete due to unfactored dead load plus live load and the yield strength of the bonded reinforcement, f_y, shall not exceed 60,000 psi (414 MPa). Bonded reinforcement shall be uniformly distributed over the precompressed tensile zone and as close as practicable to the extreme tension fiber. Its minimum length shall be one-third the clear span. If the bonded reinforcement is considered in determining the nominal strength (partially prestressed section), its length should be in accordance with the development length provisions of the ACI code.
3. In negative moment areas at column supports, the minimum area of bonded reinforcement in each direction shall be computed by:

$$A_s = 0.00075hl \tag{11.8}$$

where h is the overall slab depth and l is the length of span in the direction parallel to that of the reinforcement being determined. The above bonded reinforcement shall be distributed within a slab width between lines that are $1.5h$ outside opposite faces of the column support. At least four bars or wires shall be provided in each direction. The spacing of bonded reinforcement shall not exceed 12 in (30 cm) and its length shall be at least one-sixth the clear span on each side of the support. If the bonded reinforcement is considered in determining the nominal strength (partially prestressed section), its length should be in accordance with the development length provisions of the ACI code.

11.10 FLAT PLATES: DESIGN FOR SHEAR

1. Concrete Shear Capacity

Shear is likely to be critical in concrete flat plates supported directly on columns. The shear strength of prestressed concrete flat plates is governed by the more severe of the following two conditions: (1) *beam-type* shear behavior with potential diagonal cracking failure along the plane of principal tension (Fig. 11.17a) and (2) *punching shear* behavior with potential diagonal cracking failure along a truncated pyramid- (or cone-) shaped surface around the column (Fig. 11.17b).

For beam-type shear, the plate is considered to act as a wide beam and the related shear design provisions are identical to those of beams covered in Sec. 6.8. In particular, the first critical section is taken at a distance $h/2$ from the face of the column and no shear reinforcement is needed unless v_u exceeds ϕv_c.

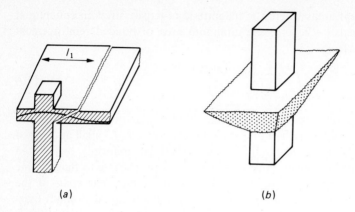

(a)   (b)

Figure 11.17 (a) Beam-type shear failure. (b) Punching shear failure.

When punching shear is considered, the critical section for design is taken at a periphery around the column distant $d_p/2$ (or $d/2$ if both prestressed and nonprestressed reinforcement are used) from the face of the column. For such a case, the shear resistance of the concrete v_c is given by Eq. (6.35), the same as Eq. (11.10), and shear reinforcement is provided if v_u exceeds ϕv_c.

A review of available data on punching shear tests of two-way prestressed concrete slabs led ACI Committee 423 (Ref. 11.8) to make the following recommendation for slabs with average prestress in each direction not less than 125 psi (0.86 MPa): the nominal punching shear capacity of concrete v_c can be taken equal to v_{cw} given by Eq. (6.30), that is:

$$v_c = v_{cw} = 3.5\sqrt{f'_c} + 0.3\sigma_g + V_p/b_o d \qquad (11.9)$$

where σ_g is the average precompression in the direction of moment transfer, b_o is the perimeter of the critical section considered for punching shear, and V_p is the shear carried through the critical section by the tendons (vertical component of the force carried by the tendons through the critical section). V_p can be conservatively taken as zero in a preliminary design. For precompression values less than 125 psi (0.86 MPa) and for nonprestressed construction, the punching shear capacity v_c is given by Eq. (6.41), that is:

$$v_c = \left(2 + \frac{4}{\beta_c}\right)\sqrt{f'_c} \leq 4\sqrt{f'_c} \qquad (11.10)$$

where β_c is the ratio of long side to short side of column or concentrated load.

To determine the need for shear reinforcement, the concrete shear resistance v_c is to be compared with the factored vertical and torsional shear stress due to applied loads. Torsional shear stress is generated by moment transfer between column and slab as described next. Because prestressed concrete flat plates contain in general some amount of nonprestressed flexural reinforcement at the critical sections, d will be used instead of d_p. Of course, $d = d_p$ for a fully prestressed section.

2. Moment Transfer Between Column and Slab

When a rigid connection exists between column and slab, bending moments are transferred from slab to column and vice versa. Such moments are mainly generated by pattern loadings, lateral loadings due to winds and earthquakes, and temperature movements. Based on tests on reinforced concrete slabs (Ref. 11.18 and 11.19), the ACI code recognizes that a fraction of the moment is transferred by flexure across the perimeter of the critical section while the remainder is transferred by eccentricity of the shear about the centroid of the critical section. It is believed that a similar behavior prevails in prestressed concrete flat plates (Ref. 11.16). The fraction of moment assumed transferred by shear for a typical interior column is given in Sec. 11.12 of the ACI code by:

$$\gamma_v = 1 - \frac{1}{1 + \frac{2}{3}\sqrt{\frac{c_1 + d}{c_2 + d}}} \tag{11.11}$$

where $(c_2 + d)$ is the width of the face of the critical section resisting the moment and $(c_1 + d)$ is the width of the face at right angle to $(c_2 + d)$. The critical section is assumed perpendicular to the plane of the slab so that its perimeter b_o is a minimum but does not need to approach closer than $d/2$ from the perimeter of the column. A typical critical section for shear around an interior column is shown in Fig. 11.18a.

Note that the fraction of transfer moment $(1 - \gamma_v)M_u$ transferred by flexure is considered to be transferred by the ultimate resisting moment of a portion of slab taken between lines that are $1.5h$ outside opposite faces of the column (or capital, if any). This may require additional nonprestressed reinforcement in that portion of slab.

The factored shear force V_u and the factored moment M_u to be transferred to both columns meeting at an interior joint are first determined at the centroidal axis of the critical section. Then, the factored shear stresses are calculated by superposing the stresses due to both shear and moment as shown in Fig. 11.18a. For an interior column joint, the resulting maximum factored shear stress may be calculated from:

$$v_u = \text{the larger of} \begin{cases} \left| \dfrac{V_u}{A_c} + \dfrac{\gamma_v M_u c_3}{J_c} \right| & (11.12) \\[4mm] \text{and} \\[4mm] \left| \dfrac{V_u}{A_c} - \dfrac{\gamma_v M_u c_4}{J_c} \right| & (11.13) \end{cases}$$

where γ_v is given by Eq. (11.11), A_c is the area of the assumed critical section, and J_c

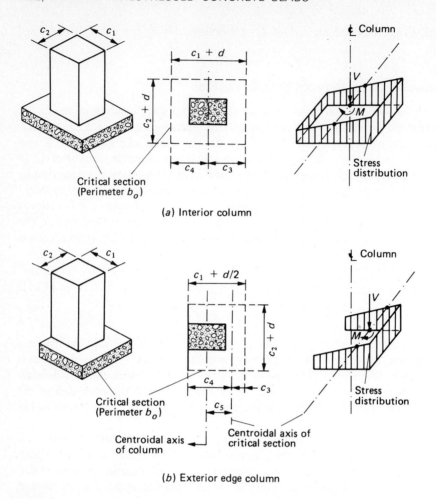

(a) Interior column

(b) Exterior edge column

Figure 11.18 Typical shear stress distribution. (a) Around an interior column. (b) Around an exterior column.

is its polar moment of inertia. Referring to Fig. 11.18a, it can be easily shown that for an interior column:

$$A_c = 2(c_1 + c_2 + 2d)d \qquad (11.14)$$

$$J_c = \frac{d(c_1 + d)^3}{6} + \frac{(c_1 + d)d^3}{6} + \frac{d(c_2 + d)(c_1 + d)^2}{2} \qquad (11.15)$$

$$c_3 = c_4 = (c_1 + d)/2 \qquad (11.16)$$

A similar set of equations can be derived for the case of exterior edge or corner columns. Referring to Fig. 11.18b, where the critical section of an exterior edge column is shown, the following expressions are obtained:

$$\gamma_v = 1 - \frac{1}{1 + \frac{2}{3}\sqrt{\dfrac{c_1 + d/2}{c_2 + d}}} \tag{11.17}$$

$$A_c = (2c_1 + c_2 + 2d)d \tag{11.18}$$

$$c_3 = \frac{d(c_1 + \tfrac{1}{2}d)^2}{A_c} \tag{11.19}$$

$$c_4 = (c_1 + \tfrac{1}{2}d) - c_3 \tag{11.20}$$

$$c_5 = c_4 - \frac{c_1}{2} \tag{11.21}$$

$$J_c = \frac{(c_1 + d/2)d^3}{6} + \frac{2d(c_3^3 + c_4^3)}{3} + d(c_2 + d)c_3^2 \tag{11.22}$$

In determining maximum factored shear stresses given by Eqs. (11.12) and (11.13) for the case of an exterior column, the values of M_u and V_u are assumed taken at the centroid of the critical section. Hence, M_u is equal to the moment at the centroid of the column minus $V_u c_5$ where c_5 is the distance from the centroid of the column to the centroid of the critical section (Fig. 11.17b).

Note that similar equations can be theoretically derived for the case of round columns. However, the use of an equivalent square column having the same cross-sectional area as the round column is generally acceptable.

3. Shear Reinforcement

According to the ACI code, when the design shear stress v_u exceeds ϕv_c, v_c should be taken equal to $2\sqrt{f_c'}$ and shear reinforcement must be provided for the excess shear. The nominal shear resistance v_n given by Eq. (6.22) shall not be taken greater than $7\sqrt{f_c'}$ when shearhead reinforcement is used and $6\sqrt{f_c'}$ when bent bar reinforcement is used.

Shearhead reinforcement consists of standard I or channel-shaped steel beams embedded in the slab and extending beyond the column faces. They serve to increase the perimeter of the critical section for shear (see ACI code commentary, Sec. 11.11.4) and contribute to the ultimate flexural resistance of the slab at the columns. Typical shearhead reinforcements are shown in Fig. 11.19a. The design of shearhead reinforcement is prescribed by Sec. 11.11.4 of the ACI code. The derivations of the corresponding equations are explained in Ref. 11.20.

When bent bar reinforcement is used, it is generally placed along the centerline of the column strip in each direction (Fig. 11.19b). Assuming an angle of inclination α at the critical section, it can easily be shown that the area of bent bar reinforcement can be obtained from the following equation, given in Sec. 11.5 of the

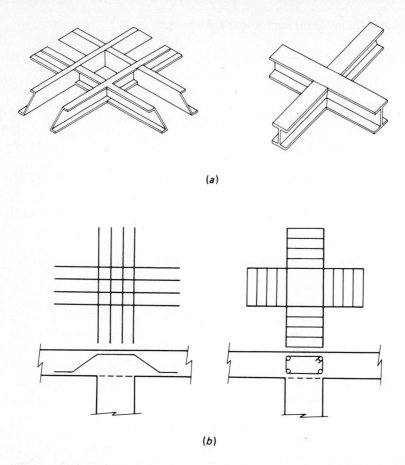

(a)

(b)

Figure 11.19 Typical shear reinforcement for two-way slabs. (a) Shearheads made out of steel channels, or I beams. (b) Bent bars and beam stirrups.

ACI code:

$$A_v = \frac{V_s}{f_y \sin \alpha} \qquad (11.23)$$

where f_y is the yield strength of the bent bar reinforcement and V_s is the shear force resisted by the reinforcement. It can be determined from:

$$V_s = \left(\frac{v_u}{\phi} - v_c\right) A_c \qquad (11.24)$$

in which A_c is given by Eq. (11.14) or Eq. (11.18) and $\phi = 0.85$ for shear. Only the center three-fourths of the inclined portion of bent bars are considered effective and adequate anchorage of the bars is required.

11.11 DEFLECTION OF FLAT PLATES

Although deflection may often govern the design in reinforced concrete two-way flat plates, it is likely to be less critical in prestressed concrete flat plates subjected to relatively light live loads (less than 100 psf (4.8 kN/m²)) and having span-to-depth ratios within the range recommended in Sec. 11.8. This is because generally (1) due to load balancing, the plate has zero deflection under dead load, hence, only the live load produces deflection and (2) the concrete section is assumed uncracked, thus leading to higher stiffness and lower deflection.

Deflection limitations specified in the ACI code and described for one-way slabs and beams in Table 7.2, also apply to two-way flat plates. In using Table 7.2, the shorter span of the panel center-to-center of the supports (column centerlines) is generally considered. This is because, when the ratio of long to short span increases, the slab tends to behave as a one-way slab in the shorter direction for which the shorter span applies.

A number of approximate methods are available to estimate deflections in flat plates. The most common ones are reviewed in a state-of-the-art report by ACI Committee 435 (Ref. 11.21). Two of them are briefly described next: the first is essentially based on the elastic analysis of thin plates and the second is based on the equivalent-frame method.

1. Elastic Solution

Timoshenko and Woinowsky-Krieger proposed a simplified solution to estimate deflections in typical rectangular panels having various support conditions and various ratios of long to short span (Ref. 11.22). Assuming a typical column-supported panel with a column size considered small in comparison to the column spacing, they derived the following expression for a uniformly loaded flat plate system:

$$\Delta_i = k \frac{w l_a^4}{E_c h^3} \tag{11.25}$$

where Δ_i = instantaneous deflection assuming short-term loading
l_a = longer span of panel center-to-center of supports
k = coefficient depending on both the ratio of long span l_a to short span l_b of the panel and the Poisson's ratio of the concrete.

The above equation must be dimensionally correct. Hence, if E_c is in pounds per square inch, h in inches, and l in inches, w should be given in pounds per square inch. For a continuous flat plate supported by columns, the maximum deflection, assuming alternate bay loading, leads to a value of k of the order of 0.11. Tests on four-panel prestressed concrete flat plates with various boundary conditions have indicated actual values of k of 0.148 (Ref. 11.23) and 0.095 (Ref. 11.24). These are not too different from the theoretical value of 0.11. Moreover, one can use as a first approximation the following rule of thumb: k decreases linearly to about 50 percent of its value when the long-to-short-span ratio of the panel increases from one to two. Equation (11.21) may represent, in some cases, a very rough approximation. However, a more accurate analysis will be needed only if the calculated deflection is close to the allowable deflection.

2. Equivalent Frame Approach

The deflection of flat plates can also be estimated from the equivalent-frame method which essentially applies to all types of two-way slabs reinforced or prestressed. The method is described in more detail in Refs. 11.4, 11.21, 11.25, and 11.26. The main steps of the procedure are summarized as follows. A typical slab panel bounded by column centerlines is considered. The panel is analyzed in each of the main directions as if it was part of a one-way slab system (or equivalent frame) having nonyielding support lines along column centerlines. The midpanel deflections, one for each direction, are computed and added to obtain the total deflection. The assumed deflected shape of the panel in each direction and the resulting deflected surface are illustrated in Figs. 11.20a to c. Although a panel bounded by column centerlines was used to explain the procedure, actual computations for each direction are undertaken using the same slab strips and corresponding distribution of moments as defined in Sec. 11.6 for the equivalent-frame method. Thus, the midpanel deflection is obtained as the sum of the midspan deflection of the column strip in one direction and the midspan deflection of the middle strip in the other direction, that is:

$$\Delta_i = \Delta_{cx} + \Delta_{my} \tag{11.26}$$

or

$$\Delta_i = \Delta_{cy} + \Delta_{mx} \tag{11.27}$$

where Δ_i is the instantaneous elastic short-term deflection, and the subscripts c, m, x, and y are for column strip, middle strip, x direction, and y direction, respectively (Fig. 11.19c). The midpanel deflection should essentially be the same whether calculated from Eq. (11.26) or Eq. (11.27). However, a difference will generally exist due to the approximate nature of the analysis. As the main contribution to the deflection comes from the longer span, it is recommended to use, for a typical interior panel, the column strip in the long direction and the middle strip in the short direction (Ref. 11.25). This is shown schematically in Fig. 11.21. As the equivalent-frame method leads to the moments in one direction or the other, deflection components such as Δ_{cx} and Δ_{my} can easily be calculated using the second moment area theorem (see Sec. 7.3) and assuming parabolic moment diagrams between supports.

The deflections calculated from Eqs. (11.25) to (11.27) are the instantaneous elastic short-term deflections assuming a uniformly applied load. For a sustained load, the additional long-term deflection due to creep must also be considered in a manner similar to beams and one-way slabs. Commonly, for two-way slabs, the

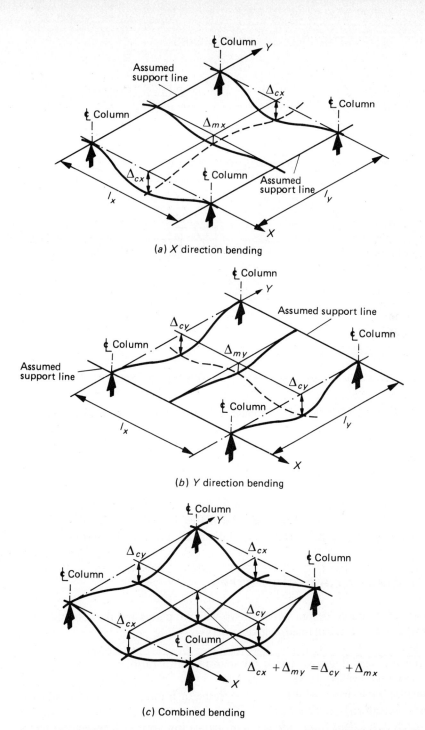

(a) X direction bending

(b) Y direction bending

$$\Delta_{cx} + \Delta_{my} = \Delta_{cy} + \Delta_{mx}$$

(c) Combined bending

Figure 11.20 Deflected surface of flat plate. (*a*) Assuming bending in the *x* direction only. (*b*) Assuming bending in the *y* direction only. (*c*) Assuming bending in two directions.

(Ref. 11.21, Courtesy American Concrete Institute.)

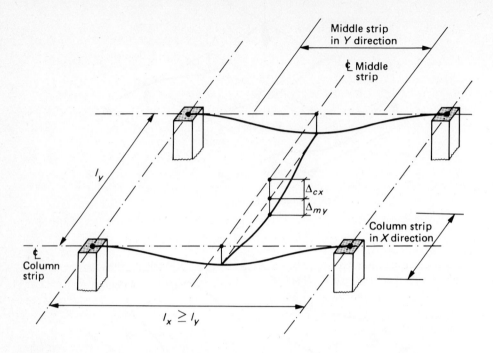

Figure 11.21 Determination of midpanel deflection by the equivalent-frame method.

additional long-term deflection is taken equal to the instantaneous elastic deflection multiplied by a factor of 2.

An example illustrating deflection computations is covered in Sec. 11.13. Experimental data on prestressed concrete flat plates, including deflection measurements, can be found in Refs. 11.16, 11.17, 11.23, 11.24, 11.27, 11.28, and 11.29.

11.12 SUMMARY OF DESIGN STEPS

A summary of the major design steps for two-way flat plates and the corresponding sections where they are explained, is given next:

1. Proportion slab thickness based on span-to-depth ratios and fire endurance requirements (see Sec. 11.8).
2. Define for each principal direction the frame or frames equivalent to the slab system considered (see Sec. 11.6).
3. For each equivalent frame use the load-balancing approach to determine the prestressing force. The load to be balanced should preferably include the entire dead load. Compute average prestress and compare with the average prestress values suggested in Sec. 11.8 (see Sec. 10.13).

4. Analyze the equivalent frame for the unbalanced load and determine corresponding moments and stresses.
5. Superimpose to the actual average prestress the stresses due to the unbalanced load and compare the resulting stresses with allowable stresses (see Sec. 11.9).
6. Determine minimum nonprestressed ordinary reinforcement, if required.
7. Detail tendon and bar layout and check if spacing and cover requirements are satisfied (see Sec. 11.7).
8. Check ultimate flexural strength requirements (see Secs. 11.9 and 10.14).
9. Check shear, both one-way shear and punching shear, and provide shear reinforcement if needed (see Sec. 11.10).
10. Compute deflection and compare with deflection limitations (see Sec. 11.11).

It was pointed out earlier that the effect of smoothing the tendon profile over the supports after load balancing is generally neglected in slabs. It may be necessary to estimate this effect in some cases. For this, the equivalent load due to prestressing, for the new tendon profile, is determined and the equivalent frame is analyzed for that load. The moments obtained are compared with the primary moments due to prestressing. The difference, if any, at any section represents the secondary moments. If the secondary moments are relatively significant (say more than 5 to 10 percent of the primary moment), they should then be accounted for in both the computation of stresses and the strength design moment where they are included with a factor of 1.

If in the above steps one of the design criteria is not satisfied, the design should be revised or refined as needed.

11.13 EXAMPLE: DESIGN OF A TWO-WAY FLAT PLATE

Consider the prestressed concrete flat plate floor system with square panels shown in Fig. 11.22. It is proposed for an office building and utilizes unbonded tendons. The columns are 30 ft apart, center-to-center in each direction. Their cross section is given in Fig. 11.22. The story height is typically 12 ft center-to-center of slab floor. A design live load of 80 psf is specified. The superimposed dead load due to flooring and partitions is assumed equivalent to a uniform load of 25 psf.

Let us provide a preliminary design for a typical interior frame taken along the short direction of the slab (shaded area of Fig. 11.22). This frame is the most critical for the slab shown. The following information is provided: normal weight concrete with $\gamma_c = 150$ pcf, $f'_c = 5000$ psi, $f'_{ci} = 4000$ psi, $E_c = 4.28 \times 10^6$ psi; unbonded strands with a diameter of 0.6 in, $f_{pu} = 270$ ksi, $f_{pe} = 160$ ksi, cross section area of one strand = 0.217 in^2; concrete cover to center of strands = 1.25 in.

Using a span-to-depth ratio of 40 for a slab without drop panels leads to a slab depth $h = 9$ in with a dead weight $w_G = 112.5$ psf.

Analysis for moment and shear. The moment distribution method is used to determine moments and shears. Member stiffness and distribution factors at each joint are needed.

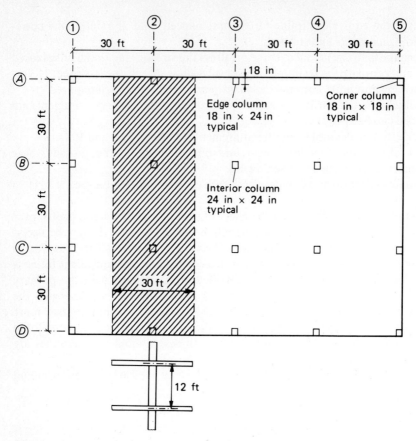

Figure 11.22 Flat plate of Example, Section 11.13.

Theoretical values will be used, assuming center-to-center spans. The stiffnesses are given by:

Slab
$$K_s = \frac{4E_c I}{l} = \frac{4E_c \times 12 \times 30 \times 9^3}{12 \times 12 \times 30} = 243E_c$$

Interior column
$$K_c = \frac{4E_c I}{l} = \frac{4E_c \times 24 \times 24^3}{12 \times 12 \times 12} = 768E_c$$

Edge column
$$K_c = \frac{4E_c \times 24 \times 18^3}{12 \times 12 \times 12} = 324E_c$$

The stiffness of the torsional member (slab strip) attached to an interior column is given by Eq. (11.4) in which the torsional constant C is given by Eq. (11.5). Hence:

$$C = \left(1 - 0.63 \frac{x}{y}\right) \frac{x^3 y}{3} = \left(1 - 0.63 \frac{9}{24}\right) \frac{9^3 \times 24}{3} = 4454.19 \text{ in}^4$$

$$K_t = \sum \frac{9E_c c}{l_2(1 - c_2/l_2)^3} = \frac{2 \times 9 \times E_c \times 4454.19}{30 \times 12(1 - 24/(30 \times 12))^3} = 273.92E_c$$

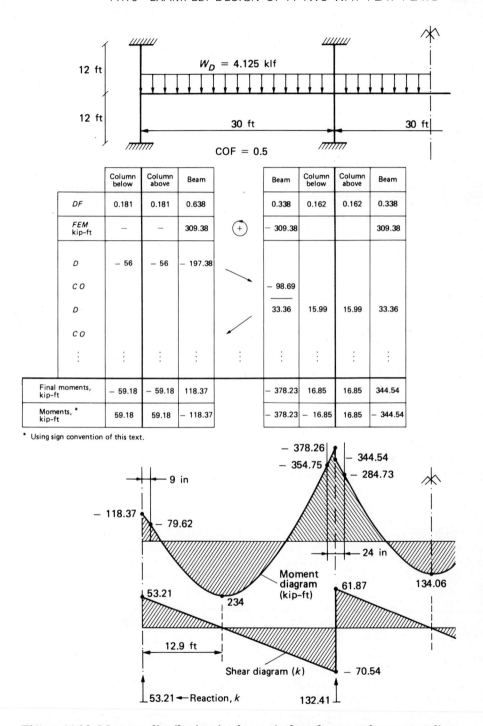

	Column below	Column above	Beam			Beam	Column below	Column above	Beam
DF	0.181	0.181	0.638			0.338	0.162	0.162	0.338
FEM kip-ft	—	—	309.38			− 309.38			309.38
D	− 56	− 56	− 197.38						
CO						− 98.69			
D						33.36	15.99	15.99	33.36
CO									
	⋮	⋮	⋮		⋮	⋮	⋮	⋮	⋮
Final moments, kip-ft	− 59.18	− 59.18	118.37			− 378.23	16.85	16.85	344.54
Moments, * kip-ft	59.18	59.18	− 118.37			− 378.23	− 16.85	16.85	− 344.54

* Using sign convention of this text.

Figure 11.23 Moment distribution in the equivalent frame and corresponding shear forces and reactions under dead load.

The stiffness of the equivalent interior column is obtained from Eq. (11.3):

$$\frac{1}{K_{ec}} = \frac{1}{\sum K_c} + \frac{1}{K_t} = \frac{1}{2 \times 768E_c} + \frac{1}{273.92E_c}$$

from which:

$$K_{ec} = 232.46E_c$$

K_{ec} is the stiffness equivalent to the two columns above and below the slab. Similar calculations for the edge column lead to:

$$C = 2996.19 \text{ in}^4 \qquad K_t = 174.73E_c* \qquad \text{and} \qquad K_{ec} = 137.62E_c$$

The distribution factors at each joint defined by the ratio of stiffness $K_{ij}/\sum K_{ik}$ are calculated and shown in Fig. 11.23, where the moment distribution is run assuming a uniform dead load which includes the weight of the slab and the superimposed dead load. Its value is:

$$w_D = (0.1125 + 0.025)30 = 4.125 \text{ klf}$$

The corresponding fixed-end moment is given by $-w_D l^2/12 = -309.38$ kips-ft. The moment distribution is also run assuming various pattern loadings of live load (Fig. 11.24). The live load per linear foot width of frame is given by:

$$w_L = 0.080 \times 30 = 2.4 \text{ klf}$$

and the corresponding fixed-end moment is given by $-w_L l^2/12 = -180$ kips-ft.

The results of the analysis for the uniform dead load (loading 1) and for four different patterns of live load (loadings 2 to 5) are summarized in Figs. 11.23 and 11.25. Note that not only the moment at the centerline of the column but also the moments at the face of columns as well as the shear forces and the reactions are given. They provide a needed source of data for the design.

A summary of most critical effects due to combinations of loadings is given in Table 11.2. The unbalanced service moments (assuming unbalanced live load) are needed in the load-balancing method. Strength design moments are needed for ultimate strength analysis and are taken at the face of the supports where applicable. The factored transfer moments and the corresponding loading reactions are taken at the centerline of columns. They are needed for shear design.

Load balancing. Let us balance the entire dead load, that is $w_b = w_D = 4.125$ klf. Assuming a cover of 1.25 in to the centroid of the tendons leads to a maximum sag $\delta_1 = 4.6$ in in the exterior span and $\delta_2 = 6.5$ in in the interior span (Fig. 11.26a). The required prestressing force for load balancing is controlled by the exterior span and is given by:

$$F = \frac{w_b l^2}{8\delta_1} = \frac{4.125 \times 30^2}{8 \times 4.6/12} = 1210.6 \text{ kips}$$

Although a smaller value could be used for the interior span, it seems more economical to run the same tendons throughout the length of the slab, because there is only one interior span. Hence, to maintain load balancing, the sag in the interior span is reduced to $\delta_2 = \delta_1 = 4.6$ in. The ideal steel profile is shown in Fig. 11.26a while the actual profile can be as shown in Fig. 11.26b.

---

* This value of K_t is erroneous and should be $184.26E_c$. If it is corrected, slight differences will result in the numerical values of K_{ec}, the distribution factors, and the moments obtained from the moment distribution analysis. However, the design of the slab remains essentially the same.

Loading 1, dead load or loading 2, live load

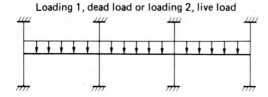

Loading 3, live load

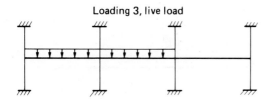

Loading 4, live load

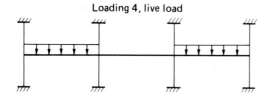

Loading 5 live load

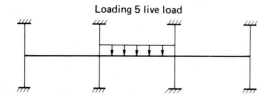

Figure 11.24 Pattern live loadings for the equivalent frame of Example slab, Section 11.13.

Assuming a required force of 1210.6 kips, leads to the following number of strands:

$$N = \frac{1210.6}{0.217 \times 160} = 34.87 \simeq 35 \text{ strands}$$

Thirty-five strands provide a final prestressing force of 1215.2 kips. To achieve about 75 percent distribution in the column strip, 25 strands are placed in the 15-ft-wide column strip at about 7.5-in spacing and 10 strands in the 15-ft-wide middle strip at about 18-in spacing (Fig. 11.26c). The corresponding average prestress values are:

$$(\sigma_g)_c = \frac{25 \times 0.217 \times 160}{15 \times 12 \times 9} = 536 \text{ psi}$$

$$(\sigma_g)_m = \frac{10 \times 0.217 \times 160}{15 \times 12 \times 9} = 214 \text{ psi}$$

Note that the average prestress for the whole slab strip is $\sigma_g = 375$ psi and is within practical range.

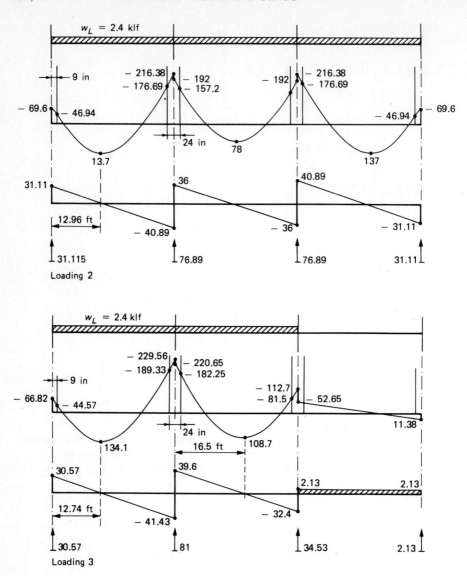

Figure 11.25 Moment diagrams, shear diagrams, and reactions in the equivalent frame for different live loadings.

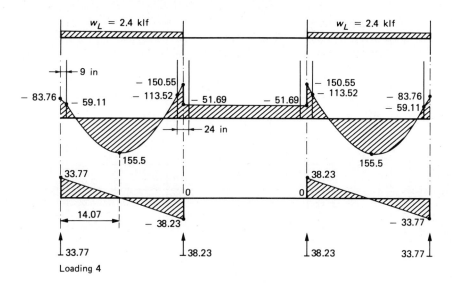

Loading 4

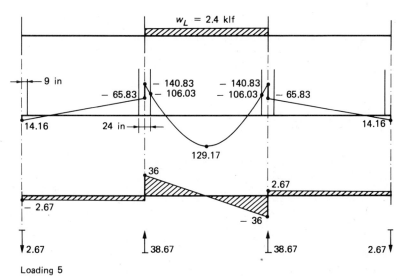

Loading 5

Table 11.2 Maximum moments and support reaction of slab Example Sec. 11.13

		Exterior span			℄ first interior support	Interior span	
	℄ exterior support	Face left support	Near midspan	Face right support		Face left support	Midspan
Unbalanced moment, kips-ft							
Loading 3: M_{nb}	....	−59.11	....	−189.33	....	−182.25	....
Loading 4: M_{nb}	....	....	+155	....	....	....	....
Loading 5: M_{nb}	....	....	....	....	....	....	+129.17
Factored moment, kips-ft							
Loading 1 + 3: M_u	....	|−211.95|	....	|−755.51|	....	|−708.45|	....
Loading 1 + 4: M_u	....	....	591.1	....	....	....	....
Loading 1 + 5: M_u	....	....	....	....	....	....	387
Transfer moment, kips-ft, and reaction, kip							
Loading 1 + 3, M_u	279				62.31		
Loading 1 + 3, R_u	126				323		
Loading 1 + 4, M_u	308				215.23		
Loading 1 + 4, R_u	132				250.36		

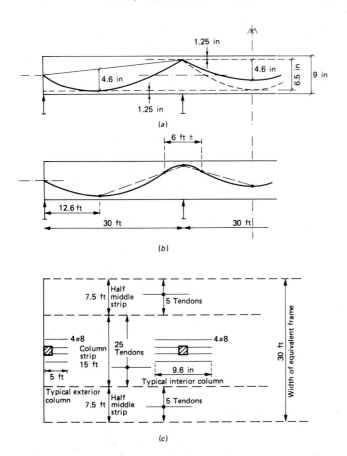

Figure 11.26 Example flat plate. (*a*) Ideal tendon profile. (*b*) Actual tendon profile. (*c*) Reinforcement layout in one direction.

Service stresses. Flexural stresses are due to the uniform stress induced by load balancing and stresses generated by the unbalanced live load. It will be assumed that 75 percent of the moments is distributed to the column strip and the remainder to the middle strip. The section moduli for either strip are given by:

$$Z_t = Z_b = \frac{bh^2}{6} = \frac{15 \times 12 \times 9^2}{6} = 2430 \text{ in}^3$$

Critical values of the unbalanced moment are given in Table 11.2. Stresses at the corresponding critical sections can be calculated separately for the column strip and the middle strip. For the column strip, they are given by:

Exterior span, near midspan

$$\sigma_t = (\sigma_g)_c + \frac{M}{Z_t} = 536 + \frac{0.75 \times 155 \times 12,000}{2430} = 1110 \text{ psi} < 0.3f'_c \qquad \text{OK}$$

$$\sigma_b = (\sigma_g)_c - \frac{M}{Z_t} = 536 - 574 = -38 \text{ psi} > -2\sqrt{f'_c} \qquad \text{OK}$$

First interior support

$$\sigma_t = (\sigma_g)_c + \frac{M}{Z_b} = 536 - \frac{0.75 \times 189.33 \times 12{,}000}{2430} = -165 \text{ psi} > -6\sqrt{f'_c} \qquad \text{OK}$$

$$\sigma_b = 536 + 701 = 1237 \text{ psi} < 0.3f'_c \qquad \text{OK}$$

Interior span, midspan

$$\sigma_t = (\sigma_g)_c + \frac{M}{Z_t} = 536 + \frac{0.75 \times 129.17 \times 12{,}000}{2430} = 1014.4 \text{ psi} < 0.3f'_c \qquad \text{OK}$$

$$\sigma_b = 536 - 478.4 = 57.6 > -2\sqrt{f'_c} \qquad \text{OK}$$

Similar calculations for the middle strip lead to smaller stresses and are not given here. Note that, because the magnitude of tensile stresses in positive moment regions does not exceed $2\sqrt{f'_c}$, there is no need to add nonprestressed bonded reinforcement. However, nonprestressed bonded reinforcement is needed at the supports and is given by Eq. (11.8):

$$A_s = 0.00075hl = 0.00075 \times 9 \times 30 \times 12 = 2.43 \text{ in}^2$$

This can be achieved using four #8 bars, leading to an actual $A_s = 3.16$ in^2. They are placed within $1.5h$ on each side of the column faces at a spacing of about 15 in. Their length should be at least one-sixth the clear span on each side of the support, that is, a total of about 9.5 ft (Fig. 11.26c). Similarly, four #8 bars are placed at the exterior columns projecting about five feet into the slab.

Ultimate flexural strength. The stress in the prestressing steel at nominal moment resistance can be estimated (on the safe side) from Eq. (11.6):

$$f_{ps} = f_{pe} + 15 = 160 + 15 = 175 \text{ ksi}$$

Let us check ultimate strength requirements at the most critical section, that is, the first interior support. For such calculations, the whole slab width is considered using the following variables:

$$A_{ps} = 35 \times 0.217 = 7.60 \text{ in}^2$$

$$A_s = 3.16 \text{ in}^2 \text{ (see previous section)}$$

$$d_p = d_s = d = 7.75 \text{ in}$$

$$f_y = 60 \text{ ksi}$$

The reinforcing index is given by:

$$q = \frac{A_{ps}f_{ps} + A_s f_y}{bdf'_c} = \frac{7.6 \times 175 + 3.16 \times 60}{12 \times 30 \times 7.75 \times 5} = 0.1089$$

and the design nominal moment is obtained from (Fig. 5.13):

$$\phi M_n = \phi f'_c bd^2 q(1 - 0.59q)$$

$$= 0.9 \times 5000 \times 12 \times 30 \times 7.75^2 \times 0.1089(1 - 0.59 \times 0.1089)/12{,}000$$

$$\simeq 826.3 \text{ kips-ft}$$

The strength design moment M_u at the face of the support is given in Table 11.2 and is equal in magnitude to 755.51 kips-ft. As $M_u < \phi M_n$, ultimate strength is satisfactory at the support. It can be shown that the design nominal moment resistance ϕM_n is equal to

729 kips-ft at the near midspan section of the exterior span, 540 kips-ft at the midspan of the interior span, and 405 kips-ft at the exterior support (considering prestressed reinforcement only) and is satisfactory.

Shear strength. It can be easily shown that beam-type shear is largely satisfactory for this slab. Let us consider punching shear around the first interior column. The area of the critical section and its polar moment of inertia are given by Eqs. (11.14) and (11.15), namely:

$$A_c = 2(c_1 + c_2 + 2d)d = 2(24 + 24 + 15.5)7.75 = 984.25 \text{ in}^2$$

$$J_c = \frac{7.75(24 + 7.75)^3}{6} + \frac{(24 + 7.75)7.75^3}{6} + \frac{7.75(24 + 7.75)(24 + 7.75)^2}{2}$$

$$= 167,827 \text{ in}^4$$

The fraction of moment assumed transferred by shear for a typical interior column is given by Eq. (11.11) and, for a square column, leads to $\gamma_v = 0.40$.

Shear stresses are induced by the shear force at the critical section and the fraction of moment transferred by shear to the columns. The shear force at the critical section is obtained from the ultimate reaction at the supporting column minus the factored load applied to the slab inside a square area limited by the perimeter b_o of A_c and equal $(c_1 + d)(c_2 + d) = 7.0 \text{ ft}^2$. Referring to Table 11.2, it can be seen that two loading combinations may be critical for shear. Considering the combination of loadings 1 and 3, we have:

$$V_u = 323 - 7(1.4 \times 0.1375 + 1.7 \times 0.08) = 320.7 \text{ kips}$$

The corresponding value of transfer moment is given by $M_u = 62.31$ kips-ft. The maximum shear stress is obtained from Eq. (11.12) as:

$$v_u = \frac{V_u}{A_c} + \frac{\gamma_v M_u c_3}{J_c} = \frac{320,700}{984.25} + \frac{0.4 \times 62.31 \times 12,000 \times 15.875}{167,827} = 354.12 \text{ psi}$$

For the other loading combination (1 + 4) at the interior column, we have (Table 11.2):

$$V_u = 250.36 - 7(0.3285) = 248 \text{ kips}$$

and

$$M_u = 215.23 \text{ kips-ft}$$

and the maximum shear stress is given by:

$$v_u = \frac{248,000}{984.25} + \frac{0.4 \times 215.23 \times 12,000 \times 15.875}{167,827} = 349.7 \text{ psi}$$

Hence, for design, we will consider the controlling value of $v_u = 354.12 \simeq 354$ psi. The shear resistance of the concrete is estimated from Eq. (11.9), that is:

$$v_c = v_{cw} = 3.5\sqrt{f'_c} + 0.3\sigma_g + \frac{V_p}{b_o d_p}$$

The average prestress in the column strip will be used for σ_g. Assuming same prestress in both directions of the slab and a steel profile as described in Fig. 11.26b, it can be shown that $V_p/b_o d$ is small and of the order of 15 psi. Hence, the value of v_c is given by:

$$v_c = 3.5\sqrt{5000} + 0.3 \times 536 + 15 \simeq 424 \text{ psi}$$

and

$$\phi v_c = 0.85 \times 423 \simeq 360 \text{ psi}$$

As $v_u < \phi v_c$, shear reinforcement is not required at the interior column.

However, to illustrate the use of bent bars, let us compute the area of shear reinforcement as if it was required. Assuming an angle of inclination $\alpha = 30°$, $f_y = 60$ ksi, and using Eqs. (11.24) and (11.23), we have:

$$V_s = \left(\frac{v_u}{\phi} - v_c\right) A_c = \left(\frac{354}{0.85} - 2\sqrt{5000}\right) \frac{984.25}{1000} = 270.72 \text{ kips}$$

$$A_s = \frac{V_s}{f_y \sin \alpha} = \frac{270.72}{60 \times 0.5} = 9.02 \text{ in}^2$$

This reinforcement is to be distributed across the four sides of the critical section, as shown in Fig. 11.19b. It can be provided by four #7 bars crossing at right angles in each direction and leading to $A_v = 4 \times 4 \times 0.6 = 9.6$ in².

A similar procedure is followed throughout to investigate punching shear at the exterior column. The following values are obtained from Eqs. (11.17) to (11.22):

$$A_c = 585.125 \text{ in}^2$$

$$J_c = 32,275 \text{ in}^4$$

$$c_3 = 6.338 \text{ in}$$

$$c_4 = 15.537 \text{ in}$$

$$c_5 = 6.537 \text{ in}$$

$$\gamma_v = 0.356$$

The critical loading combination (Table 11.2) corresponds to loadings 1 and 4. The transfer moment is to be taken at the centroid of the critical section distant c_5 from the column centerline and is given by:

$$M_u = 308 - 132 \times \frac{6.537}{12} = 236 \text{ kips-ft}$$

Hence, the design shear stress is given by:

$$v_u = \frac{132,000}{585.125} + \frac{0.356 \times 236 \times 12,000 \times 6.338}{32,275} = 225.59 + 197.98 = 423.6 \text{ psi}$$

As $v_u > \phi v_c$, shear reinforcement is needed at the exterior column. In such a case, the concrete contribution v_c is reduced to $2\sqrt{f'_c}$. If bent bar reinforcement is to be used, the nominal resistance v_v is limited to $6\sqrt{f'_c}$ (424 psi) which is smaller than what is needed ($v_u/\phi = 498$ psi). Several solutions can be considered, including the use of (1) shearhead reinforcement acceptable for v_n up to $7\sqrt{f'_c} = 495$ psi, (2) beam stirrups, (3) a drop panel or a column capital, and (4) a larger size column. Let us illustrate the use of the beam stirrups described in Fig. 11.19b, assuming #4 closed stirrups. The critical shear section around the exterior column has a periphery $b_o = 75.5$ in and has three sides. Hence, the cross-sectional area of three stirrups should be considered. That is:

$$A_v = 3 \times 2 \times 0.20 = 1.20 \text{ in}^2$$

The corresponding spacing is given by Eq. (6.37):

$$s = \frac{A_v f_y}{(v_u/\phi - v_c)b_o} = \frac{1.2 \times 60,000}{(423.6/0.85 - 2\sqrt{5000})75.5} = 2.67 \text{ in}$$

This is a relatively heavy shear reinforcement. The beam stirrups are to be extended by an appropriate distance beyond the faces of the column until punching shear stresses are reduced below v_c. About two feet are sufficient here.

It can be easily shown that there is sufficient moment resistance within $1.5h$ of the faces of both interior and exterior columns to resist the portion of moment $(1 - \gamma_v)M_u$ transferred by flexure. At the exterior column, the contribution of the nonprestressed tensile reinforcement will be found necessary.

Deflection. No deflection is induced under dead load because of load balancing and, as the slab is level, no additional long-term deflection should theoretically be recorded. Let us estimate the elastic deflection due to live load from Eq. (11.21), assuming pattern loading and a value of $k = 0.11$. That is:

$$\Delta_i = k \frac{w l_a^4}{E_c h^3} = 0.11 \frac{80 \times (12 \times 30)^4}{144 \times 4.28 \times 10^6 \times 9^3} = 0.33 \text{ in}$$

This value is less than $l/360 = 1$ in, allowed by the code (Table 7.2).

Additional comments. The above preliminary design was as close as possible to a final design. Some difficulties were encountered for punching shear resistance at the exterior column. As mentioned earlier, a drop panel or a column capital or a thicker slab could have been used in a revised design to avoid costly shear reinforcement. It seems that the best way to overcome punching shear is to provide a larger critical section or a thicker slab, or both. Note that a significant portion of the shear stress around the exterior column was due to the transfer moment. In determining the value of the transfer moment, the dead load, which is equal to the balanced load, was included. It can be argued that the slab is level under balanced load and prestressing. Thus, in effect, no transfer moment exists under dead load. If only the transfer moment due to the unbalanced load is considered in the computations, the punching shear at the exterior column would have been satisfactory. However, it is preferable to be on the safe side when dealing with punching shear. This may explain why, in practice, the entire transfer moment is considered.

REFERENCES

11.1. G. D. Nasser: "A Look at Prestressed Flat Plate Design and Construction," *PCI Journal*, vol. 14, no. 6, December 1969, pp. 62–77.

11.2. R. Park and W. L. Gamble: *Reinforced Concrete Slabs*, John Wiley & Sons, New York, 1980, 618 pp.

11.3. F. M. Ferguson: *Reinforced Concrete Fundamentals*, 4th ed., John Wiley & Sons, New York, 1979, 724 pp.

11.4. G. Winter and A. H. Nilson: *Design of Concrete Structures*, 9th ed., McGraw-Hill Book Company, New York, 1979, 647 pp.

11.5. C. K. Wang and G. Salmon: *Reinforced Concrete Design*, 3d ed., Thomas Y. Crowell Company, New York 1979, 918 pp.

11.6. P. F. Rice and E. S. Hoffman: *Structural Design Guide to the ACI Building Code*, 2d ed., Van Nostrand-Reinhold, 1979, 470 pp.

11.7. ACI-ASCE Joint Committee 423: "Tentative Recommendations for Prestressed Concrete Flat Plates," *ACI Journal*, vol. 71, no. 2, February 1974, pp. 61–71.

11.8. ————: "Recommendations for Concrete Members Prestressed with Unbonded Tendons," Draft Report, 1980.

11.9. K. Saether: "The Structural Membrane Theory Applied to the Design of Flat Slabs," *PCI Journal*, vol. 8, no. 5, October 1963, pp. 68–79.

11.10. G. I. N. Rozvany and A. J. K. Hampson: "Optimum Design of Prestressed Flat Plates," *ACI Journal*, vol. 60, no. 8, August 1968, pp. 1065–1082.

11.11. C. H. Wang: "Direct Design Method for Prestressed Concrete Slabs," *PCI Journal*, vol. 13, no. 3, June 1968, pp. 62–72.

11.12. A. L. Parme: "Prestressed Flat Plates," *PCI Journal*, vol. 13, no. 6, December 1968, pp. 14–32.

11.13. T. Y. Lin: "Load-Balancing Method for Design and Analysis of Prestressed Concrete Structures," *ACI Journal*, vol. 60, no. 6, June 1963, pp. 719–742.

11.14. W. G. Corley and J. O. Jirsa: "Equivalent Frame Analysis for Slab Design," *ACI Journal*, vol. 67, no. 11, November 1970, pp. 875–884.

11.15. L. L. Gerber and N. H. Burns: "Ultimate Strength Tests of Post-Tensioned Flat Plates," *PCI Journal*, vol. 16, no. 6, November/December 1971, pp. 40–58.

11.16. *Design of Post-Tensioned Slabs*, Post-Tensioning Institute, Phoenix, Arizona, 1977, 52 pp.

11.17. N. H. Burns and R. Hemakom: "Test of Scale Model Post-Tensioned Flat Plate," *Journal of the Structural Division, ASCE*, vol. 103, no. ST6, June 1977, pp. 1237–1255.

11.18. N. W. Hanson and J. M. Hanson: "Shear and Moment Transfer Between Concrete Slabs and Columns," Portland Cement Association, Development Department Bulletin D129, 1968, 16 pp.

11.19. N. M. Hawkins: "Shear Strength of Slabs with Moments Transferred to Columns," Shear in Reinforced Concrete, vol. 2, SP-42, American Concrete Institute, Detroit, 1974, pp. 817–846.

11.20. W. G. Corley and N. M. Hawkins: "Shearhead Reinforcement for Slabs," *ACI Journal*, vol. 65, no. 10, October 1968, pp. 811–824.

11.21. ACI Committee 435: "Deflection of Two-Way Reinforced Concrete Floor Systems: State-Of-The-Art Report," ACI Manual of Concrete Practice, pt 2, American Concrete Institute, Detroit, Michigan, 1979, pp. 435–83, 435–106.

11.22. S. P. Timoshenko and S. Woinowsky-Kreiger: *Theory of Plates and Shells*, 2d ed., McGraw-Hill Book Company, New York, 1959, 580 pp.

11.23. A. C. Scordelis, T. Y. Lin, and R. Itaya: "Behavior of a Continuous Slab Prestressed in Two Directions," *ACI Journal*, vol. 56, no. 6, December 1959, pp. 441–459.

11.24. E. G. Nawy and P. Chakrabarti: "Deflection of Prestressed Concrete Flat Plates," *PCI Journal*, vol. 21, no. 2, March/April, 1976, pp. 86–102.

11.25. A. H. Nilson and D. B. Walters: "Deflection of Two-Way Floor Systems by the Equivalent Frame Method," *ACI Journal*, vol. 72, no. 5, May 1975, pp. 210–218.

11.26. ————: *Design of Prestressed Concrete*, John Wiley and Sons, New York, 1978, 526 pp., Chap. 10.

11.27. J. F. Brotchie: "Some Australian Research on Flat Plate Structures," *ACI Journal*, vol. 77, no. 1, January/February 1980, pp. 3–11.

11.28. ⸻: "Experimental Studies of Prestressed Thin Plate Structures," *ACI Journal*, vol. 77, no. 2, March/April 1980, pp. 87–95.

11.29. N. M. Hawkins: "Lateral Load Resistance of Unbonded Post-Tensioned Flat Plate Construction," *PCI Journal*, vol. 26, no. 1, January/February 1981, pp. 94–115.

PROBLEMS

11.1 Redesign the slab of Example 11.3 assuming that the same prestressing force needed for the exterior span is used throughout the length of the slab.

11.2 Read the provisions of the ACI code and its commentary on shearhead reinforcement. Draw the corresponding critical sections for punching shear.

11.3 Determine the value of A_c, the critical section for punching shear, and its polar moment of inertia J_c for (a) a rectangular corner column and (b) a round interior column.

11.4 Read the provisions of the ACI code and its commentary related to openings in slabs near the columns. How do openings influence the critical section for punching shear?

11.5 Go back to the slab of Example 11.13 and design the exterior equivalent frame in each direction. All the input data are assumed the same.

11.6 Assuming that the required reinforcement in the slab of Example 11.13 is the same for both the longitudinal and the transverse directions, provide a reinforcement layout of the slab using a banded tendon distribution.

11.7 The effect of smoothing the tendon profile over the supports, after load balancing, was disregarded in Example 11.13. For the final tendon profile and prestressing force obtained, analyze the equivalent frame and determine the secondary moments at the supports. Compare their values with the primary moments and conclude on whether they are in effect negligible.

11.8 Go back to the slab of Example 11.13 and Fig. 11.22 and assume that beams having the same width as the columns are placed between the columns along the numbered grid lines. Hence, a one-way slab system along the lettered grid lines is obtained. Design this one-way slab using the same input data as for Example 11.13 and a slab depth of 10 in.

Five million gallon (190,000 m³) prestressed concrete water tank in Pinnellas County, Florida. *(Courtesy Portland Cement Association.)*

CHAPTER 12

ANALYSIS AND DESIGN OF TENSION MEMBERS

12.1 TYPES OF TENSION MEMBERS

Prestressed concrete tension members are structural elements predominantly subjected to axial tension. They are mostly linear, circular, or parabolic (catenary) in shape. Linear tension elements, commonly called ties, include restraining ties for arch bridges, soil anchors for retaining structures, and truss members. Circular elements are part of any figure of revolution and include cylindrical tanks, silos, and pressure vessels. Parabolic prestressed concrete tension members, often described as stress ribbons, follow the same principles as catenary steel cables. They are used in inverted suspension bridges. Typical examples of tension members are shown in Fig. 12.1 where only the tensile member is emphasized in its position with respect to the rest of the structure, schematically described by a line configuration.

Prestressed linear tension members are similar to compressive members, except that they carry a much higher level of prestress. They are generally concentrically prestressed. However, when a combination of flexure (self-weight) and tension exists, the prestress may be slightly eccentric to balance the weight and keep a uniform state of stress in the section.

Although experimental data and design information on prestressed concrete tension members are scarce (perhaps because they are simpler to analyze than other elements), they are widely used in actual applications. In some examples they represent the main element that confers to the structure its innovative aspects.

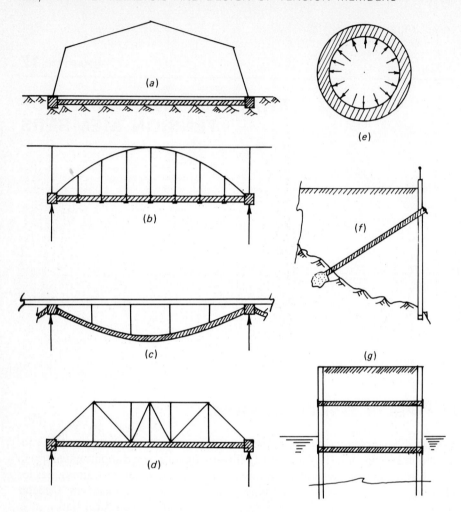

Figure 12.1 Typical examples of tension members.

Morandi (Ref. 12.1) built several bridges featuring prestressed concrete tensile elements, used in a cable-stayed configuration. Finsterwalder (Ref. 12.2) presented an innovative design for the Bosphorus Bridge (not built) where a shallow funicular-shaped prestressed concrete stress-ribbon tensile member spanning 190 m (620 ft) was proposed. The ribbon was to serve as tensile element and deck in the central portion of the bridge's main span of 408 m (1340 ft). Lin (Ref. 12.3) designed and built the inverted suspension Rio Colorado Bridge in Costa Rica, where a catenary-shaped prestressed concrete ribbon spanning 146 m (480 ft) was used to support the horizontal deck. Such bridges may represent a comparatively low-cost solution for crossing deep valleys (Refs. 12.4 and 12.5).

Figure 12.2 Prestressed concrete truss railway bridge: Iwahana Bridge, Japan.
(Ref. 12.7, Courtesy Prestressed Concrete Institute.)

Precast prestressed concrete tensile and compressive elements can successfully be used in trusses where they are assembled on site and connected by post-tensioning (Refs. 12.6 and 12.7). Several truss railway bridges have been built in Japan (Ref. 12.7). A typical example is shown in Fig. 12.2.

Some of the advantages of prestressed concrete tensile members are discussed next.

12.2 ADVANTAGES OF PRESTRESSED CONCRETE TENSION MEMBERS

The use of concrete in pure tensile members, also called ties, may seem paradoxical, as concrete is weak in tension. However, a prestressed concrete element can be treated as a single composite material that can sustain tension, with no reference to its components, steel and concrete (Fig. 4.5). Prestressed concrete tension members can present a number of salient advantages over their counterparts made with either steel of reinforced concrete:

1. A prestressed concrete tie can be designed not to crack under normal service loads and acceptable levels of service overloads.

2. Because it is crack-free, it offers an excellent corrosion protection to the steel reinforcement.
3. The use of concrete offers an inherent fire resistance which, in some instances, may be an important design factor.
4. The most significant advantage of using a prestressed concrete tension member is that the total deformation necessary to develop the full resistance to the applied external load can be controlled practically to any desired degree. This is because the member behaves essentially as a linear elastic crack-free material and a change in its cross section leads to a proportional change in its deformation under load. The total deformation (elongation or shortening) under load is a very important design variable. An excessive deformation may lead to distorsional distress and failure in the structure itself or in some of its elements.

Everything else being equal, the deformation of a prestressed concrete tie can be more than an order of magnitude smaller than that of a steel tie. This is illustrated in the following example.

1. Example: Relative Deformation of Tension Members

Consider a tensile force N of magnitude 100 kips to be carried over a span of 100 ft. Several design proposals are to be evaluated including a high-strength steel cable made of prestressing strands, a structural steel member, a prestressed concrete member, and a reinforced concrete member. A schematic representation of the problem and typical formula are given in Fig. 12.3.

The steel cable (Fig. 12.3a) has an area $A_{ps} = 0.918$ in^2 and an elastic modulus $E_{ps} = 28.14 \times 10^6$ psi. It can be shown that its elongation under load is given by $\Delta l_{ps} = Nl/E_{ps} A_{ps} \simeq 4.64$ in.

The structural steel member (Fig. 12.3b) has a cross-sectional area $A_s = 5$ in^2 and same elastic modulus as the cable, i.e., $E_s = 28.14 \times 10^6$ psi. Using the same formula as above leads to $\Delta l_s = 0.85$ in.

The prestressed concrete tie (Fig. 12.3c) has the same area of prestressing steel as the steel cable, i.e., $A_{ps} = 0.918$ in^2, a gross sectional area $A_g = 100$ in^2, and a concrete elastic modulus $E_c = 4.69 \times 10^6$ psi. The transformed area of the section is $A_t = 104.59$ in^2. The member's instantaneous elongation under the same load N is given by $|\Delta l_c| = |Nl/E_c A_t| = 0.245$ in.

If we keep the same prestressing steel reinforcement and double the concrete area (Fig. 12.3d), the elongation is halved to approximately 0.125 in. This value is less than three percent of that obtained if the free steel cable were used alone. In practice, long-term effects must also be considered and generally counteract the effect of external loads. In the above comparison the steel cable was chosen to have the same cross-sectional area as the prestressing tendons in the prestressed concrete tie. Of course, in practice a free cable will have a substantially lower allowable stress or higher area than those assumed.

The design of a reinforced concrete tie is based on a cracked concrete section. The corresponding area of steel reinforcement is significantly higher than that of a prestressed tie. Everything else being the same, its elongation under load would be higher than that of a prestressed tie, the difference being mainly due to the sum of crack widths along the member (Fig. 12.3e). For the same steel area and steel stress, the elongation of a reinforced concrete tie can be expected to be less than that of a structural steel member.

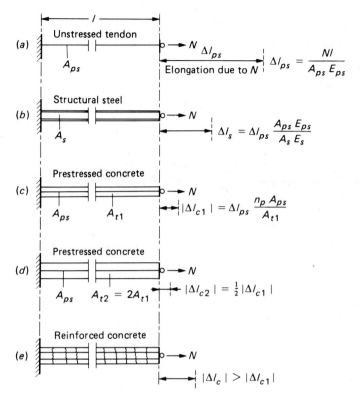

Figure 12.3 Relative deformation in tension members of different materials.

12.3 BEHAVIOR OF PRESTRESSED CONCRETE TENSION MEMBERS

Because prestressed concrete ties are simple structural elements that are expected to perform in the linear elastic uncracked range of behavior, they have generated little need for experimental research. A recent investigation by Wheen (Ref. 12.8) has clarified the influence of many variables on the load-deformation response of the composite and allows for some comparison and correlation with reinforced concrete ties.

When a prestressed concrete tensile member is subjected to a monotonically increasing tensile load, its load elongation curve is characterized by an initial linear elastic portion up to first structural cracking. The increase in external load is accompanied by a slow stress increase in the steel and a fast stress decrease in the concrete which eventually leads to tension in the concrete and subsequent cracking. Wheen reports that the occurrence of first cracking is dramatic and is invariably accompanied by a loud bang (Ref. 12.8). Significant changes in characteristics occur after cracking. They include a very large reduction in the stiffness of the composite and a sudden increase in steel stress (stress jump) as the steel must resist

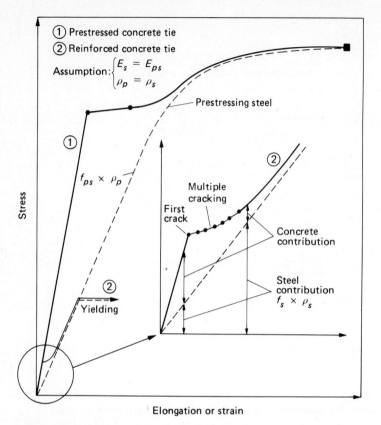

Figure 12.4 Typical stress-elongation curves of reinforced and prestressed concrete tension members.

the additional force released by the concrete. Such a stress jump may often lead to a steel stress in the nonlinear range of behavior. It may cause serious steel debonding on either side of the crack, thus leading to unusually wide cracks. If the member is less than minimally reinforced, that is if the cracking load is larger than the ultimate resistance of the tendons alone, cracking will also lead to failure. For normally reinforced members, the composite resistance after cracking is reduced to that of the reinforcement alone and the composite load elongation curve is essentially reduced to that of the reinforcement.

A typical stress elongation curve of a prestressed concrete tie is plotted in Fig. 12.4 and compared to that of a reinforced concrete tie having the same reinforcement ratio. It can be observed that the reinforced concrete tie will crack in the very early stages of loading. A slight increase in load after cracking leads to multiple crack formations shortly followed by crack stabilization. No further cracks form thereafter but the widths of existing cracks increase with the load and the composite response tends toward that of the reinforcing steel alone. A reinforced concrete tie is designed to resist service loads after cracking, while a prestressed concrete tie is mostly serviceable before cracking. The contribution of the reinforcement to the composite response, represented by $\rho_p f_{ps}$ or $\rho_s f_s$, is also plotted in Fig. 12.4 versus the elongation. It is assumed, for clarity, that both the prestressing steel and the reinforcing steel have the same modulus of elasticity. The difference in ordinate between the composite curve and the reinforcement curve represents the average contribution of the concrete. Such contribution is substantial prior to cracking. After cracking it decreases with the elongation and tends to vanish. Simultaneously, the composite's response tends toward that of the reinforcement alone. Note that, because of the extent of concrete contribution prior

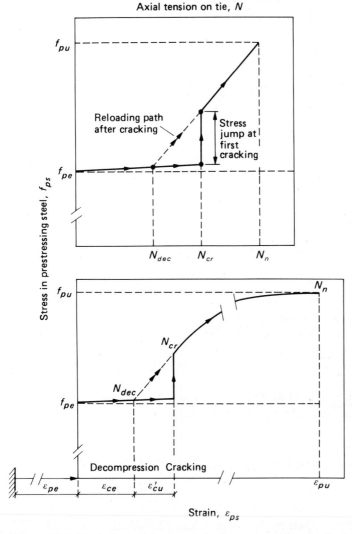

Figure 12.5 Variation of the stress in the prestressing steel with loading.

to cracking, prestressing takes substantial advantage of the presence of the concrete.

If, due to an overload, a prestressed tensile member with bonded reinforcement cracks, its response under subsequent loading would be similar to that of the first loading except that existing cracks will start opening at decompression instead of at the cracking load. A smoother transition in the steel stress will replace the stress jump otherwise encountered at first cracking, as illustrated in Fig. 12.5. After cracking, the stress in the prestressing steel increases in direct proportion to the applied load N until failure (Fig. 12.5), but the strain increases at a much faster rate, according to the stress-strain relationship of the steel. In Fig. 12.5, ε_{ce} rep-

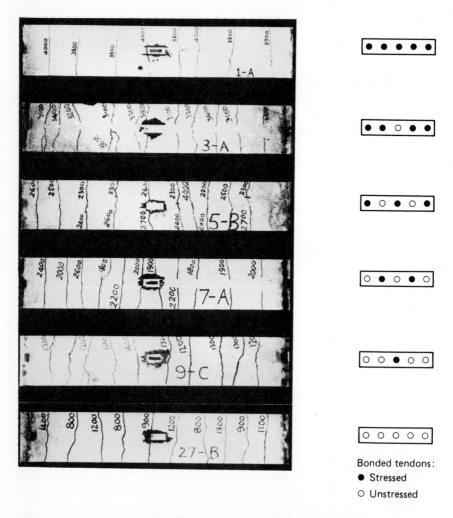

Bonded tendons:
● Stressed
○ Unstressed

Figure 12.6 Typical crack patterns in prestressed tension members.
(Courtesy R. J. Wheen, Ref. 12.8.)

resents the concrete strain under effective prestress and ε'_{cu} the tensile fracture strain of concrete.

If the reinforcement is unbonded and cracking occurs, generally only one crack develops and subsequent reloading will see the stiffness and the response of the composite reduced to those of the reinforcement alone. Because of this dramatic reduction in stiffness, the use of unbonded reinforcement in prestressed tensile members is not recommended.

Typical crack patterns in prestressed concrete tension members using bonded stressed and unstressed tendons, as reported by Wheen, are shown in Fig. 12.6 (Ref. 12.8). Note that cracking in these members is very similar to that generally observed using reinforced concrete.

12.4 ANALYSIS OF TENSION MEMBERS

The analysis of prestressed concrete tensile members presents little mathematical difficulty. It is generally assumed that the member behaves in a linear elastic manner under service loads and that both the prestressing force and the external tensile load are concentric to the axis of the member, thus leading to uniform stresses in the section.

In addition to developing the basic analytic equations for prestressed tension members, criteria usually checked in the "review or investigation" process will also be covered in this section. They include: maximum allowable compression, ultimate strength, safety against cracking, safety against decompression, minimum reinforcement, and maximum instantaneous and long-term deformations.

Let us consider a prestressed concrete tension member subjected to an axial force N (Fig. 12.7). According to our sign convention, compression is positive for concrete and tension is positive for the steel. As N is applied to the concrete, it is negative. Standard notation will be used whenever possible.

1. Service Stresses, Cracking, and Ultimate Load

The net area of the concrete section subjected to the prestressing force is defined by:

$$A_n = A_g - A_{ps} \qquad (12.1)$$

where A_g is the gross sectional area of the concrete section and A_{ps} the area of the prestressing steel.

An external force N applied to the member causes an equal elongation in the steel and the concrete and is resisted by both. Equivalently, it can be shown that N is resisted by the transformed cross-sectional area of the member defined by:

$$A_t = A_n + n_p A_{ps} = A_g + (n_p - 1)A_{ps} \qquad (12.2)$$

where $n_p = E_{ps}/E_c$ and E_{ps} and E_c are the moduli of elasticity of the prestressing steel and the concrete.

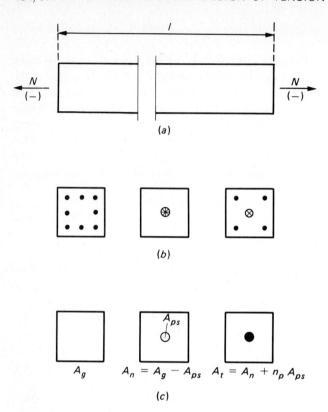

Figure 12.7 (*a*) **Prestressed tension member: sign convention.** (*b*) **Pretensioned section, posttensioned section, or both.** (*c*) **Gross, net, and transformed sections.**

The uniform compressive stress in the concrete due to the prestressing force at release is given by:

$$\sigma_{gi} = \frac{F_i}{A_n} = \frac{A_{ps} f_{pi}}{A_n} \tag{12.3}$$

and the effective stress after all losses is given by:

$$\sigma_g = \frac{F}{A_n} = \frac{A_{ps} f_{pe}}{A_n} \tag{12.4}$$

Note that, in Chap. 4, the gross sectional area A_g was used instead of A_n to determine σ_g and σ_{gi} in beams.

If allowable stresses are not to be exceeded, the following condition should be satisfied:

$$\sigma_{gi} \le \bar{\sigma}_{ci} \tag{12.5}$$

$$\sigma_g \le \bar{\sigma}_{cs}$$

where $\bar{\sigma}_{ci}$ and $\bar{\sigma}_{cs}$ are allowable compressive stresses in the concrete under initial and effective or final prestress. As the compression is uniform, their magnitude is generally not the same as for flexure where extreme fiber stress is considered. The ACI code does not have special provisions for uniform compression. However, in its alternative design method, App. B, it recommends a maximum bearing stress of $0.30f'_c$. In practice it seems reasonable not to exceed such a value in order to minimize creep strains and long-term deformation. Note that ACI Committee 344 suggests the following allowable stresses for tanks: $\bar{\sigma}_{cs} = 0.45f'_c$ and $\bar{\sigma}_{ci} = 0.55f'_{ci}$. These are relatively high stresses for uniform compression. However, actual stresses under service loads, which occur when the tank is full, are much smaller (Ref. 12.9).

An external tensile force applied to the member induces a stress change of magnitude N/A_t. The resulting uniform stress in the concrete due to the combined action of F and N is:

$$\sigma_c = \frac{F}{A_n} + \frac{N}{A_t} = \frac{A_{ps}\,f_{pe}}{A_n} + \frac{N}{A_t} \tag{12.6}$$

where N is negative. The corresponding stress in the prestressing steel is given by:

$$f_{ps} = f_{pe} - n_p\,\frac{N}{A_t} \tag{12.7}$$

Equation (12.6) can be used to predict the cracking load N_{cr} for which the stress σ_c becomes equal to the tensile strength of the concrete f'_{tc}. It leads to:

$$N_{cr} = A_t\left(f'_{tc} - \frac{F}{A_n}\right) = A_t\left(f'_{tc} - \frac{A_{ps}\,f_{pe}}{A_n}\right) \tag{12.8}$$

The margin of safety against cracking m_{cr} is defined by:

$$m_{cr} = \frac{N_{cr}}{N} \tag{12.9}$$

A minimum margin of safety is generally required in design. When prior cracking has occurred, subsequent loading of the prestressed member will see cracks opening at a stress $\sigma_c = 0$ instead of f'_{tc}. This stage is called decomposition. The tensile load at decompression can be computed from:

$$N_{dec} = -\frac{A_t\,F}{A_n} = -\frac{A_t\,A_{ps}\,f_{pe}}{A_n} \tag{12.10}$$

The margin of safety against decompression is defined by:

$$m_{dec} = \frac{N_{dec}}{N} \tag{12.11}$$

A minimum value of m_{dec} is often required in design. The stress in the prestressing steel just before cracking is given by Eq. (12.4) in which N is replaced by N_{cr}.

That is:

$$(f_{ps})^*_{cr} = f_{pe} - n_p \frac{N_{cr}}{A_t} \tag{12.12}$$

The stress in the prestressing steel just after first cracking is given by:

$$(f_{ps})_{cr} = \frac{-N_{cr}}{A_{ps}} = \frac{A_t f_{pe}}{A_n} - \frac{A_t f'_{tc}}{A_{ps}} \tag{12.13}$$

It should be less than f_{pu} if failure is not to accompany cracking. This can be satisfied by setting a minimum reinforcement criterion as shown in Sec. 12.5.

The stress in the prestressing steel just after decompression in a previously cracked member is obtained from:

$$(f_{ps})_{dec} = \frac{-N_{dec}}{A_{ps}} = \frac{A_t f_{pe}}{A_n} \tag{12.14}$$

At ultimate loading the nominal tensile resistance of the cracked section is equal to that of the steel alone. Thus:

$$N_n = -A_{ps} f_{pu} \tag{12.15}$$

Using the ultimate strength design approach prescribed by ACI, the following condition must be satisfied:

$$|N_u| \le |\phi N_n| \tag{12.16}$$

where N_u is the required factored load or design strength (for instance, $N_u = 1.4 N_D + 1.7 N_L$). Absolute values were used in order to keep Eq. (12.16) in accordance with the general form used for other types of loading. Otherwise, as the tensile forces are negative, we would have algebraically $N_u \ge \phi N_n$.

2. Short- and Long-Term Deformations

The instantaneous shortening of a concrete element of length l, subjected to a concentric prestressing force F_i, is given by:

$$\Delta l_i = \frac{F_i l}{A_n E_{ci}} \tag{12.17}$$

A similar expression can be obtained if the final prestressing force F and E_c are considered.

The instantaneous elongation of a prestressed concrete member subjected to an axial tension N is given by:

$$\Delta l_i = \frac{N l}{A_t E_c} \tag{12.18}$$

It is to be used for short-term loads such as live loads.

Long-term deformations in tensile members depend on many factors such as the construction method, the stressing sequence, if any, the age at loading, and the loading history. If, for instance, the member is pretensioned and stored in plant for a reasonable period of time, a substantial portion of the additional time-dependent shortening would have occurred when the member is attached to the rest of the structure. Subsequent external loading would release part of the compression and reduce further any additional long-term effect thereafter.

If the tensile member is posttensioned on site, the prestress may be applied in stages depending on the progress of the construction of the structure and the application of the loads. Stage stressing may be designed for an average resulting compressive stress of the same order as the final sustained stress for which instantaneous and long-term deformations can be estimated.

In such a case, a shortening due to the combined effects of prestressing and external load can be computed in function of the average resulting compressive stress σ_c. The instantaneous shortening can be obtained from:

$$\Delta l_i = l \frac{\sigma_c}{E_c} \qquad (12.19)$$

and the additional long-term shortening can be taken as the instantaneous value multiplied by the ultimate creep coefficient or, depending on the particular circumstances, an adjusted creep coefficient.

In summary, there is no general solution to estimating long-term deformations, and engineering judgment must be exercised depending on the particular circumstances. The reader may want to review Chap. 7 on deflections and Chap. 8 on prestress losses for further information. Note that the computations of prestress losses are easier for tensile members than for flexural members, even if the time-interval approach is used. This is because a uniform stress exists in the section and corresponding computations due to creep are greatly simplified.

3. Example: Investigation of a Tension Member

Consider the prestressed concrete tensile member of length 100 ft shown in Fig. 12.8. It is subjected to an axial tensile force in service $N = N_D + N_L = -60,000 - 40,000 = -100,000$ lb. The following material properties are assumed: $f'_c = 6000$ psi, $f'_{ci} = 4500$ psi, $f'_{tc} = -4\sqrt{f'_c} = -310$ psi, $E_c = 4.69 \times 10^6$ psi, $E_{ci} = 4.06 \times 10^6$ psi, $\bar{\sigma}_{ci} = 2000$ psi, $\bar{\sigma}_{cs} = 1500$ psi, $f_{pi} = 175,000$ psi, $f_{pe} = 145,000$ psi, $f_{pu} = 270,000$ psi, $E_{ps} = 28.14 \times 10^6$ psi, $n_p = E_{ps}/E_c = 6$, ultimate creep coefficient of concrete $C_{CU} = 2$.

In the investigation or review process the following unknowns are provided: $A_g = 100$ in^2 and $A_{ps} = 0.918$ in^2. A_{ps} corresponds to six half-inch diameter prestressing strands; thus $F = A_{ps} f_{pe} = 133,110$ lb and $F_i = A_{ps} f_{pi} = 160,650$ lb.

Let us use the equations and criteria in the order presented in Sec. 12.4, Service Stresses, Cracking, and Ultimate Load.

Equation (12.1): $A_n = 100 - 0.918 \simeq 99.08$ in^2

Equation (12.2): $A_t = 100 + (6 - 1)0.918 \simeq 104.59$ in^2

Equation (12.3): $\sigma_{gi} = \dfrac{160,650}{99.08} = 1621$ psi

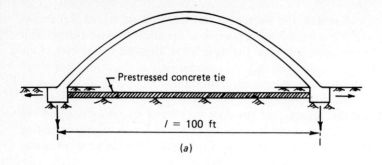

Figure 12.8 Investigation example. (*a*) Tie position within structure. (*b*) Tie cross section.

Equation (12.4): $\sigma_g = \dfrac{133,110}{99.08} = 1343$ psi

Equation (12.5): $\begin{cases} \sigma_{gi} < \bar{\sigma}_{ci} = 2000 \text{ psi} \\ \sigma_g < \bar{\sigma}_{cs} = 1500 \text{ psi} \end{cases}$

Thus, the allowable stress criterion is satisfied. The stress in the concrete under combined final prestressing force and external tension N is given by Eq. (12.6):

$$\sigma_c = \frac{133,110}{99.08} - \frac{100,000}{104.59} \simeq 387 \text{ psi}$$

The external tensile force that would produce first cracking in the member is given by Eq. (12.8):

$$N_{cr} = 104.59\left(-310 - \frac{133,110}{99.08}\right) = -172,935 \text{ lb}$$

The corresponding margin of safety against cracking is given by Eq. (12.9):

$$m_{cr} = \frac{N_{cr}}{N} \simeq 1.73$$

and seems sufficient.

The external load that would lead to decompression is given by Eq. (12.10):

$$N_{dec} = -\frac{104.59 \times 133,110}{99.08} = -140,512 \text{ lb}$$

and the margin of safety against decompression (Eq. (12.11)) is:

$$m_{dec} = \frac{N_{dec}}{N} \simeq 1.40$$

The stress in the prestressing steel just before first cracking is given by Eq. (12.12):

$$(f_{ps})^*_{cr} = 145,000 + \frac{6 \times 172,935}{104.59} = 154,920 \text{ psi}$$

The stress in the prestressing steel just after cracking is given by Eq. (12.13):

$$(f_{ps})_{cr} = \frac{172,935}{0.918} = 188,382 \text{ psi}$$

and is less than f_{pu}, the ultimate strength of the steel, with an allowance for an adequate safety margin.

Note that the stress jump at cracking can be computed from:

$$(f_{ps})_{cr} - (f_{ps})^*_{cr} = 188,382 - 154,920 = 33,462 \text{ psi}$$

This is a relatively high value if debonding on either side of the crack is to be limited. However, unless cracking is artificially produced, such stress jumps cannot be avoided in tensile members.

The stress in the steel just after decompression in a previously cracked member is given by Eq. (12.14):

$$(f_{ps})_{dec} = \frac{140,512}{0.918} = 153,063 \text{ psi}$$

The stress in the prestressing steel just before decompression in a cracked member, or at decompression in an uncracked member, can also be obtained from Eq. (12.7) in which N is replaced by N_{dec}, thus

$$(f_{ps})_{dec} = f_{pe} - n_p \frac{N_{dec}}{A_t} = 145,000 + \frac{6 \times 140,512}{104.59} \simeq 153,061 \text{ psi}$$

That is essentially the same as the previously calculated value indicating that the stress jump is eliminated once first cracking has occurred.

The nominal resistance of the member at ultimate is given by:

$$N_n = -A_{ps} f_{pu} = -0.918 \times 270,000 = -247,860 \text{ lb}$$

Thus

$$\phi N_n = -0.9 \times 247,860 = -223,074 \text{ lb}$$

Using the ACI load factors, the design strength is given by:

$$N_u = -1.4 \times 60,000 - 1.7 \times 40,000 = -152,000 \text{ lb}$$

As $|N_u| < |\phi N_n|$ ultimate strength requirements are satisfied.

To illustrate deformation computations let us consider two cases:

1. Assume the member is posttensioned on site and immediately thereafter scaffolding is removed under the structure and the dead load N_D is applied. Using an average

prestressing force $(F + F_i)/2 = 146{,}880$ lb, and an average modulus $(E_c + E_{ci})/2 = 4.375 \times 10^6$ psi, the resulting stress in the concrete can be computed from Eq. (12.6):

$$\sigma_c = \frac{146{,}880}{99.08} - \frac{60{,}000}{104.59} \simeq 909 \text{ psi}$$

The corresponding value of average instantaneous shortening (to be used in design) is given by Eq. (12.19):

$$\Delta l_i = \frac{100 \times 12 \times 909}{4.375 \times 10^6} \simeq 0.25 \text{ in}$$

Additional long-term shortening can be obtained, as a first approximation, by multiplying Δ_i by $C_{CU} = 2$. Thus:

$$\Delta l_{\text{add}} = 2 \times 0.25 = 0.50 \text{ in}$$

Note that the live load would produce an elongation given by Eq. (12.18):

$$\Delta l_i = \frac{-40{,}000 \times 100 \times 12}{104.59 \times 4.69 \times 10^6} \simeq -0.1 \text{ in}$$

Hence, long-term deformation will be most critical and its effect on the rest of the structure must be assessed.

2. Assume the member is pretensioned, stored, and attached to the structure only at one year of age. Also assume that the dead load N_D is applied immediately thereafter. Most of the long-term shortening of the member would have taken place prior to attaching it to the rest of the structure. The application of N_D would produce an instantaneous elongation given by Eq. (12.18)

$$\Delta l_i = \frac{-60{,}000 \times 100 \times 12}{104.59 \times 4.69 \times 10^6} \simeq -0.15 \text{ in}$$

Some creep recovery will also occur. However, the resulting stress in the concrete remains a compression and is given by Eq. (12.6):

$$\sigma_c = \frac{133{,}100}{99.08} - \frac{60{,}000}{104.59} \simeq 770 \text{ psi}$$

One can estimate an additional long-term shortening due to this sustained stress, say only about 20 percent of what it would be if prior creep has not occurred over one year of age. Its value may be estimated from:

$$\Delta l_{\text{add}} = 0.20 C_{CU} \left(\frac{l\sigma_c}{E_c} \right) = 0.4 \left(\frac{1200 \times 770}{4.69 \times 10^6} \right) \simeq 0.08 \text{ in}$$

This value balances in part the instantaneous elongation due to N_L. Thus, little long-term shortening is expected for the sustained loading. Deformation is not critical in this case.

12.5 OPTIMUM DESIGN OF TENSION MEMBERS

Unless the various design criteria are clearly understood and accounted for, the design of prestressed concrete tension members may often err on the unsafe side and may lead to inconsistent results. Most common design criteria have been mentioned in Sec. 12.4 on analysis. Depending on the particular problem, other criteria may be added. For instance, in partially prestressed members, a maximum crack width criterion may be necessary. Each criterion leads to an analytic relationship to be satisfied by the design. Only two unknowns are needed in practice, the area of prestressing steel A_{ps} and the gross sectional area of the concrete A_g. Other variables may be assumed at first and revised later, if necessary. Generally, it is not possible to know a priori which criterion or set of criteria will control a particular design.

A systematic procedure is proposed here, which would guarantee a range of satisfactory designs and, if desired, an optimum design. In this procedure each design criterion is expressed in function of the two unknown variables A_{ps} and A_g and a feasibility domain is sought in a way similar to what was done in Chap. 4 for F_i and e_o. Any point within the domain should provide an acceptable design, while generally only one point provides an optimum (minimum cost) design. The various criteria and the corresponding analytic relationships are given next.

1. Design Criteria

(a) Maximum compressive stress criterion. It stipulates that the maximum uniform compressive stress in the concrete should not exceed the corresponding allowable stress. From Eqs. (12.3) to (12.5) we have:

$$\begin{cases} \dfrac{A_{ps} f_{pi}}{A_n} \le \bar{\sigma}_{ci} \\[2ex] \dfrac{A_{ps} f_{pe}}{A_n} \le \bar{\sigma}_{cs} \end{cases} \tag{12.20}$$

Replacing A_n by its value from Eq. (12.1) leads to the following inequality conditions:

$$\begin{cases} A_{ps} \le \dfrac{A_g}{1 + f_{pi}/\bar{\sigma}_{ci}} \\[2ex] A_{ps} \le \dfrac{A_g}{1 + f_{pe}/\bar{\sigma}_{cs}} \end{cases} \tag{12.21}$$

These equations, at equality, are represented by straight lines on a graph with A_{ps} as ordinate and A_g as abcissa. Each inequality is satisfied for any point below the corresponding line. A typical example is shown in Fig. 12.9 where the lines are marked (a') and (a).

(b) Margin of safety against cracking criterion. Because cracking is such a dramatic event in prestressed tensile members, it is generally wise to design for a minimum margin of safety against cracking, $\bar{m}_{cr}$, the value of which should depend on the type of structure, the probability of overload, and the risk associated with

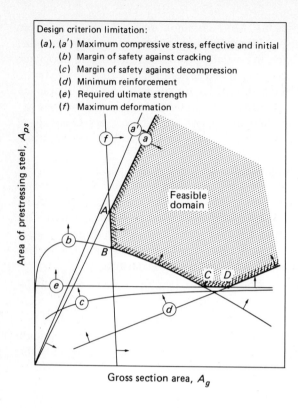

Design criterion limitation:

(a), (a') Maximum compressive stress, effective and initial
(b) Margin of safety against cracking
(c) Margin of safety against decompression
(d) Minimum reinforcement
(e) Required ultimate strength
(f) Maximum deformation

Figure 12.9 Geometric representation of the various design criteria and the feasible domain.

cracking (i.e., a water tank versus a nuclear power containment vessel; a short bridge versus a long bridge). Thus, we have the following condition:

$$\frac{N_{cr}}{N} \geq \bar{m}_{cr} \tag{12.22}$$

Using Eq. (12.8) for N_{cr} and noting that N is negative leads to:

$$\bar{m}_{cr} N \geq A_t \left(f'_{tc} - \frac{A_{ps} f_{pe}}{A_n} \right) \tag{12.23}$$

Replacing A_n and A_t by their values from Eqs. (12.1) and (12.2) in Eq. (12.23) leads to the following quadratic equation in A_{ps}:

$$A_{ps}^2 (n_p - 1)(f_{pe} - f'_{tc}) - A_{ps}(A_g f_{pe} - A_g(n_p - 2)f'_{tc} - \bar{m}_{cr} N)$$
$$+ A_g(\bar{m}_{cr} N - A_g f'_{tc}) \geq 0 \tag{12.24}$$

At equality, Eq. (12.24) has generally two roots of opposite signs. The positive root of interest here is given by:

$$A_{ps} = -\left(\frac{b}{2a}\right) + \sqrt{\left(\frac{b}{2a}\right)^2 - \frac{c}{a}} \tag{12.25}$$

in which:

$$\frac{b}{2a} = \frac{-A_g[f_{pe} - (n_p - 2)f'_{tc}] + \bar{m}_{cr} N}{2(n_p - 1)(f_{pe} - f'_{tc})}$$ (12.26)

$$\frac{c}{a} = \frac{A_g(\bar{m}_{cr} N - A_g f'_{tc})}{(n_p - 1)(f_{pe} - f'_{tc})}$$ (12.27)

Note that A_{ps} can be plotted in function of A_g on the same x-y graph as it was done for the other criteria. It separates the plane into two regions. The minimum required margin of safety is obtained for any point above the curve. A typical such curve is plotted in Fig. 12.9 and marked b.

(c) Margin of safety against decompression criterion. Similarly to the minimum margin of safety against cracking, a minimum margin of safety against decompression, $\bar{m}_{dec}$, may represent an important safety measure. Thus, the following condition can be set:

$$\frac{N_{dec}}{N} \geq \bar{m}_{dec}$$ (12.28)

The treatment of this case is exactly similar to case b above, except that f'_{tc} is replaced by zero. It leads to:

$$A_{ps} = -\left(\frac{b}{2a}\right) + \sqrt{\left(\frac{b}{2a}\right)^2 - \frac{c}{a}}$$ (12.25)

in which:

$$\frac{b}{2a} = \frac{-A_g f_{pe} + \bar{m}_{dec} N}{2(n_p - 1)f_{pe}}$$ (12.29)

$$\frac{c}{a} = \frac{A_g \bar{m}_{dec} N}{(n_p - 1)f_{pe}}$$ (12.30)

A_{ps} is plotted versus A_g and the points of the plane above the curve satisfy the required margin of safety against decompression. A typical such plot is shown as curve c in Fig. 12.9.

(d) Minimum reinforcement criterion. This criterion would ensure that failure does not occur immediately after cracking. Thus, we can write the following condition:

$$\phi A_{ps} f_{pu} \geq |\alpha N_{cr}| \qquad \text{or} \qquad \leq -\alpha N_{cr}$$ (12.31)

where α is a factor not less than one. The ACI code prescribes a similar condition for flexure where $\alpha = 1.2$. Replacing N_{cr} by its value from Eq. (12.8) leads to:

$$\phi A_{ps} f_{pu} \leq -\alpha A_t\left(f'_{tc} - \frac{F}{A_n}\right)$$ (12.32)

where A_n and A_t can be replaced by their values from Eqs. (12.1) and (12.2). The resulting equation is lengthy but not impossible to solve. Because this condition seldom controls the design and a high level of accuracy is not needed for a minimum level of reinforcement, the following approximation is proposed: assume $A_t \simeq A_n \simeq A_g$; Eq. (12.32) would lead to:

$$A_{ps} \geq \left(\frac{-\alpha f'_{tc}}{\phi f_{pu} - \alpha f_{pe}} \right) A_g \tag{12.33}$$

Typically, it is represented at equality by a straight line marked d in Fig. 12.9. Any point above the line should satisfy the minimum reinforcement criterion.

(e) Required ultimate strength criterion. It guarantees an ultimate resistance not less than the factored strength design load. Using Eqs. (12.15) and (12.16), we have:

$$A_{ps} \geq \frac{-N_u}{\phi f_{pu}} \tag{12.34}$$

which is represented at equality by a straight line marked e in Fig. 12.9.

(f) Maximum deformation criterion. This criterion garantees an allowable deformation $\overline{\Delta l}$ (elongation or shortening) under a given load or a combination of loadings. It should be tailored to the particular problem at hand. If, for instance, the instantaneous elongation under live load N_L is to be limited to $\overline{\Delta l}$ (negative value), we will have the following condition

$$\frac{-N_L l}{E_c A_t} \leq -\overline{\Delta l} \quad \text{or} \quad \frac{N_L l}{E_c A_t} \geq \overline{\Delta l} \tag{12.35}$$

and replacing A_t by its value from Eq. (12.2) leads to:

$$A_{ps} \geq -\frac{A_g}{n_p - 1} + \frac{N_L l}{(n_p - 1)E_c \overline{\Delta l}} \tag{12.36}$$

which, at equality, is represented by a straight line marked f in Fig. 12.9.

A similar approach can be used for limiting the additional long-term deformation by estimating the instantaneous deformation first and multiplying it by an appropriate factor. A relation similar to Eq. (12.36) is generally obtained and can be plotted on an A_{ps} versus A_g graph.

A geometric representation of the feasible sets of values of A_{ps} and A_g that satisfy all the above criteria is shown in Fig. 12.9 and is defined as the feasible domain. A wide range of possibilities is available. The choice of the design point should be in a region where both A_{ps} and A_g are minimized (somewhere between B and C). A more accurate minimum cost solution can be sought as shown later in this section—3. Minimum Cost Solution.

2. Design Approximations

Two design approximations will be given: one is a rule of thumb and should be used with care and the other greatly simplifies the geometric solution.

(a) *Rule of thumb for pure tensile members.* Use a value of F 20 percent higher than the tensile load in service and determine A_g based on a uniform compressive stress under F of $0.25f'_c$, that is $A_g \simeq F/0.25f'_c$. A_{ps} is determined from F/f_{pe}. One can then quickly analyze the member obtained, check other criteria, and revise the design accordingly.

(b) *Linearization of the safety criteria against cracking and decompression.* All design criteria covered in the preceding subsection led to straight lines on Fig. 12.9, except the two criteria dealing with safety against cracking and decompression. The computations would gain enormous speed if these were approximated by straight lines. This can be simply done by assuming for these criteria only that $A_t \simeq A_n \simeq A_g$ over the range of interest. Equation (12.23) would then lead to the following conditions:

For the margin of safety against cracking

$$A_{ps} \geq \frac{f'_{tc}}{f_{pe}} A_g - \frac{\bar{m}_{cr} N}{f_{pe}}$$ (12.37)

and for the margin of safety against decompression

$$A_{ps} \geq - \frac{\bar{m}_{dec} N}{f_{pe}}$$ (12.38)

Equations (12.37) and (12.38) would replace Eqs. (12.25) to (12.30) and therefore the geometric feasibility domain, bound entirely by straight lines, can be easily built. In addition, the solution for minimum cost becomes readily available and requires only one additional step as shown next.

3. Minimum Cost Solution

The cost of the tensile member per unit length, defined by Z, can be expressed as follows:

$$Z = U_p W_{ps} + U_c V_c$$ (12.39)

where W_{ps} = weight of prestressing steel per unit length
U_p = unit cost of prestressing steel in place
V_c = volume of concrete per unit length
U_c = unit cost of concrete in place.

Using U.S. units, W_{ps} will be given in pounds per foot, U_p in dollars per pound, V_c in cubic yards per foot, and U_c in dollars per cubic yard. Thus, Z is the cost in dollars per linear foot of member and the coefficients in the following equations are valid for the U.S. system only. Theoretically, they can be developed for any other system.

W_{ps} and V_c can be expressed, respectively, in functions of A_{ps} and A_g, leading to:

$$Z = U_p\left(\frac{\gamma_s A_{ps}}{144}\right) + U_c\left(\frac{A_g}{144 \times 27}\right) \tag{12.40}$$

where γ_s is the unit weight of steel in pounds per cubic foot. Equation (12.40) is a linear equation in A_{ps} and A_g.

The following general optimization problem can be formulated:

$$\text{Minimize cost } Z = \left(\frac{\gamma_s U_p}{144}\right) A_{ps} + \left(\frac{U_c}{3888}\right) A_g \tag{12.41}$$

subject to a number of constraints given by Eqs. (12.21), (12.33), (12.34), (12.36), (12.37), and (12.38) and representing the various design criteria. As the above objective function and the constraints are linear, the problem is a linear optimization problem. It can be solved by any linear programming algorithm. In such a case it is known that the optimum solution corresponds to one of the vertices of the feasible geometric domain, such as points A, B, C, or D of Fig. 12.10. The minimum cost solution can be found by inspection or by plotting on the same graph as the feasible domain the objective function (Eq. (12.41)), written in the following form:

$$A_{ps} = -\left(\frac{U_c}{27\gamma_s U_p}\right) A_g + \frac{144Z}{\gamma_s U_p} \tag{12.42}$$

The representative line can be moved graphically parallel to itself from $Z = 0$ to a

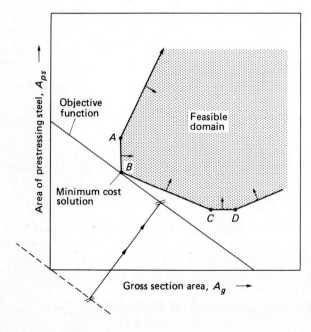

Figure 12.10 Graphical solution for the minimum-cost design problem.

minimum Z for which only one point of the line belongs to the feasible domain. As we mentioned earlier, that point is a vertex and gives the minimum cost solution. Other approaches or special cases particular to linear programming problems apply here as well. The method is graphically illustrated in Fig. 12.10.

4. Example

Find the values of A_{ps} and A_g for the Example in Sec. 12.4, assuming they are not given. All other information is the same.

The design has to accommodate the following minimum requirements: $\bar{m}_{cr} = 1.5$, $\bar{m}_{dec} = 1.2$, ultimate resistance at least 20 percent higher than the cracking load, and an instantaneous elongation not to exceed $\overline{\Delta l} = -0.15$ in for an overload 50 percent higher than the live load.

Let us write the relationships between A_{ps} and A_g for the various design criteria and build the corresponding feasible domain graphically.

(a) Maximum compressive stress criterion. Using Eq. (12.21) leads to:

$$A_{ps} \leq \frac{A_g}{1 + 175,000/2000} = \frac{20}{1770} A_g$$

and

$$A_{ps} \leq \frac{A_g}{1 + 145,000/1500} = \frac{15}{1465} A_g$$

The corresponding lines at equality are plotted in Fig. 12.11 and marked a' and a, respectively. Note, a is more critical than a'.

(b) Margin of safety against cracking. Using the linear approximation given by Eq. (12.37) leads to:

$$A_{ps} \geq \frac{-310}{145,000} A_g + \frac{1.5 \times 100,000}{145,000}$$

that is:

$$A_{ps} \geq -\frac{31}{14,500} A_g + \frac{150}{145}$$

The corresponding line at equality is plotted in Fig. 12.11 and marked b. Also plotted as a dashed line marked b' is the exact solution corresponding to Eq. (12.25). It gives the reader the opportunity to check the exact solution. *Note:* The approximation obtained using Eq. (12.37) is on the safe side.

(c) Margin of safety against decompression. Using the linear approximation given by Eq. (12.38) leads to:

$$A_{ps} \geq \frac{1.2 \times 100,000}{150,000} = 0.80 \text{ in}^2$$

The corresponding line at equality is parallel to the x axis and marked c in Fig. 12.11.

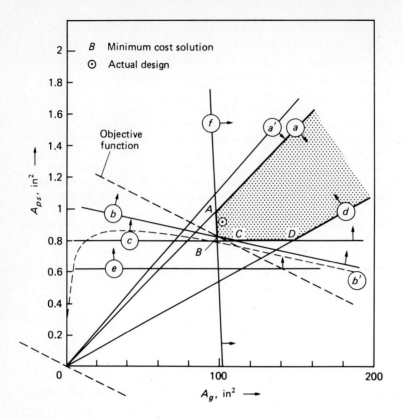

Figure 12.11 Minimum-cost solution for Example Problem.

(d) Minimum reinforcement criterion. Using Eq. (12.33) leads to:

$$A_{ps} \geq \frac{1.2 \times 310}{0.9 \times 270,000 - 1.2 \times 145,000} A_g = \frac{372}{69,000} A_g$$

The corresponding line at equality is plotted on Fig. 12.11 and marked d.

(e) Ultimate strength criterion. Using Eq. (12.34), we have

$$A_{ps} \geq \frac{152,000}{0.9 \times 270,000} \simeq 0.625 \text{ in}^2$$

and the corresponding line at equality is parallel to the x axis and marked e on Fig. 12.11.

(f) Maximum deformation criterion. Using Eq. (12.36) for an overload 50 percent higher than the live load ($-60,000$ lb) and a limit $\overline{\Delta l} = -0.15$ in, we have:

$$A_{ps} \geq -\frac{A_g}{5} + \frac{60,000 \times 100 \times 12}{5 \times 4.69 \times 10^6 \times 0.15}$$

that is:

$$A_{ps} \geq -\frac{A_g}{5} + 20.47$$

The corresponding line at equality is plotted in Fig. 12.11 and marked f.

The feasible domain is thus identified in Fig. 12.11 and any point inside the domain is a feasible solution. Let us try to approach a minimum-cost solution. Assuming $U_c = \$80$ per cubic yard (this value is assumed to include the cost of forms for this type of structure), $U_p = \$1.25$ per pound, $\gamma_s = 490$ pcf and using these values in Eq. (12.42) leads to:

$$A_{ps} \simeq -0.0048A_g + 0.235Z$$

A first dashed line is plotted for $Z = 0$ and gives the needed slope of -0.0048. A parallel that touches the domain at point B gives the minimum intercept leading to the minimum cost Z. Thus, point B corresponds to the minimum-cost solution with $A_{ps} \simeq 0.825$ in^2 and $A_g \simeq 98.2$ in^2. However, in rounding off the numbers, a value of $A_g = 100$ in^2 and $A_{ps} = 0.915$ in^2 (corresponding to six strands) is selected and leads to the same results as used in the Example in Sec. 12.4. Note that the coordinates of point B can also be obtained from the point of intersection of lines b and f. The minimum cost Z is then calculated from Eq. (12.41) in which A_{ps} and A_g are replaced by their optimum values.

12.6 CIRCULAR STRUCTURES: TANKS AND PRESSURE VESSELS

Because it can sustain tensile stress and remain crack-free, prestressed concrete is an ideal material for tanks, reservoirs, pipes, pressure vessels, and containers in general. Although early applications were mostly for water tanks, they have expanded to accommodate oil, gas, chemicals, slurries, liquids at cryogenic temperatures, and granular materials (silos). Nuclear containment vessels are among the latest and largest scale applications. Typical elevation cross sections of a nuclear containment vessel and an open water tank are shown in Fig. 12.12.

Several methods can be used to build cylindrical water tanks. Most popular in the United States is the wire-winding technique in which a concrete core (the tank wall) is built first and a tendon (mostly wires) is wound around it under stress, thus creating the necessary prestress in the wall. Once the prestressing operation is completed, the wires are covered with a layer of mortar generally applied by "guniting" or "shotcreting" (Fig. 12.13a). Practical provisions for the design and construction of this type of structures can be found in Ref. 12.9. Alternatively, as in posttensioned members, the prestressing tendons can be replaced in ducts within the tank wall and posttensioned after the concrete reaches a sufficient resistance (Fig. 12.13b). The ducts are then grouted to ensure proper bond. Tensioning is achieved from several buttresses (commonly 3 to 6), distributed along the periphery of the tank, at which the tendons overlap as shown in Fig. 12.13c. Unbonded tendons can also be used for convenience, provided the criteria for ultimate strength and stiffness after cracking are satisfied. A more recent technique of tank construction consists of using precast wall units, jointed on site to form the wall

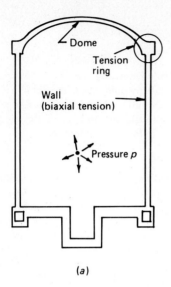

(a)

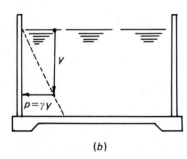

(b)

Figure 12.12 (a) Typical nuclear containment structure. (b) Typical cylindrical water tank.

of the tank, then posttensioned by wire-winding or regular posttensioning (Fig. 12.13d).

Open tanks that are restrained at their base may also need structural reinforcement (in addition to shrinkage and temperature reinforcement) in the vertical direction in the form of reinforcing bars or prestressed tendons. Pressure vessels need biaxial prestressing to balance tensile stresses in both the circumferential and meridional directions.

1. Analysis of Stresses

Most prestressed concrete tanks and vessels are figures of revolution, that is, they have a circular cross section. In the simplest cases they are subjected to pure tensile stresses in one or two directions. However, other effects, such as end restraints, creep, shrinkage, temperature changes, and even swelling under water should in general be considered in the analysis and lead to a more complex state of stress.

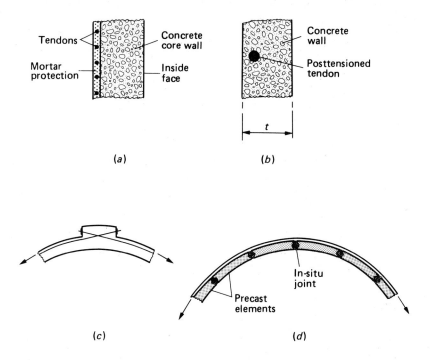

Figure 12.13 (*a*) **Tank wall section using wire-winding technique.** (*b*) **Posttensioned wall.** (*c*) **Posttensioning buttress.** (*d*) **Wall made of precast elements.**

For instance, restraining the base of the wall of an open tank causes significant bending stresses in the vertical (meridional) direction. These must be rationally evaluated and properly resisted by reinforcing or prestressing in the vertical direction. Depending on the extent of the restraint at the base, significant reduction in hoop or ring forces is achieved. As illustrated in Fig. 12.14, a fully fixed base leads to zero ring force at the base, while a free sliding base leads to a maximum ring force. A reduction in the ring force relative to that obtained in the free sliding state leads to an increased load and subsequent bending in the vertical element.

The analysis of prestressed concrete tanks and vessels should be based on accepted methods of shell analysis (generally elastic) where stresses and deformations can be determined. Such methods are covered in detail elsewhere and will not be repeated here (Refs. 12.11 to 12.13).

Once the tensile stresses have been determined, the analysis and design of the tensile elements follow the approaches developed in Secs. 12.4 and 12.5. This is illustrated next for the simplest case.

Let us consider a free circular ring (Fig. 12.15) with thickness t, small relative to its radius R and height h. Such a ring can be considered to represent a slice of a cylindrical tank with unit height $h = 1$. If the ring is subjected to an internal

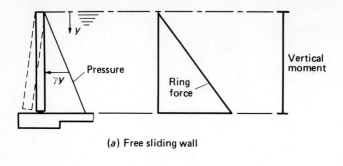

(a) Free sliding wall

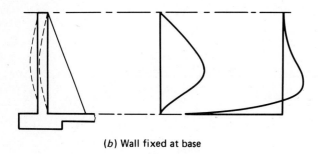

(b) Wall fixed at base

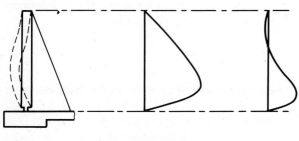

(c) Wall hinged at base

Figure 12.14 Influence of base restraints on ring forces and bending moments in tank walls.

pressure p, a tensile hoop stress (also called tangential or circumferential stress) develops normally to the cross section of the ring and is given by:

$$\sigma_1 = -\frac{pR}{t} \tag{12.43}$$

The corresponding tensile force in the ring is:

$$N = \sigma_1 th = -pRh \tag{12.44}$$

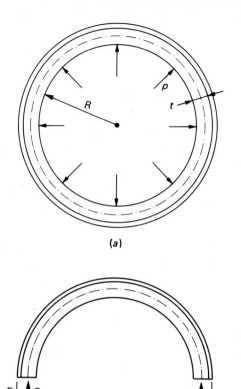

(a)

(b)

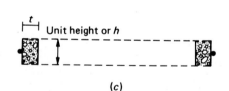

Unit height or h

(c)

Figure 12.15 (*a*) Circumferential cross section of tank or ring. (*b*) Prestressing steel location and corresponding C line. (*c*) Ring dimensions.

and for $h = 1$

$$N = -pR \qquad (12.45)$$

To balance such tensile force, circular prestressing is used around the ring. Note that a circular prestressing force F, placed anywhere in the ring section, including its outer periphery, produces a C line or pressure line that coincides with the centroid of the ring section. This is because the ring is a statically indeterminate structure and the linear transformation theorem explained in Sec. 10.6 for continuous beams is applicable to a ring. That is, a tendon along the centroid of the ring section is a concordant tendon. A tendon parallel to it with a smaller or larger diameter can be considered linearly transformed and thus produces the same C line as the concordant tendon. Thus, the prestressing force F can be considered to act

Table 12.1 Stresses in thin wall containers

Shape	Loading	Stresses†
Long cylinder closed ends radius, R wall thickness, t	Uniform internal pressure, p	$\sigma_1 = -\dfrac{pR}{t}$ $\sigma_2 = -\dfrac{pR}{2t}$
Closed sphere radius, R wall thickness, t	Uniform internal pressure, p	$\sigma_1 = \sigma_2 = -\dfrac{pR}{2t}$
Open cylinder radius, R wall thickness, t	Liquid pressure, γh	$\sigma_1 = -\dfrac{\gamma h R}{t}$ $\sigma_2 = 0$

† σ_1 = tangential, circumferential or hoop stress. (Does not account for the effects of end restraints.) σ_2 = meridian stress (direction normal to σ_1).

along the centroid of the ring section. The resulting stress, due to F and N, is given by Eq. (12.6):

$$\sigma_c = \frac{F}{A_n} + \frac{N}{A_t} \tag{12.6}$$

where N is obtained from Eqs. (12.44) and (12.45).

For a spherical ring or for closed tanks where a meridian tensile stress σ_2 develops in addition to σ_1 and for cases where p is replaced by a liquid pressure, the values of σ_1 and σ_2 are summarized in Table 12.1. The corresponding values of N in either direction can be readily determined and the resulting stresses due to F and N can be calculated from Eq. (12.6). Subsequent analysis of tension members follows the approach covered in Sec. 12.4.

2. Design

All the design criteria covered in Sec. 12.5 apply to circular structures. Because excessive cracking and subsequent leakage can be considered a failure state in tanks and pressure vessels, the criterion specifying the margin of safety against cracking is often critical in the design. If the maximum allowable compression criterion is also assumed binding (for economic reasons), these two criteria can be used to determine at least in a preliminary design, the two unknowns of the problem, A_{ps} and A_g. A_{ps} gives the prestressing force F and A_g gives the wall thickness t when a unit height is assumed. The solution for such a case is covered next.

Let us define:

$$\lambda = \text{the larger of } \frac{f_{pi}}{\bar{\sigma}_{ci}} \quad \text{or} \quad \frac{f_{pe}}{\bar{\sigma}_{cs}} \tag{12.46}$$

If maximum allowable compression is attained under prestressing, we have from Eqs. (12.3) to (12.5):

$$A_n = \lambda A_{ps} \tag{12.47}$$

If the margin of safety against cracking is satisfied, we have from Eq. (12.23):

$$\frac{F}{A_n} + \frac{\bar{m}_{cr} N}{A_t} = f'_{tc} \tag{12.48}$$

Replacing, in Eq. (12.48), F by $A_{ps} f_{pe}$, A_t by $A_n + n_p A_{ps}$, A_n by λA_{ps}, and solving for A_{ps} leads to:

$$A_{ps} = \frac{\bar{m}_{cr} N}{(\lambda + n_p)(f'_{tc} - f_{pe}/\lambda)} \tag{12.49}$$

in which N is the applied tensile force. For a given A_{ps}, A_n can be computed from Eq. (12.47) and A_g from:

$$A_g = A_n + A_{ps} \tag{12.1}$$

For a cylindrical ring of unit depth $h = 1$ subjected to an internal pressure p, Eq. (12.49) gives:

$$A_{ps} = \frac{-\bar{m}_{cr} pR}{(\lambda + n_p)(f'_{tc} - f_{pe}/\lambda)} \tag{12.50}$$

and the corresponding wall thickness is given by

$$t = \frac{A_g}{h} = \frac{A_g}{1} \tag{12.51}$$

The above solution by Eq. (12.50) was first proposed by Lin (Ref. 12.14). The use of Eq. (12.50) leads to a very fast design. However, other criteria will have to be checked according to Sec. 12.5 before finalizing the design. An example illustrating the use of Eq. (12.50) in the preliminary dimensioning of a vessel and a tank is given next.

3. Example

(a) Determine the wall thickness and the prestressing force in the hoop direction of a cylindrical nuclear containment vessel, assuming the following information is given: $f'_c = 5000$ psi, $\bar{\sigma}_{cs} = 1500$ psi, $\bar{\sigma}_{ci} = 2000$ psi, $f_{pi} = 190{,}000$ psi, $f_{pe} = 150{,}000$ psi, $f'_{tc} = -283$ psi, $n_p = 6.54$, $R = 70$ ft. The design calls for an accidental internal pressure $p = 60$ psi for which a margin of safety against cracking $\bar{m}_{cr} = 1.5$ is specified.

Consider a ring slice of wall having $h = 1$ in in height and located sufficiently far away from the ends of the vessel to be assumed free and thus subjected to the full ring forces induced by p.

Using Eq. (12.46)

$$\lambda = \text{the larger of } \frac{190,000}{2000} \quad \text{or} \quad \frac{150,000}{1500}$$

thus

$$\lambda = 100$$

Using Eq. (12.50) gives us:

$$A_{ps} = \frac{-\bar{m}_{cr}\, pR}{(\lambda + n_p)(f'_{tc} - f_{pe}/\lambda)} = \frac{-1.5 \times 60 \times 70 \times 12}{(100 + 6.54)(-283 - 150,000/100)}$$

or

$$A_{ps} \simeq 0.40 \text{ in}^2$$

and the corresponding prestressing force per inch of height is:

$$F = 150,000 \times 0.40 = 60,000 \text{ lb} \quad \text{or} \quad 60 \text{ kips}$$

From Eq. (12.47):

$$A_n = \lambda A_{ps} = 100 \times 0.40 = 40 \text{ in}^2$$

Thus

$$A_g = A_n + A_{ps} = 40.4 \text{ in}^2$$

and the required thickness of the wall, Eq. (12.51), will be:

$$t = \frac{A_g}{1} = 40.4 \text{ in}$$

The actual design will probably use $t = 42$ in or 3.5 ft.

For comparison let us assume that, instead of prestressed concrete, a steel vessel is proposed. Using an allowable stress in the steel of 15,000 psi will lead to a steel thickness:

$$(t \times h) \times 15,000 = N = pR$$

$$t = \frac{60 \times 70 \times 12}{1 \times 15,000} = 3.36 \text{ in}$$

Of course, such a thickness is not possible for a large structure where extensive welding would be required.

(b) Consider exactly the same data as for (a) above, except that a 30-ft-high water tank with a free sliding base is considered.

The pressure at the base of the tank is given by:

$$p = \gamma H = 62.5 \times 30 = 1875 \text{ psf} \simeq 13 \text{ psi}$$

The required area of prestressing steel per inch of height is given by Eq. (12.50):

$$A_{ps} = \frac{-1.5 \times 13 \times 70 \times 12}{(100 + 6.54)(-283 - 150,000/100)} = 0.086 \text{ in}^2$$

or a prestressing force:

$$F = 0.086 \times 150,000 = 12,900 \text{ lb} = 12.9 \text{ kips/in}$$

The corresponding net concrete area is given by Eq. (12.47):

$$A_n = \lambda A_{ps} = 100 \times 0.086 = 8.6 \text{ in}^2$$

For which:

$$A_g = A_n + A_{ps} \simeq 8.68 \text{ in}^2$$

Thus, the required wall thickness at the base will be:

$$t = \frac{A_g}{1} = 8.68 \text{ in}$$

Actual design will probably use $t = 9$ in.

12.7 COMBINED TENSION AND BENDING

Throughout the analysis developed in this chapter, it was assumed that only axial loads were applied. It is however common to have some bending moment in linear tensile elements in addition to the axial load, such as induced from their own weight.

If the bending moment M is small, there may be no need to have an eccentric prestress to counteract its effects. In such a case the stresses on the top and bottom fibers are obtained by adding the value of M/Z_t or $-M/Z_b$ to the uniform stress of Eq. (12.6). The smaller value is used in the design procedure to determine A_{ps}. In computing Z_t and Z_b, the moment of inertia of the transformed section should be used.

If the bending moment is significant, it may be more appropriate to use eccentric prestress. For this the eccentricity e_o of the prestressing force is selected such that the moment due to prestressing is equal in magnitude to the external moment M, that is:

$$Fe_o = M$$

or

$$e_o = \frac{M}{F}$$

The resulting stress in the section becomes again uniform, as the effect of moment is eliminated. Let us illustrate this result by expressing the stress on the bottom fiber of the concrete due to the combined effects of F, N, and a positive moment M:

$$\sigma_b = \frac{F}{A_n} + \frac{Fe_o}{Z_b} - \frac{M}{Z_b} + \frac{N}{A_t}$$

As $Fe_o = M$, σ_b is reduced to the value of σ_c given by Eq. (12.6). Note that one assumption is implied here and is acceptable: the Z_b obtained for the net section and used in Fe_o/Z_b is the same as that used in M/Z_b for the transformed section.

REFERENCES

12.1. R. Morandi: "Some Types of Tied Bridges in Prestressed Concrete," First International Symposium on Concrete Bridge Design, American Concrete Institute, Special Publication SP-23, Detroit, Michigan, 1969, pp. 447–465.

12.2. U. Finsterwalder: "Prestressed Concrete Bridge Construction," *ACI Journal*, vol. 62, no. 9, September 1965.

12.3. T. Y. Lin and F. Kulka: "Construction of Rio Colorado Bridge," *PCI Journal*, vol. 18, no. 6, November/December 1973, pp. 92–101.

12.4. Anonymous: "Inverted Suspension Span is Simple and Cheap," *Engineering News Record*, May 11, 1972, pp. 27–31.

12.5. H. Matsushita and M. Sato: "The Hayahi-No-Mine Prestressed Bridge," *PCI Journal*, vol. 24, no. 2, March/April 1979, pp. 90–109.

12.6. W. T. Carroll, F. W. Beaufait, and R. H. Byran: "Prestressed Concrete Trusses," *ACI Journal*, vol. 75, no. 8, August 1978.

12.7. Ben C. Gerwick, Jr.: "Prestressed Concrete Developments in Japan," *PCI Journal*, vol. 23, no. 6, November/December 1978, pp. 66–76.

12.8. R. J. Wheen: "Prestressed Concrete Members in Direct Tension," *Journal of the Structural Division, ASCE*, vol. 105, no. ST7, July 1979, pp. 1471–1487.

12.9. ACI Committee 344: "Design and Construction of Circular Prestressed Concrete Structures," *ACI Journal*, vol. 67, no. 9, September 1970.

12.10. S. P. Timoshenko: *Theory of Plates and Shells*, 2d ed., McGraw-Hill Book Company, New York, 1959.

12.11. L. R. Creasy: *Prestressed Concrete Cylindrical Tanks*, John Wiley & Sons, New York, 1961.

12.12. D. P. Billington: *Thin Shell Concrete Structures*, McGraw-Hill Book Company, New York, 1965.

12.13. A. Ghali: *Circular Storage Tanks and Silos*, E. & F. N. Spon, Publisher, London, Methuen, Inc., Distributor, N.J., 210 pp.

12.14. T. Y. Lin and N. Burns: *Design of Prestressed Concrete Structures*," 3d ed., John Wiley & Sons, New York, 1981.

PROBLEMS

12.1 Go back to subsection **(b)** of the Example in Sec. 12.6. Assume the tank is closed at its top by a spherical dome-shaped roof with a prestressed concrete tension ring at its base. The ring is supported on the tank wall by elastomeric pads (Fig. P12.1) and, thus, is free to move. The dome radius is $R^* = 140$ ft and its opening angle θ is 30°. Assume the load on the dome is uniform over a horizontal projection of the dome surface and leads to a total load $W = W_D + W_L = 770 + 462 = 1232$ kips. Provide a complete design for the ring assuming same materials properties and requirements as for the tank. As a first approximation consider either a square cross section or a rectangular cross section with a ratio of depth to width equal 2.

12.2 Go back to subsection **b** of the Example in Sec. 12.6. Assume that the base of the tank is partially fixed to its foundation and that the ring forces and vertical bending moments in the wall are the arithmetic average of those for which the tank is either free at the base or fully fixed. (*a*) Design the tank wall, that is, determine its thickness and the corresponding prestressing in the hoop direction. (*b*) Determine the vertical prestressing needed and the

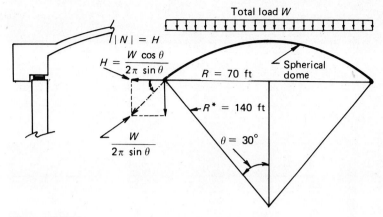

Total load W

$|N| = H$

$H = \dfrac{W \cos \theta}{2\pi \sin \theta}$ $R = 70$ ft

Spherical dome

$R^* = 140$ ft

$\dfrac{W}{2\pi \sin \theta}$ $\theta = 30°$

Figure P12.1

corresponding eccentricity. Revise the design if necessary. Use the same design information as given in the above-mentioned subsection of the Example in Sec. 12.6.

12.3 Because of limited clearance under a planned railway bridge, a truss bridge is proposed as shown in Fig. P12.3. Several solutions are considered and include a truss made out of

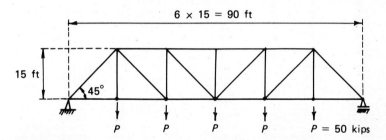

$6 \times 15 = 90$ ft

15 ft

45°

P P P P $P = 50$ kips

Figure P12.3.

prestressed concrete elements. Assume that the lower cord of the truss is to be designed as a single precast prestressed concrete element and that connection details with other elements have been worked out. Maximum joint loadings in service are shown in Fig. P12.3. Fifty percent of service load is due to dead loads and the remainder to live loads. Provide a design for the lower cord assuming the following information is given: $f'_c = 9000$ psi, $f'_{ci} = 6300$ psi, $f'_{tc} = -4\sqrt{f'_c} \simeq -380$ psi, $E_c = 5.75 \times 10^6$ psi, $E_{ci} = 4.81 \times 10^6$ psi, $\bar{\sigma}_{ci} = 3150$ psi, $\bar{\sigma}_{cs} = 2700$ psi, $f_{pi} = 175{,}000$ psi, $f_{pe} = 145{,}000$ psi, $f_{pu} = 270{,}000$ psi, $E_{ps} = 24 \times 10^6$ psi, $C_{CU} = 2$, $\bar{m}_{cr} = 1.5$, $\bar{m}_{dec} = 1.2$. Make any other reasonable assumptions if needed.

Towers of the Statfjord B Condeep Offshore Oil Platform. *(Courtesy Norwegian Contractors.)*

CHAPTER 13

ANALYSIS AND DESIGN OF COMPRESSION MEMBERS

13.1 TYPES AND ADVANTAGES

Compression members are structural elements mostly of linear shape, such as columns, poles, and foundation piles. On a larger scale, TV towers and shafts of offshore structures are treated as compression members. The latter types are compressed not only in the longitudinal direction but also, because of hydrostatic pressure, in the circumferential direction.

It may seem irrational to prestress, that is precompress, a compression member. However, compression members are seldom subjected to pure compression only. Columns, for instance, must be capable of resisting a variety of loads, including lateral loads from wind and earthquakes, from shearing forces transmitted by beams and slabs, and, if precast, from transportation and erection. Moreover, code provisions generally imply the existence of a minimum eccentricity, thus bending, even when pure compression is theoretically considered. In most cases the most critical loading combination of compression members involve substantial bending.

The use of prestressed versus nonprestressed steel in a column leads to a small reduction in its resistance to pure compression but increases significantly its resistance to first cracking. Consequently, its deflection in the uncracked state is substantially reduced and its performance in service is improved. Prestressing allows the use of precasting and, thus, offers its related benefits, such as savings on forms and the use of high-strength concrete. As a column's capacity in compression is directly proportional to the concrete strength, this may be a very substantial

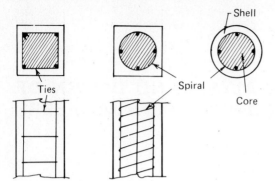

Figure 13.1 Typical cross sections of prestressed columns.

advantage. Precast prestressed columns used in building structures are often designed to span several stories. They are connected in place by posttensioning or other standard jointing techniques. The cost of connections is an important factor to consider in the early stages of design. Typical columns' cross sections are shown in Fig. 13.1 and are identical in shape to reinforced concrete columns.

Because they can be made of a single element, prestressed concrete piles are very efficient structural members. They are widely used for marine structures and building foundations. Lengths of up to 120 ft (36 m) are common. The longest length reported is 2606ft (78 m) for a single piece (Ref. 13.1). Prestressed concrete piles offer a number of important advantages that have made them competitive in

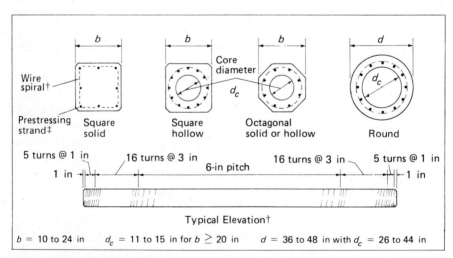

† Wire spiral varies with pile size.
‡ Strand pattern may be circular or square.

Figure 13.2 Typical cross sections of prestressed piles.
(Ref. 13.3, Courtesy Prestressed Concrete Institute.)

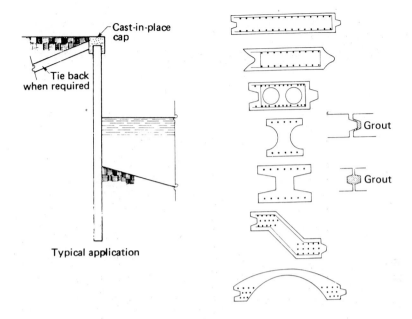

Figure 13.3 Typical cross sections of prestressed sheet piles.
(Adapted from Refs. 13.1 and 13.3.)

all applications requiring piling. These advantages include durability, high load-moment resistance, ability to take uplift (tension), ability to penetrate hard strata, ease of handling and transportation, and economy (Ref. 13.1). Their use has been extended to fenders and sheet piling for waterfront bulkheads. Typical cross sections of piles and sheet piles are shown in Figs. 13.2 and 13.3.

Prestressed concrete poles are used for lighting, electric and telephone transmission lines, antenna masts, and the like. They are highly suited for urban installation (Ref. 13.1). Because they are often subjected to torsion in addition to compression and bending, their cross section is generally selected to achieve good torsional resistance. Typical cross sections of prestressed concrete poles are shown in Fig. 13.4.

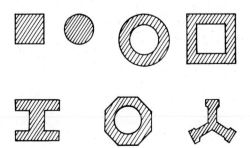

Figure 13.4 Typical cross sections of prestressed poles.

This chapter covers the general principles of analysis and design of compression members subjected to a combination of axial force and bending and, thus, is mostly concerned with columns. Particular design aspects related to piles and poles can be found elsewhere (Refs. 13.2 and 13.3).

13.2 BEHAVIOR OF COLUMNS

1. Load-Deformation Response

Prestressed concrete columns have in general a low level of prestress. The ACI code recommends a minimum effective uniform prestress of 225 psi (1.55 MPa). This is insignificant in comparison to the compressive strength of concrete. Prestressed concrete columns, subjected to monotonically increasing axial compression, are expected to behave similarly to reinforced concrete columns (Fig. 13.5). Ferguson reports that the ascending part of the load-deformation response reaches a pseudoyield point at about 85 percent of the corresponding ultimate resistance of the concrete, after which the behavior of the column depends on whether it is transversely reinforced with ties or spirals (Ref. 13.4). Substantial ductility and considerable increase in energy absorption before failure are achieved when spirals are used. This is particularly important in earthquake-prone regions. Shortly after the pseudoyield point, spalling of the concrete cover starts and essentially only the core of the column is left to resist loading.

2. Classification

It was assumed in the preceding section that only an axial load was applied and that buckling did not control the behavior of the column. Assuming everything else

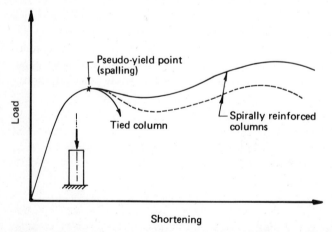

Figure 13.5 Typical load-shortening curves of tied and spirally reinforced short columns.

is the same, the criticality of buckling increases continuously with the length of the column. However, for the purpose of analysis and design, columns are essentially classified into three categories (Ref. 13.5), namely, short columns, medium columns, and long columns. The exact delineation between the three categories is clarified in Sec. 13.8. In brief, a short column can be analyzed or designed from its cross section only; a medium column is essentially designed as a short column with due account to slenderness effects; and the design of long columns is governed by instability criteria.

3. Load-Moment Interaction Diagram

A plot of the column axial load capacity P_n versus the moment it can simultaneously carry M_n is called a column interaction diagram. A typical load-moment interaction diagram is shown in Fig. 13.6 and covers the range of pure compression to pure tension. Several failure points of interest can be identified on the diagram: the point of pure compression for which a uniform compressive stress exists in the section; the point of zero tension above which no tensile stress exists in the concrete section; the balanced point for which the maximum compressive strain in the concrete is attained simultaneously with the yield strain in the steel; the point of pure bending or flexure where the axial force is zero; and the point of pure tension. These are important points of the diagram, as the diagram can be fairly well approximated by joining these points either by a continuous curve or by straight-line segments. Any point inside the area limited by the diagram is theoretically within the capacity of the column.

The following notation will be adopted to describe the coordinates of the major points of the diagram:

$$(P_{n,o}, 0) \qquad \text{for the point of pure compression}$$
$$(P_{n,ot}, M_{n,ot}) \quad \text{for the point of zero tension}$$
$$(P_{n,b}, M_{n,b}) \quad \text{for the balanced point}$$
$$(0, M_{n,f}) \qquad \text{for the point of pure bending or flexure}$$

Any combination of axial load and moment (P, M) acting on the column can be resolved into the same load P acting at an eccentricity $e = M/P$. Radial lines passing by the origin correspond to a constant value of eccentricity and have a slope equal to $1/e$ (Fig. 13.6). Thus, the lines corresponding to "zero tension" eccentricity e_{ot} or "balanced" eccentricity e_b can be easily identified. Given e, an increase in P leads to a proportional increase in M (assuming no buckling) and the representative point moves along the corresponding line until it reaches the interaction diagram where failure occurs. An increase in eccentricity leads to a decrease in the slope of the representative line.

It is beneficial to understand how the interaction diagram is influenced by various parameters. Typical effects are illustrated in Fig. 13.7. Everything else being equal, it can be observed that an increase in concrete compressive strength (Fig. 13.7a) substantially increases the region where compression prevails but leads to little improvement in the region where bending prevails (flexural capacity is

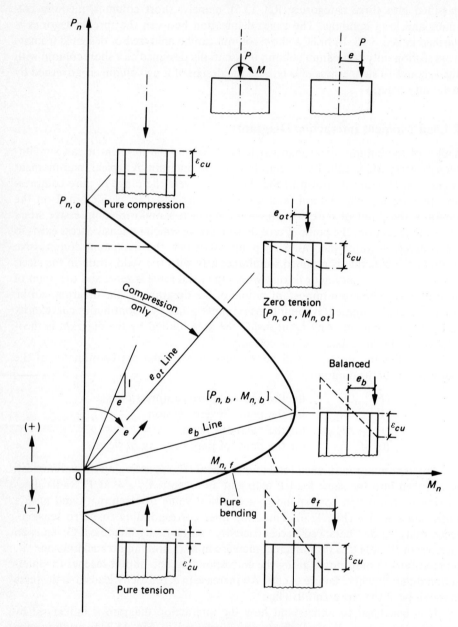

Figure 13.6 Column load-moment interaction diagram.

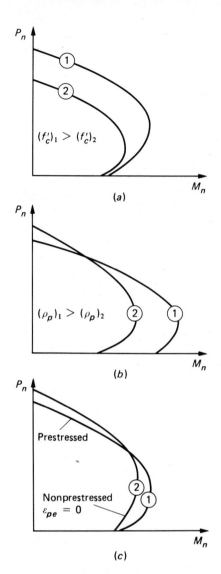

Figure 13.7 Typical effects of (*a*) **compressive strength,** (*b*) **reinforcement ratio, and** (*c*) **effective prestress on the interaction diagram.**

controlled mostly by the steel). An increase in the amount of prestressing steel (Fig. 13.7*b*) leads to a small reduction in pure compression capacity but a significant increase in pure bending resistance. An increase in pure compression capacity is obtained (Fig. 13.7*c*) if the steel is left nonprestressed while the region where bending prevails is reduced.

4. ACI Code Design Interaction Diagram

For the analysis and design of concrete columns, the ACI code recommends the use of an interaction diagram derived from the nominal interaction diagram as

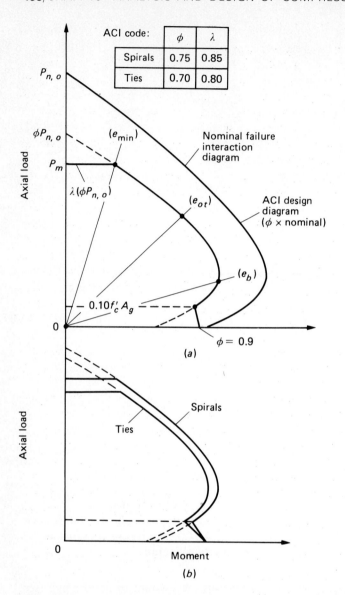

Figure 13.8 (*a*) **ACI code design interaction diagram.** (*b*) **Diagram of tied versus spirally reinforced column.**

follows (Fig. 13.8a): a diagram similar to the (P_n, M_n) diagram is obtained using the origin as center and a similarity factor equal ϕ; the strength reduction factor ϕ is equal to 0.75 for spirally reinforced columns and 0.70 for tied columns. The design diagram so obtained is truncated at its top by an upper limit corresponding to a maximum compressive force P_m equal to $\lambda \phi P_{n,o}$ where $\lambda = 0.85$ for spirally reinforced columns and 0.80 for tied columns. Note that the cutoff limit P_m indirectly sets a minimum eccentricity limit for design. The lower limit of the diagram is extended by a transition segment from the point where the compressive force equals $0.10f'_c A_g$ to the point of pure bending for which $\phi = 0.9$. This is to be consistent with flexural design where the ϕ factor is different from that used for compression. Typical interaction diagrams for the same column laterally reinforced with spirals or ties are shown in Fig. 13.8b. The fact that spirals induce higher ductility is reflected in the results. According to the ACI code, any combination of factored axial load and moment, represented by a point falling inside the region limited by the design diagram, is an acceptable loading combination.

13.3 ANALYSIS OF SHORT COLUMNS

1. Assumptions

The analysis and design of prestressed concrete columns are based on ultimate strength requirements and are approached much in the same way as for reinforced concrete columns (Refs. 13.6 to 13.8). Because of the relatively low level of prestress in columns, little is gained by a stress analysis under service loads. However, such an analysis may be necessary for piles where the level of prestress can be high. Moreover, stresses induced during handling and transportation of precast elements, especially wall panels, must be assessed according to the criteria described in Chap. 4.

The analysis of columns is generally reduced to determining the load-moment interaction diagram of the column and checking if the diagram provides an upper bound to factored loading combinations. The following assumptions, pertinent to the ultimate strength design approach of the ACI code, are made:

1. Plane sections remain plane under loading. Thus, the strain in the concrete is proportional to the distance from the neutral axis.
2. Perfect bond exists between steel and concrete. Thus, the same strain change develops in the steel and in the concrete at the level of the steel.
3. The tensile strength of concrete is neglected.
4. The ACI representation of the rectangular stress block in the concrete at ultimate is assumed.
5. The concrete compressive strain at failure is $\varepsilon_{cu} = 0.003$.

2. Interaction Diagram

The load-moment interaction diagram of a given column is generally determined point by point. The procedure for each point consists of two major steps:

1. Select a location of neutral axis. Draw a linear strain diagram passing by the neutral axis and showing an extreme fiber compressive strain ε_{cu}.
2. Find the values of P_n and M_n for which internal equilibrium and strain compatability are satisfied. Derive the design values ϕP_n and ϕM_n.

The procedure is repeated for each point. In practice, only few points are needed, such as the pure compression, the zero tension, the balanced and the pure flexure points. The diagram is approximately plotted through these points. When additional accuracy is needed, more points are determined.

3. Basic Equations

Once a strain diagram is selected, two equations describing the solicitations at the onset of failure can be written, one for the sum of forces and one for the sum of moments. It is also assumed that strain compatibility holds and that the stress-strain relation of the steel is known. If the calculated strains in the various layers of steel are less than the proportional limit strain, the corresponding stresses are

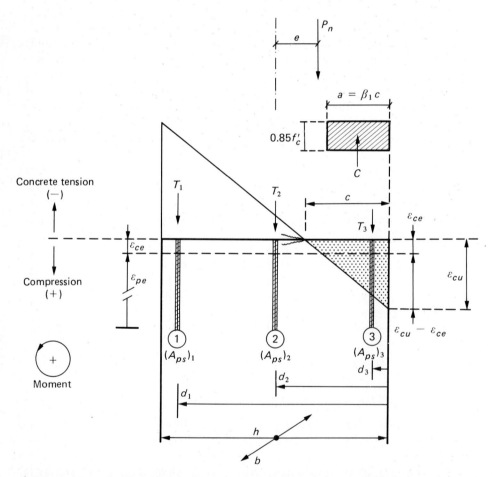

Figure 13.9 Typical strain diagram and corresponding forces at ultimate in a column subjected to compression and bending.

obtained using Hooke's law. Otherwise the stress-strain curve (or representative relationship) must be used to determine the stresses and the corresponding forces in the steel.

Let us consider a typical strain diagram at ultimate (Fig. 13.9) and determine for it the various expressions of interest. The following notation is used:

ε_{ce} = uniform compressive strain in the concrete under effective prestress

$(A_{ps})_i$ = area of prestressing steel in layer i (Fig. 13.9 shows three layers numbered 1, 2, and 3)

d_i = distance from extreme compressive fiber to centroid of $(A_{ps})_i$

b = column width

h = column depth

c = distance from extreme compressive fiber to location of neutral axis

T_{ip} = tensile force due to $(A_{ps})_i$. The subscript p is not used in Fig. 13.10, as nonprestressed reinforcement may also be present in layer i as described in the next section.

Assuming materials and sectional properties are given, the following equations can easily be derived:

$$\varepsilon_{ce} = \frac{A_{ps}f_{pe}}{A_n E_c} = \frac{A_{ps}f_{pe}}{(A_g - A_{ps})E_c} \tag{13.1}$$

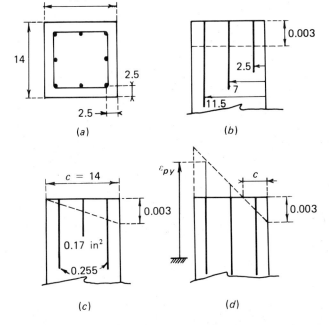

Figure 13.10 Example column. (*a*) Cross section. (*b*) Strain diagram for uniform compression. (*c*) Zero tension. (*d*) Balanced state.

where $A_{ps} = \sum_i (A_{ps})_i$. It is assumed that all tendons have the same effective pre-stress.

The strain change in any layer of steel i is given by:

$$(\Delta \varepsilon_{ps})_i = \varepsilon_{ce} + \varepsilon_{cu} \left(\frac{d_i - c}{c} \right) \tag{13.2}$$

where d_i can be smaller or larger than c. For $d_i > c$, $(\Delta \varepsilon_{ps})_i$ is positive and corresponds to a strain increase in the steel. The strain in any layer of steel i is given by:

$$(\varepsilon_{ps})_i = \varepsilon_{pe} + (\Delta \varepsilon_{ps})_i \tag{13.3}$$

When $(\varepsilon_{ps})_i$ is less than ε_{pp} (the proportional limit strain), the stress in the prestressing steel layer i can be computed from:

$$(f_{ps})_i = E_{ps}(\varepsilon_{ps})_i \tag{13.4}$$

For $(\varepsilon_{ps})_i > \varepsilon_{pp}$ the stress in the steel should be determined from its stress-strain relation. The tensile force in any layer i of prestressing steel is then given by:

$$T_{ip} = (A_{ps})_i (f_{ps})_i \tag{13.5}$$

The compressive resistance of the concrete at ultimate is obtained from the rectangular stress block as:

$$C = 0.85 f'_c \, ba = 0.85 f'_c \, b \beta_1 c \tag{13.6}$$

It is assumed in the above equation that the area occupied by the steel is negligible. It could be accounted for in the analysis. However, it leads to almost no difference in the final results.

Summing up the axial forces on the section leads to the nominal force resistance:

$$P_n = C - \sum_i T_{ip} \tag{13.7}$$

In order to sum up the moments, a sign convention for moments is set. In Fig. 13.9 a positive moment is assumed counterclockwise. Summing up the moments with respect to the centerline of the column leads to the nominal moment resistance:

$$M_n = C \left(\frac{h}{2} - \frac{a}{2} \right) + \sum_i T_{ip} \left(d_i - \frac{h}{2} \right) \tag{13.8}$$

The corresponding eccentricity is given by:

$$e = \frac{M_n}{P_n} \tag{13.9}$$

For design, the ultimate resistance is characterized by the point of the interaction diagram with eccentricity e and coordinates ϕP_n and ϕM_n.

4. Partially Prestressed Columns

Prestressed concrete columns may contain a substantial amount of nonprestressed conventional reinforcement whose contribution can add significantly to the ultimate resistance. In each layer of steel the reinforcement can be separated into prestressed and nonprestressed reinforcement carrying a force:

$$T_i = T_{ip} + T_{is} \tag{13.10}$$

where T_{ip} is defined in Eq. (13.5). In order to determine T_{is} and its contribution to the moment resistance, the strains in the nonprestressed steel at the level of each layer i are needed. Let us define ε_{se} as the compressive strain in the nonprestressed steel due to effective prestress only. It can be easily shown that:

$$\varepsilon_{se} = -\frac{A_{ps} f_{pe}}{A_t E_c} \tag{13.11}$$

in which

$$A_t = A_n + n_s A_s \tag{13.12}$$

A_s is the total area of nonprestressed reinforcement in the section and n_s the moduli ratio. Note that, in the presence of nonprestressed reinforcement, the value of A_n in Eq. (13.1) should be replaced by A_t leading to $\varepsilon_{ce} = |\varepsilon_{se}|$.

For a given location of neutral axis and strain diagram, the strain change (from the zero strain reference state) in the nonprestressed steel of layer i is, similarly to Eq. (13.2), given by:

$$(\Delta \varepsilon_s)_i = \varepsilon_{cu} \left(\frac{d_i - c}{c} \right) \tag{13.13}$$

The resulting strain in each steel layer is:

$$(\varepsilon_s)_i = \varepsilon_{se} + (\Delta \varepsilon_s)_i \tag{13.14}$$

When the absolute value of $(\varepsilon_s)_i$ is less than ε_y, the stress in the nonprestressed steel of layer i can be computed from:

$$(f_s)_i = E_s(\varepsilon_s)_i \tag{13.15}$$

When the magnitude of $(\varepsilon_s)_i$ is larger than ε_y, the stress is given by:

$$(f_s)_i = \pm E_s \varepsilon_y \tag{13.16}$$

The plus sign holds for $(\varepsilon_s)_i$ positive, that is tension, and the minus sign for $(\varepsilon_s)_i$ negative, that is compression.

The tensile force in the nonprestressed steel of layer i is given by:

$$T_{is} = (A_s)_i (f_s)_i \tag{13.17}$$

The two equations leading to the resulting nominal force and nominal moment resistance for the strain diagram considered become:

$$P_n = C - \sum_i (T_{ip} + T_{is}) \tag{13.18}$$

$$M_n = C\left(\frac{h}{2} - \frac{a}{2}\right) + \sum_i (T_{ip} + T_{is})\left(d_i - \frac{h}{2}\right) \tag{13.19}$$

The corresponding point of the design load-moment interaction diagram is then given by ϕP_n and ϕM_n.

13.4 EXAMPLE: COLUMN INTERACTION DIAGRAM

Construct the load-moment interaction diagram of the tied column, the section of which is shown in Fig. 13.10. The column is prestressed with eight $\frac{3}{8}$-in-diameter Grade 270 strands. The following information is given: $f'_c = 5$ ksi, $E_c = 4.28 \times 10^3$ ksi, $f_{pu} = 270$ ksi, $f_{pe} = 150$ ksi, $f_{py} = 243.5$ ksi, $\varepsilon_{py} = 0.010$, $f_{pp} = 196$ ksi, $\varepsilon_{pp} \simeq 0.007$, $E_{ps} = 27,890$ ksi, $A_g = 14 \times 14 = 196$ in^2, $A_{ps} = 8 \times 0.085 = 0.68$ in^2.

The stress-strain curve of the prestressing steel is assumed linear up to the proportional limit after which it can be represented by the following relationship:

$$f_{ps} = 27,890\varepsilon_{ps}\left[0.01174 + \frac{0.98826}{[1 + (107.8715\varepsilon_{ps})^{7.344}]^{1/7.344}}\right]$$

where f_{ps} is given in kips per square inch. The above relationship is described in more detail in the chapter on nonlinear analysis of Vol. 2.

The strand distribution is symmetrical with respect to either the x or the y axis leading to the same resistance in each direction. The strands can be separated in three layers. Referring to the above data and to the notation described in Fig. 13.9 and Sec. 13.3, the following quantities can be derived:

$$(A_{ps})_1 = (A_{ps})_3 = 3 \times 0.085 = 0.255 \text{ in}^2$$

$$(A_{ps})_2 = 2 \times 0.085 = 0.17 \text{ in}^2$$

$$d_1 = 11.5 \text{ in}$$

$$d_2 = 7 \text{ in}$$

$$d_3 = 2.5 \text{ in}$$

$$\phi = 0.70$$

$$\lambda = 0.80$$

$$\beta_1 = 0.80$$

$$A_n = A_g - A_{ps} = 196 - 0.68 = 195.32 \text{ in}^2$$

$$\varepsilon_{ce} = \frac{0.68 \times 150}{4.28 \times 10^3 \times 195.32} = 0.000122$$

$$\varepsilon_{pe} = \frac{150}{27,890} = 0.005378$$

Four main points of the load-moment interaction diagram will be determined, namely, the point of pure compression, the point of zero tension, the balanced point, and the point of pure bending.

(a) Point of pure compression. A uniform strain diagram exists. It can be generated, assuming $c = \infty$ (Eq. (13.2) remains valid). Failure occurs when the strain reaches $\varepsilon_{cu} = 0.003$ (Fig. 13.10b). The compression force in the concrete is given by Eq. (13.6) in which ba is replaced by A_n:

$$C = 0.85f'_c A_n = 830.1 \text{ kips}$$

Note that the use of A_g instead of A_n as a first approximation would have led to $C = 833$ kips, an acceptable result.

For the three layers of steel we have:

Eq. (13.2): $\Delta\varepsilon_{ps} = \varepsilon_{ce} - \varepsilon_{cu} = 0.000122 - 0.003 = -0.002878$

Eq. (13.3): $\varepsilon_{ps} = \varepsilon_{pe} + \Delta\varepsilon_{ps} = 0.005378 - 0.002878 = 0.00250$

As ε_{ps} is less than the proportional limit strain ε_{pp}, we have:

Eq. (13.4): $f_{ps} = E_{ps}\varepsilon_{ps} = 27,890 \times 0.00250 = 69.725$

Eq. (13.5): $\sum_i T_{ip} = A_{ps}f_{ps} = 0.68 \times 69.725 = 47.41 \text{ kips}$

The nominal resistance in pure compression is given by Eq. (13.7):

$$P_{n, o} = C - \sum_i T_{ip} = 830.1 - 47.41 \simeq 783 \text{ kips}$$

The corresponding point of the design diagram is given by:

$$\phi P_{n, o} = 0.7 \times 783 = 548.1 \text{ kips}$$

and the maximum ACI acceptable cutoff compressive force for a tied column is given by:

$$P_m = \lambda(\phi P_{n, o}) = 0.80 \times 548.1 \simeq 438.5 \text{ kips}$$

(b) Point of zero tension in the concrete. The corresponding strain diagram is shown in Fig. 13.10c and leads to $c = 14$ in and $a = \beta_1 c = 11.2$ in. The corresponding force in the concrete compression block is given by Eq. (13.6):

$$C = 0.85f'_c \, ba = 0.85 \times 5 \times 14 \times 11.2 = 666.4 \text{ kips}$$

In order to determine the tensile forces in the three layers of steel, the following quantities are needed:

From Eq. (13.2)

$$(\Delta\varepsilon_{ps})_1 = \varepsilon_{ce} + \varepsilon_{cu}\left(\frac{d_1 - c}{c}\right) = 0.000122 + 0.003\left(\frac{11.5 - 14}{14}\right) = -0.000414$$

$$(\Delta\varepsilon_{ps})_2 = 0.000122 + 0.003\left(\frac{7 - 14}{14}\right) = -0.001378$$

$$(\Delta\varepsilon_{ps})_3 = 0.000122 + 0.003\left(\frac{2.5 - 14}{14}\right) = -0.002342$$

From Eq. (13.3)

$$(\varepsilon_{ps})_1 = \varepsilon_{pe} + (\Delta\varepsilon_{ps})_1 = 0.005378 - 0.000414 = 0.004964$$

$$(\varepsilon_{ps})_2 = 0.005378 - 0.001378 = 0.0040$$

$$(\varepsilon_{ps})_3 = 0.005378 - 0.002342 = 0.00304$$

As all these strains are less than the proportional limit strain, the stresses are obtained from Eq. (13.4):

$$(f_{ps})_1 = E_{ps}(\varepsilon_{ps})_1 = 27,890 \times 0.004964 = 138.50 \text{ ksi}$$

$$(f_{ps})_2 = 27,890 \times 0.004 = 111.56$$

$$(f_{ps})_3 = 27,890 \times 0.00304 = 84.78 \text{ ksi}$$

The corresponding tensile force in each layer of steel is given by Eq. (13.5):

$$T_{1p} = (A_{ps})_1(f_{ps})_1 = 0.255 \times 138.5 = 35.32 \text{ kips}$$

$$T_{2p} = 0.17 \times 111.5 = 18.96 \text{ kips}$$

$$T_{3p} = 0.255 \times 84.78 = 21.62 \text{ kips}$$

Summing up the forces as per Eq. (13.7) leads to the nominal resistance at the zero tension point:

$$P_{n,\,ot} = C - \sum_i T_{ip} = 666.4 - 35.32 - 18.96 - 21.62 = 590.50 \text{ kips}$$

The corresponding nominal moment is given by Eq. (13.8):

$$M_{n,\,ot} = C\left(\frac{h}{2} - \frac{a}{2}\right) + \sum_i T_{ip}\left(d_i - \frac{h}{2}\right)$$

$$= 666.4(7 - 5.6) + 35.32(11.5 - 7)$$

$$+ 18.96(7 - 7) + 21.62(2.5 - 7)$$

$$= 932.96 + 158.94 + 0 - 97.29$$

$$M_{n,\,ot} = 994.61 \text{ kips-in}$$

The corresponding eccentricity is given by:

$$e_{ot} = \frac{M_{n,\,ot}}{P_{n,\,ot}} = \frac{944.61}{590.50} = 1.68 \text{ in}$$

For design, we will use:

$$\phi P_{n,\,ot} = 0.7 \times 590.50 = 413.35 \text{ kips}$$

$$\phi M_{n,\,ot} = 0.7 \times 994.61 = 696.23 \text{ kips-in}$$

The representative point is shown in Fig. 13.11.

(c) Balanced point. The corresponding strain diagram is shown in Fig. 13.10d. The strain in the extreme tensile layer of steel is assumed equal to the yield strain $\varepsilon_{py} = 0.010$. From

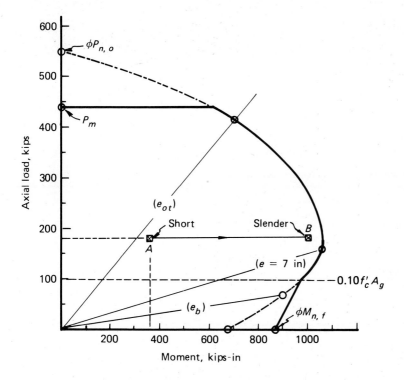

Figure 13.11 Load-moment interaction diagram of example column.

similar triangles, it can be easily shown that:

$$\frac{c}{d_1 - c} = \frac{\varepsilon_{cu}}{\varepsilon_{py} - \varepsilon_{ce} - \varepsilon_{pe}}$$

from which $c \simeq 4.60$ in. Thus, $a = \beta_1 c = 3.68$ in.

The corresponding force in the concrete compression block is given by:

$$C = 0.85 \times 5 \times 14 \times 3.68 = 218.96 \text{ kips}$$

Using Eq. (13.2), it can be shown that:

$$(\Delta\varepsilon_{ps})_1 = 0.00462$$

$$(\Delta\varepsilon_{ps})_2 = 0.00169$$

$$(\Delta\varepsilon_{ps})_3 = -0.001247$$

The corresponding strain values are:

$$(\varepsilon_{ps})_1 = 0.010$$

$$(\varepsilon_{ps})_2 = 0.00707$$

$$(\varepsilon_{ps})_3 = 0.004133$$

The first strain value is equal to the yield strain of the prestressing steel. The corresponding stress is given by:

$$(f_{ps})_1 = 243.5 \text{ ksi}$$

Using the given stress-strain relation for the prestressing steel leads to:

$$(f_{ps})_2 = 192 \text{ ksi}$$

The stress in the third layer of steel is less than the proportional limit, thus:

$$(f_{ps})_3 = E_{ps}(\varepsilon_{ps})_3 = 115.27 \text{ ksi}$$

The following quantities are then easily derived:

$$T_{1p} = 62.09 \text{ kips}$$
$$T_{2p} = 33.52 \text{ kips}$$
$$T_{3p} = 29.39 \text{ kips}$$
$$P_{n, b} = 93.96 \text{ kips}$$
$$M_{n, b} = 1276.98 \text{ kips-in}$$
$$e_b = 13.59 \text{ in}$$
$$\phi P_{nb} = 65.77 \text{ kips}$$
$$\phi M_{nb} = 895.88 \text{ kips-in}$$

The corresponding point is plotted in Fig. 13.11. It is interesting to note that the balanced point for this column falls below the transition point of ordinate $0.10 f_c' A_g = 98$ kips. In order to have a more accurate representation above the value of 98 kips, another point of the diagram corresponding to $c = 7$ in was determined and is also shown in Fig. 13.11. The fact that the balanced point falls below the point where the diagram changes direction seems to characterize prestressed columns in comparison to reinforced concrete columns.

(d) Point of pure bending. Pure bending occurs when the sum of forces acting on the section leads to $P_n = 0$. The location of neutral axis for this case is unknown. A trial-and-error approach is used in which the value of c is first assumed and then the forces are computed as for the other cases described above. If their sum is not nil, a new converging value of c is selected, and so on.

Few iterations are necessary. It can be shown, for this example, that the pure bending point corresponds to a value of $c = 3$ in. The reader may want to check the following results:

$$C = 142.8 \text{ kips}$$

$(\varepsilon_{ps})_1 = 0.01398$	$(f_{ps})_1 = 258.4 \text{ ksi}$	$T_{1p} = 65.9 \text{ kips}$
$(\varepsilon_{ps})_2 = 0.00948$	$(f_{ps})_2 = 238.1 \text{ ksi}$	$T_{2p} = 40.47 \text{ kips}$
$(\varepsilon_{ps})_3 = 0.00498$	$(f_{ps})_3 = 138.95 \text{ ksi}$	$T_{3p} = 35.43 \text{ kips}$

$$P_{n, f} = 142.8 - 65.9 - 40.47 - 35.43 = 1 \text{ kip} \simeq 0$$
$$M_{n, f} = 965.35 \text{ kips-in}$$

The corresponding design value is given by:

$$\phi M_{n,f} = 0.9 \times 965.35 = 868.82 \text{ kips-in}$$

For a ϕ factor of 0.7, we would have obtained 675.75 kips-in.

The point corresponding to pure bending is shown on Fig. 13.11 where the other main points of the diagram are joined by a continuous curve with due account to the ACI design cutoff point and the transition region.

13.5 ACI CODE AND OTHER DESIGN CONSIDERATIONS

Some of the ACI code provisions related to the interaction diagram and ϕ factors were given in the previous sections. A number of additional design requirements are summarized below.

1. *Minimum longitudinal reinforcement.* Prestressed concrete compressive members (columns and bearing walls) should have an average effective prestress not less than 225 psi (1.55 MPa). This provision indirectly sets a minimum reinforcement ratio. Compressive members with lower prestress shall, like ordinary reinforced concrete, have a minimum nonprestressed reinforcement ratio of at least 1 percent.
2. *Lateral reinforcement.* Except for walls, for which Sec. 10.15 of the ACI code applies, members with average prestress equal to or greater than 225 psi (1.55 MPa) shall have all prestressing tendons enclosed by lateral ties or spirals in accordance with the following (ACI code Sec. 18.11):

   a. Lateral ties shall be at least #3 in size or welded wire fabric of equivalent area with lateral spacings not to exceed 48 tie bar or wire diameters or least dimension of compression member. For reinforced concrete or partially prestressed concrete the lateral spacing is also limited by 16 longitudinal bar diameters. Ties shall be arranged so that every corner and alternate longitudinal bar, wire, or strand has lateral support provided by the corner of a tie.
   b. Spiral reinforcement, when used instead of ties, shall have a reinforcement ratio that satisfies the following relation:

$$\rho_s \geq 0.45 \left(\frac{A_g}{A_{co}} - 1 \right) \frac{f'_c}{f_y} \tag{13.20}$$

   where A_{co} is the area of concrete core of spirally reinforced compression member measured to outside diameter of spiral (Fig. 13.12) and f_y is the specified yield strength of spiral reinforcement but is not more than 60 ksi (414 MPa). ρ_s is the ratio of volume of spiral reinforcement to the total

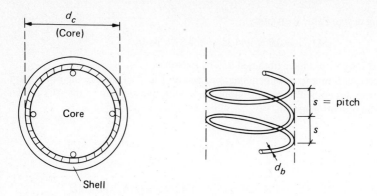

Figure 13.12 Spiral reinforcement.

volume of the core. Calling s the pitch of the spiral (center to center), it can be shown that:

$$\rho_s \simeq \frac{\pi A_{sp}(d_c - d_b)}{A_{co}\, s} \qquad (13.21)$$

where A_{sp} is the cross section area of the spiral steel. Replacing A_{co} by $\pi d_c^2/4$ and neglecting d_b relative to d_c, leads to:

$$\rho_s \simeq \frac{4A_{sp}}{sd_c} \qquad (13.22)$$

Inversely, given a minimum value of ρ_s from Eq. (13.20), the area of spiral steel at a pitch s can be determined from Eq. (13.22) as:

$$A_{sp} = \frac{\rho_s\, sd_c}{4} \qquad (13.23)$$

The clear spacing between spirals, or clear pitch $(s - d_b)$, shall not exceed 3 in (75 mm) nor be less than 1 in (25 mm). For cast-in-place construction bar size for spirals shall not be less than #3.

c. For walls with average prestress not less than 225 psi (1.55 MPa), minimum lateral reinforcement may be waived where analysis shows adequate strength and stability.

Besides its practical role of holding the longitudinal steel together, lateral reinforcement provides a confinement that increases strength and significantly improves ductility. It increases the shear resistance of columns and limits the buckling of longitudinal bars.

3. *Minimum size of columns.* Contrary to previous editions of the code, no minimum cross-sectional sizes are set for columns in the current ACI code. However, slenderness effects, lateral deflections, and other practical considerations limit the size of prestressed concrete columns. In practice, a cross section of less than 8×8 in $(20 \times 20$ cm$)$ is not desirable.

Additional design recommendations related to columns, bearing walls, and piles can be found in Refs. 13.3 and 13.5. The PCI committee on prestressed concrete columns and bearing walls recommends a number of maximum allowable deflections under service loads. Although the wording is adapted to compression members, the prescribed limits are essentially identical to those given in Table 7.2 for flexural members.

13.6 SLENDER COLUMNS: THEORETICAL BACKGROUND

Slender columns are columns for which the effect of buckling (or lateral instability) is significant. Their strength is less than that of their cross section and must be evaluated in function of their length, cross section dimensions, mechanical characteristics, lateral deformation under load, and restraint conditions at their ends. This section gives the very essential background needed to account for slenderness in columns. Specialized publications (Refs. 13.9 and 13.10) must be consulted to extend the concepts and handle special problems requiring exact stability analysis.

Euler's derivation of the critical buckling load P_{cr} of an elastic column subjected to axial compression and having its two ends hinged and restricted against lateral movement leads to the following formula:

$$P_{cr} = \frac{\pi^2 EI}{l_u^2} \qquad (13.24)$$

where l_u is the column's length between hinges or unbraced length, E the elastic modulus of the column's material, and I the moment of inertia of its cross section.

Bracing implies restriction of lateral movement or joint translation at the ends of a column. Lateral movement is often described as "sidesway." Most concrete structures are intentionally or unintentionally braced against sidesway by walls, elevator shafts, staircases, and the like.

Equation (13.24) can be generalized to account for different end conditions, such as for a fixed-ends column or a cantilever column. It leads to:

$$P_{cr} = \frac{\pi^2 EI}{(kl_u)^2} \qquad (13.25)$$

where k is called the effective length factor and l_u is the actual unbraced length of the column between its two ends. The term kl_u, defined as the effective column length, represents in effect the distance between the two points of inflections or zero moments (equivalent hinges) of the column. For a column with two fixed ends, the points of zero moment are at a quarter length, thus $kl_u = 0.5l_u$, or $k = 0.5$. It can easily be shown that for a cantilever column $kl_u = 2l_u$. (The column is considered half a fictitious column, having two hinged ends.)

Theoretical values of k for several typical end conditions are shown in Fig. 13.13. Also shown are corresponding design values recommended by AISC for steel columns where ideal conditions are approximate, and values implied in the ACI code approximate method described in Sec. 13.8.

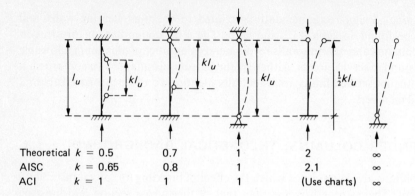

Theoretical	$k = 0.5$	0.7	1	2	∞
AISC	$k = 0.65$	0.8	1	2.1	∞
ACI	$k = 1$	1	1	(Use charts)	∞

Figure 13.13 Deformed shapes and effective length factors of columns with various end restraints.

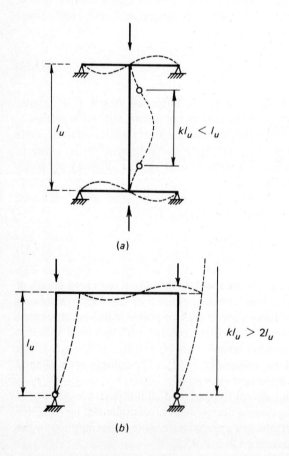

(a)

(b)

Figure 13.14 Typical deformed shapes of (a) braced column and (b) unbraced column.

Perfectly hinged or perfectly fixed connections are seldom encountered in real structures. The effective length depends on the degree of fixity of the column at its ends. The degree of fixity depends on the relative stiffness of the column to the connecting members. The column, whether braced or unbraced, can be modeled as shown in Fig. 13.14 and the evaluation of k in such cases is explained in Sec. 13.8.

The critical buckling load at instability given by Eq. (13.25) can be rewritten as follows:

$$P_{cr} = \frac{\pi^2 E I A}{(k l_u)^2 A} = \frac{\pi^2 E A}{[k(l_u/r)]^2} \tag{13.26}$$

where r is the radius of gyration of the section and A its cross-sectional area in general (transformed area, or as a first approximation, gross area). The ratio l_u/r is called slenderness ratio and $k l_u/r$ is called the effective slenderness ratio. It can be seen from Eq. (13.26) that, for a given column cross section and materials properties, the critical load is inversely proportional to the square of the effective slenderness ratio and, thus, decreases significantly with an increase in $k l_u/r$. Note that the slenderness ratio of a column can easily be reduced by bracing the column at intermediate points, which is equivalent to reducing l_u.

13.7 SLENDERNESS EFFECTS: ACI CODE PHILOSOPHY

The ACI code does not specifically cover slenderness effects in prestressed concrete columns. It is important to realize that, although prestressed concrete columns are subjected to a compressive force F in addition to external loading, they are not more vulnerable to buckling than reinforced concrete columns. This is because the tendons do not change position within the cross section, even when a lateral displacement is induced. Thus, contrary to Euler's case, a lateral displacement does not generate an additional moment due to F in the section.

As at ultimate, both reinforced and prestressed concrete columns show similar behavior, the provisions given in the ACI code for reinforced concrete can be somewhat extended to prestressed concrete. This was essentially done in the PCI committee report on prestressed columns (Ref. 13.5) in which some ACI code provisions were further simplified to better accommodate prestressed columns and walls in accordance with recent research results (Refs. 13.11 and 13.12).

The ACI code philosophy regarding the design of slender columns can be summarized as follows:

1. A comprehensive analysis of the structure, accounting for lateral deflection, variable stiffness, and duration of loading is generally preferred. Such an analysis, also called second-order analysis, is required when the effective slenderness ratio $k l_u/r$ exceeds 100 (that is, for long columns). Procedures for carrying out a second-order analysis are explained and documented in Refs. 13.13 to 13.16.

Among the minimum requirements (stated in Sec. 10.10 of the ACI code commentary) for an adequate analysis is a realistic estimate of the moment-curvature or moment-end rotation relationships. This implies a reasonably accurate estimate of the stiffness or EI values. The main difficulty is that, although the EI values are used in an elastic analysis, they should be representative of the ultimate limit state for which columns are designed.
2. An approximate method is proposed when the effective slenderness ratio is less than 100. The method treats each column individually but takes into account the effect of lateral bracing of the entire structure and the relative stiffness between a column and the beams at its ends.

In the approximate method the approach consists of designing the column for the applied axial load P and for a magnified moment δM. This is illustrated graphically in Fig. 13.15 where the load-moment interaction diagram of a column is plotted. For a given eccentricity e, the load P on a short column can be increased and the corresponding loading path is represented by line $0A$, that is, a proportional increase in moment is induced ($M = Pe$). For a slender column, for which the lateral deformation y is significant, the moment generated at any load P is given by ($Pe + Py$). The loading path follows curve $0B$ until it reaches the interaction diagram at B, where failure occurs. Lines $0A$ and $0B$ can also be seen as the loading paths followed by the end section and the midsection of the column, respectively.

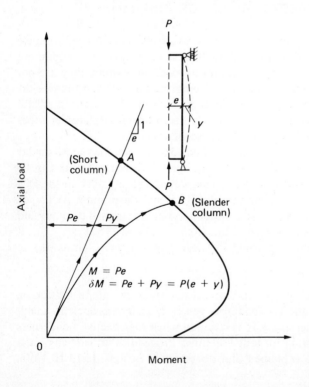

Figure 13.15 Effect of slenderness on loading path.

The moment along $0B$ can be expressed in function of the moment along $0A$ as:

$$\delta M = Pe + Py = P(e + y) \tag{13.27}$$

where δ is called the moment magnification factor and is theoretically given by:

$$\delta = \frac{e + y}{e} \tag{13.28}$$

The value of δ should depend on the slenderness of the column and the applied load. The determination of slenderness ratio and the design value of δ for factored loads, as recommended by the ACI code, are explained in the next section.

13.8 ACI CODE DESIGN PROVISIONS FOR SLENDER COLUMNS

The approximate design method of the ACI code can be reduced to five major steps (Ref. 13.17) as follows:

1. Determine if frame is braced against sidesway
2. Determine effective length factor k
3. Determine effective slenderness ratio and conditions for which slenderness must be accounted for
4. Determine design moment magnification factor δ
5. Design for factored axial load and magnified factored moment

These steps are explained in detail below. A flow chart summarizing the overall approach is given in Sec. 13.10 and Fig. 13.21.

1. Bracing Condition

Bracing influences the value of the effective length factor k. In order to determine if the frame to which the column considered belongs is braced against sidesway, the stiffness of any wall or any bracing in the structure is compared to that of parallel frames. If the stiffness of walls and lateral bracing is larger than six times that of parallel frames, the frames can be considered braced against sidesway. MacGregor and Hage (Ref. 13.16) have proposed a more rational approach to evaluate if bracing prevails for a compression member by using a parameter called the stability index. Their approach is summarized in Sec. 10.11.2 of the code commentary and, thus, is a code acceptable alternative. The stability index μ is defined as:

$$\mu = \frac{\sum P_u \Delta_u}{H_u h_s} \tag{13.29}$$

where $\sum P_u$ = sum of factored axial loads in a given story

Δ_u = elastically computed first-order lateral deflection due to H_u (neglecting $P\Delta$ effects) at the top of the story relative to the bottom of the story

H_u = total factored lateral force within the story

h_s = height of story center to center of floors or roofs

A frame is considered braced if the stability index does not exceed 0.04.

The stability index provides a simple means to estimate the moment magnification factor δ when μ is larger than 0.04 and does not exceed 0.22, namely:

$$\delta = \frac{1}{1 - \mu} \tag{13.30}$$

More on δ can be found later in this section under the heading 4. Moment Magnification Factor.

2. Effective Length Factor k

The ACI code recommends the following values:

$$\begin{cases} k = 1 & \text{for braced frames} \\ k \geq 1 & \text{for unbraced frames} \end{cases} \tag{13.31}$$

In order to determine k, for unbraced columns, the stiffness of columns and beams at the two ends of each column and the corresponding end restraint coefficients are needed. Section 10.11 of the ACI code commentary suggests a procedure to be associated with Sec. 10.11 of the code as described next.

The stiffness of flexural members can be computed, assuming a moment of inertia of the cracked transformed section equal to half the gross moment of inertia. The stiffness of columns can be estimated using Eq. (10.9) of the code, in which β_d (described in Eq. (13.40)) is eliminated. That is:

$$(EI)_{\text{column}} = \frac{E_c I_g}{5} + E_s I_s \tag{13.32}$$

where the subscript s is for the steel reinforcement (prestressed and nonprestressed).

The end-restraint coefficient ψ is defined as follows:

$$\psi = \frac{(\sum EI/l_u)_{\text{columns}}}{(\sum EI/l)_{\text{beams}}} \tag{13.33}$$

where the summation applies to all elements at the end considered. Particular values called ψ_1 and ψ_2 are calculated for each end of a column.

The effective length factor k is determined from ψ_1 and ψ_2, either by using the well-known alignment charts of Jackson and Moreland, reproduced in Fig. 13.16, or by using the following approximation:

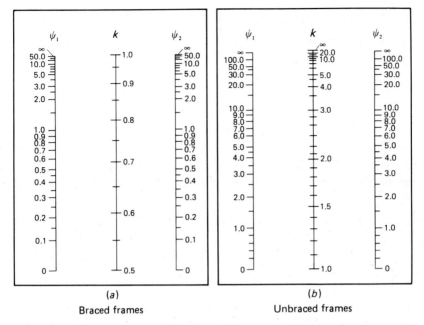

ψ = ratio of ΣK of compression members to ΣK of flexural members in a plane
   at one end of a compression member
k = effective length factor

Figure 13.16 Alignment charts for the effective length factor k. **(a) Braced frames.
(b) Unbraced frames.**

Call ψ_m the average value of ψ_1 and ψ_2 for the column considered:
For $\psi_m < 2$

$$k = \frac{20 - \psi_m}{20}\sqrt{1 + \psi_m} \tag{13.34}$$

For $\psi_m \geq 2$

$$k = 0.9\sqrt{1 + \psi_m} \tag{13.35}$$

For unbraced compression members hinged at one end, k can be taken as:

$$k = 2 + 0.3\psi \tag{13.36}$$

where ψ is for the restrained end.

In practice and for most structures, ψ, at the base of a column, should not be taken less than 1.0. For a base assumed hinged ψ can be taken equal to 10 (Ref. 13.3).

3. Effective Slenderness Ratio and Slenderness Condition

Once k is determined, the effective slenderness ratio (kl_u/r) can be calculated. Section 10.11.4 of the ACI code sets the following conditions to evaluate slenderness:

1. For compression members braced against sidesway, the effects of slenderness may be neglected when

$$\frac{kl_u}{r} < 34 - 12\frac{M_{u1}}{M_{u2}} \qquad (13.37)$$

where M_{u1} and M_{u2} are the factored end moments of the column and where M_{u1} is smaller in magnitude than M_{u2}. M_{u2} is taken always positive and M_{u1} is taken positive if the member is bent in single curvature and negative if the member is bent in double curvature. Figure 13.17 illustrates what is meant by single and double curvature.

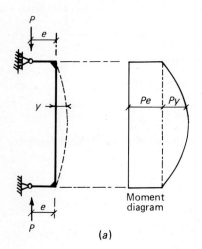

(a)

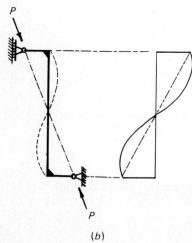

(b)

Figure 13.17 Typical deformed shapes having (a) single curvature and (b) double curvature.

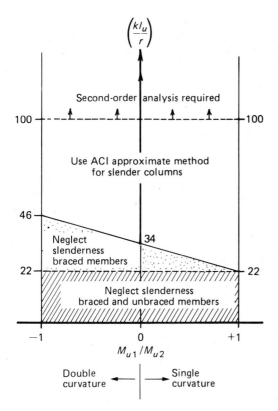

Figure 13.18 Summary of ACI code slenderness conditions.

2. For compression members not braced against sidesway, the effects of slenderness can be neglected when $(kl_u/r) \leq 22$.
3. For all compression members with (kl_u/r) greater than 100, a comprehensive second-order analysis shall be made.

The above conditions are graphically illustrated in Fig. 13.18. It can be observed that, for all practical purposes, three regions of interest are identified: one, where slenderness is neglected, leading essentially to a "short column" analysis; one, where slenderness must be accounted for by a comprehensive second-order analysis, leading to "a long-column" analysis; and an in-between region, where slenderness can be accounted for using the approximate method of the code, leading to a "medium-column" analysis.

4. Moment Magnification Factor

The ACI code gives the following value of the moment magnification factor δ:

$$\delta = \frac{C_m}{1 - (P_u/\phi P_{cr})} \geq 1 \tag{13.38}$$

where P_u = factored axial load obtained from conventional frame analysis
$\quad P_{cr}$ = critical buckling load (see Eq. (13.25))
$\quad C_m$ = reduction factor explained below
$\quad \phi$ = 0.7 for tied columns and 0.75 for spirally reinforced columns

The reduction factor C_m (also used in steel design) relates the actual moment diagram along the member to an equivalent uniform moment diagram. For members braced against sidesway and without transverse loads between supports, the value of C_m may be taken as:

$$C_m = 0.6 + 0.4 \frac{M_{u1}}{M_{u2}} \geq 0.4 \qquad (13.39)$$

where M_{u1} and M_{u2} are as defined in Sec. 13.8. For all other cases, C_m shall be taken as 1.

In calculating P_{cr} from Eq. (13.25) and in lieu of a more exact analysis, the value of EI may be taken either as:

$$EI = \frac{E_c I_g/5 + E_s I_s}{1 + \beta_d} \qquad (13.40)$$

or conservatively:

$$EI = \frac{E_c I_g/2.5}{1 + \beta_d} \qquad (13.41)$$

where I_s = moment of inertia of the reinforcement (prestressed and nonprestressed) about the centroidal axis of the section
$\quad \beta_d$ = ratio of maximum factored dead load moment to maximum factored total load moment, always positive.

β_d essentially accounts for the reduction of elastic modulus of concrete under sustained load.

The PCI committee on prestressed columns (Ref. 13.5) prefers the use of Eq. (13.41) to the use of Eq. (13.40) and suggests values of a parameter λ to replace the divider 2.5 for members with and without compression flange (T walls). Also, a slightly modified value of C_m is suggested. It is possible that additional research will bring more provisions particular to prestressed columns in the future (Ref. 13.12).

Note that the value of δ holds only for one direction of moments. If the analysis is run in two directions, a separate value of δ must be calculated for each axis.

5. Design

Compression members are designed using a factored axial load P_u and a magnified factored moment M_{um} defined by:

$$M_{um} = \delta M_{u2} \geq M_{u2} \qquad (13.42)$$

where M_{u2} is the value of the larger factored end moment. This is achieved by designing a column with a design load-moment interaction diagram that envelops the point of coordinates P_u and M_{um}. Many such points representing different loading combinations are in general present and must also be designed for.

13.9 EXAMPLE: SLENDER COLUMN

Consider the same column cross section and corresponding materials properties described in Example Sec. 13.4 and in Fig. 13.10. Assume the column is fixed at its base, hinged at its top, and braced against sidesway. Its unbraced length is $l_u = 24$ ft (Fig. 13.19a). The column is to be designed for the following load combination:

$$P_u = 1.4P_D + 1.7P_L = 1.4 \times 80 + 1.7 \times 40 = 180 \text{ kips}$$

acting at an eccentricity $e = 2$ in (Fig. 13.19b). The corresponding moments M_{u1} and M_{u2} at the end of the column are equal in magnitude.

Furthermore, as the column is bent in single curvature, M_{u1} and M_{u2} are both positive. Thus:

$$M_{u1} = M_{u2} = 1.4(P_D \times 2) + 1.7(P_L \times 2) = 360 \text{ kips-in}$$

The point representing the loading P_u and $M_u = M_{u1}$ or M_{u2} is plotted in Fig. 13.11 as point A. It can be observed that it falls well inside the column interaction diagram. If the column was a short column, the load would be an acceptable combination with a wide safety margin. However, it is very likely that slenderness effects will prevail, leading to a more critical loading.

Let us follow the flow chart of Fig. 13.21 and the steps described in Sec. 13.8.

Determine the radius of gyration r of the section:

$$r = \sqrt{\frac{I_g}{A_g}} = 4.04 \text{ in}$$

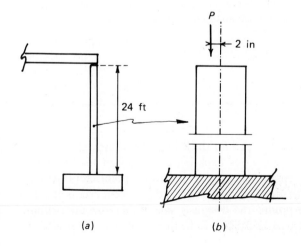

(a)

(b)

Figure 13.19 Example of slender column. (a) Elevation. (b) Eccentric loading.

As the column is braced against sidesway, $k = 1$ and the effective slenderness ratio is

$$\frac{kl_u}{r} = \frac{1 \times 24 \times 12}{4.04} = 71.3$$

It is less than 100 and larger than 22. Thus, the approximate ACI procedure to account for slenderness effects can be used.

The load ratio β_d is given by:

$$\beta_d = \frac{\text{factored } M_D}{\text{factored } M_{\text{total}}} = \frac{1.4 \times 80 \times 2}{360} = 0.62$$

The stiffness is estimated from Eq. (13.41):

$$EI = \frac{0.4 E_c I_g}{1 + \beta_d} = \frac{0.4 \times 4.28 \times 10^3 \times 3201.3}{1 + 0.62} = 3,383,102 \text{ kips-in}^2$$

and the critical buckling load is obtained from Eq. (13.25):

$$P_{cr} = \frac{\pi^2 EI}{(kl_u)^2} = \frac{\pi^2 3,383,102}{(1 \times 24 \times 12)^2} = 402.5 \text{ kips}$$

As the member is braced against sidesway, the factor C_m is given by Eq. (13.39):

$$C_m = 0.6 + 0.4 \frac{M_{u1}}{M_{u2}} = 1$$

The moment magnification factor is determined from Eq. (13.38):

$$\delta = \frac{C_m}{1 - P_u/\phi P_{cr}} = \frac{1}{1 - 180/(0.7 \times 402.5)} \doteq 2.77$$

and the magnified moment is:

$$M_{um} = \delta M_{u2} = 2.77 \times 360 \simeq 997 \text{ kips-in}$$

The effect of slenderness leads us to design for the same load $P_u = 180$ kips and a magnified moment M_{um} substantially higher than M_u. The point of coordinate P_u and M_{um} is plotted on Fig. 13.11 as point B. It can be observed that it is much closer to the interaction diagram than point A and, thus, leads to a substantially smaller safety margin.

13.10 DESIGN AIDS

No closed-form solution has been developed for the design of columns. Generally a trial column is selected based on previous experience. Its design load-moment interaction diagram is then determined and compared to all points representing the various loading combinations (due account being made for slenderness effects). If the diagram envelops all the points, the design is safe. Otherwise the column section or the amount of prestress or both have to be increased and the procedure repeated. If, on the other hand, it is found that the margin of safety is too large, the column section can be decreased to achieve a more economical design.

In dimensioning a column cross section, the following rule of thumb can be used as a first approximation: select the load combination with the largest specified axial load P_u and dimension the concrete cross section so that $0.95P_m \simeq P_u$; neglect the effect of reinforcement in determining P_m. Thus, we have:

$$P_u = 0.95P_m = 0.95\lambda\phi P_{n,o} \simeq 0.95\lambda\phi(0.85f'_c A_g) \tag{13.43}$$

from which we derive:

$$A_g \simeq \frac{P_u}{0.95\lambda\phi 0.85f'_c} \tag{13.44}$$

Eq. (13.44) becomes:
For a tied column

$$A_g \simeq \frac{P_u}{0.45f'_c} \tag{13.45}$$

For a spirally reinforced column

$$A_g \simeq \frac{P_u}{0.53f'_c} \tag{13.46}$$

Once the cross section is determined, at least a minimum reinforcement is provided and the problem is essentially transformed from a design problem to an analysis or review problem. Revision of the cross section characteristics may be necessary after a first design evaluation. Common range of average prestress in columns varies from the minimum of 225 psi (1.55 MPa) to about 900 psi (6.2 MPa). In applications for building structures, the most likely range is between 300 to 400 psi (2 to 3 MPa).

To speed up the design, load-moment interaction diagrams have been extensively developed for reinforced concrete columns and cover different cross-section configurations, reinforcement ratios, reinforcement properties, and the like. The PCI handbook offers a very limited number of interaction graphs for precast prestressed square columns with minimum prestress. Nondimensionalized load-moment interaction diagrams for prestressed concrete columns are shown in Fig. 13.20. They apply to square columns prestressed with 270 ksi (1860 MPa) strands, assumed having an effective prestress $f_{pe} = 150$ ksi (1035 MPa). The steel configuration is as shown in the figure. Four reinforcement ratios are used, corresponding to four levels of average prestress. The smallest value of 0.015 corresponds to the minimum average prestress of 225 psi (1.55 MPa) recommended by the ACI code for prestressed concrete columns, and the largest is four times the minimum value. A ϕ factor of 0.7 was used, hence assuming tied columns. Although the range of variables covered is limited, these diagrams are very convenient for the preliminary dimensioning of prestressed concrete columns.

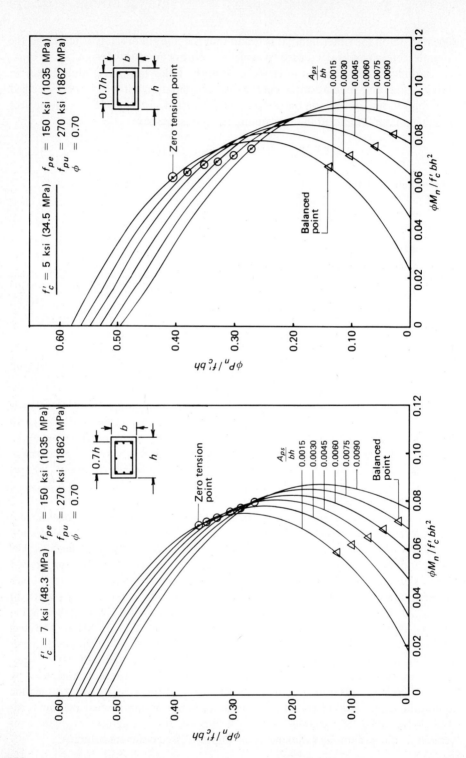

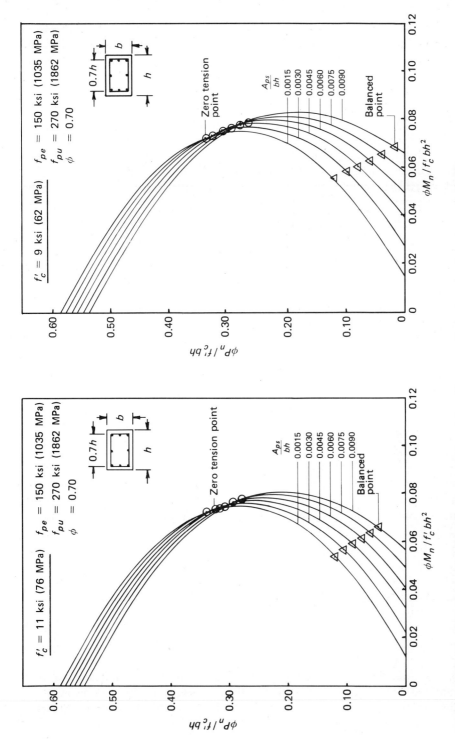

Figure 13.20 Nondimensionalized load-moment interaction diagrams for prestressed concrete columns.

Once the load-moment interaction diagram is determined, the effects of slenderness have to be checked, as described in Sec. 13.8. A flow chart summarizing the steps according to the ACI code procedure is given in Fig. 13.21 and should cover the design in a majority of cases.

For column economy, it is better to use the highest possible compressive strength of concrete. However, this may not help if the column is subjected to relatively large moments. In such cases, the amount of prestressing steel has to be increased. Because of the minimum prestress requirement of 225 psi, an increase in column cross section also leads to an increase in the amount of prestressing steel resisting flexure, and the corresponding moment resistance is also increased.

Because the failure of columns can be catastrophic in terms of damage to the whole structure and to human lives, it is this author's opinion that columns should be designed with a sufficient margin of safety, even exceeding code recommendations. This is also justifiable on the basis that creep and shrinkage, which are generally ignored in everyday design of columns, can substantially affect their performance. Finally, future modifications of the structure, such as addition of a story, can be easily accommodated with slightly overdesigned columns.

13.11 BIAXIAL BENDING

When biaxial bending exists, the load-moment interaction diagram can be derived for each of the two principal axes, x or y, or any axis in between which can be characterized by the ratio of moments. In general, an interaction failure surface can be developed (Fig. 13.22). Such a surface can also be generated by a family of contour curves, each defining the geometric lieu of a characteristic point and corresponding to a constant value of P_n. Contour curves comprise the zero tension contour, balanced contour, pure flexure contour, and the like. Bresler proposed a general nondimensional interaction relationship to approximate these contour curves (Ref. 13.19). It has the following form:

$$\left(\frac{M_{nx}^*}{M_{nx}}\right)^\alpha + \left(\frac{M_{ny}^*}{M_{ny}}\right)^\beta = 1 \tag{13.47}$$

where M_{nx} and M_{ny} are the nominal *uniaxial* moment strengths in the directions of the x and y axes, separately, and M_{nx}^* and M_{ny}^* are the nominal *biaxial* moment strengths in the x and y directions, respectively. The exponents α and β are functions of the dimensions of the column, the properties of the steel and concrete, and the amount, distribution, and location of the reinforcement. They can be adjusted to simulate the behavior of a particular type of column. Bresler indicates that it is reasonably accurate to assume $\alpha = \beta$ for reinforced concrete columns. He also

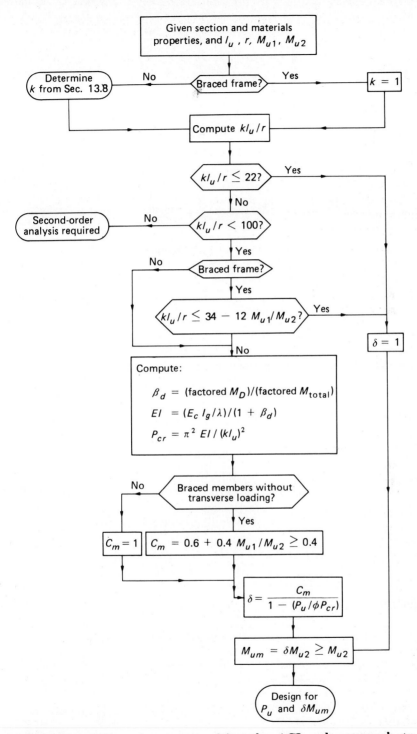

Figure 13.21 Flow chart summarizing the ACI code approach to evaluate slenderness.

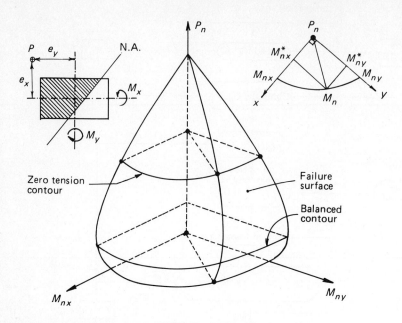

Figure 13.22 Biaxial load-moment interaction surface.

suggests that, for mostly square and rectangular columns with uniformly distributed reinforcement, a value of $\alpha \simeq 1.5$ leads to a reasonably accurate prediction of the contour curves. Thus, Eq. (13.43) becomes, as a first approximation:

$$\left(\frac{M_{nx}^*}{M_{nx}}\right)^{1.5} + \left(\frac{M_{ny}^*}{M_{ny}}\right)^{1.5} = 1 \qquad (13.48)$$

The above equation is expected also reasonably to apply to prestressed concrete columns.

REFERENCES

13.1. Ben C. Gerwick, Jr.: *Construction of Prestressed Concrete Structures*, Wiley-Interscience, New York, 1971, Chaps. 11 and 17.

13.2. Shu-t'ien Li and V. Ramakrishnan: "Optimum Prestress, Analysis and Ultimate Strength Design of Prestressed Concrete Sheet Piles," *PCI Journal*, vol. 16, no. 3, May/June 1971, pp. 60–74.

13.3. *PCI Design Handbook—Precast and Prestressed Concrete*, 2d ed., Prestressed Concrete Institute, Chicago, 1978.

13.4. P. M. Ferguson: *Reinforced Concrete Fundamentals*, 4th ed., John Wiley & Sons, New York, 1979.

13.5. PCI Committee on Prestressed Concrete Columns: "Recommended Practice for the Design of Prestressed Concrete Columns and Bearing Walls," *PCI Journal*, vol. 21, no. 6, November/December 1976, pp. 16–45. Also, discussion by L. D. Martin and Committee Closure in PCI Journal, vol. 23, no. 1, January/February 1978, pp. 92–94.

13.6. T. Y. Lin and T. R. Lakhwara: "Ultimate Strength of Eccentrically Loaded Partially Prestressed Columns," *PCI Journal*, vol. 11, nol 3, June 1966, pp. 37–49.

13.7. P. Zia and E. C. Guillermo: "Combined Bending and Axial Load in Prestressed Concrete Columns," *PCI Journal*, vol. 12, no. 3, June 1967, pp. 52–59.

13.8. A. R. Anderson and S. E. Moustafa: "Ultimate Strength of Prestressed Concrete Piles and Columns," *ACI Journal*, vol. 67, no. 8, August 1970, pp. 620–635.

13.9. S. P. Timoshenko and J. M. Gere: *Theory of Elastic Stability*, 2d ed., McGraw-Hill Book Company, New York, 1961.

13.10. A. Chajes: *Principles of Structural Stability Theory*, Prentice-Hall, Inc., Englewood Cliffs, N.J., 1974.

13.11. N. D. Nathan: "Applicability of ACI Slenderness Computations to Prestressed Concrete Sections," *PCI Journal*, vol. 20, no. 3, May/June 1975, pp. 68–85.

13.12. W. J. Alcock and N. D. Nathan: "Moment Magnification Tests of Prestressed Concrete Columns," *PCI Journal*, vol. 22, no. 4, July/August 1977, pp. 50–61.

13.13. B. R. Wood, D. Beaulieu, and P. F. Adams: "Column Design by P-Delta Method," *Journal of the Structural Division, ASCE*, vol. 102, no. ST2, February 1976, pp. 411–427.

13.14. ———, ———, and ———: "Further Aspects of Design by P-Delta Method," *Journal of the Structural Division, ASCE*, vol. 102, no. ST3, March 1976, pp. 487–500.

13.15. J. G. MacGregor, V. H. Oelhafen, and S. E. Hage: "A Reexamination of the EI Value for Slender Columns," in *Reinforced Concrete Columns*, SP-50, American Concrete Institute, Detroit, 1975, pp. 1–40.

13.16. ——— and S. E. Hage: "Stability Analysis and Design of Concrete," *Journal of the Structural Division, ASCE*, vol. 103, no. ST10, October 1977, pp. 1953–1970.

13.17. P. F. Rice and E. S. Hoffman: *Structural Design Guide to the ACI Building Code*, 2d ed., Van Nostrand-Reinhold Company, New York, 1979.

13.18. J. E. Breen, J. G. MacGregor, and E. O. Pfrang: "Determination of Effective Length Factors for Slender Concrete Columns," *ACI Journal*, vol. 69, no. 11, November 1972, pp. 669–672.

13.19. B. Bresler: "Design Criteria for Reinforced Concrete Columns Under Axial Load and Biaxial Bending," *ACI Journal*, vol. 57, November 1960, pp. 481–490.

PROBLEMS

13.1 Repeat the example of Sec. 13.4, assuming the prestressed steel used is not prestressed. That is $f_{pe} = 0$. Compare the load-moment interaction diagram obtained with that of Fig. 13.12. Plot the nominal load ϕP_n versus curvature ($\Phi = \varepsilon_{cu}/c$) for each case ($f_{pe} = 0$ and $f_{pe} = 150$ ksi) and each of the main points of the interaction diagram. Draw conclusions on the effects of prestressing.

13.2 In a spirally reinforced column, the failure strain of the concrete is substantially higher than the value of 0.003 assumed by the ACI code. Repeat the example of Sec. 13.4, assuming the column is spirally reinforced (by square spirals) and $\varepsilon_{cu} = 0.006$. Everything else is kept the same.

13.3 Go back to the equations developed in Sec. 13.3 and point out the particular modifications needed to analyze the columns of round or triangular cross sections.

13.4 You are considering the use of a prestressed concrete column-pile as a pier for an elevated guideway. You are given the following information: $P_D = 40$ kips, $P_L = 45$ kips; horizontal load at top of column due to vehicle braking $H_L = 6$ kips (considered live load). Assume that the column is fixed at its base and free at its upper end as shown in Fig. P13.4.

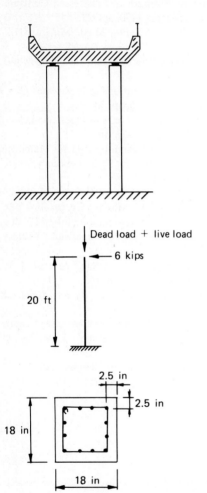

Figure P13.4

A column has been suggested by the local prestressed concrete manufacturer with the following properties: $f'_c = 6000$ psi, cross section 18×18 in, tied column; prestressing steel: 12 strands, $\frac{1}{2}$-in-diameter, 270 ksi strength; $f_{pe} = 150$ ksi. You want to check its feasibility.

(a) Plot the load-moment interaction diagram and show where actual and design loading combinations fit within that diagram assuming no buckling effect. (b) If the slenderness of the column is considered, would it still be safe for use? Justify your answer. Show corresponding representation on the diagram. Make any reasonable assumption when needed.

Columbia River Bridge built in balanced cantilevers with up to 600-ft (183-m) long spans. *(Courtesy Engineering News Record.)*

CHAPTER 14

PRESTRESSED CONCRETE BRIDGES

14.1 SCOPE

Bridges are structures that generally perform a single but major function: that of providing a simple means to "cross" or "reach" between two points separated by a deep valley, a river, a highway, or the like. There are many advantages in using prestressed concrete for bridges. Among them are minimum maintenance, increased durability, good aesthetics, and, when factory precast elements are used, assured plant quality, fast and easy construction, and low initial cost.

In common structural applications, prestressed concrete usually complements reinforced concrete at moderate-span lengths and competes with structural steel at high-span lengths. However, for bridge decks and where factory precast products are available, prestressed concrete competes with reinforced concrete throughout the span range. Hence, it is establishing, at least in the United States, a very strong dominance in bridge applications.

In spite of their unique common purpose, bridges are each characterized by particular site conditions and other factors that may dictate the type of design and construction solution selected. Such factors include the span length and size of the structure, types of loading, clearance, access, available technologies of construction or fabrication, site profile, importance of the bridge, and cost.

Most bridges are designed to carry vehicles and people for which they offer a flat riding or walking surface called the deck. In its simplest form, the deck of a bridge, as in the case of a one-way slab bridge, acts as a simple flexural element.

Increased analytical difficulties arise depending on the design requirements, the type of construction, and the construction sequence. Examples include a continuous span versus a simple span or a series of simple spans, a statically indeterminate structure versus a statically determinate one, a skewed or a curved deck where torsion becomes critical versus a straight deck, a variable-depth bridge versus a constant-depth bridge and a composite versus a noncomposite structure. The type of construction may by itself bring additional constraints. For instance, factory precast elements often have to sustain transportation and erection stresses more severe than service stresses. In segmentally built bridges, each segment acts as a cantilever during construction and as part of the continuous deck during service. Hence, it may be subjected to large stress reversals. The selection of a final solution, using prestressed concrete, implies the evaluations of various design and construction alternatives in which solutions involving cast-in-place posttensioned structures are often compared to solutions involving factory precast pretensioned elements or site precast elements.

The design and construction of bridges is a specialty by itself. It involves not only the deck but also other essential elements, such as the piers, abutments, foundations, connections and bearings, and the like. Building a bridge may take a long time and can be a unique experience. Each project generates particular problems and corresponding solutions which add to the sum of existing knowledge. A comprehensive approach to the design of bridges, even limited to prestressed concrete, cannot be covered without great length. In the following sections only essential aspects and peculiarities related to the design of bridge decks are addressed. They should provide a sufficient background for the design of relatively small and simple bridges. However, a great deal of knowledge can be gained from the experience of others. Some recent references are given at the end of this chapter and should provide a useful source of information.

14.2 TYPES OF BRIDGES

For all practical purposes, bridges can be classified according to their span length, namely short, medium, and long spans. A clear cut does not exist between the above categories as they generally overlap. Moreover, such classification is subjective and relative. For instance, a span considered long for a factory precast element may be considered medium for a cast-in-place element. Similarly, the maximum length for a factory precast element using land transportation may be substantially smaller than an on-site precast solution using floating equipment.

Bridges spanning up to about 50 ft (15 m) are generally considered short. Spans of between 50 and 100 ft (15 and 30 m) are considered moderate, while spans above 100 ft are considered long when factory precast elements are used. The length range for long-span bridges can again be divided into long and very long spans that can exceed 1000 ft (305 m).

Typical bridge deck sections, either cast-in-place and posttensioned or using precast pretensioned elements are shown in Fig. 14.1.

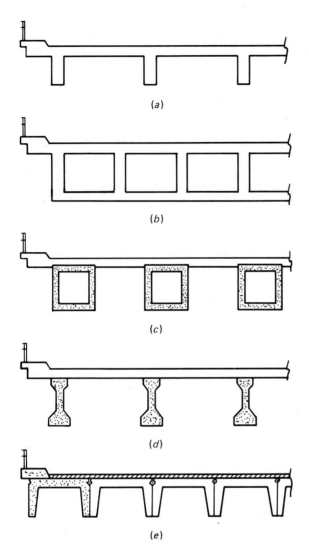

Figure 14.1 Typical cross-sections of bridge decks.
(*a*) T beams. (*b*) Box beams.
(*c*) Spread box beams.
(*d*) I beams. (*e*) Adjacent
channel beams.

1. Short-Span Bridges

In the United States short-span bridge decks are usually built with precast pre-tensioned beams that are transported to site and erected. A typical isometric view of this construction procedure is shown in Fig. 14.2 (Ref. 14.1). The beams which have builtin shear keys are generally placed adjacent to each other, as shown in Fig. 14.1*e*. The shear keys are then filled with a mortar grout to provide for lateral resistance and load transfer. A topping of bituminous concrete or equivalent, of about 2-in (5-cm) thickness, is generally added to provide a wearing surface and for the purpose of leveling. If a concrete topping is used, the deck will act as a composite deck as described in Chap. 9.

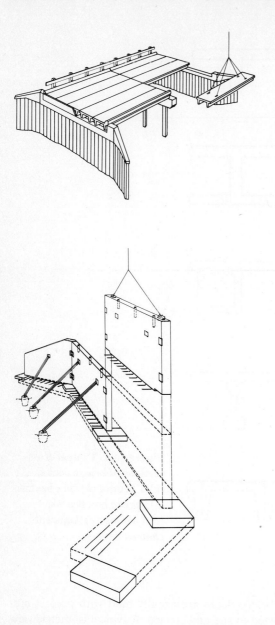

Figure 14.2 Isometric view of a bridge deck construction using precast prestressed elements.
(Ref. 14.1, Courtesy of the Prestressed Concrete Institute.)

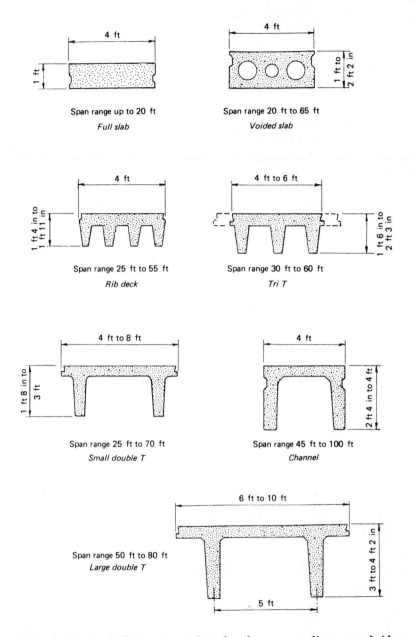

Figure 14.3 Typical precast sections for short- to medium-span bridges.
(Ref. 14.2, Courtesy of the Prestressed Concrete Institute.)

Typical cross sections of precast pretensioned elements used for bridges and the corresponding common ranges of dimensions and spans are shown in Fig. 14.3 (Ref. 14.2). They include solid slabs for very short spans of less than about 25 ft (8 m), voided slabs, rib decks, channels, small double Ts, and tri Ts.

2. Medium- and Long-Span Bridges Using Precast Beams

The decks of medium-span bridges using precast pretensioned beams are generally built either in a manner similar to short-span bridges where the beams are placed adjacent to each other (Fig. 14.1e) or as composite decks where the beams are transversely spaced and a cast-in-place slab is added to provide for lateral continuity (Figs. 14.1c and d).

Typical cross sections and common span ranges are shown in Figs. 14.3 and 14.4. They include box beams, channels, single Ts, double Ts, and bulb Ts. Load tables and charts are available to help the designer achieve an acceptance design. A typical example for box beams taken from the PCI manual on short-span bridges (Ref. 14.1) is shown in Fig. 14.5.

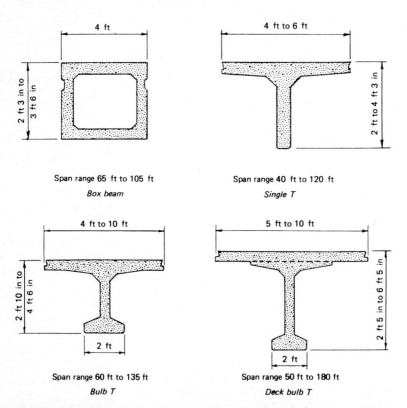

Figure 14.4 Typical precast sections for medium- to long-span bridges.
(Ref. 14.2, Courtesy of the Prestressed Concrete Institute.)

The box beams shown are the former AASHO-PCI standard sections. They can be used either as adjacent units with or without an added wearing surface or spaced apart in which case the deck slab is cast-in-place. Box beams for railway loadings have been standardized by AREA.

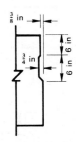

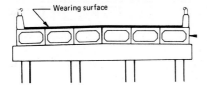

Wearing surface

Typical longitudinal section

Typical section properties

Type	Width, ft	Depth, in	Weight, lb/ft	Net area, in^2	I in^4	y_b, in	Z_b, in^3	Z_t, in^3
B I-36	3	27	584	561	50,334	13.35	3770	3687
B II-36	3	33	647	621	85,153	16.29	5227	5096
B III-36	3	39	709	681	131,145	19.25	6813	6640
B IV-36	3	42	740	711	158,644	20.73	7653	7459
B I-48	4	27	722	693	65,941	13.37	4932	4838
B II-48	4	33	784	753	110,499	16.33	6767	6629
B III-48	4	39	847	813	168,367	19.29	8728	8542
B IV-48	4	42	878	843	203,088	20.78	9773	9571

Typical keyway detail

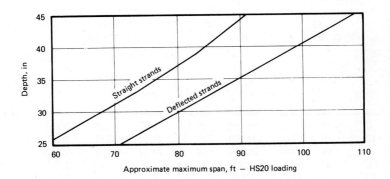

Approximate maximum span, ft — HS20 loading

Figure 14.5 Typical precast box beams for bridges.
(Ref. 14.1, Courtesy of the Prestressed Concrete Institute.)

Many of the sections shown in Figs. 14.3 to 14.5 can be used in combination with a cast-in-place slab to achieve composite action. An example described as the spread box-beam configuration is shown in Fig. 14.1c. However, the most common composite bridge deck in the United States consists of precast prestressed I beams, such as those especially developed and standardized by AASHTO-PCI for highway bridges, with a structural cast-in-place concrete slab on top. Their dimensions, properties, and span ranges are shown in Fig. 14.6. Type I girder is shown for

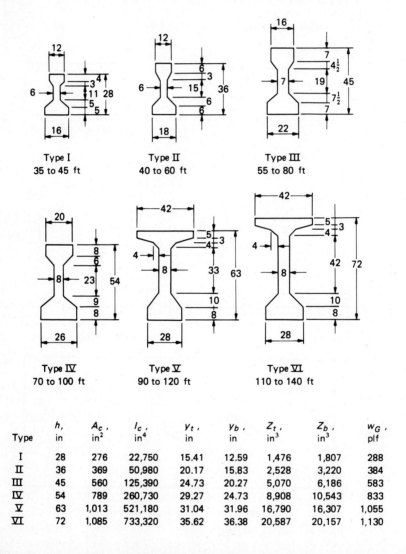

Type	h, in	A_c, in²	I_c, in⁴	y_t, in	y_b, in	Z_t, in³	Z_b, in³	w_G, plf
I	28	276	22,750	15.41	12.59	1,476	1,807	288
II	36	369	50,980	20.17	15.83	2,528	3,220	384
III	45	560	125,390	24.73	20.27	5,070	6,186	583
IV	54	789	260,730	29.27	24.73	8,908	10,543	833
V	63	1,013	521,180	31.04	31.96	16,790	16,307	1,055
VI	72	1,085	733,320	35.62	36.38	20,587	20,157	1,130

Figure 14.6 Standard AASHTO-PCI bridge I beams: section properties and span range.

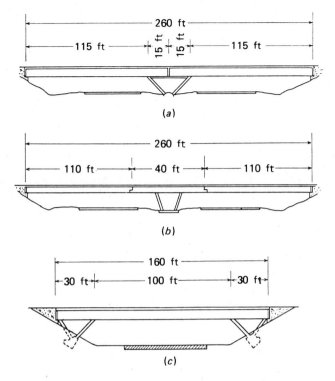

Figure 14.7 Typical effective span-shortening systems.
(Ref. 14.3, Courtesy of the Prestressed Concrete Institute.)

completeness. It is seldom used because, for its span range, it is not as cost effective as the noncomposite bridge sections described earlier for short-span bridges. Note that cast-in-place bridge beams of medium and not too long span lengths tend to have cross sections quite similar to the precast single T or the bulb T shown in Fig. 14.4.

There are many ways to increase the effective span range of a precast prestressed beam when its capacity is attained or when transportation and erection constraints prevail. Typical solutions are shown in Fig. 14.7 and are worth considering (Ref. 14.3).

3. Long- and Very Long Span Bridges

Long-span prestressed concrete bridges are generally cast-in-place or site precast and posttensioned. Although length is relative, spans above 165 ft (50 m) are considered long. Very long spans are spans above about 500 ft (152 m). Long span bridges are often built by a segmental construction technique. In segmental construction the deck is built by segments, one at a time. Segments can be precast or cast-in-place. Box beams are considered best suited for this type of construction.

Bridge (and maximum span)	Cross section (dimensions in meters)	Segment length	Maximum segment wt. (tons)

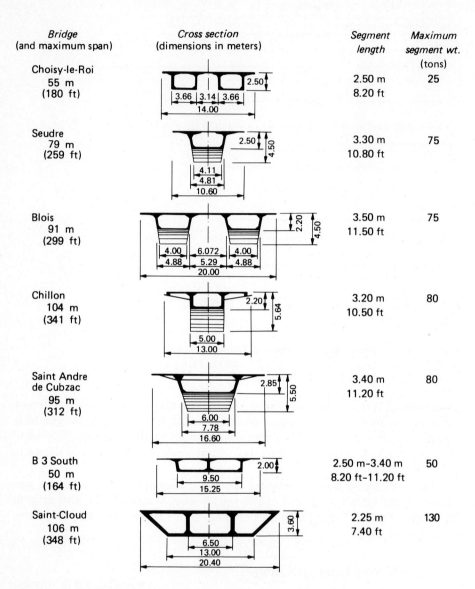

Figure 14.8 Evolution of typical sections for segmental bridges.
(Ref. 14.1, Courtesy of the Prestressed Concrete Institute.)

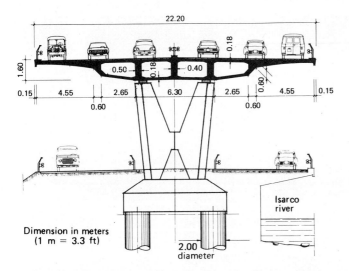

Figure 14.9 Typical cross section of an elevated motorway.
(Ref. 14.5, Courtesy of the Prestressed Concrete Institute.)

Figure 14.10 The segmentally constructed Grand-Mère Bridge in Quebec, Canada.
(Ref. 14.6, Courtesy of the Prestressed Concrete Institute.)

Figure 14.11 The Pasco-Kennewick cable-stayed bridge in Washington, U.S.A.
(Ref. 14.8, Courtesy Arvid Grant & Associates, Consulting Engineers.)

They offer superior torsional rigidity and stability during construction and in service. Typical cross sections and their evolution are shown in Fig. 14.8 where actual dimensions are also given for several segmental bridges built in France (Ref. 14.4). It is observed that, for spans above about 200 ft (60 m), sections tend to have a variable depth.

A bridge deck width of about 35 ft (10.67 m) can be achieved with a single box, while several individual boxes or a box with several webs or openings can be used for larger widths. An example is shown in Fig. 14.9 for a motorway built in Bolzano, Italy, in which the shape of the cross section has been smoothed for aesthetic purposes (Ref. 14.5).

A typical example of a segmentally constructed bridge is the Grand-Mère Bridge built in Quebec, Canada, with a main span of 595 ft (181.36 m) (Ref. 14.6). It is shown in Fig. 14.10. A record main span of 790 ft (241 m) using a similar technique was achieved with the Koror-Babelthuap Bridge in the Pacific Islands (Ref. 14.7).

Because the depth of the deck structure tends to increase substantially for very long spans, a cable-stayed solution can be considered as an alternative to segmental cantilever construction. In such a solution the prestressed concrete deck ribbon,

Figure 14.12 The suspended half of the Pasco-Kennewick Bridge.
(Ref. 14.8, Courtesy Arvid Grant & Associates, Consulting Engineers.)

Figure 14.13 Conceptual design of the Ruck-A-Chucky Bridge in California, U.S.A.
(Ref. 14.9, Courtesy of T. Y. Lin International/Hanson Engineers.)

generally of constant depth, is suspended from sloping stay cables emanating from
a high tower. The sloping cables are stiff and dissipate all anchorage forces in the
deck girder which results in a beneficial longitudinal compressive force in the deck.
The longest cable-stayed bridge in the United States is the Pasco-Kennewick
Intercity Bridge which was completed in 1978 (Figs. 14.11 and 14.12 and Ref. 14.8).
It has a main span of 981 ft (299 m) and a total length of 2503 ft (763 m). Each of
the precast deck segments is 80 ft (24.4 m) wide, 27 ft (8.2 m) long, 7 ft (2.1 m) deep,
and weighs 300 tons. Another bridge that pushes even farther the limits of prestressed concrete is being currently planned in California and deserves mention. It
is the Ruck-A-Chucky Bridge, designed by T. Y. Lin International/Hanson Engineers. It has a main curved span of 1300 ft (396 m) and is suspended by cables from
the walls of the canyon it crosses (Fig. 14.13 and Ref. 14.9).

14.3 INNOVATIVE CONSTRUCTION TECHNIQUES
FOR BRIDGES

A large number of imaginative solutions have been developed for the design and
construction of bridges. As described earlier, some were aimed at the use of factory

precast elements in short- and medium-span bridges. Other solutions, such as for cable-stayed bridges, were derived for long-span bridges and geared at achieving the highest spans at least cost. In some instances, solutions offering a new alternative in bridge design and construction as applied to prestressed concrete were generated.

Some innovative construction techniques and corresponding examples of application are described next, starting with segmental construction. A great deal can be learned from reviewing the related references.

1. Segmental Construction

There are currently four erection techniques for constructing segmental bridges (Fig. 14.14): the balanced cantilever method using a launching girder, the balanced cantilever method using travelling forms, the span-by-span method, and the progressive placement in one direction method (Refs. 14.4 and 14.10).

In the balanced cantilever method using a launching girder (Fig. 14.14a) an overhead truss or launching girder riding above the superstructure places the segments alternating from one cantilever to the other to balance the loads.

In the balanced cantilever method with travelling forms (Fig. 14.14b), traveling forms at the tips of the cantilevers move in opposite directions at the same rate to cast new segments, thereby maintaining balance and stability until midspan is reached. Then a closure is made with a previous half-span cantilever from the preceding pier. To account for possible imbalance, a moment-resisting pier or temporary clamps between the deck and the pier are provided. An example of a bridge built by the segmental method with traveling forms is the Quebec Grand-Mère Bridge shown in Fig. 14.10. Its construction sequence is described in Fig. 14.15. Note that the cantilever was built on the main span side only and stability was maintained by ballasting the end spans.

The span-by-span segmental construction method (Fig. 14.14c) is used mostly for long viaduct structures when crossing over water. The technique was initially developed by the German firm of Dyckerhoff and Widmann and features the execution of a superstructure in one direction, span by span. Generally, a barge crane places the segments on a truss spanning between piers. The segments are placed progressively from one end to the other in span increments, then post-tensioning tendons are installed and stressed.

The progressive placing in one direction method is similar to the span-by-span method (Fig. 14.14d). In this method the precast segments are placed continuously from one end of the structure to the other in successive cantilevers on the same side of the various piers. A crane, placed at the tip of the cantilever, lowers them into position. Temporary stay cables are used to maintain the necessary balance. As the cantilever grows, additional temporary cables are installed for stability. A typical isometric view of the procedure is shown in Fig. 14.16 (Ref. 14.4).

More information on the design and construction of segmentally built bridges can be found in Refs. 14.11 to 14.13.

Representative Erection Techniques

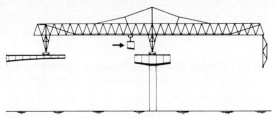

(a) **Balanced Cantilever Using Launching Girder.** *An overhead truss or "launching girder," riding above the superstructure, places segments — alternating from one cantilever to the other — to balance loads.*

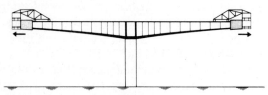

(b) **Balanced Cantilever with Travelling Forms.** *Travelling forms at the tips of the cantilevers move in opposite directions at the same rate to cast new segments, thereby maintaining balance and stability.*

(c) **Span-by-Span Method.** *A barge crane places segments on a truss spanning between piers, then post-tensioning is installed and stressed.*

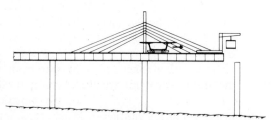

(d) **Progressive Placement in One Direction.** *Segments are transported out to the tip of the cantilever where a crane places them in position. As the cantilever grows, additional temporary cables are installed for stability.*

Figure 14.14 Representative erection techniques for segmental construction.
(Courtesy of Figg and Muller, Engineers Inc., Florida.)

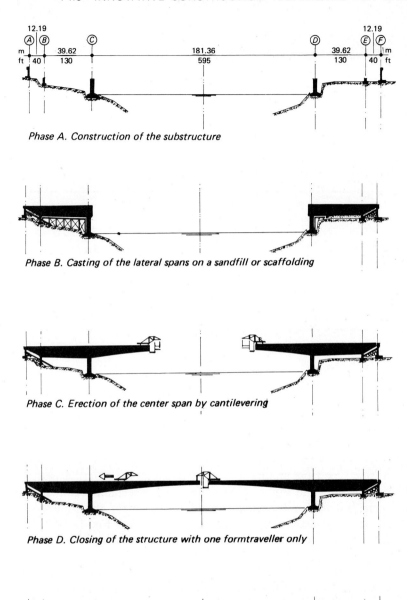

Phase A. Construction of the substructure

Phase B. Casting of the lateral spans on a sandfill or scaffolding

Phase C. Erection of the center span by cantilevering

Phase D. Closing of the structure with one formtraveller only

Phase E. Placing of parapets and asphalt

Figure 14.15 Construction sequence of the Grand-Mère Bridge.
(Ref. 14.6, Courtesy of the Prestressed Concrete Institute.)

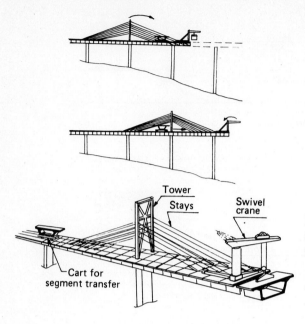

Figure 14.16 Isometric view of the progressive placing technique.
(Ref. 14.4, Courtesy of the Prestressed Concrete Institute.)

2. Truss Bridges

The use of prestressed concrete for truss bridges is not common. However, some recent applications, among which are the Iwahana Bridge in Japan (Ref. 14.14) and the Rip Bridge in Australia (Ref. 14.15), deserve to be mentioned.

The Iwahana Bridge is the first prestressed concrete truss railway bridge in Japan. It is a Warren-type truss and has a span of about 80 m (260 ft). A photograph of the bridge is shown in Fig. 12.2. The construction technique consists of using precast elements (cords and diagonals) match-cast in plant, then joined on site by posttensioning. The construction sequence is described in Fig. 14.17. Note that high-strength concrete of compressive strength of about 12 ksi (83 MPa) was used for the precast elements.

The Rip Bridge (Fig. 14.18) is a three-span, 1082-ft- (330-m) long structure with a central span of 600 ft (183 m). It spans the entrance to Brisbane Waters, north of Sydney, Australia (Ref. 14.15). It consists of a three-arch-shaped cantilever truss system assembled from large precast concrete elements joined together on site by posttensioning. The construction sequence is described in Fig. 14.19 for each of the two cantilever trusses. Note that the upper chord forming the deck essentially acts as a tensile member while the lower chord acts as a compressive member. Each panel of the upper and lower chords of the truss was assembled from five precast elements across the width. The cross sections of these elements were adequately designed so as to create the formwork for the in-situ cast-in-place concrete used for joining the elements and for embedding part of the posttensioning tendons.

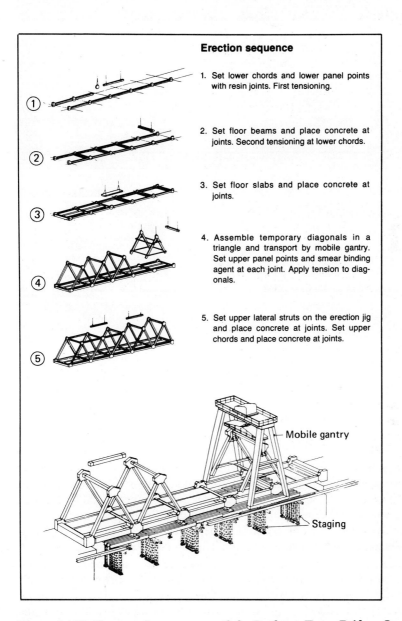

Erection sequence

1. Set lower chords and lower panel points with resin joints. First tensioning.

2. Set floor beams and place concrete at joints. Second tensioning at lower chords.

3. Set floor slabs and place concrete at joints.

4. Assemble temporary diagonals in a triangle and transport by mobile gantry. Set upper panel points and smear binding agent at each joint. Apply tension to diagonals.

5. Set upper lateral struts on the erection jig and place concrete at joints. Set upper chords and place concrete at joints.

Mobile gantry

Staging

Figure 14.17 Construction sequence of the Iwahana Truss Bridge, Japan. *(Ref. 14.14, Courtesy of the Prestressed Concrete Institute.)*

Figure 14.18 The Rip Bridge, Australia.
(Ref. 14.15, Courtesy Department of Main Roads, New South Wales, Australia and American Concrete Institute.)

3. Inverted Suspension Bridges

The concept of using prestressed concrete to carry the load mostly in tension instead of in flexure has always been considered as a most desirable alternative (see Chap. 12). However, little has been done, in practice, to take advantage of such potential. Recent applications include trusses as described above and inverted suspension bridges. Two such bridges have recently been built: the Rio Colorado Bridge in Costa Rica (Ref. 14.16) and the Hayahi-No-Mine Bridge in Japan (Ref. 14.17).

A photograph of the completed Rio Colorado Bridge is shown in Fig. 14.20. The bridge is 669 ft (204 m) long with a main span of 380 ft (116 m). The construction sequence is described in Fig. 14.21 and comprises the following stages: (1) the foundations, the vertical piers, and the abutments for the approach spans are cast; (2) the precast T beams of the approach spans are erected and the main piers are cast vertically on hinges; (3) the main piers are leaned downward 30°, by adjusting anchorage cables, and their base is fixed; (4) the main span cables are strung and the topping of the approach spans completed; (5) precast sections forming the

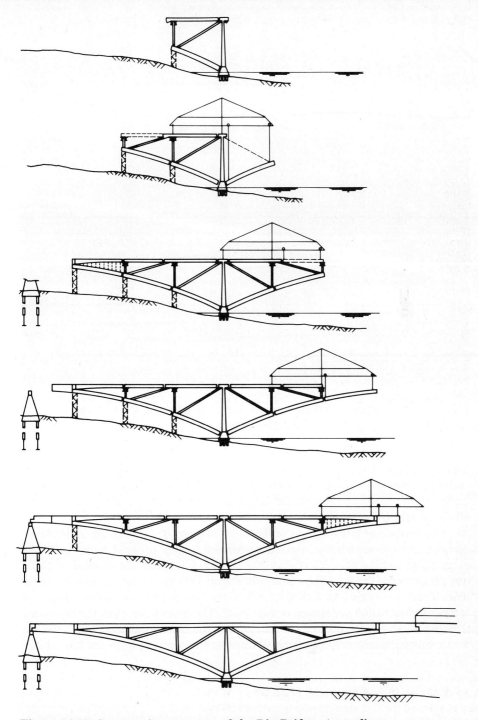

Figure 14.19 Construction sequence of the Rip Bridge, Australia.
(Ref. 14.15, Courtesy Department of Main Roads, New South Wales, Australia and American Concrete Institute.)

Figure 14.20 The Rio Colorado inverted suspension Bridge, Costa Rica.
(Ref. 14.16, Courtesy of T. Y. Lin International.)

lower platform of the center span are fastened to the cables symmetrically from each end, then vertical column bents are attached to the cables; (6) the precast T beams of the roadway are lowered into place and posttensioned for continuity; (7) the structure is completed.

The design and construction of the Rio Colorado Bridge has led to the development of a self-anchoring inverted suspension bridge believed to be very economical for a span range of between 200 and 400 ft (61 and 122 m) (Ref. 14.18). It was proposed for the Shai-Dant-Shuey Bridge in Taiwan and is mentioned here for its innovative concept. An artist's sketch of the bridge and corresponding typical section and elevation are shown in Fig. 14.22. The main features of the bridge are that the deck is fully utilized as the main compression member and only one posttensioned tensile element resists all the tension acting like a suspension cable.

14.4 DESIGN SPECIFICATIONS

In the United States highway bridges are generally designed according to the Standard Specifications for Highway Bridges of the American Association of State Highway and Transportation Officials (AASHTO) (Ref. 14.19). Railway bridges

Stage 1

Stage 2

Stage 3

Stage 4

Stage 5

Stage 6

Stage 7

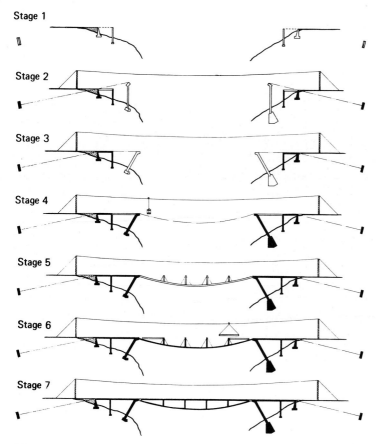

Figure 14.21 Construction sequence of the Rio Colorado Bridge, Costa Rica.
(Ref. 14.16, Courtesy of T. Y. Lin International.)

are designed according to the provisions of the Manual of Railway Engineering of
the American Railway Engineering Association (AREA) (Ref. 14.20). Other local
and regional codes must also be considered, in particular for secondary and
special-purpose structures. The design provisions of the AASHTO and AREA
specifications are very similar to those of the ACI code. Although, in their basic
design philosophy, they favor the working stress design over the ultimate strength
design approach preferred by ACI, such difference does not much materialize when
prestressed concrete is considered. The report of ACI Committee 343 on the analysis and design of reinforced concrete bridge structures provides a valuable docu-
ment where both the AASHTO and ACI philosophies are accommodated at best
(Ref. 14.21).

Allowable stresses recommended by the AASHTO specifications are given in
App. E. Loading combinations, the corresponding percentages of basic unit
stress recommended for working stress design, and load factors for the ultimate

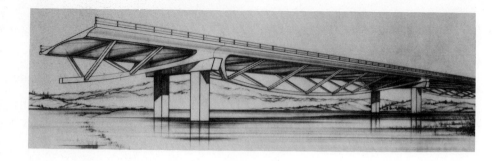

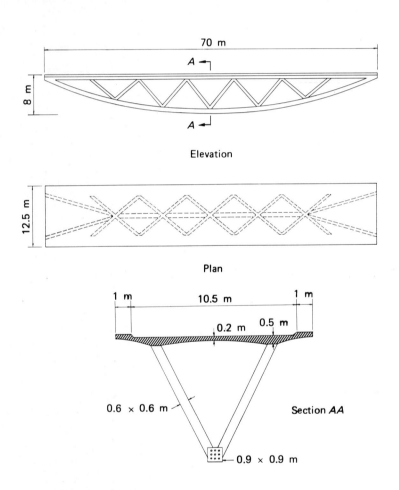

Figure 14.22 Proposed Shai-Dant-Shuey Bridge, Taiwan.
(Ref. 14.18, Courtesy of T. Y. Lin International.)

strength design are detailed in Ref. 14.19. For instance, the one frequent loading combination for ultimate strength design in flexure which correlates with that given by ACI in Table 3.9, item 1, leads to:

$$1.3[D + \tfrac{5}{3}(L + I) + (CF + E + B + SF)] \qquad (14.1)$$

where D = dead load
$L + I$ = live load plus live load impact
CF = centrifugal force
E = vertical earth pressure
B = buoyancy
SF = stream flow pressure

$$Mu = 1.3\left[M_D + \tfrac{5}{3} M_{L+I}\right]$$

Other load combinations include the effects of wind, earthquakes, temperature shrinkage, and the like. It is, of course, not within the scope of this text to cover in depth the application of the AASHTO specifications to bridge design. However, it is essential to briefly describe the live loads and their application to bridge super-structures, as bridge live loads are very different from live loads generally considered for other structures. Live loads applied to the substructure by the super-structure and other loads directly applied to the substructure will not be addressed here. Once the effects of live loads have been determined, the design of prestressed concrete bridge superstructures should proceed in a manner very similar to that used for other prestressed concrete structures.

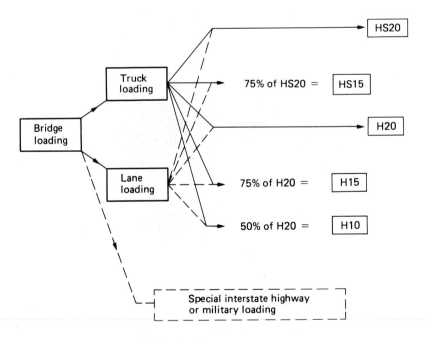

Figure 14.23 AASHTO live load systems.

14.5 LIVE LOADS

Two standard systems of live loads are recommended by AASHTO for bridges: truck loading and lane loading. Five classes of loadings, having loads of different magnitudes, are defined within each system (Fig. 14.23). For each class of truck loading, there is a corresponding class of lane loading. Both are assumed applicable for a given structure and the loading which produces maximum stress governs the design. These live loads are applicable for ordinary highway bridges with spans of up to 500 ft (152.4 m).

1. Truck Loadings

Two basic types of trucks are considered, the H type and the HS type. The H type is a truck with two axles and the HS type is a tractor truck with a semitrailer having a total of three axles.

Three classes of loadings are associated with the H truck: H20-44, H15-44, and H10-44. The first number following the H, such as 20, represents the weight of the truck in U.S. tons (1 U.S. ton = 2 kips), while the second number 44 refers to the year 1944 in which these trucks were standardized. (This second number will be ignored in this text, as it is the same for all truck loadings.) H trucks are described in Fig. 14.24. Note that they have the same dimensions and differ only in their weight. Hence, the H15 truck is 75 percent of the weight of the H20 truck and the H10 truck is 50 percent of the weight of the H20 truck. The front axle is assumed to carry 20 percent of the weight and the rear axle the remaining 80 percent. The number in parentheses near each truck designation, such as (M18), represents the total weight of the truck in metric tons.

Two classes of loadings are associated with the HS truck: the HS20-44 and the HS15-44. They are described in Fig. 14.25. Again, the number 44 refers to the year 1944 and will be dropped. The HS15 truck represents 75 percent of the weight of the HS20 truck. The number 20 in HS20 represents the total weight in U.S. tons of the first two axles of the tractor truck. The third axle is assumed to carry a load equal to that of the second axle, that is 80 percent of the weight of the tractor truck. Hence, the HS20 truck has a total weight of 36 U.S. tons or 72 kips. Note that, although the spacing between the first two axles is assumed fixed, a variable spacing of 14 to 30 ft (4.266 to 9.144 m) is assumed between the second and third axles and is to be selected so that it will produce maximum stresses.

2. Lane Loadings

Five classes of lane loadings are associated with the five classes of truck loadings and are given the same designation (Fig. 14.23). Lane loadings simulate truck trains and each consists of a uniform load per linear foot (meter) of lane and a single concentrated load. The concentrated load is given a different magnitude depending on whether shears or moments are computed and is to be placed so as to produce maximum stresses. Lane loadings are described in Fig. 14.26.

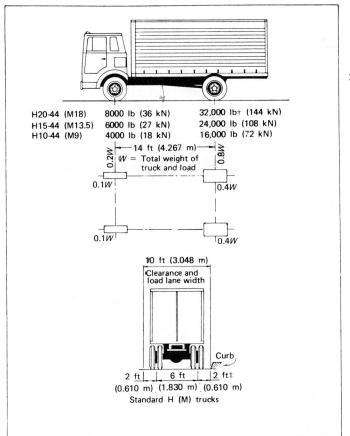

M18 = Total wt. of
truck in
metric tons.

H20 = 20 U.S. Tons
= 2 kips

H20-44 (M18)	8000 lb (36 kN)	32,000 lb† (144 kN)
H15-44 (M13.5)	6000 lb (27 kN)	24,000 lb (108 kN)
H10-44 (M9)	4000 lb (18 kN)	16,000 lb (72 kN)

14 ft (4.267 m)

0.2W

W = Total weight of truck and load

0.8W

0.1W

0.4W

0.1W

0.4W

10 ft (3.048 m)

Clearance and load lane width

Curb

2 ft | 6 ft | 2 ft‡
(0.610 m) (1.830 m) (0.610 m)

Standard H (M) trucks

† In the design of timber floors and orthotropic steel decks (excluding transverse beams) for H20 (M18) loading, one axle load of 24,000 lb (108 kN) or two axle loads of 16,000 lb (72 kN) each spaced 4 ft (1.219 m) apart may be used, whichever produces the greater stress, instead of the 32,000-lb (144-kN)-axle shown.

‡ For slab design, the centerline of wheels shall be assumed to be 1 ft (0.305 m) from face of curb.

Figure 14.24 Standard H truck loadings.
(Ref. 14.19, Courtesy AASHTO.)

In continuous beams, for maximum positive moment, only one concentrated load shall be used per lane combined with as many spans loaded uniformly (continuously or discontinuously) as required to produce maximum moment. However, for maximum negative moment, another concentrated load of equal weight as that described in Fig. 14.26 shall be placed in one other span in the series in such a position as to produce maximum negative moment. The loading procedure is illustrated for a two-equal-span continuous bridge for maximum positive moment in Fig. 14.27a and for maximum negative moment in Fig. 14.27b.

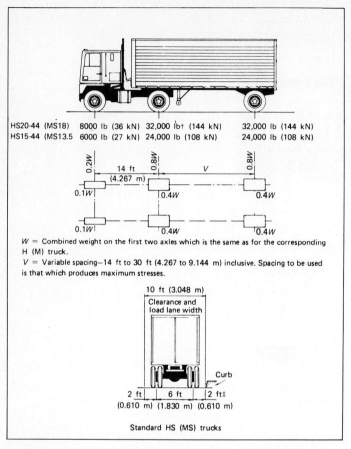

| HS20-44 (MS18) | 8000 lb (36 kN) | 32,000 lb† (144 kN) | 32,000 lb (144 kN) |
| HS15-44 (MS13.5) | 6000 lb (27 kN) | 24,000 lb (108 kN) | 24,000 lb (108 kN) |

W = Combined weight on the first two axles which is the same as for the corresponding H (M) truck.

V = Variable spacing—14 ft to 30 ft (4.267 to 9.144 m) inclusive. Spacing to be used is that which produces maximum stresses.

Standard HS (MS) trucks

† In the design of timber floors and orthotropic steel decks (excluding transverse beams) for H20 (M18) loading, one axle load of 24,000 lb (108 kN) or two axle loads of 16,000 lb (72 kN) each, spaced 4 ft (1.219 m) apart may be used, whichever produces the greater stress, instead of the 32,000-lb (144-kN)-axle shown.

‡ For slab design, the centerline of wheels shall be assumed to be 1 ft (0.305 m) from face of curb.

Figure 14.25 Standard HS truck loadings.
(Ref. 14.19, Courtesy AASHTO.)

For simply supported structures, the H and HS lane loadings are more critical than the corresponding truck loadings for spans above 56 ft (17.07 m) and 140 ft (42.67 m), respectively.

3. Application of Live Loadings

Standard truck or lane loadings are assumed to occupy a loaded width of 10 ft (3.048 m). These loads shall be placed in 12-ft- (3.65-m) wide-design traffic lanes spaced across the entire bridge roadway width in number and position required to

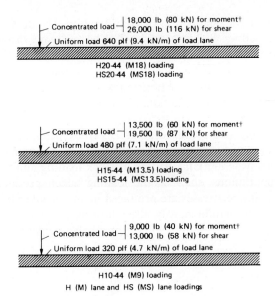

Figure 14.26 Standard lane loading.
(Ref. 14.19, Courtesy AASHTO.)

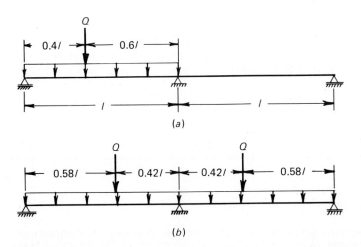

Figure 14.27 Typical lane loading of a two-equal-span continuous bridge. (*a*) For maximum positive moment. (*b*) For maximum negative moment.

produce the maximum stress in the member under consideration. The uniform and concentrated load of a lane loading shall be considered to be uniformly distributed over a 10-ft (3.048-m) width on a line normal to the centerline of the lane. In computing stresses, each 10-ft (3.048-m) lane loading or single standard truck shall be considered as a unit that can occupy any position within its individual traffic lane, so as to produce maximum stress. Fractional load lane widths or fractional trucks shall not be used.

For continuous spans, only one standard H or HS truck per lane shall be considered and placed so as to produce maximum positive or negative moments.

The type of loading used, whether lane loading or truck loading and whether the spans are simple or continuous, shall be the loading which produces the maximum stress. Where maximum stresses are produced in any member by loading any number of traffic lanes simultaneously, the following percentages of the resultant live load stresses shall be used in view of improbable coincident maximum loading:

One or two lanes	100%
Three lanes	90%
Four lanes or more	75%

Note that the combined design load capacity of all the beams in a bridge span shall not be less than that required to support the total live and dead loads of the span.

4. Impact

To allow for the increase in stresses due to the sudden application of live loads (truck or lane loadings), the following live load increment shall be used:

$$I = \frac{50}{l + 125} \leq 0.30 \qquad (14.2)$$

where I = impact fraction or coefficient (maximum 30 percent)

l = length in feet of the portion of the span which is loaded to produce the maximum stress in the member (it is generally taken as equal to the span length)

If the metric system is used, Eq. (14.2) becomes

$$I = \frac{15.24}{l + 38} \leq 0.30 \qquad (14.3)$$

where l is in meters.

5. Other Requirements

Some other requirements given in the AASHTO specifications are summarized below:

1. A minimum live load corresponding to HS15 shall be used for trunk highways or other highways which carry or may carry heavy truck traffic.

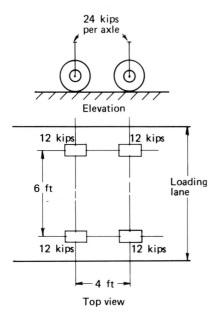

Figure 14.28 Special interstate highway loading.

2. Bridges supporting interstate highways shall be designed for HS20 loading or an alternative military loading of two axles 4 ft (1.219 m) apart with each axle weighing 24,000 lb (108 kN), whichever produces the greater stress (Fig. 14.28). This special loading will govern the design of simply supported structures up to spans of about 37 ft (11.28 m).
3. If loadings of weights other than those designated are desired, they shall be obtained by proportionately changing the weights shown for both the standard truck and the corresponding lane loading. When applicable, special military loadings shall be obtained from the appropriate military manual.
4. Sidewalks shall be designed for a live load of 85 psf (4070 Pa).
5. Provision shall be made for the effect of a longitudinal force of five percent of the live load, without the effect of impact, in all lanes assumed to carry traffic headed in the same direction.

14.6 DISTRIBUTION OF LIVE LOADS

When a concentrated load is placed on a bridge deck, be it a monolithic slab or a slab on beams, the load is distributed over an area larger than the actual contact area. Hence, a larger portion of the deck will assist in resisting the load. This is described as load distribution and should be taken into account in design.

Several methods of analysis are available to determine the wheel load distribution in bridge decks among which is the renowned method of Guyon-Massonet (Refs. 14.22 to 14.25). They are based on elastic theory and assume the deck to be idealized by a structure forming a grid system of interconnected beams, or an orthotropic plate, or a system of thin plate elements and beams. Computerized procedures using finite element are well suited for such analysis. However, in lieu of an elaborate analysis, the approximate approach recommended by AASHTO can be used.

Following are some of the AASHTO specifications related to load distribution. In general, no load distribution is assumed in the longitudinal or span direction (also traffic direction) for the computations of moments and shears. Lateral or transverse distribution of wheel loadings always involves a line of wheels (front and rear) associated with a standard truck. The distribution of wheel loadings takes on a different form depending on the type of bridge deck and, if applicable, whether an interior beam or an exterior beam is considered.

AASHTO recommends that the bending moment for exterior roadway beams be determined by applying to the beam the reaction of wheel loads obtained by assuming the flooring to act as a simple span between beams. In no case shall an exterior beam have less carrying capacity than an interior beam.

The distribution of wheel loads for interior beams and other common types of bridge decks using reinforced or prestressed concrete is covered next.

1. T Beams, I Beams, and Box Girders

Bridge decks included in this category comprise monolithically built bridges and composite decks of the type shown in Figs. 14.1a to d. In calculating end shears and end reactions, lateral distribution of the wheel load shall be that produced by assuming the flooring to act as a simple span between stringers or beams. For loads in other positions on the span, the distribution for shear shall be determined by the method prescribed for moments.

The live load bending moment for each interior longitudinal beam shall be determined by applying to the beam a fraction DF of a wheel load (line of wheels: both front and rear) described in Table 14.1 in which S is the average beam spacing in feet.

For the case of prestressed concrete spread box beams (Fig. 14.1c) the distribution factor DF shall be determined by the following equation:

$$DF = \frac{2n_L}{n_B} + k\,\frac{S}{l} \tag{14.4}$$

where n_L = number of design traffic lanes
  n_B = number of beams ($4 \le n_B \le 10$)
  S = beam spacing in feet ($6.57 \le S \le 11$) or in meters ($2.257 \le S \le 3.353$)
  l = span length in feet (meters)
  $k = 0.07W - n_L(0.10n_L - 0.26) - 0.20n_B - 0.12$
  W = numerical value of the roadway width between curbs expressed in feet ($22 \le W \le 66$)

The live load bending moment for outside beams shall be determined by applying to the beam the reaction of the wheel load obtained by assuming the flooring to act as a simple span between beams.

Table 14.1 AASHTO wheel load distribution factor DF for interior beams (*adapted from Ref. 14.19.*)

Type of bridge deck	Bridge designed for one traffic lane	Bridge designed for two or more traffic lanes
Concrete slab on prestressed concrete girders or steel I-beam stringers	$\dfrac{S}{7}, \left(\dfrac{S}{2.134}\right)$ If S exceeds 10 ft (3.048 m) use footnote 1	$\dfrac{S}{5.5}, \left(\dfrac{S}{1.676}\right)$ If S exceeds 14 ft (4.267 m) use footnote 1
Concrete T beams	$\dfrac{S}{6.5}, \left(\dfrac{S}{1.829}\right)$ If S exceeds 6 ft (1.829 m) use footnote 1	$\dfrac{S}{6}, \left(\dfrac{S}{1.829}\right)$ If S exceeds 10 ft (3.048 m) use footnote 1
Concrete box girders[2]	$\dfrac{S}{8}, \left(\dfrac{S}{2.438}\right)$ If S exceeds 12 ft (3.658 m) use footnote 1	$\dfrac{S}{7}, \left(\dfrac{S}{2.134}\right)$ If S exceeds 16 ft (4.877 m) use footnote 1
Spread box beams	See Sec. 14.6	See Sec. 14.6
Multibeam adjacent precast concrete beams	See Sec. 14.6	See Sec. 14.6

1. In this case, the load on each beam shall be the reaction of the wheel loads assuming the flooring between the beams to act as a simple span beam.
2. The sidewalk live load shall be omitted for interior and exterior box girders designed in accordance with the wheel load distribution indicated herein.

The combined design load capacity of all the beams in a bridge span shall not be less than that required to support the total live and dead loads of the span.

2. Multibeam Adjacent Precast Concrete Beams

A multibeam bridge is constructed with precast reinforced or prestressed concrete beams placed side by side on the supports (Fig. 14.1e). The interaction between the beams is developed by continuous longitudinal shear keys and lateral bolts which may or may not be prestressed.

The fraction of wheel load or distribution factor DF for each beam is determined from the following expression:

For $C \le 3$

$$DF = \frac{(12n_L + 9)/n_B}{5 + (n_L/10) + [3 - (2n_L/7)][1 - (C/3)]^2} \tag{14.5}$$

And for $C > 3$

$$DF = \frac{(12n_L + 9)/n_B}{5 + (n_L/10)}$$

(14.6)

where n_L = total number of traffic lanes

n_B = number of longitudinal beams

$C = K(W/l)$, a stiffness parameter

W = overall width of bridge, feet (meters)

l = span length, feet (meters)

The value of K used in $K(W/l)$ is given by:

$K = 0.7$ for nonvoided rectangular beams

$K = 0.8$ for rectangular beams with circular voids   (14.7)

$K = 1$ for box beams

$K = 2.2$ for channel beams

3. Floor Beams Transverse To Traffic

In calculating moments in floor beams that are placed normally to the direction of traffic, no transverse distribution of the wheel loads shall be assumed. If longitudinal stringers are omitted and the concrete floor is supported directly on floor beams, the fraction of wheel load assigned to each floor beam is given by $S/6$, where S is the spacing of beams in feet. If S exceeds six feet (1.83 meters), the load on the beam shall be the reaction of wheel loads assuming the flooring between beams to act as a simply supported beam.

4. Concrete Slabs

For simple span slabs, the span length l is defined by the center-to-center distance of supports but should not exceed the clear span plus the thickness of the slab. For slabs monolithic with beams or walls, the span length l shall be the clear span. Two cases are considered:

Case a. Main reinforcement perpendicular to traffic (spans 2 to 24 ft (0.61 to 7.315 m), inclusive)

The maximum live load moment for simple spans shall be determined by the following formula (impact not included):

$$M_L = \frac{l + 2}{32} P$$

(14.8)

where M_L = moment in foot-pounds per foot width of slab

P = 16,000 pounds for HS20 and 12,000 for HS15

l = span defined at the beginning of this section

In slabs continuous over three or more supports, a continuity factor of 0.8 shall be applied to the above formula for both positive and negative moments. If the SI system is used, Eq. (14.8) leads to:

$$M_L = \frac{l + 0.61}{9.75} P \tag{14.9}$$

where M_L = moment in kilonewton-meters per meter width of slab
 P = 72 kN for HS20 and 54 kN for HS15

In addition to the main reinforcement, reinforcement transverse to the main reinforcement shall be provided at the bottom of the slabs to ensure lateral distribution of concentrated loads. It should correspond to a percentage of the main reinforcement not to exceed 67 percent and is given by:

$$\begin{cases} \dfrac{220}{\sqrt{l}} & \text{where } l \text{ is in feet} \\[2ex] \text{or} \quad \dfrac{121}{\sqrt{l}} & \text{where } l \text{ is in meters} \end{cases} \tag{14.10}$$

At least 50 percent of the transverse reinforcement should be placed in the outer quarters of the slab span.

The above provision (Eq. (14.10)) covers slabs that are either reinforced or prestressed in both directions. If a slab is prestressed in one direction only and reinforced in the other direction, the above percentage should be adjusted accordingly. The author suggests that the value given by Eq. (14.10) be then multiplied by f_{pu}/f_y.

Case b. Main reinforcement parallel to traffic
The distribution of wheel loads (both front and rear) is taken over a distance:

$$\begin{cases} E = 4 + 0.06l & \text{in feet} \\ \text{or} \quad E = 1.219 + 0.06l & \text{in meters} \end{cases} \tag{14.11}$$

with a maximum value of 7 ft (2.134 m). Lane loadings are distributed over a width $2E$. This is equivalent to having a distribution factor per unit width of slab equal to:

$$DF = \frac{1}{E} \tag{14.12}$$

where E is in feet or in meters and the unit width is one foot or one meter.

For simple spans, the maximum live load moment in foot-pounds per foot width of slab can be approximated by:

For HS20

$$\begin{cases} M_L = 900l & \text{for } l \leq 50 \text{ ft} \\ M_L = 1000(1.3l - 20) & \text{for } 50 < l \leq 100 \text{ ft} \end{cases} \tag{14.13}$$

Use 75 percent of the above values for HS15. For continuous spans, the moments shall be determined by suitable analysis using the truck or appropriate lane loading. If the SI system is used, Eq. (14.13) leads to the moments in kilonewton-meters per meter given by:

$$\begin{cases} M_L = 13.14l & \text{for } l \le 15.24 \text{ m} \\ M_L = 14.6(1.3l - 6.1) & \text{for } 15.24 < l \le 30.48 \text{ m} \end{cases} \tag{14.14}$$

Reinforcement transverse to the main reinforcement shall be provided at the bottom of slabs to ensure lateral distribution of concentrated load. The percentage recommended by AASHTO is not to exceed 50 percent of the main reinforcement and is given by:

$$\begin{cases} \dfrac{100}{\sqrt{l}} & \text{where } l \text{ is in feet} \\ \\ \text{or} \quad \dfrac{55}{\sqrt{l}} & \text{where } l \text{ is in meters} \end{cases} \tag{14.15}$$

Here too, if nonprestressed reinforcement is used instead of prestressed reinforcement, the author recommends multiplication of Eq. (14.15) by f_{pu}/f_y.

Furthermore, edge beams shall be provided for all slabs having main reinforcement parallel to traffic. The edge beam may consist of a slab section additionally reinforced, a beam integral with and deeper than the slab, or an integral reinforced section of slab and curb. For simple spans it shall be designed to resist a live load moment of at least $0.10Pl$, in which P and l are defined above (Eq. (14.8)). A reduction of 20 percent is suggested for continuous spans unless a greater reduction results from a more exact analysis.

5. Slabs Supported on Four Sides

In the case of slabs supported along four edges and reinforced in both directions, the proportion p of the load carried by the short span a shall be assumed as given by the following equations:

For uniformly distributed load

$$p = \frac{b^4}{a^4 + b^4} \tag{14.16}$$

For load concentrated at center

$$p = \frac{b^3}{a^3 + b^3} \tag{14.17}$$

in which p = proportion of load carried by short span
a = length of short span of slab
b = length of long span of slab

Where the length of the slab exceeds $1\frac{1}{2}$ times its width, the entire load shall be assumed to be carried by the transverse reinforcement. Hence, the design is essentially reduced to that of a one-way slab.

The distribution width E for the load taken by either span shall be determined as provided for other slabs (Eq. (14.11)). Moments obtained shall be used in designing the center half of the short and long slabs. The reinforcement steel in the outer quarters of both short and long spans may be reduced by 50 percent.

14.7 DESIGN AIDS FOR LIVE LOAD MOMENTS AND SHEARS

Although there is no conceptual difficulty in determining the moments and shears due to live loads, substantial time can be saved by using existing design aids and solutions. The AASHTO specifications provide in their App. A (Ref. 14.19) tables giving absolute maximum moments and end reactions in simply supported beams due to bridge loading. The values are for a whole truck or corresponding lane loading, whichever controls. When the lane loading (in which a uniform load and a single concentrated load are used) controls, the maximum bending moment occurs at midspan. This is not true, however, for truck loadings, which are represented by a group of concentrated loads. For such a case it can be shown that: "*the absolute maximum moment in the span occurs under the load closest to the resultant force and placed in such a way that the centerline of the span bisects the distance between that load and the resultant.*"

The application of this theorem to the HS and the H trucks is shown in Figs. 14.29a and b, respectively. The corresponding simple span absolute maximum moments are given by:

For the HS truck

$$(M)_{max} = P\left(\frac{9}{8}l + \frac{24.5}{l} - 17.50\right) \tag{14.18}$$

For the H truck

$$(M)_{max} = P\left(\frac{5}{8}l + \frac{4.9}{l} - 3.5\right) \tag{14.19}$$

where $P = 16$ kips for HS20 and H20, 12 kips for HS15 and H15, and 10 kips for H10.

In the above equations l is in feet and M in kips-feet.

While absolute maximum moments are given by AASHTO, moments at other sections are not. For the HS20 loading, a chart originally developed by G. K. Gillan and printed in *Engineering News-Record* of October 7, 1954, can be used (Fig. 14.30). It gives the maximum moment at any section x of a simple span due to

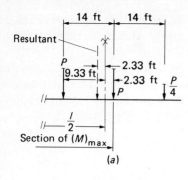

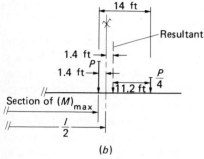

Figure 14.29 Section of maximum moment in simple spans due to (a) HS truck, (b) H truck.

the truck or corresponding lane loading, whichever governs. The moment is determined from an equivalent concentrated load Q obtained from the chart and assumed to be acting at x. It is given by:

$$M_{lane} = Q\,\frac{x(l-x)}{l} \tag{14.20}$$

where x is less than $l/2$. For the HS15 loading, the moment obtained above is multiplied by 75 percent.

The above-described design aids do not provide answers to many practical design cases. Hence, equations giving maximum moments and shears at any section x of a simple span are developed here for all standard loading types. Loadings and load configurations are described in Fig. 14.31. They include the five classes of truck loadings, the corresponding lane loadings, and the special interstate highway loading. Two load configurations are considered for the HS trucks, as each can control depending on the value of x. Expressions for maximum moments and shears at any section x for each of the cases described in Fig. 14.31 are given in Table 14.2 with some remarks on their limit of application. Also given is the absolute maximum moment for the span. Note that the values shown are for a single loading lane (that is, a whole truck or corresponding lane loading) and for simple spans. For continuous spans, influence lines can be used to obtain maximum effects due to the application of concentrated loads.

In designing a typical beam (or unit slab width) of a bridge deck, the maximum

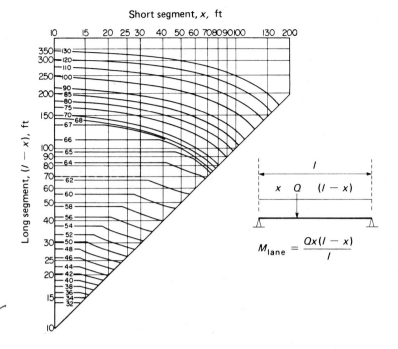

Figure 14.30 Bending moments chart for HS20 loading.
(Courtesy McGraw-Hill Book Company.)

live load moments and shears at any section x can be determined from the moments and shears due to a single live loading lane. If the effect of impact is considered, the following expressions can be derived at any section x:

$$M_{L+I} = (1 + I)DF \frac{M_{\text{lane}}}{2} \qquad (14.21)$$

$$V_{L+I} = (1 + I)DF \frac{V_{\text{lane}}}{2} \qquad (14.22)$$

where M_{L+I} = beam maximum moment at x due to live load plus impact
$\quad V_{L+I}$ = beam maximum shear at x due to live load plus impact
$\qquad I$ = impact coefficient (Eq. (14.2))
$\quad DF$ = wheel load distribution factor (see Sec. 14.6)
M_{lane} = maximum moment at x due to a single live loading lane (truck or lane loading, whichever governs)
$\quad V_{\text{lane}}$ = maximum shear at x due to a single live loading lane (truck or lane loading, whichever governs).

For interstate highway bridges M_{lane} and V_{lane} may also be controlled by the special highway interstate loading or other military loading, if applicable.

Case	Loading	Load configuration†
A	HS20 HS15	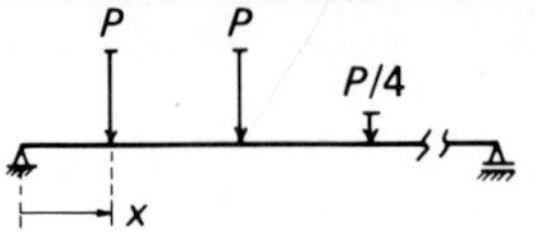
B	HS20 HS15	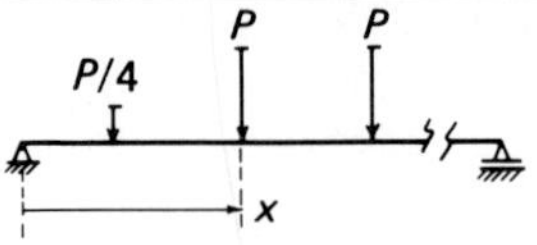
C	H20 H15 H10	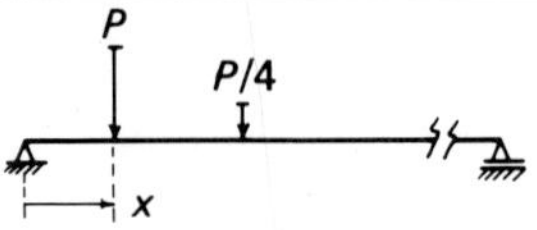
D	Special interstate highway loading	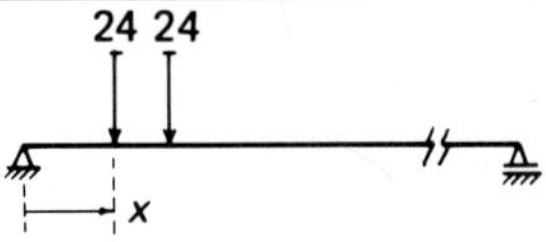
E	Lane loading	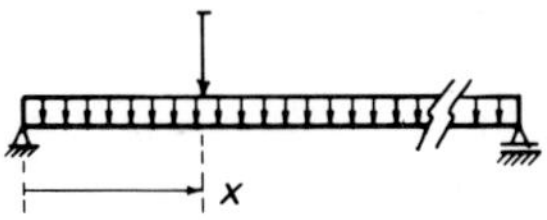

† One line of wheels is shown when applicable.

Figure 14.31 Loading cases for maximum effects in simple spans.

14.8 DESIGN CHARTS FOR BRIDGE BEAMS

Because a large proportion of bridges and highway overpasses are in a span range below about 150 ft (46 m), many attempts were made in the United States and elsewhere to standardize bridge deck sections for such bridges. In the United States this effort was particularly fostered by the precast prestressed concrete industry, the U.S. Department of Transportation, AASHTO, and many state departments of transportation. Cost savings and other benefits can be substantial. Computer programs were written for the analysis and design of bridges and were used to generate design charts for typical bridge deck configurations using standardized beams such as box beams (Fig. 14.5), I beams, T beams, and the like (Ref. 14.1). Several studies have dealt with the use of composite decks made with precast prestressed concrete beams and a cast-in-place concrete slab (Refs. 14.26, 14.27). The use of the AASHTO I beams, described in Fig. 14.6, was particularly extensive. A typical design chart for selecting a typical interior beam for such bridges is shown in

Table 14.2 Simple span moments and shears due to a single truck or lane loading (equivalent to two lines of wheels)

Case	Moments (kips-ft) and shears (kips)†	Loading and limitations (x and l in feet)
A	$(M)_{max} = P\left(\dfrac{9}{8}l + \dfrac{24.5}{l} - 17.50\right)$ $M(x) = Px\left[4.5\left(1 - \dfrac{x}{l}\right) - \dfrac{42}{l}\right]$ $V(x) = P\left[4.5\left(1 - \dfrac{x}{l}\right) - \dfrac{42}{l}\right]$	HS20 & HS15 $M_A \geq M_B$ for: $l > 28$ $x \leq l/3$ $x + 28 \leq l$ $V_A > V_B$ for any x
B	$(M)_{max} = P\left(\dfrac{9}{8}l + \dfrac{24.5}{l} - 17.50\right)$ $M(x) = Px\left[4.5\left(1 - \dfrac{x}{l}\right) - \dfrac{21}{l} - \dfrac{7}{x}\right]$ $V(x) = P\left(4 - 4.5\dfrac{x}{l} - \dfrac{21}{l}\right)$	HS20 & HS15 $M_B \geq M_A$ for: $l > 28$ $x > l/3$ $14 \leq x \leq l/2$
C	$(M)_{max} = P\left(\dfrac{5}{8}l + \dfrac{4.9}{l} - 3.5\right)$ $M(x) = Px\left[2.5\left(1 - \dfrac{x}{l}\right) - \dfrac{7}{l}\right]$ $V(x) = P\left[2.5\left(1 - \dfrac{x}{l}\right) - \dfrac{7}{l}\right]$	H20, H15, & H10 Equations apply for: $l > 14$ $x \leq l/2$ $x + 14 \leq l$
D	$(M)_{max} = 48\left(\dfrac{l}{4} + \dfrac{1}{l} - 1\right)$ $M(x) = 48x\left(1 - \dfrac{x}{l} - \dfrac{2}{l}\right)$ $V(x) = 48\left(1 - \dfrac{x}{l} - \dfrac{2}{l}\right)$	Special interstate highway loading: Is more severe than HS20 for $l \leq 37$ ft
E	$(M)_{max} = Pl\left(0.005l + \dfrac{9}{32}\right)$ $M(x) = Px\left[0.02(l - x) + \dfrac{9}{8}\left(1 - \dfrac{x}{l}\right)\right]$ $V(x) = P\left[0.02l + \dfrac{13}{8}\left(1 - \dfrac{x}{l}\right)\right]$	Lane loading: All classes

† $P = 16$ kips for HS20 and H20, 12 kips for HS15 and H15, and 8 kips for H10. x is defined in Fig. 14.31.

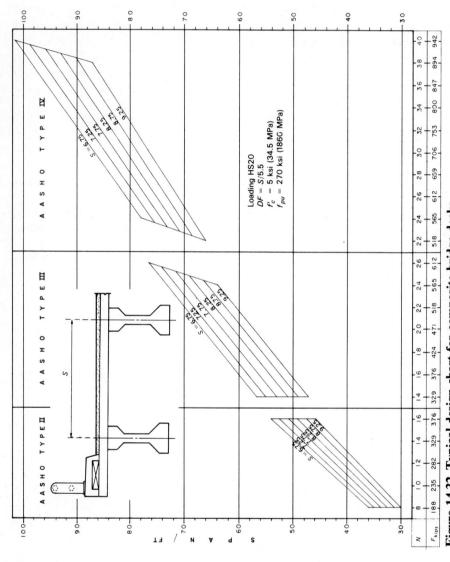

Figure 14.32 Typical design chart for composite bridge decks. (*Ref. 14.27, Courtesy of the Prestressed Concrete Institute.*)

574

Fig. 14.32 (Ref. 14.27). The chart was developed assuming an 8-in- (20-cm-) thick cast-in-place concrete slab with a specified compressive strength of 4000 psi (27.6 MPa). For the precast prestressed beams, the following strengths were specified: $f'_c = 5000$ psi and $f'_{ci} = 4000$ psi (34.5 and 27.6 MPa). For a given span and beam spacing, the chart leads to the type of AASHTO beam, the final prestressing force, and the corresponding number of half-inch diameter strands assumed having a specified strength of 270 ksi (1860 MPa). The tendon's eccentricity at midspan is to be within the maximum practical range. Similar studies should be consulted whenever the need arises, especially during the preliminary planning and design of bridges where standardized solutions can be utilized.

14.9 EXAMPLE OF COMPUTATION OF LIVE LOAD MOMENTS AND SHEARS

1. Composite Bridge

Determine the maximum moments and shears due to live load plus impact for a typical interior span of the composite bridge deck described in Sec. 9.11 and Fig. 9.15. The following information is given: span $l = 80$ ft, beam spacing $S = 6.75$ ft, HS20 loading.

Let us determine first the moment due to a single loading lane. The maximum absolute moment due to the HS20 truck is given by Eq. (14.18):

$$(M)_{max} = P\left(\frac{9}{8}l + \frac{24.5}{l} - 17.50\right)$$

$$= 16\left(\frac{9 \times 80}{8} + \frac{24.5}{80} - 17.50\right) = 1165 \text{ kips-ft}$$

The maximum moment due to HS20 lane loading occurs at midspan and is given in Table 14.2, case E by:

$$(M)_{max} = Pl\left(0.005l + \frac{9}{32}\right)$$

$$= 16 \times 80\left(0.005 \times 80 + \frac{9}{32}\right) = 872 \text{ kips-ft}$$

As the moment due to the truck loading is larger than that due to the lane loading, it governs. Hence, the maximum moment due to a single loading lane is given by:

$$M_{lane} = 1165 \text{ kips-ft}$$

Although the maximum live load moment occurs near midspan, for all practical purposes it is treated as if it occurs at midspan. Thus, it could be combined with maximum dead load moments at midspan.

The wheel load distribution coefficient for a typical interior I beam is given in Table 14.1 by:

$$DF = \frac{S}{5.5} = \frac{6.75}{5.5} \simeq 1.2272$$

The impact coefficient is given by Eq. (14.2)

$$I = \frac{50}{125 + l} = \frac{50}{125 + 80} \simeq 0.244$$

The maximum moment on the typical interior composite beam due to live load plus impact is obtained from Eq. (14.21):

$$M_{L+I} = (1 + I)DF \frac{M_{\text{lane}}}{2}$$

$$= 1.244 \times 1.2272 \times \frac{1165}{2} \simeq 889.19 \text{ kips-ft}$$

This value is shown in Table 9.3. Similarly, using the equations for shear given in Table 14.2, cases A and E, it can be shown that the truck loading controls over the lane loading. Thus, the maximum shear force due to a whole truck at the first critical section located a distance $x = 3$ ft from the center of the support is given by:

$$V_{\text{lane}} = V(x) = P\left[4.5\left(1 - \frac{x}{l}\right) - \frac{42}{l}\right]$$

$$= 16\left[4.5\left(1 - \frac{3}{80}\right) - \frac{42}{80}\right] = 60.90 \text{ kips}$$

The corresponding shear on a typical interior beam is given by Eq. (14.22):

$$V_{L+I} = (1 + I)DF \frac{V_{\text{lane}}}{2} \simeq 46.49 \text{ kips}$$

This value is shown in Table 9.3. The design of the beam is pursued in Sec. 9.11.

2. Two-Span Continuous Bridge

Determine the maximum negative moment due to HS20 live loading for the two-equal-span continuous bridge deck shown in Fig. 14.33 assuming two traffic lanes and using the AASHTO specifications.

The maximum negative moment occurs at the intermediate support B when the uniform lane loading is applied throughout with an additional concentrated load placed so as to create maximum effect. It can easily be shown that a concentrated load Q produces a maximum moment at B when it is placed at $x = l/\sqrt{3} = 0.58l$ from the end support. The loading configuration is shown in Fig. 14.27b.

Let us determine the moment at B due to one loading lane, that is, for a uniform load $w = 0.64$ klf and a value of concentrated load $Q = 18$ kips. Note, two concentrated loads Q are used, one on each span. The moment at B is given by:

$$M_{\text{lane}} = -\left[w\frac{l^2}{8} + 2Q\frac{x}{4}\left(1 - \left(\frac{x}{l}\right)^2\right)\right]$$

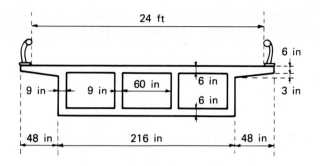

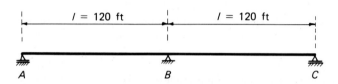

Figure 14.33

where $x = 0.58l$. Hence:

$$M_{\text{lane}} = -\left[0.64 \frac{120^2}{8} + 2 \times 18 \times \frac{0.58 \times 120}{4}(1 - 0.58^2) \right]$$

$$= -1567.68 \text{ kips-ft}$$

The impact coefficient is given by:

$$I = \frac{50}{l + 125} = 0.204$$

The bridge is designed for two traffic lanes. Thus, let us consider the whole bridge deck as a single beam subjected to two loading lanes. This is equivalent to having a distribution factor $DF = 4$ (that is, four lines of wheels or two trucks). The maximum moment at B due to live load plus impact is given by Eq. (14.21):

$$M_{L+I} = (1 + I)DF \frac{M_{\text{lane}}}{2}$$

$$= -1.204 \times 4 \times \frac{1567.68}{2} \simeq -3775 \text{ kips-ft}$$

For this problem, it is not clear if the maximum positive moment will be produced by the truck loading or the lane loading, such as shown in Fig. 14.27a. The use of influence lines would indicate that the truck loading governs here for maximum positive moment.

Note that the section of the bridge deck shown in Fig. 14.33 is the same as that shown in Fig. 10.23, except that it has a cantilever slab added on each side. If the dead load moment due to these cantilever slabs is accounted for and if a wearing surface of 25 psf is considered, the maximum service moment at B for this example would not be too different from the maximum service moment at B of Example Sec. 10.15. The reader may want to refer to Sec. 10.15 to see how the design can be pursued.

14.10 PRELIMINARY DESIGN HINTS

In case a standard bridge deck configuration is not feasible, several design rules of thumb can be used to facilitate the quick dimensioning of bridge beams. Some are summarized in Table 14.3 where common ranges of values related to the depth h, the beam spacing S, the top slab thickness h_f, and the web thickness b_w are given.

Note that the cross-sectional shape of the selected beam can play an important role as it greatly influences the geometric efficiency of the section in resisting

Table 14.3 Common range of main design variables in bridge decks

Variable	Design range		
Depth h	Slab	$\begin{cases}\text{Simple span:} & 0.03 \ \le h/l \le 0.04 \\ \text{Continuous:} & 0.025 \le h/l \le 0.035\end{cases}$	
	Beams (T, box)	$\begin{cases}\text{Simple span:} & 0.035 \le h/l \le 0.055 \\ \text{Continuous:} & 0.03 \ \le h/l \le 0.045\end{cases}$	
	I beams (composite)	$\begin{cases}\text{Simple span:} & 0.045 \le h/l \le 0.06 \\ \text{Continuous:} & 0.040 \le h/l \le 0.055\end{cases}$	
Beam spacing S	Range:	$2 \le S \le 16$ ft	
	Most common range:	$5 \le S \le 10$ ft	
Top slab thickness h_f	Range:	$\dfrac{S}{15} \le h_f \le \dfrac{S}{12}$	
	Most common range:	$6 \le h_f \le 9$ in	
Web thickness b_w	Pretensioned beams	$\begin{cases}\text{Minimum value:} & b_w \simeq 4 \text{ in} \\ \text{Common range:} & 5 \le b_w \le 8 \text{ in} \\ b_w \text{ is also controlled by type of} \\ \quad \text{vibration and by shear stresses}\end{cases}$	
	Posttensioned beams	$\begin{cases}\text{Precast:} & b_w \simeq' \phi + 5 \text{ in} \\ \text{Cast-in-place:} & b_w \simeq \phi + 8 \text{ in} \\ \phi = \text{outside diameter of tendon duct}\end{cases}$	

flexure and, hence, can lead to significant weight and cost savings. In arriving at an acceptable solution, several remarks are in order:

1. All bridges must provide a top surface to carry people and vehicles.
2. Given a uniform load, such as the dead load, the bending moment varies in function of the square of the span, while the shear varies in direct proportion to the span. Hence, the moments in long-span bridges can be very large and necessitate the use of a large size top flange to resist compressive stresses. As the shear does not increase at the same rate as the moment, a webbed type section, such as a T section, becomes desirable for long-span bridges. This also has the advantage of reducing the weight and the dead load moment.
3. The ratio of live load to dead load shear in short-span bridges is relatively large. Hence, a relatively large shear area is needed. This suggests that a slab or a hollow-cored slab is preferable for short-span bridges.
4. The ratio of live load to dead load moment is relatively large in short spans and small in long spans. Hence, the effect of impact due to live loads decreases with an increase in span length.

Note that, once a beam or deck cross section has been arrived at in a preliminary design, it should be further modified and refined according to the final design requirements. These may include not only mechanical and code requirements but also minimum weight and/or cost considerations (Ref. 14.28). A discussion of minimum weight and minimum cost design of prestressed concrete beams is given in Vol. 2.

14.11 EXAMPLE: PRELIMINARY DESIGN OF TWO BRIDGE DECKS

A simply supported bridge deck spanning 50 ft is to be designed according to the AASHTO specifications, assuming two traffic lanes and HS20 loading. Two alternatives are considered: the first consists of a cast-in-place posttensioned slab and the other consists of precast pretensioned beams placed adjacent to each other (Fig. 14.34). The following information common to the two alternatives is provided: $f'_c = 6000$ psi, $f'_{ci} = 4500$ psi, $\bar{\sigma}_{ti} = -201$ psi, $\bar{\sigma}_{ci} = 2700$ psi, $\bar{\sigma}_{ts} = -232$ psi, $\bar{\sigma}_{cs} = 2400$ psi. A wearing surface weighing 25 psf is to be placed on top of the deck. The following live load impact coefficient applies to the two alternatives (Eq. (14.2)):

$$I = \frac{50}{l + 125} = \frac{50}{175} = 0.2857$$

1. Cast-In-Place Posttensioned Slab

A typical deck cross section is shown in Fig. 14.34. Using the suggested quick dimensioning rules for preliminary design (Table 14.3), let us try a depth:

$$h \simeq 0.035l = 0.035 \times 50 \times 12 = 21 \text{ in}$$

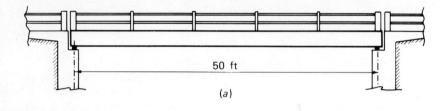

(a)

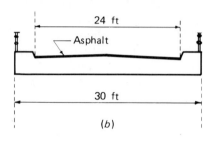

(b)

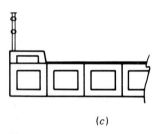

(c)

Figure 14.34 Example Bridge. (a) Longitudinal deck profile. (b) Transverse section for the slab solution. (c) Transverse section for the adjacent box beams solution.

Assuming normal-weight concrete, it leads to a unit weight of $w_G = 0.2625$ klf per foot width. Geometric properties per foot width of slab are: $A_c = 252$ in^2, $Z_b = Z_t = 882$ in^3, $k_b = 3.5$ in, $k_t = -3.5$ in.

Let us determine the live load moment. The main reinforcement is parallel to traffic. Referring to Section 14.6, 4. Concrete Slabs, case b, the distribution of wheel load is to be taken over a distance E given by Eq. (14.11):

$$E = 4 + 0.06l = 4 + 0.06 \times 50 = 7 \text{ ft}$$

The distribution factor per unit width of slab is given by Eq. (14.12):

$$DF = \frac{1}{E} = \frac{1}{7} \text{ per foot}$$

The maximum live load moment per foot width of slab due to the HS20 loading can be obtained as a first approximation from Eq. (14.13):

$$M_L = 900l = 900 \times 50 = 45,000 \text{ lb-ft} = 45 \text{ kips-ft}$$

Hence, the maximum moment due to live load plus impact is given by:

$$M_{L+I} = 45 \times (1 + 0.2857) \simeq 57.857 \text{ kips-ft}$$

Note that we could have used the exact equations given in Table 14.2 and Eq. (14.21) to determine the above moment per foot width of slab. They lead to:

$$M_{L+I} = (1 + I)DF \, \frac{M_{\text{lane}}}{2}$$

$$= (1 + I)DF \times \frac{P}{2} \left(\frac{9}{8} l + \frac{24.5}{l} - 17.50 \right) = 57.678 \text{ kips-ft}$$

which is very close to the result obtained above.

Prestressing force. The minimum and maximum service moments per foot width of slab are given by:

$$M_{\min} = M_G = w_G \, \frac{l^2}{8} = 0.2625 \, \frac{50^2}{8} = 82.031 \text{ kips-ft}$$

$$M_{\max} = M_G + M_{SD} + M_{L+I} = 82.031 + 0.025 \, \frac{50^2}{8} + 57.857$$

$$= 147.700 \text{ kips-ft}$$

Assuming a value of $(e_o)_{mp} = h/2 - 3 \simeq 7.5$ in and using stress condition IV of Table 4.2 for $e_o = (e_o)_{mp}$, leads to:

$$F = \frac{M_{\max} + \bar{\sigma}_{ts} z_b}{e_o - k_t}$$

$$= \frac{147.7 \times 12,000 - 232 \times 882}{7.5 + 3.5} = 142,525 \text{ lbs} = 142.525 \text{ kips}$$

The Dywidag single-bar posttensioning system is selected (App. C). It offers a bar with a nominal diameter of $1\frac{3}{8}$ in and an area of 1.58 in^2 that can carry a force of 142.2 kips at $0.60 f_{pu}$. Hence, one such single-bar tendon is selected per foot width of slab. Assuming a value of $\eta \simeq 0.85$, it can easily be shown that the other three stress conditions of Table 4.2 are largely satisfied.

Ultimate moment. The strength design moment using AASHTO is given by Eq. (14.1):

$$M_u = 1.3(M_D + \tfrac{5}{3} M_{L+I})$$

$$= 1.3(89.8435 + \tfrac{5}{3} \times 57.857) \simeq 242 \text{ kips-ft}$$

Referring to the flow chart Fig. 5.13, let us determine the nominal resistance of the slab at ultimate behavior, assuming a one-foot-wide section:

$$\rho_p = \frac{A_{ps}}{bd_p} = \frac{1.58}{12 \times 18} = 0.007315$$

$$f_{ps} = f_{pu}\left(1 - 0.5\rho_p \frac{f_{pu}}{f'_c}\right) = 160\left(1 - 0.5\rho_p \frac{160}{6}\right) = 144.4 \text{ ksi}$$

$$q = \omega_p = \frac{A_{ps} f_{ps}}{bd_p f'_c} = 0.176 < 0.30$$

$$\phi M_n = \phi f'_c bd_p^2 q(1 - 0.59q)$$

$$= 0.9 \times 6 \times 12 \times 18^2 \times 0.176(1 - 0.59 \times 0.176)$$

$$\simeq 3311.5 \text{ kips-in} \simeq 276 \text{ kips-ft}$$

As M_u is less than ϕM_n, ultimate strength requirements are satisfied. It can also be shown that ϕM_n is more than 20 percent larger than the cracking moment.

Shear strength. Let us check shear at the first critical section assumed at $x = 1.25$ ft from the center of the support. The dead load shear per foot width of slab is given by:

$$V_D = V_G + V_{SD} = (0.2625 + 0.025)\left(\frac{l}{2} - x\right) = 0.2875(25 - 1.25) = 6.828 \text{ kips}$$

The shear force per foot width of slab due to live load plus impact is given by Eq. (14.22) in which V_{lane} is taken from Table 14.2, case A which controls in this case:

$$V_{L+I} = (1 + I)DF \frac{P}{2}\left[4.5\left(1 - \frac{x}{l}\right) - \frac{42}{l}\right]$$

$$\simeq 5.2126 \text{ kips}$$

Using the AASHTO specifications, the specified factored shear at ultimate is given by:

$$V_u = 1.3(V_D + \tfrac{5}{3}V_{L+I}) = 20.17 \text{ kips}$$

Assuming $0.8h$ is larger than d_p at the first critical section, the factored design shear is obtained from (flow chart, Fig. 6.13):

$$v_u = \frac{V_u}{b_w d_p} = \frac{V_u}{b_w \times 0.8h} = \frac{20,170}{12 \times 0.8 \times 21} \simeq 100 \text{ psi}$$

The shear resistance v_c of the concrete is at least equal to $2\sqrt{f'_c} \simeq 154.9$ psi. As $v_u = 100$ psi is less than $\phi v_c = 131.7$ psi, and as we have a slab, no shear reinforcement is needed at this section. A similar conclusion is drawn for other sections.

Edge beam. According to AASHTO an edge beam must be provided for slabs. Let us assume that the edge beam is made of the last 3 ft width of slab cast monolithically with the sidewalk. Its depth is assumed equal to $h = 21 + 11 = 32$ in. The corresponding weight of the edge beam is $w_G = 1.2$ klf and the dead load moment is given by:

$$M_G = 1.2 \frac{50^2}{8} = 375 \text{ kips-ft}$$

The edge beam should resist a live load moment at least equal to:

$$M_L = 0.10Pl = 0.10 \times 50 \times 16 = 80 \text{ kips-ft}$$

This live load moment is about equal to that generated by a live load of 85 psf on the sidewalk. However, it is smaller than the live load moment of 57.857 kips-ft per foot width for which the rest of the slab is designed. As a truck may accidentally ride on the sidewalk and as an "exterior" beam should not have less capacity than an "interior" beam, the live load moment on the edge beam is taken as equal to:

$$M_{L+I} = 3 \times 57.857 = 173.57 \text{ kips-ft}$$

Hence, the maximum service moment is:

$$M_{\max} = M_G + M_L = 548.57 \text{ kips-ft}$$

The corresponding value of the required prestressing force is obtained from stress condition IV of Table 4.2, assuming $e_o = h/2 - 3 = 13$ in, that is:

$$F = \frac{M_{\max} + \bar{\sigma}_{ts} Z_b}{e_o - k_t}$$

$$= \frac{548.57 \times 12,000 - 232 \times 6144}{13 + 5.33} = 281,366 \text{ lb} \simeq 281 \text{ kips}$$

It can be largely achieved using two bars of the same type as used for the rest of the slab. Other requirements related to allowable stresses, ultimate strength, and shear resistance of the edge beam are also found satisfactory.

Transverse reinforcement. Reinforcement transverse to the main reinforcement must also be provided in slabs. According to Section 14.6, 4. Concrete Slabs, case b, the area of transverse reinforcement should correspond to a percentage of the area of the main reinforcement equal to:

$$\frac{100}{\sqrt{l}} = \frac{100}{\sqrt{50}} = 14.14\%$$

Hence, we should provide $0.1414 \times 1.58 \simeq 0.223$ in^2 of prestressing steel per foot width of slab in the transverse direction. This can be achieved by using a Dywidag bar, $\frac{5}{8}$ in in diameter, every 15 in. Its cross-sectional area is 0.28 in^2.

If we do not want to prestress the slab in the transverse direction, the required area of conventional grade 60 reinforcing bars per foot of slab would be:

$$0.223 \frac{f_{pu}}{f_y} = 0.223 \times \frac{160}{60} = 0.594 \text{ in}^2$$

It can be achieved using a #7 bar every 12 inches or a #8 bar every 15 inches.

It can also be shown, for this example, that deflection limitations are largely satisfied. Note that, under the effect of sustained loading and final prestressing force, the slab is almost uniformly compressed. This essentially eliminates long-term camber or deflection problems.

2. Adjacent Precast Pretensioned Box Beams

According to the quick dimensioning rules suggested in Table 14.3, the depth of a simple span box beam can be estimated from $h/l \simeq 0.045$ or $h \simeq 27$ in. Referring to Fig. 14.5, where standard precast prestressed box beams are described, we find that beam BI-36 has a depth

$h = 27$ in, weighs 0.584 klf, and is 3 ft wide. Its geometric properties are: $A_c = 561$ in^2, $y_b = 13.35$ in, $y_t = 13.65$ in, $Z_b = 3770$ in^3, $Z_t = 3687$ in^3, $k_t = -6.72$ in, $k_b = 6.57$ in, $I_g = 50{,}334$ in^4. Ten such beams can be placed adjacent to each other, totalling 30 ft width, and can form the bridge deck. Let us check for a typical interior beam if the design is acceptable.

Let us first determine the moment due to live load plus impact. Using Eq. (14.5), the wheel load distribution factor is given by:

$$DF = \frac{(12n_L + 9)/n_B}{5 + n_L/10 + (3 - 2N_L/7)(1 - C/3)^2}$$

in which $C = KW/l = 1 \times 30/50 = 0.6 < 3$.

Hence:

$$DF = \frac{(12 \times 2 + 9)/10}{5 + 2/10 + (3 - 4/7)(1 - 0.6/3)^2} \simeq 0.4886$$

The maximum moment on the beam, due to live load plus impact, is given by Eq. (14.21) in which M_{lane} is obtained from Table 4.2, case A:

$$M_{L+I} = (1 + I)DF \, \frac{M_{\text{lane}}}{2}$$

$$= 1.2857 \times 0.4886 \times \frac{16}{2}\left(\frac{9}{8} \, 50 + \frac{24.5}{50} - 17.50\right)$$

$$\simeq 197.20 \text{ kips-ft}$$

The minimum and maximum service moments for the beam are given by:

$$M_{\text{min}} = M_G = 0.584 \, \frac{50^2}{8} = 182.5 \text{ kips-ft}$$

$$M_{\text{max}} = M_G + M_{SD} + M_{L+I} = 182.5 + 3 \times 0.025 \, \frac{50^2}{8} + 197.20$$

$$= 403.1375 \text{ kips-ft}$$

Prestressing force. Assuming a value of $(e_o)_{mp} = y_b - 3 = 10.35$ in and using stress condition IV of Table 4.2 for $e_o = (e_o)_{mp}$ leads to:

$$F = \frac{M_{\text{max}} + \bar{\sigma}_{ts} Z_b}{e_o - k_t}$$

$$= \frac{403.1875 \times 12{,}000 - 232 \times 3770}{10.35 + 6.72} = 232{,}197 \text{ lb} = 232.2 \text{ kips}$$

Let us assume that the tendons consist of $\frac{1}{2}$-in diameter strands with $f_{pu} = 270$ ksi, $f_{pe} = 145$ ksi, and an area per strand equal 0.153 in^2. We would need:

$$N = \frac{F}{0.153 \times f_{pe}} = 10.5 \simeq 11 \text{ strands}$$

Thus, the final prestressing force provided is equal to:

$$F = 11 \times 0.153 \times 145 = 244.035 \text{ kips}$$

Assuming a value of $\eta = 0.83$ it can easily be shown that the other three stress conditions of Table 4.2 are largely satisfied.

Ultimate moment. The strength design moment using AASHTO is given by:

$$M_u = 1.3(205.9375 + \tfrac{5}{3} \times 197.25) = 695.09 \text{ kips-ft}$$

Referring to the flow chart, Fig. 5.13, let us determine the nominal moment resistance of the beam:

$$\rho_p = \frac{A_{ps}}{bd_p} = \frac{11 \times 0.153}{36 \times 24} = 0.00195$$

$$f_{ps} = 270\left(1 - 0.5\rho_p \frac{270}{6}\right) = 258.16 \text{ ksi}$$

$$q = \omega_p = \rho_p \frac{f_{ps}}{f'_c} = 0.0839$$

$$\frac{1.18qd_p}{\beta_1} = \frac{1.18 \times 0.0839 \times 24}{0.75} = 3.168 \text{ in} < h_f = 5.5 \text{ in}$$

Hence, we have rectangular section behavior. As $q < 0.30$, the design nominal moment resistance is given by:

$$\phi M_n = 0.90 \times 6 \times 36 \times 24^2 \times 0.0839(1 - 0.59 \times 0.0839)$$

$$= 8929.6 \text{ kips-in} = 744.1 \text{ kips-ft}$$

Thus:

$$M_u = 695.09 < \phi M_n = 744.1 \text{ kips-ft} \qquad\qquad \text{OK}$$

Let us determine the cracking moment, using Eq. (4.38):

$$M_{cr} = F(e_o - k_t) - f_r Z_b$$

$$= 244.035(10.35 + 6.72) + 0.581 \times 3770$$

$$= 6356 \text{ kips-in} = 529.67 \text{ kips-ft}$$

Hence

$$\phi M_n = 743.3 > 1.2M_{cr} = 635.6 \text{ kips-ft} \qquad\qquad \text{OK}$$

Shear. Let us check shear at the first critical section assumed at $x = 1.5$ ft from the center of the support. The shear due to dead load is given by:

$$V_D = V_G + V_{SD} = (0.584 + 3 \times 0.025)\left(\frac{l}{2} - x\right) = 15.486 \text{ kips}$$

The shear due to live load plus impact is given by Eq. (14.22) in which V_{lane} is obtained from Table 14.2, case A:

$$V_{L+I} = (1 + I)DF \frac{P}{2}\left[4.5\left(1 - \frac{x}{l}\right) - \frac{42}{l}\right]$$

$$\simeq 17.715 \text{ kips}$$

The AASHTO specified strength design shear is given by:

$$V_u = 1.3(15.486 + \tfrac{5}{3} \times 17.715) \simeq 58.514 \text{ kips}$$

It leads to a design shear stress

$$v_u = \frac{V_u}{b_w d_p} = \frac{58,514}{10 \times 24} = 243.8 \text{ psi}$$

Referring to the flow chart, Fig. 6.13, let us determine the concrete contribution v_c using the conservative method, that is:

$$v_c = 0.6\sqrt{f'_c} + 700 \frac{V_u d_p}{M_u}$$

in which $V_u d_p/M_u$ is taken at $x = 1.5$ ft. It can be shown that $V_u d_p/M_u$ is larger than 1, hence use 1. Thus:

$$v_c = 0.6\sqrt{f'_c} + 700 = 746.5$$

but v_c is limited by $5\sqrt{f'_c} = 387.3$ psi, hence use $v_c = 387.3$ psi.

As v_u is less than $\phi v_c = 329.2$ psi, a minimum shear reinforcement is required. It is found that a #3 U-shaped stirrup at a spacing $s = 18$ in is sufficient. Note that the more elaborate method to calculate v_c also leads to minimum shear reinforcement.

Note that the cross section of this beam is more than sufficient to resist both bending and shear for the span considered. In revising the design, a smaller depth such as $h \simeq 24$ in could be considered.

14.12 OTHER DESIGN CONSIDERATIONS

In the preliminary design of bridge decks, an idealized representation is generally assumed to speed up the analysis and quickly arrive at an acceptable solution. However, several additional details that require special attention have to be integrated in the final design. They include the design of diaphragms, bearing pads or expansion joints, joints between consecutive spans or between spans and abutments, waterproofing, transverse sloping or crowning of the deck to allow for drainage, and the like. Some information on diaphragms and bearing pads is given next.

Diaphragms are thin reinforced or prestressed walls placed transversely to the beams of bridge decks to improve lateral rigidity and ensure good lateral load distribution. The AASHTO specifications recommend to place diaphragms between the girders at span ends and within the span at intervals not exceeding 40 ft (12.19 m) for T-beam constructions and 60 ft (18.29 m) for box-beam constructions. Diaphragms may be omitted where tests or structural analysis show adequate strength. Diaphragm spacing for curved beams shall be given special consideration.

Bearing pads are generally used at the supports to transmit and distribute vertical reactions. In addition, they must accommodate (1) changes in the length of the span due to variations in temperature and (2) end rotations. A number of suitable materials are available for use in bearing pads, among which are several elastomeric (neoprene) materials. They are commonly characterized by a relatively small shear modulus in comparison to their elastic modulus. Sections 12 of Division I and 25.2 of Division II of the AASHTO specifications deal with structural grade neoprene pads. For high compressive stresses and large horizontal deformations, laminated pads made of various layers of elastomer bonded between steel sheets can be used. The acceptable shear deformation is a function of the thickness or number of layers of the total assembly. Typical information on a commercially available bearing pad, satisfying the AASHTO specifications, and an example of bearing pad selection are given in App. D.

14.13 FORECAST OF BRIDGE ENGINEERING

A study was undertaken to forecast the nature of bridge engineering and construction for the years 1980 to 2000 (Ref. 14.29). Some of the findings specifically related to prestressed concrete are summarized below:

1. There will be more standardization for short- and medium-span bridges and prestressed concrete will be used more frequently than reinforced concrete. For longer spans, there will be growing use of prestressed concrete segmental box-girder bridges and cable-stayed bridges. In industrialized countries, including the United States, there will be increased concern for aesthetics and environmental harmony.
2. As to materials, higher strength-to-weight ratios will be utilized. This means higher-strength concretes and higher-strength steels. More durable concrete decks (possibly polymerized) and more corrosion resistant tendons (possibly nonmetallic) are expected. Efforts at using recycled materials will increase.
3. The maintenance, rehabilitation, revitalization, or replacement of old bridges will assume a significant percentage of bridge work. New materials and new methods for repair work will emerge. Old bridges will be subjected to rigorous inspection.
4. Electronic computers will penetrate farther into almost every aspect of bridge analysis, design, construction, and management. Analysis and design procedures will become more exact and complex. Secondary factors such as temperature, fatigue, and creep will be increasingly assessed. Instrumentation and automatic recording equipment to determine more exactly actual loads, stresses, deformations, cracking, and their evolution with time will be commonly used. There will be increased use of load factor design, probabilistic design, and optimum design.
5. The cost of construction will rise dramatically in both the material and labor sectors. There will be more prefabrication, better and more automatic construction equipment, and increased simplification of construction procedures.

The above findings suggest that design will be simultaneously more global and more refined. Although very powerful analytic tools will be available, one has to recognize that there are inherent idealizations in the design assumptions and inherent defects in the materials used and in the structures built. Hence, the following advice is offered: "It does no good to design to a level of refinement that cannot tolerate inherent materials and structural defects" (Ref. 14.29).

REFERENCES

14.1. *Precast Prestressed Concrete Short Span Bridges—Spans to 100 Feet*, 2d ed., Prestressed Concrete Institute, Chicago, Illinois, 1981.

14.2. R. Tokerud: "Precast Prestressed Concrete Bridges for Low-Volume Roads," *PCI Journal*, vol. 24, no. 4, July/August 1979, pp. 42–56.

14.3. F. J. Jacques: "Study of Long Span Prestressed Concrete Bridge Girders," *PCI Journal*, vol. 16, no. 2, March/April 1971, pp. 24–42.

14.4. J. Muller: "Ten Years of Experience in Precast Segmental Construction," *PCI Journal*, vol. 20, no. 1, January/February 1975, pp. 28–61.

14.5. B. Gentillini and L. Gentillini: "Precast Prestressed Segmental Elevated Urban Motorway in Italy," *PCI Journal*, vol. 20, no. 5, September/October 1975, pp. 26–43.

14.6. Editor: "Quebec's Grand-Mère Bridge—935-Ft Long Post-Tensioned Segmental Structure," *PCI Journal*, vol. 24, no. 1, January/February 1979, pp. 94–99.

14.7. A. A. Yee: "Record Span Box Girder Bridge Connects Pacific Islands," *Concrete International*, vol. 1, no. 6, June 1979, pp. 22–25.

14.8. A. Grant: "The Pasco-Kennewick Intercity Bridge," *PCI Journal*, vol. 24, no. 3, May/June 1979, pp. 90–109.

14.9. "Ruck-A-Chucky Bridge," First Prize of the 26th Progressive Architecture Award, *Progressive Architecture*, January 1979, pp. 68–69.

14.10. W. Podolny Jr.: "An Overview of Precast Prestressed Segmental Bridges," *PCI Journal*, vol. 24, no. 1, January/February 1979, pp. 56–87.

14.11. J. R. Libby: "Segmental Box Girder Bridge Superstructure Design," *ACI Journal*, vol. 73, no. 5, May 1976, pp. 279–290.

14.12. *Precast Segmental Box Girder Bridge Manual*, joint publication by the Post-Tensioning Institute and the Prestressed Concrete Institute, Chicago, Illinois, 1978, 116 pp.

14.13. *Post-Tensioned Box Girder Bridges Design and Construction*, joint publication by the Concrete Reinforcing Steel Institute and the Prestressed Concrete Institute, Chicago, Illinois, 1971, 107 pp.

14.14. G. C. Gerwick Jr.: "Prestressed Concrete Developments in Japan," *PCI Journal*, vol. 23, no. 6, November/December 1978, pp. 66–76.

14.15. R. J. Wheen: "The Rip Bridge—A Unique Australian Structure," *Concrete International*, vol. 1, no. 11, November 1979, pp. 12–15.

14.16. T. Y. Lin and F. Kulka: "Construction of Rio Colorado Bridge," *PCI Journal*, vol. 18, no. 6, November/December 1973, pp. 92–101.

14.17. H. Matsushita and M. Sato: "The Hayashi-No-Mine Prestressed Bridge," *PCI Journal*, vol. 24, no. 2, March/April 1979, pp. 90–109.

14.18. E. Loh: "Comments on Construction of Rio Colorado Bridge, by T. Y. Lin and F. Kulka," *PCI Journal*, vol. 19, no. 2, March/April 1974, pp. 131–133.

14.19. *Standard Specifications for Highway Bridges*, 12th ed., American Association of State Highway and Transportation Officials (AASHTO), Washington, D.C., 1977, 496 pp.

14.20. *Manual For Railway Engineering*, American Railway Engineering Association (AREA), Washington, D.C., 1973.

14.21. ACI Committee 343: "Analysis and Design of Reinforced Concrete Bridge Structures," Report 343R-77, American Concrete Institute, Detroit, Michigan, 1977, 118 pp.

14.22. D. Motarjemi and D. A. Van Horn: "Theoretical Analysis of Load Distribution in Prestressed Concrete Box Beam Bridges," Lehigh University, Fritz Engineering Laboratory, Report no. 315.9, October 1969.

14.23. K. J. William and A. C. Scordelis: "Computer Program for Cellular Structures of Arbitrary Plan Geometry," Report No. UC SESM 70-10, University of California, Berkeley, 1970.

14.24. A. E. Cusens and Y. C. Loo: "Applications of the Finite Strip Method in the Analysis of Concrete Box Bridges," Proceedings of the Institution of Civil Engineers, vol. 57, pt 2, Research and Theory, June 1974, pp. 251–273.

14.25. J. R. Libby and N. D. Perkins: *Modern Prestressed Concrete Highway Bridge Superstructures: Design Principles and Construction Methods*, Grantville Publishing Co., San Diego, 1976, 254 pp.

14.26. C. L. Freyermuth: "Computer Program for Analysis and Design of Simple-Span Precast Prestressed Highway or Railway Bridges," *PCI Journal*, vol. 13, no. 3, June 1968, pp. 28–39.

14.27. A. E. Naaman: "A Computer Program for Selection and Design of Simple Span Prestressed Concrete Highway Girders," *PCI Journal*, vol. 17, no. 1, January/February 1972, pp. 73–81.

14.28. ―――― and M. L. Silver: "Minimum Cost Design of Elevated Transit Structures," *Journal of the Construction Division, ASCE*, vol. 102, no. C01, March 1976, pp. 99–110.

14.29. W. Zuk: "A Forecast of Bridge Engineering, 1980–2000," Virginia Highway and Transportation Research Council, Report no. 79-R55, Charlottesville, Virginia, June 1979, 57 pp.

PROBLEMS

14.1 Drawing on information available in the technical literature, describe the construction sequence of a recently built prestressed concrete bridge. Point out problems that were encountered, how they influenced the design, and how they were resolved.

14.2 Determine the maximum moment due to live load plus impact on the interior beam (shaded area) of a box bridge (Fig. P14.2) assuming a simple span $l = 120$ ft and HS20 loading.

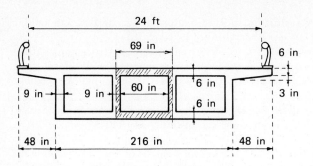

Figure P14.2

14.3 Going back to the example problem, Sec. 14.9, 2. Two-Span Continuous Bridge, determine the maximum positive moment in the span due to both truck and lane loadings and show that the truck loading governs.

14.4 Referring to the example problem in Sec. 14.11, 2. Adjacent Precast Pretensioned Box Beams, where adjacent precast pretensioned box beams are used, repeat the problem assuming a span $l = 70$ ft and everything else the same.

14.5 Referring to the design chart, Fig. 14.32, justify the feasibility of using AASHTO type IV beams for a 90-ft-span simply supported composite bridge deck. Use the following variables: $S = 8$ ft, slab thickness = 8 in, $f'_c = 5000$ for the precast prestressed beams. Check, not only the value of the prestressing force, but also ultimate strength requirements and shear. Make any other reasonable assumptions when needed.

LIST OF SYMBOLS

NOTATION: ENGLISH LETTERS

a depth of equivalent rectangular stress block ($= \beta_1 c$ if ACI code)

a parameter, distance, coefficient

A area in general

A used as subscript for age, anchorage set, additional weight

A_c area of concrete at the cross section considered in general (depending on the particular case, it may be the net area, the gross area, or the transformed area)

A_{cc} area of concrete composite section

A_{co} area of concrete core of a spirally reinforced column measured to outside diameter of spiral

A_g gross area of concrete at the cross section considered

A_f area of forms

A_i area of part i of a section

A_l total area of longitudinal reinforcement to resist torsion

A_n net area of concrete at the cross section considered

A_{pf} area of prestressed reinforcement required to develop the ultimate compressive strength of the overhanging portions of the flange of a flanged section

A_{pi} cross-sectional area of ith tendon

A_{ps} area of prestressed reinforcement in tension zone

A_{pw} $A_{ps} - A_{pf}$, area of prestressed reinforcement associated with the web of a flanged member at nominal moment resistance

A_s area of nonprestressed tension reinforcement

A_{sp} area of spiral reinforcement

A_s' area of compression reinforcement

A_t transformed area of concrete at section considered (depending on the particular case it may represent a cracked or uncracked section)

A_t area of one leg of closed stirrups used as torsion reinforcement within distance s

A_v area of shear reinforcement within a distance s

A_{vf} area of shear-friction reinforcement

b used as a subscript to indicate "bottom fiber"

b width of compression face of member

b_e effective flange width

b_o perimeter of critical section for slabs or footings

b_{tr} transformed flange width

b_v width of cross section at contact surface being investigated for horizontal shear

b_w web width of a flanged member

c used as a subscript to describe "concrete" or "composite" section

c distance from extreme compression fiber to neutral axis

$c_1, c_2, \ldots, c_i$ various dimensions or distances

C resulting compressive force in the concrete section due to the prestressing force and applied external forces

C cross-sectional constant to define torsional properties

C used as a subscript to describe effect of creep

$C_C(t)$ creep coefficient at time t

C_{CU} ultimate creep coefficient, or creep coefficient at end of service life

C line   geometric lieu of the compressive force in a member

C_m factor relating actual moment diagram to equivalent uniform moment diagram

d distance from extreme compression fiber to combined centroid of tensile force when prestressed and nonprestressed tension reinforcement are used

d_b nominal diameter of bar

d_c diameter of core of spirally reinforced column

d_c concrete cover measured from the extreme tension fiber to centroid of reinforcement

d_p distance from extreme compression fiber to centroid of prestressing steel

d_{pc} distance from extreme compression fiber of cast-in-place slab to centroid of prestressing steel, in a composite beam

d_s distance from extreme compression fiber to centroid of nonprestressed tension reinforcement

d_s' distance from extreme compression fiber to centroid of compressive reinforcement

D dead loads or their internal moments and forces

DF distribution factor of wheel load or concentrated load

e base of napierian logarithms

e_c eccentricity of the C force in the concrete section measured from the centroid of the section

e_{cl} lower eccentricity limit of the C line

e_{cu} upper eccentricity limit of the C line

e_i eccentricity of ith tendon, or eccentricity of the steel at section i

e_o eccentricity of the prestressing force at the section considered measured from the centroid of the section

$e_o(x)$ eccentricity of the prestressing force at section x

e_{oA}, e_{oB} left and right support eccentricities of the prestressing steel in a typical span AB

$e_{oc}(x)$ eccentricity of the Zero-Load-C line at section x

e_{ol} lower eccentricity limit of the prestressing steel

e_{ou} upper eccentricity limit of the prestressing steel

$(e_o)_{mp}$ maximum practically feasible eccentricity

E load effects of earthquakes or their related internal moments and forces; modulus of elasticity, in general

E_c modulus of elasticity of concrete

$E_{ce}(t)$ effective or equivalent modulus of elasticity of concrete at time t

E_{ci} modulus of elasticity of concrete at time of initial prestress

E_o tangent modulus of elasticity measured at the origin of the stress-strain curve

E_m secant modulus measured at the maximum or peak stress

E_{ps} modulus of elasticity of prestressing steel

E_s modulus of elasticity of nonprestressed steel

EI flexural stiffness of compression members

ES elastic shortening

f used as stress in general, preferably for the steel and occasionally for concrete when a symbol is widely used

f_b stress on bottom fiber of concrete section (also σ_b)

f_{cgs} stress in the concrete at the centroid of prestressing steel

$f_{cgs}(t_i)$ stress in the concrete at the centroid of the prestressing steel at time t_i at section considered

f_{cr} stress range in the concrete

f_c' specified compressive strength of concrete

f'_{ci} compressive strength of concrete at time of initial prestress

f_{pe} effective stress in the prestressing steel, after losses, at section considered

f_{pi} initial stress in the prestressing steel at section considered

f_{pJ} stress in the prestressing steel just before transfer at end of jacking

f_{pp} proportional limit stress of the prestressing steel

f_{ps} calculated stress in prestressing steel at section considered and loading considered

$f_{ps}(t)$ stress in the prestressing steel at time t at section considered and for the loading considered

f_{pu} specified tensile strength of prestressing steel

f_{py} specified yield strength of prestressing steel

f_r modulus of rupture of concrete

f_{sr} stress range in the steel

f_s stress in the nonprestressed tensile reinforcement

f'_s stress in the compressive reinforcement

$\bar{f}_s$ allowable stress in the steel

f'_{tc} direct tensile strength (stress) of concrete

f_y specified yield strength of nonprestressed tensile reinforcement

f'_y specified yield strength of compressive reinforcement

F final or effective prestressing force (after all losses) at section considered

F used as a subscript to describe effect of friction

F_i initial prestressing force at time of transfer at section considered

F_J prestressing force at end of jacking

F_n tensile force in the prestressing steel at the nominal moment resistance of the section

g used as a subscript for center of gravity

$g(t)$ time function

G used as a subscript for gravity load or self-weight

h overall thickness or depth of member

h_c overall depth of composite member

h_f flange thickness of a flanged member

H relative humidity, percent; also used to describe loading due to earth pressure

H_u total horizontal shear force at the interface between the precast section and the cast-in-place slab of a composite beam

i used as a subscript to describe "initial" conditions or ith element

I effect of impact or impact coefficient; moment of inertia, in general

I_c moment of inertia of uncracked section resisting externally applied load (it represents the inertia of either the net or the gross section depending on the particular case)

I_{cc} moment of inertia of uncracked composite section

I_{cr} moment of inertia of cracked section (transformed to concrete)

I_e effective or equivalent moment of inertia for computation of deflections after cracking

I_g moment of inertia of gross concrete section about the centroidal axis, neglecting the reinforcement

I_{gc} gross moment of inertia of composite section

k effective length factor for compression members

k_b distance from centroid of concrete section to the lower (bottom) limit of central kern

k_r reduction factor of additional long-term deflection due to the presence of nonprestressed reinforcement

k_t distance from centroid of concrete section to the upper (top) limit of central kern

k'_b distance from centroid of concrete section to the lower (bottom) limit of the limit kern

k'_t distance from centroid of concrete section to the upper (top) limit of the limit kern

K wobble friction coefficient per unit length of prestressing steel

K flexural stiffness of member; moment per unit rotation

K_{CA} age at loading factor for creep

K_{CH} humidity correction factor for creep

K_{CS} shape and size factor for creep

K_{SH} humidity correction factor for shrinkage

K_{SS} shape and size factor for shrinkage

l span length of member generally center-to-center of supports

l used as a subscript for lifetime, or lower

l_a, l_b longer and shorter span of a slab panel

l_c height of column center-to-center of floors or roofs

l_d development length

l_i ith span of a continuous beam or one-way slab

l_n clear span measured face to face of supports

l_t transfer length

l_u unbraced length of column or column length between hinges

l_x, l_y spans in the x and y direction for a two-way slab system

l_1, l_2 span in the direction being analyzed for bending and span or width transverse to l_1 (used in the equivalent-frame method for two-way slabs), measured center to center of supports

L live loads or their internal moments and forces

m_A, m_B fixed end moments

$M, M(x)$ moment in general at section considered or at section x

M_a maximum absolute moment in member at stage deflection is computed

M_c maximum moment on section while acting as composite section

M_{cr} cracking moment

M_D moment due to dead load

M_F, $M_F(x)$	total moment due to prestressing at section considered or at section x
M_{FA}, M_{FB}	total moment due to prestressing at supports A and B of a typical span AB
M_G	bending moment due to self-weight of member
M_L	moment due to live load
M_{lane}	maximum moment due to a single live loading lane at section considered
M_{L+I}	moment due to live load plus impact at section considered
M_{max}	maximum bending moment at section considered under service load conditions
$(M)_{max}$	maximum absolute moment in a span due to truck loading
M_{min}	minimum bending moment at section considered under service load conditions
M_n	nominal moment resistance
M_{nf}	nominal moment resistance due to the overhanging portion of the flange of a T section
M_{nw}	nominal moment resistance due to the web of a T section
M_p	maximum moment on the precast prestressed section of a composite member
M_S	moment due to cast-in-place slab in a composite member at section considered
M_{SD}	moment due to superimposed dead load at section considered
M_u	strength design moment or factored moment at section considered
M_{um}	magnified factored moment
M_{uw}	strength design moment for the web of a T section
M_{u1}	value of smaller factored end moment on compression member, positive if member is bent in single curvature, negative if bent in double curvature
M_{u2}	value of larger factored end moment on compressive member, always positive
M_1, $M_1(x)$	primary moment due to prestressing in a continuous structure at section considered or at section x
M_2, $M_2(x)$	secondary moment due to prestressing in a continuous structure at section considered or at section x
M_{2A}, M_{2B}	secondary moments due to prestressing at supports A and B or at sections A and B
n	used as a subscript for nominal
n_c	modular ratio of concrete of cast-in-place slab to concrete of precast section of a composite beam
n_p	modular ratio E_{ps}/E_c
n_{pi}	initial modular ratio E_{ps}/E_{ci}
n_s	modular ratio E_s/E_c
N	axial load acting on member, preferably used for tension

N_{cr} tensile load leading to cracking of prestressed member

N_{dec} tensile load leading to decompression of prestressed member

N_n nominal resistance of prestressed member under axial tension

N_u factored design tensile load acting on member

p used as a subscript for prestressing; also used for pressure or percentage

P concentrated external load in general

P axial load acting on member, preferably used for compression

P_{cr} critical buckling load

P_m maximum cutoff compressive force on column allowed by code

P_n nominal axial load capacity, in general, at a given eccentricity

$P_{n,b}$ nominal axial load capacity at balanced conditions

$P_{n,o}$ nominal axial load capacity of compression member subject to pure compression

$P_{n,ot}$ nominal axial load capacity leading to zero tension on extreme fiber of column or wall

P_u factored axial compressive load at given eccentricity

PPR partial prestressing ratio

q $= \omega_p + \omega - \omega' =$ reinforcing index

q_b reinforcing index corresponding to balanced conditions

q_{max} $= 0.75 \times q_b =$ maximum recommended value of the reinforcing index

q_{min} minimum recommended value of the reinforcing index

q_w reinforcing index associated with the web of a flanged member, same as q except that b is replaced by b_w and A_{ps} by A_{pw}

Q first static moment with respect to centroid of the portion of section above the shear plane considered

Q concentrated external load

r radius of gyration of cross section $= \sqrt{I/A}$

R used as a subscript to describe effect of steel relaxation

R radius of circular, cylindrical, or curved element

s curvilinear abscissa; also used as a subscript for "steel" reinforcement

s spacing of stirrups, or ties, or bent-up bars in direction parallel to longitudinal reinforcement

s pitch of spiral reinforcement

S effect of cast-in-place slab in a composite beam

S used as a subscript to describe effect of shrinkage

S transverse spacing center-to-center of beams or girders in a deck or slab structure

S_n transverse clear spacing of beams or girders

SD superimposed dead load or its related internal moments and forces

t time

t torsional shear stress

t used as a subscript to indicate "top fiber" or "tension"

t wall thickness

t_A age at loading

t_c torsional shear stress contributed by concrete after cracking; also used when member is subjected to torsion alone

t_c^* torsional strength (stress) of concrete under combined torsion and flexural shear

t_{cr} torsional shear stress at cracking

t_i, t_j particular values of time, mostly used to define the beginning and the end of a time interval

t_l design lifetime of member

t_t time at transfer or at release of prestress

t_o time at jacking of prestressing steel

t_u factored design torsional strength (stress)

T tensile force in the steel; also used as subscript for total to describe cumulative effects; for temperature loading; for torque or torsional moment; for T section

T_c nominal torsional moment resistance contributed by concrete after cracking

T_c^* torsional shear strength of concrete under combined torsion and flexural shear

T_{cr} cracking torsional moment

T_{ip} tensile force in the ith layer of prestressing steel

T_{is} tensile or compressive force in the ith layer of reinforcing steel

T_n nominal torsional moment resistance of section

T_{nf} tensile force in the steel balancing the compression force in the overhanging portion of the flange of a T section at nominal moment capacity

T_{nw} nominal tensile force in the steel balancing the compressive force in the web of a T section at nominal moment capacity

T_s nominal torsional moment resistance contributed by torsion reinforcement

T_u factored design torsional moment at section considered

u used as subscript for "factored effects" or design specified values at ultimate capacity

U required strength to resist factored loads or related internal moments and forces

U unit cost; subscript c holds for concrete, f for forms, p for prestressing steel, and s for reinforcing steel

v shear stress in general

v_c permissible shear stress carried by concrete

v_c^* shear strength (stress) of concrete in presence of torsion

v_{ci} nominal shear strength (stress) provided by concrete when diagonal cracking results from combined shear and moment

v_{cw} nominal shear strength (stress) provided by concrete when diagonal cracking results from excessive principal tensile stresses in the web

v_n — nominal shear strength (stress) at section considered

v_{nh} — nominal horizontal shear strength (stress)

v_s — nominal shear strength (stress) provided by shear reinforcement

v_u — factored design shear strength (stress) at section considered

v_{uh} — factored horizontal shear strength (stress)

$V, V(x)$ — shear force in general at section considered or at section x

V_D — shear force due to unfactored dead load at section considered

V_{lane} — maximum shear force due to a single live loading lane at section considered

V_{L+I} — shear force due to unfactored live load plus impact at section considered

V_n — nominal shear strength (force) at section considered

V_p — vertical component of effective prestressing force at section considered

V_{SD} — unfactored shear force due to superimposed dead load

V_u — factored design shear force at section considered

w — unfactored load per unit length of beam or per unit area of slab

w_b — balanced load

w_D — dead load per unit length of beam or per unit area of slab

w_G — self-weight of member per unit length or per unit area

w_L — live load per unit length of beam or per unit area of slab

w_{nb} — nonbalanced load or unbalanced load

w_u — factored load per unit length of beam or per unit area of slab

W — weight; wind load or related internal moments and forces; crack width

x — abcissa along the x axis; also x represents, in general, an unknown

x — shorter overall dimension of rectangular part of cross section

x_1 — shorter center-to-center dimension of closed rectangular stirrup

X — abcissa of section of tendon beyond which the stress loss due to anchorage set is zero

y — ordinate along the y axis

y — longer overall dimension of rectangular part of cross section

y_b — distance from centroidal axis of section to extreme bottom fiber

y_{bc} — distance from centroidal axis of composite section to extreme bottom fiber

y_t — distance from centroidal axis of section to extreme top fiber

y_{tc} — distance from centroidal axis of composite section to extreme top fiber

y'_{tc} — distance from centroidal axis of composite section to extreme top fiber of the precast member

y_1 — longer center-to-center dimension of closed rectangular stirrup

z — ordinate along the z axis

Z objective function in an optimization problem

Z_b section modulus with respect to extreme bottom fiber $= I_c/y_b$

Z_{bc} section modulus with respect to extreme bottom fiber for a composite section $= I_{cc}/y_{bc}$

Z_t section modulus with respect to extreme top fiber $= I_c/y_t$

Z_{tc} section modulus with respect to extreme top fiber for a composite section $= I_{cc}/y_{tc}$

Z'_{tc} section modulus with respect to extreme top fiber of precast elements of a composite section $= I_{cc}/y'_{tc}$

NOTATION: GREEK LETTERS

α angle in general or factor in general

α total angular change of prestressing steel profile in radians between two points

α angle between inclined stirrups and longitudinal axis of member

β_c ratio of long side to short side of concentrated load, reaction, or column

β_d ratio of maximum factored dead load moment to maximum factored total load moment, always positive

β_1 factor used to define the depth of the equivalent rectangular stress block at ultimate as a function of the location of the neutral axis. According to ACI:
$\beta_1 = 0.85 f'_c$ for $f'_c \leq 4000$ psi
$\beta_1 = 0.85 - 5 \times 10^{-5}(f'_c - 4000)$ for $4000 \leq f'_c \leq 8000$ psi
$\beta_1 = 0.65$ for $f'_c \geq 8000$ psi

γ geometric efficiency; unit weight in general; factor in general

γ_c unit weight of concrete

γ_s unit weight of steel

γ_v fraction of unbalanced moment transferred by eccentricity of shear at slab-column connections

δ differential increase in variable considered

δ moment magnification factor for columns

δ amount of anchorage set or slippage

δ, δ_i sag in span considered or in span i

Δ deflection in general, positive for deflection and negative for camber

Δ difference or differential amount between two values of variable that follows the Δ

Δ_{add} additional long-term deflection

Δ_D deflection due to dead load

Δ_G deflection due to self-weight

Δ_L deflection due to live load

Δ_i	initial, instantaneous elastic deflection
Δ_u	life deflection for the sustained loading considered
$\Delta_1, \Delta_2, \Delta_3$	deflections at different times or loading stages
Δf_{pA}	total stress loss in the prestressing steel due to anchorage set at section considered
$\Delta f_{pC}, \Delta f_{pC}(t_i, t_j)$	total stress loss in the prestressing steel during service life due to creep of concrete and stress loss during a time interval (t_i, t_j) at section considered
Δf_{pES}	total stress loss in the prestressing steel due to elastic shortening at time of transfer or release
Δf_{pF}	total stress loss in the prestressing steel due to friction at section considered
$\Delta f_{pR}, \Delta f_{pR}(t_i, t_j)$	total stress loss in the prestressing steel during service life due to relaxation of the tendons and stress loss during a time interval (t_i, t_j) at section considered
$\Delta f_{pS}, \Delta f_{pS}(t_i, t_j)$	total stress loss in the prestressing steel during service life due to shrinkage of concrete and stress loss during a time interval (t_i, t_j) at section considered
$\Delta f_{pT}, \Delta f_T(t_i, t_j)$	total stress loss in the prestressing steel during service life due to all sources of loss and stress loss during a time interval (t_i, t_j) at section considered
$\overline{\Delta f_{ps}}$	average stress loss in the prestressing steel
ΔM	moment amplitude $= M_{max} - M_{min}$ at section considered
ΔM_{cr}	moment in excess of self-weight moment, causing flexural cracking in the precompressed tensile fiber at section considered
ΔM_u	factored bending moment due to superimposed dead load plus live load at section considered
ΔV_u	factored shear force due to superimposed dead load plus live load at section considered
$\Delta \sigma$	differences between two stresses or stress amplitude
$\Delta \bar{\sigma}$	permissible stress amplitude
ε	strain in general
$\varepsilon_C(t)$	creep strain at time t
ε_{cb}	strain in concrete bottom fiber
ε_{ce}	concrete strain at the centroid of prestressing steel due to effective prestress
ε_{ci}	initial elastic instantaneous strain in concrete
ε_{ct}	strain in concrete top fiber; also used for tensile strain in concrete
ε_{cu}	strain in extreme compression fiber of concrete at nominal resistance of the section
ε_{CU}	ultimate creep strain or creep strain at end of life of member
ε_m	strain at maximum or peak stress of the stress-strain curve
ε_{pe}	strain in prestressing steel under effective stress f_{pe}

ε_{ps} strain in prestressing steel at section considered and loading considered

ε_{pu} ultimate failure strain of prestressing steel

ε_{py} yield strain of prestressing steel

ε_{SU} ultimate shrinkage strain or shrinkage strain at end of life of member

$\varepsilon_S(t)$ shrinkage strain at time t

ε_y strain at onset of yielding of reinforcing steel

η F/F_i ratio of final prestressing force to initial prestressing force or ratio of corresponding stresses

η torsional coefficient

θ angle in general

λ coefficient used in prestress losses

λ_i ith constant or parameter

μ curvature friction coefficient; also coefficient of friction or simply coefficient

ν Poisson's ratio

π 3.14159

ρ A_s/bd, ratio of non-prestressed tension reinforcement

ρ' A_s'/bd, ratio of compression reinforcement

ρ_b reinforcement ratio producing balanced condition

ρ_{min} minimum specified reinforcement ratio

ρ_{max} maximum specified reinforcement ratio

ρ_p A_{ps}/bd, ratio of prestressed reinforcement

ρ_s ratio of volume of spiral reinforcement to total volume of core of spirally reinforced compression member

σ stress in general; preferably used for concrete unless another symbol is widely accepted

σ_b stress on bottom fiber; bearing stress

$\sigma_{ci}, \bar{\sigma}_{ci}$ actual extreme fiber compressive stress in the concrete immediately after prestress transfer and code allowable limit

$\sigma_{cs}, \bar{\sigma}_{cs}$ actual extreme fiber compressive stress in the concrete at service loads and code allowable limit

$(\sigma_{cs})_{slab}, (\bar{\sigma}_{cs})_{slab}$ actual extreme fiber compressive stress in composite slab if any at service loads and code allowable limit

σ_g stress at the centroid of the concrete section due to the final prestressing force

σ_{gi} stress at the centroid of the concrete section due to the initial prestressing force

σ_m maximum or peak stress

σ_t stress on top fiber

$\sigma_{ti}, \bar{\sigma}_{ti}$ actual extreme fiber (initial) tensile stress in the concrete immediately after transfer and code allowable limit

$\sigma_{ts}, \bar{\sigma}_{ts}$ actual extreme fiber tensile stress in the concrete at service loads and code allowable limit

σ_x axial stress in the x direction

σ_y axial stress in the y direction

σ_1 principal tensile stress; also used for hoop stress

σ_2 principal compressive stress; also used for meridian stress

τ bond stress in general

$\bar{\tau}$ allowable bond or shear stress

Υ torsional constant

ϕ strength reduction factor

Φ curvature of section

ψ end restraint coefficient

ψ_m average value of restraint coefficient of column considered

ψ_1, ψ_2 restraint coefficients at extreme ends of column considered

ω $\rho f_y / f'_c$

ω' $\rho' f'_y / f'_c$

ω_p $\rho_p f_{ps} / f'_c$

$_w, \omega_{pw}, \omega'_w$ reinforcement indices for flanged sections computed as for ω, ω_p, and ω' except that b shall be the web width, and the steel area shall be that required to develop the compressive strength of the web only

ABBREVIATIONS:

cf   cubic foot

cgc   centroid of concrete section

cgs   centroid of the prestressing tendons or of the steel

cm   centimeter

ft   foot

in   inch

kips   kilopounds

kip-ft   kip × foot (unit of moment)

kip-in   kip × inch (unit of moment)

klf   kips per linear foot

kN/m^2 kilonewtons per square meter

ksi   kips per square inch

lb   pound (pounds)

m   meter

mm   millimeter

MPa   megapascal

N/mm^2 newtons per square millimeter

pcf   pounds per cubic foot

plf   pounds per linear foot

psi   pounds per square inch

psf   pounds per square foot

ABBREVIATIONS FOR PROFESSIONAL ORGANIZATIONS

AASHTO	American Association of State Highway and Transportation Officials
ACI	American Concrete Institute
ANSI	American National Standards Institute
ASCE	American Society of Civil Engineers
ASTM	American Society for Testing and Materials
CEB	Comité Européen du Béton
CRSI	Concrete Reinforcing Steel Institute
FIP	Fédération Internationale de la Précontrainte
PCA	Portland Cement Association
PCI	Prestressed Concrete Institute
PTI	Post-Tensioning Institute

SI UNIT CONVERSIONS AND
EQUIVALENT DESIGN EQUATIONS

CONVERSIONS TO SI UNITS

Length:

$1 \text{ in} = 25.4 \text{ mm}$
$1 \text{ ft} = 0.3048 \text{ m}$
$1 \text{ yd} = 0.914 \text{ m}$

Area:

$1 \text{ in}^2 = 6.452 \text{ cm}^2 = 645.2 \text{ mm}^2$
$1 \text{ ft}^2 = 0.0929 \text{ m}^2$

Volume:

$1 \text{ in}^3 = 16.39 \text{ cm}^3 = 16390 \text{ mm}^3$
$1 \text{ ft}^3 = 0.0283 \text{ m}^3$
$1 \text{ yd}^3 = 0.765 \text{ m}^3$
$1 \text{ oz} = 29.57 \text{ ml}$
$1 \text{ gal} = 3.785 \text{ l}$

Inertia:

$1 \text{ in}^4 \simeq 41.62 \text{ cm}^4 = 416{,}200 \text{ mm}^4$

Density:

$1 \text{ lb/ft}^3 = 16.03 \text{ kg/m}^3$

Unit weight:

$1 \text{ lb/ft}^3 = 0.1575 \text{ kN/m}^3$

Stress and modulus: $1 \ \text{lb/in}^2 = 1 \ \text{psi} \simeq 0.006895 \ \text{N/mm}^2$
$1 \ \text{kip/in}^2 = 1 \ \text{ksi} \simeq 6.895 \ \text{N/mm}^2 = 6.895 \ \text{MPa}$

Mass: $1 \ \text{lb} = 0.454 \ \text{kg}$
$1 \ \text{oz} = 28.35 \ \text{gr}$

Loads: $1 \ \text{lb} = 4.448 \ \text{N}$
$1 \ \text{kip} = 4.448 \ \text{kN}$
$1 \ \text{kip/ft} = 1 \ \text{klf} \simeq 14.59 \ \text{kN/m}$
$1 \ \text{lb/ft}^2 = 1 \ \text{psf} \simeq 0.0479 \ \text{kN/m}^2 = 47.9 \ \text{Pa}$
$1 \ \text{kip/ft}^2 = 1 \ \text{ksf} \simeq 47.9 \ \text{kN/m}^2$

Moment or torque: $1 \ \text{lb-ft} = 1.356 \ \text{N-m}$
$1 \ \text{lb-in} = 0.113 \ \text{N-m}$
$1 \ \text{kip-in} = 0.113 \ \text{kN-m}$
$1 \ \text{kip-ft} = 1.356 \ \text{kN-m}$

Temperature: $°C = (°F - 32)/1.8$

SI METRIC EQUIVALENT OF SOME DESIGN EQUATIONS NOT GIVEN IN TEXT

U.S. customary	SI—metric
Units: in	Units: mm
in^2	mm^2
psi	MPa
lb	N
$\sqrt{f'_c}$	$0.083 \sqrt{f'_c}$
$v_c = 2\sqrt{f'_c}$	$0.17 \sqrt{f'_c}$
$f_r = 7.5\sqrt{f'_c}$	$0.62 \sqrt{f'_c}$
$f_{ps} = f_{pe} + 10{,}000 + \dfrac{f'_c}{100\rho_p}$	$f_{pe} + 69 + \dfrac{f'_c}{100\rho_p}$
$(A_v)_{\min} = \dfrac{50 b_w s}{f_y}$	$0.35 \dfrac{b_w s}{f_y}$
$v_c = \left(2 + \dfrac{4}{\beta_c}\right)\sqrt{f'_c}$	$0.083\left(2 + \dfrac{4}{\beta_c}\right)\sqrt{f'_c}$
$v_c = 0.6\sqrt{f'_c} + 700\,\dfrac{V_u d_p}{M_u}$	$0.05\sqrt{f'_c} + 4.8\,\dfrac{V_u d_p}{M_u}$

$$v_{ci} = 0.6\sqrt{f'_c} + \frac{V_D + (\Delta V_u \times \Delta M_{cr}/\Delta M_u)}{b_w d_p} \quad \bigg| \quad 0.05\sqrt{f'_c} + \frac{V_D + (\Delta V_u \times \Delta M_{cr}/\Delta M_u)}{b_w d_p}$$

$$v_{cw} = 3.5\sqrt{f'_c} + 0.3\sigma_g + \frac{V_p}{b_w d_p} \quad \bigg| \quad 0.29\sqrt{f'_c} + 0.3\sigma_g + \frac{V_p}{b_w d_p}$$

$$A_l = \left\{\frac{400xs}{f_y}\left[\frac{T_u}{T_u + (V_u/3C_t)}\right] - 2A_t\right\}\frac{x_1 + y_1}{s} \quad \bigg| \quad \text{Replace “400” with “2.76”}$$

1000 PSI = 6.9 MPA

6.9 MPA = 70.36 kg/cm²

1 MPA = 10.1979 kg/cm².

1 kN/m³ ≈ 6.36 lb/ft³

1 kN/m² ≈ 21.43 lb/ft².

1 lb/ft³ = 16.03 kg/m³.

POSTTENSIONING SYSTEMS

THE FREYSSINET MONOGROUP K SYSTEM

(*Courtesy Freyssinet International, Inc.*)

The Monogroup System is the latest develpment of Freyssinet International. It offers a wide variety of prestressing forces within the range 130 to 1030 metric tons ultimate strength.

The Monogroup System is basically a combination of the extremely reliable and well-proven Monostrand method of anchoring with individual grips and the traditional simultaneous tensioning technique used in the multiwire and multi-strand Freyssinet Systems.

The advantage of such a combination is that a variety of very large, concentrated, low "pull-in" cables are available to the designer. These cables can be tensioned in one operation and are therefore suitable for the full range of prestressed structures, including the large tendons required by recent design for nuclear reactor pressure vessels and containment structures.

All tendons in the "K" system can either be premade and pulled into the sheath or the strands pushed one by one into the sheath, before or after concreting, to suit the construction sequence.

It is not necessary to produce a tendon of precise length, as the anchorages are easily assembled once the tendon is in position.

Strands can be omitted from the tendon sizes listed in order to achieve the greatest economy in design.

Figure C1 19K 15 Anchorage.

Figure C2 Gravelines Nuclear Power Station (France).

Table C1

Anchorage model	Force, kN (kips)			Cable weight		Sheath, i.d.		Jack model
	70%	80%	100% u.t.s.	kg/m	lb/ft	mm	in	
7K 13	899 (201)	1027 (230)	1,284 (287)	5.60	3.76	50	2	K-100
7K 15	1274 (285)	1456 (326)	1,820 (408)	7.91	5.31	65	$2\frac{1}{2}$	K-200
12K 13	1540 (345)	1761 (394)	2,201 (493)	9.60	6.45	65	$2\frac{1}{2}$	K-200
12K 15	2183 (489)	2495 (559)	3,119 (698)	13.56	9.11	85	$3\frac{3}{8}$	K-350
19K 13	2440 (546)	2788 (624)	3,485 (780)	15.20	10.21	85	$3\frac{3}{8}$	K-350
19K 15	3457 (774)	3951 (885)	4,939 (1106)	21.47	14.42	95	$3\frac{3}{4}$	K-500
27K 13	3467 (776)	3962 (887)	4,953 (1109)	21.60	14.50	95	$3\frac{3}{4}$	K-500
37K 13	4750 (1064)	5430 (1216)	6,787 (1520)	29.60	19.90	110	$4\frac{3}{8}$	K-700
27K 15	4913 (1100)	5615 (1258)	7,019 (1572)	30.51	20.50	110	$4\frac{3}{8}$	K-700
37K 15	6733 (1508)	7695 (1723)	9,619 (2155)	41.81	28.09	130	$5\frac{1}{4}$	K-1000
55K 13	7623 (1707)	8712 (1951)	10,890 (2439)	44.00	29.60	130	$5\frac{1}{4}$	K-1000

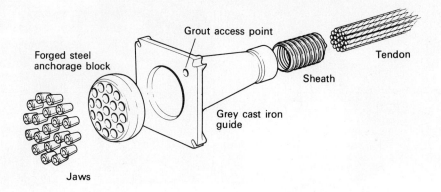

Forged steel anchorage block

Grout access point

Tendon

Sheath

Grey cast iron guide

Jaws

Figure C3 An exploded view of the 'K' range anchorage system.

Figure C4 A 'K' range anchorage.

Tendon Forces

The "K" range is designed for strands up to the following maximum breaking loads:

13 mm (0.5 in) 18.7 tonnes; 184 kN; 41.3 kips
15 mm (0.6 in) 26.5 tonnes; 260 kN; 58.6 kips

However there are corresponding anchorages for 18 mm and "drawn" strands available in those countries where these strands are marketed.

Table C2 Tendon forces

Tendon service force 70% of ultimate			Ultimate tendon force			Tendon size	System
Tonnes	kN	Kips	Tonnes	kN	Kips		
158	1545	347	225	2,208	496	12K 13 mm	K-200
223	2190	492	319	3,130	703	12K 15 mm	K-350
250	2450	550	357	3,500	786	19K 13 mm	
354	3470	780	505	4,960	1113	19K 15 mm	K-500
355	3478	782	507	4,970	1117	27K 13 mm	
488	4780	1075	697	6,830	1535	37K 13 mm	K-700
503	4910	1107	718	7,020	1582	27K 15 mm	
689	6750	1520	984	9,650	2170	37K 15 mm	K-1000
722	7080	1590	1030	10,100	2275	55K 13 mm	

Anchorages

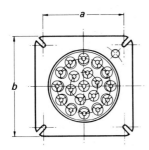

If it is necessary to reduce bearing stresses to meet local regulations, the anchorage block can be seated on a specially designed spreader plate, sometimes preferred to an anchorage guide.

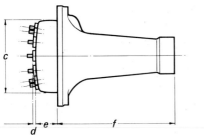

Figure C5.

Table C3 Anchorage dimensions

	12K 13	19K 13 12K 15	27K 13 19K 15	37K 13 27K 15	55K 13 37K 15
a	170	225	250	300	370
b	210	270	315	365	450
c	140	190	220	260	320
d	10	10	10	10	10
e	60	60	65	80	95
f	145	290	385	435	465
g	65	85	95	105	130
h	71	91	101	111	140

Dimensions in millimeters.

Detailing

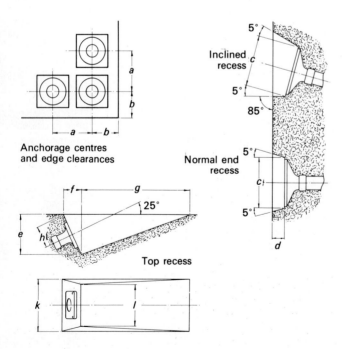

Anchorage centres
and edge clearances

Normal end
recess

Top recess

Figure C6.

Table C4 Dimensions for detailing

	12K 13	19K 13 12K 15	27K 13 19K 15	37K 13 27K 15	55K 13 37K 15
a	270	325	375	450	525
b	150	200	235	275	325
c	250	350	500	550	650
d	130	125	140	150	160
e	270	380	450	550	635
f	125	180	210	255	300
g	875	1000	1250	1500	1750
h	150	200	235	275	325
k	380	450	570	690	800
l	300	375	480	580	670

Dimensions in millimeters.

Couplers

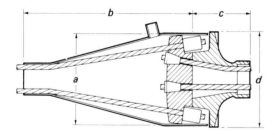

Figure C7.

Table C5 Couplers dimensions

	12K 13	19K 13 12K 15	27K 13 19K 15	37K 13 27K 15	55K 13 37K 15
a	223	253	308	Dimensions for	
b	425	490	631	larger couplers	
c	143	277	361	are available	
d	230	270	315	on request	

Dimensions in millimeters.

Blind Ends

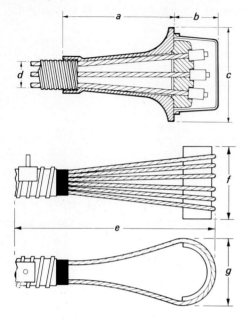

Figure C8.

Table C6 Dimensions for blind ends

	12K 13	19K 13 12K 15	27K 13 19K 15	37K 13 27K 15	55K 13 37K 15
Swaged					
a	145	290	385	435	
b	110	120	120	135	
c	210	270	315	365	
d	65	85	95	105	
Looped					
e	600	700			
f	250	250			
g	250	300			

Dimensions in millimeters.
Looped anchorages are practical only for the smaller tendons.
Details are given for the 12K 13 and 12K 15 systems.
For larger tendons the "swaged" capped anchorage should be used.

The " K " Range Jacks

The " K " range of jacks are center-hole rams of the hydraulic double-acting type with fixed cylinder and moving piston.

The attachment of the strand to the jack is by specially designed wide-angle, multiuse jaws, which are self-releasing on completion of jacking.

The system of ramming the permanent anchorages is by mechanical rubber springs which reduce the scatter of anchorage pullin valves to a minimum.

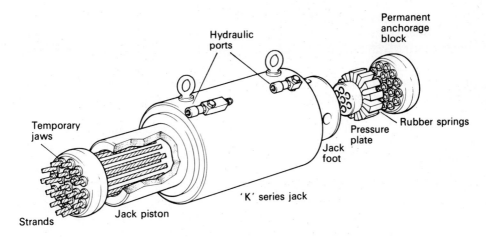

Figure C9 The 'K' range jack.

BBRV WIRE POSTTENSIONING SYSTEM

(*Courtesy Inryco, Inc., licensee of Bureau BBR Ltd., Switzerland.*)

BBRV Posttensioning was developed in 1949 by four Swiss engineers—Birkenmeier, Brandestini, Ros, and Vogt—whose initials give the system its name. Use of the BBRV system spread rapidly throughout the free world, and in the United States it is now one of the most widely used methods of posttensioning, where uncompromising reliability and stressing accuracy are required.

A BBRV tendon consists of several (or many) parallel lengths of $\frac{1}{4}$-in high-strength wire, with each end of each wire terminating in a cold-formed buttonhead after the wires pass separately through a machined anchorage fixture. The system provides for the simultaneous stressing of all the wires in a tendon, and the buttonheads allow development of the full actual ultimate strength of the prestressing steel.

Tendons are produced primarily for bonded installation, although they can be unbonded in certain applications. They can be furnished in practically any length and in a range of sizes and force capacities from 8 to 170 wires for building construction and heavy construction.

In the bonded or grout-type tendons, the wires are encased in flexible metal conduit. Wire tendons can be prefabricated in the factory as complete assemblies ready to be placed in the forms. Or the wire bundles can be pulled through the already placed conduit after concreting, a technique commonly used in bridge construction. The unbonded BBRV tendons are generally mastic-coated and wrapped in heavy paper, but in certain applications conduit-encased tendons are also used ungrouted. In these instances, the conduit is filled with grease after the tendon has been installed and stressed.

Features of the BBRV System

1. Buttonheads on every wire give positive, no-slip anchorage and eliminate seating losses, making it possible to develop the full actual ultimate strength of the prestressing wire.
2. Tendons are precisely engineered and completely shop fabricated, minimizing field labor and allowing more exact control of the quality and accuracy. Also, when using the "pull-through" method of installation, the wires are cut to exact length. One anchorage is attached in the factory, while the other anchorage is field installed using small and light field buttonheading equipment.
3. Long tendons present no problems in fabrication or in shipping, since they are furnished coiled on "lazy susan" racks 6 ft in diameter. This method of shipment also reduces job-site storage requirements, makes on-the-job handling of any length tendon easy, and minimizes exposure to transit damage.
4. Coupling of BBRV tendons is economically and easily accomplished with no loss of forces.

5. Curved or draped positioning of tendons within a structural member, often desirable from a design standpoint, is readily achieved because of the small diameter ($\frac{1}{4}$ in) of the steel elements.
6. Developed force can be closely matched to design requirements, with no waste of prestress steel, because BBRV tendons are available in a wide range of size-force units—with each single-wire increment adding seven kips of force.

BBRV TENDONS AND ANCHORS

The BBRV anchorages illustrated here are the standard types. But special types can be developed to meet unusual requirements and any anchorage can be produced with any number of wires (up to the hardware's maximum capacity) needed to produce a specified prestressing force.

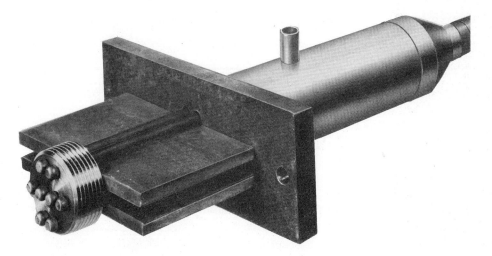

Figure C10 Type MG. Stressing anchor.

Prestressing Steel Properties

Table C7

Diameter, in	Ultimate strength, f_{pu}-ksi	Prestressing force, kips				Cross section area, in²	Weight, lb/ft
		f_{pu}	$0.80f_{pu}$	$0.70f_{pu}$	$0.60f_{pu}$		
0.25	240	11.78	9.43	8.25	7.07	0.0491	0.166

Figure C11 Type SG. Fixed anchor.

**Typical Stress Strain Curve for 0.250″ dia. High
Tensile Steel Wire — cold drawn, stress relieved**

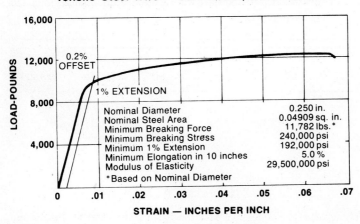

Nominal Diameter	0.250 in.
Nominal Steel Area	0.04909 sq. in.
Minimum Breaking Force	11,782 lbs.*
Minimum Breaking Stress	240,000 psi
Minimum 1% Extension	192,000 psi
Minimum Elongation in 10 inches	5.0 %
Modulus of Elasticity	29,500,000 psi

*Based on Nominal Diameter

**Figure C12 Typical stress strain curve for 0.250-in diameter high tensile steel wire—
cold drawn, stress relieved.**

STRESSING ANCHORS

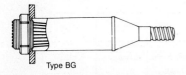

Type BG

**Figure C13 Type BG. Recessed anchor used where stressing pocket depth is limited.
Type BM also available.**

Table C8

Anchor designation		18 BG	24 BG	30 BG	38 BG	46 BG	52 BG
No. of wires (max)		18	24	30	38	46	52
Bearing plate size, in	A	$7\frac{1}{2} \times 7\frac{1}{2}$	$8\frac{1}{2} \times 8\frac{1}{2}$	$9\frac{1}{2} \times 9\frac{1}{2}$	$10\frac{1}{2} \times 10\frac{1}{2}$	$11\frac{1}{2} \times 11\frac{1}{2}$	$12\frac{1}{2} \times 12\frac{1}{4}$
	B	$6 \times 9\frac{1}{2}$	$7 \times 10\frac{1}{2}$	$8 \times 11\frac{1}{2}$	$9 \times 12\frac{1}{4}$	$10 \times 13\frac{1}{2}$	$11 \times 13\frac{1}{2}$
Trumpet o.d., in		$3\frac{3}{4}$	$4\frac{1}{4}$	5	$5\frac{1}{4}$	$5\frac{1}{2}$	$5\frac{3}{4}$
Conduit o.d., in		$1\frac{5}{8}$	$1\frac{7}{8}$	2	$2\frac{3}{8}$	$2\frac{5}{8}$	$2\frac{3}{4}$
Prestress force @ $0.6f_{su}$		127 kip	170 kip	212 kip	269 kip	325 kip	368 kip

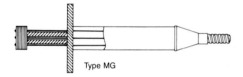

Type MG

Figure C14 Type MG. Nonrecessed type, preferred where there is no limitation on depth of stressing pocket.

Table C9

Anchor designation		18 MG	24 MG	30 MG	38 MG	46 MG
No. of wires (max)		18	24	30	38	46
Bearing plate size, in	A	5×10	$6 \times 11\frac{1}{2}$	$7 \times 12\frac{1}{2}$	$8 \times 13\frac{3}{4}$	$9 \times 14\frac{3}{4}$
	B	$6 \times 9\frac{3}{4}$	$7 \times 11\frac{1}{4}$	$8 \times 12\frac{1}{4}$	9×14	10×15
Trumpet o.d., in		3	4	$4\frac{1}{2}$	5	$5\frac{1}{2}$
Conduit o.d., in		$1\frac{5}{8}$	$1\frac{7}{8}$	2	$2\frac{3}{8}$	$2\frac{5}{8}$
Prestress force @ $0.6f_{su}$		127 kip	170 kip	212 kip	269 kip	325 kip

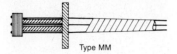
Type MM

Figure C15 Type MM. Nonrecessed type, preferred where there is no limitation on depth of stressing pocket.

Table C10

Anchor designation		18 MM	24 MM	30 MM	38 MM	46 MM
No. of wires (max)		18	24	30	38	46
Bearing plate size, in	A	5×10	$6 \times 11\frac{1}{2}$	$7 \times 12\frac{1}{2}$	$8 \times 13\frac{3}{4}$	$9 \times 14\frac{3}{4}$
	B	$6 \times 9\frac{3}{4}$	$7 \times 11\frac{1}{4}$	$8 \times 12\frac{1}{4}$	9×14	10×15
Bundle o.d., in		$1\frac{1}{2}$	$1\frac{3}{4}$	2	$2\frac{1}{8}$	$2\frac{1}{4}$
Prestress force @ $0.6f_{su}$		127 kip	170 kip	212 kip	269 kip	325 kip

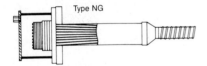
Type NG

Figure C16 Type NG.

Table C11 Data for type NG anchor

Anchor designation		90W	170W
No. of $\frac{1}{4}$-in wires (max)		90	170
Bearing plate size, in	Round	Diameter $18\frac{1}{2}$	Diameter $23\frac{1}{2}$
	Square	$16\frac{1}{2} \times 16\frac{1}{2}$	$20\frac{1}{2} \times 20\frac{1}{2}$
Trumpet o.d., in		6	7
Conduit o.d., in		$3\frac{15}{16}$	$4\frac{15}{16}$
Prestress force @ $0.6f_{su}$		636 kip	1201 kip

FIXED ANCHORS

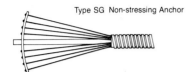

Type SG Non-stressing Anchor

Figure C17 Type SG. Nonstressing anchor.

Table C12 Data for type SG anchor

Anchor designation	18 SM	24 SM	30 SM	38 SM	46 SM	52 SM
No. of wires (max)	18	24	30	38	46	52
Bearing plate size, in	5 × 10	6 × 11½	7 × 12½	8 × 13¾	9 × 15	10 × 15
Prestress force @ $0.6f_{su}$	127 kip	170 kip	212 kip	269 kip	325 kip	368 kip

Type SM anchor unbonded installations is similar to Type SG.

STRESSING EQUIPMENT

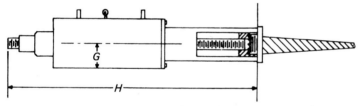

Figure C18 Jack arrangement when stressing nonrecessed anchors (Types MG and MM).

Table C13

Jack force capacity, tons	100	150	200	250	500	750
Wire capacity, number	21	32	42	53	106	160
Minimum clearance H, ft-in	5 ft	5 ft 6 in	5 ft 6 in	5 ft 6 in	6 ft	6 ft 8 in
Minimum clearance G, in	4½	5¼	6⅜	6⅜	9½	13
Jack stroke, in	10	12	12	10	8	8
Jack weight, lb	250	375	650	700	2800	4500

Total elongation must be added to minimum clearance (H) to allow for removal of jack after stressing.

DYWIDAG THREADBAR POSTTENSIONING SYSTEM

(*Courtesy Dywidag Systems International, Inc.*)

The components of Dywidag Threadbar System are manufactured in the United States exclusively by Dyckerhoff & Widmann, Inc. Used world-wide since 1965, the threadbar system provides a simple, rugged method of efficiently applying prestress force to a wide variety of structural systems including posttensioned concrete, rock and soil anchor systems.

Available in $\frac{5}{8}$-, 1-, $1\frac{1}{4}$- and $1\frac{3}{8}$-in nominal diameter, Dywidag threadbars are hot rolled and proof stressed alloy steel conforming to ASTM A 722-75.

The Dywidag threadbar prestressing steel has a continuous rolled-in pattern of threadlike deformations along its entire length. More durable than machined threads, the deformations allow anchorages and couplers to thread onto the threadbar at any point.

Exceeding the strength requirements of ACI 318-77, all Dywidag anchorages and couplers are designed to develop 100 percent of the guaranteed ultimate strength of the threadbar.

Conforming to the requirements of ASTM A 615-75, the threadbar deformations develop an effective bond with cement or resin grout. The continuous thread simplifies stressing. Lift-off readings may be taken at any time, and the prestress force increased or decreased as required.

The Dywidag Threadbar System is primarily used for grouted construction. All components of the system are designed to be fully integrated for quick and simple field assembly. Sheathing, sheathing transitions, grout sleeves, and grout tubes all feature thread-type connections.

Placing Dywidag tendons is simplified through the use of reusable plastic pocket formers. Used at each stressing end, the truncated, cone-shaped pocket former can extend through or butt up against the form bulkhead.

Available in mill lengths to 60 ft, threadbars may be cut to specified lengths before shipment to the job site. Or where circumstances warrant, the threadbars may be shipped to the job site in mill lengths for field cutting with a portable friction or band saw. Threadbars may be coupled for ease of handling or to extend a previously stressed bar.

Tendon Assembly with Bell Anchorage

Figure C19.

Prestressing Steel Properties

Table C14

Nominal threadbar diameter, in	Ultimate stress, f_{pu}-ksi	Cross section area, A_{ps}-in^2	Ultimate strength, $f_{pu}A_{ps}$	Prestressing force, kips			Weight,‡ lb/ft	Minimum elastic bending radius, ft
				$0.80f_{pu}A_{ps}$	$0.70f_{pu}A_{ps}$	$0.60f_{pu}A_{ps}$		
$\frac{5}{8}$	157	0.28	43.5	34.8	30.5	26.1	0.98	26
1	150	0.85	127.5	102.0	89.3	76.5	3.01	52
1	160†	0.85	136.0	108.8	95.2	81.6	3.01	49
$1\frac{1}{4}$	150	1.25	187.5	150.0	131.3	112.5	4.39	64
$1\frac{1}{4}$	160†	1.25	200.0	160.0	140.0	120.0	4.39	60
$1\frac{3}{8}$	150	1.58	237.0	189.6	165.9	142.2	5.56	72

† Check on availability before specifying.
‡ Shipping weight may vary.

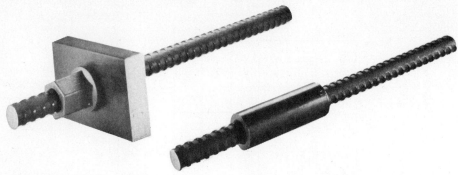

Figure C20 Plate anchorage.   **Figure C21 Coupling.**

DYWIDAG Posttensioning System Details

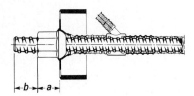

Figure C22 DYWIDAG bell anchorage.

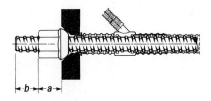

Figure C23 DYWIDAG plate anchorage.

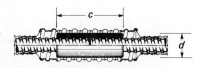

Figure C24 DYWIDAG coupler.

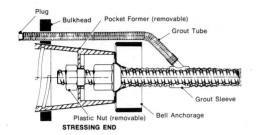

Plug
Bulkhead Pocket Former (removable)
Grout Tube
Grout Sleeve
Plastic Nut (removable) Bell Anchorage
STRESSING END

Figure C25 DYWIDAG tendon assembly.

Table C15 Anchorage details

Threadbar diameter, in		$\frac{5}{8}$	1	$1\frac{1}{4}$	$1\frac{3}{8}$
Bell anchor size, in		$3\frac{1}{4}\varnothing \times 1\frac{1}{2}$	$5\frac{1}{2}\varnothing \times 2\frac{5}{8}$	$6\frac{3}{4}\varnothing \times 2\frac{5}{8}$	$7\frac{3}{4}\varnothing \times 3\frac{1}{8}$
Anchor plate size,† in		$3 \times 3 \times \frac{3}{4}$ $2 \times 5 \times 1$	$5 \times 5\frac{1}{2} \times 1\frac{1}{4}$ $4 \times 6\frac{1}{2} \times 1\frac{1}{4}$	$6 \times 7 \times 1\frac{1}{2}$ $5 \times 8 \times 1\frac{1}{2}$	$7 \times 7\frac{1}{2} \times 1\frac{3}{4}$ $5 \times 9\frac{1}{2} \times 1\frac{3}{4}$
Nut extension, in	a	1	$1\frac{7}{8}$	$2\frac{1}{2}$	$2\frac{3}{4}$
Minimum bar protrusion, in	b	$2\frac{1}{2}$	3	$3\frac{1}{2}$	4

Table 16 Coupler details

Threadbar diameter, in		$\frac{5}{8}$	1	$1\frac{1}{4}$	$1\frac{3}{8}$
Length, in	c	$3\frac{1}{2}$	$5\frac{1}{2}$	$6\frac{3}{4}$	$8\frac{5}{8}$
Diameter, in	d	$1\frac{1}{8}$	2	$2\frac{3}{8}$	$2\frac{5}{8}$

Table C17 Sheathing details

Threadbar diameter, in	$\frac{5}{8}$	1	$1\frac{1}{4}$	$1\frac{3}{8}$
Threadbar sheathing, o.d., in	1	$1\frac{1}{2}$	$1\frac{3}{4}$	2
Threadbar sheathing, i.d., in	$\frac{3}{4}$	$1\frac{1}{4}$	$1\frac{1}{2}$	$1\frac{3}{4}$
Coupler sheathing, o.d., in	$1\frac{3}{4}$	$2\frac{3}{4}$	$3\frac{1}{4}$	$3\frac{3}{4}$
Coupler sheathing, i.d., in	$1\frac{3}{8}$	$2\frac{3}{8}$	$2\frac{7}{8}$	$3\frac{3}{8}$

Table C18 Pocket former details

Threadbar diameter, in	$\frac{5}{8}$	1	$1\frac{1}{4}$	$1\frac{3}{8}$
Length, in	$4\frac{3}{4}$	7	8	$8\frac{5}{8}$
Maximum diameter, in	$3\frac{1}{8}$	$5\frac{1}{8}$	$6\frac{1}{2}$	$6\frac{1}{2}$

† Other plate sizes available on the special order.
‡ To accommodate stressing.

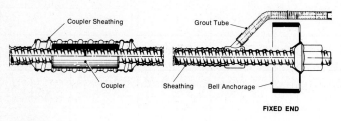

Figure C26

INRYCO CONA SINGLE-STRAND POSTTENSIONING SYSTEM

(Courtesy Inryco, Inc., Posttensioning Division)

The Cona posttensioning system adds significantly to the strength and design flexibility of concrete construction. For instance, when concrete is posttensioned:

(*a*) the depth of structural members can be reduced;
(*b*) the deflection of structural members can be controlled;
(*c*) cracks are virtually eliminated in slabs, making them virtually watertight;
(*d*) longer spans are made economically possible.

The Inryco Cona system is particularly economical in the construction of thin, one-way slabs, two-way flat plates and flat slabs, and topping slabs. As such, it is applicable to many types of structures, including high-rise office buildings, apartment complexes, parking ramps, and slabs on grade for industrial and residential construction.

Cona single-strand tendons employ cold-drawn, stress-relieved, seven-wire strands of 0.5- or 0.6-in diameter, conforming to ASTM A-416. They are normally used in unbonded construction, but can be used as bonded tendons where specifications require it.

Cona's small tendon diameters and small anchor plate dimensions meet the edge-size requirements of thin slabs and light structural shapes. The 0.5-in standard anchors can be used in slabs as thin as $3\frac{1}{2}$ in; the 0.6-in standard anchors will fit in a slab of $4\frac{1}{4}$-in thickness. When lower-strength concrete is used, requiring oversized anchorages, minimum slab thicknesses increase slightly.

The Cona wedge-anchor system exceeds code requirements for guaranteed ultimate capacity of the tendon, and seating losses are reduced to a minimum by the two-piece wedge design. Stressing equipment is small and light—easily handled by one man.

The installation of Cona single-strand tendons is simple. There are practically no length limitations, which eliminates the need to couple tendons at the site and, by stressing at intermediate points between partial slabs, tension can be applied separately to successive sections of the same continuous tendon.

There are no large pockets to form and patch. The Cona system provides plastic pocket formers as part of standard anchor hardware. Attached to formwork for quick positioning of anchors, they create small, clean stressing voids that are easily patched.

Cona single-strand tendons are "built" at the job site from coated and sheathed strands. Intermediate and fixed anchorages are installed in the factory; stressing anchorages are put on in the field. Wedges for fixed anchorages are hydraulically seated for positive grip in the factory.

Consult your Inryco representative for specific application recommendations.

Table C19 Prestressing steel properties

System	Diameter, in	Ultimate strength, f_{pu}-ksi	Prestressing force, kips				Cross section area, in^2	Weight, lb/ft
			f_{pu}	$0.80f_{pu}$	$0.70f_{pu}$	$0.60f_{pu}$		
Strand	0.5	270	41.3	33.1	28.9	24.8	0.153	0.52
	0.6	270	58.6	46.9	41.0	35.2	0.217	0.74

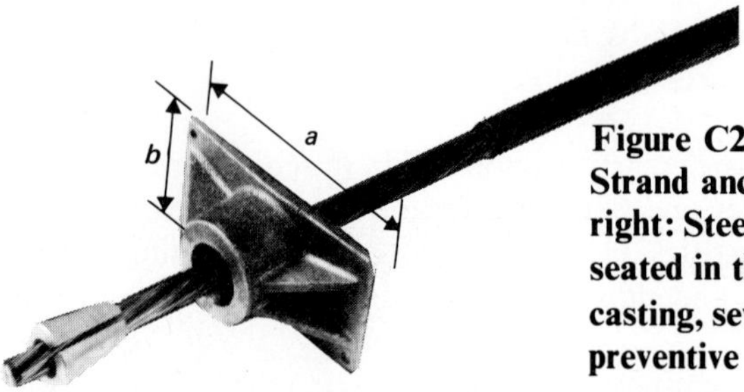

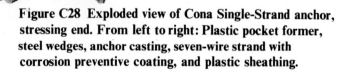

Figure C27 Exploded view of Cona Single-Strand anchor, fixed end. From left to right: Steel wedges—hydraulically power seated in the factory for firm grip, anchor casting, seven-wire strand with corrosion preventive coating, and plastic sheathing.

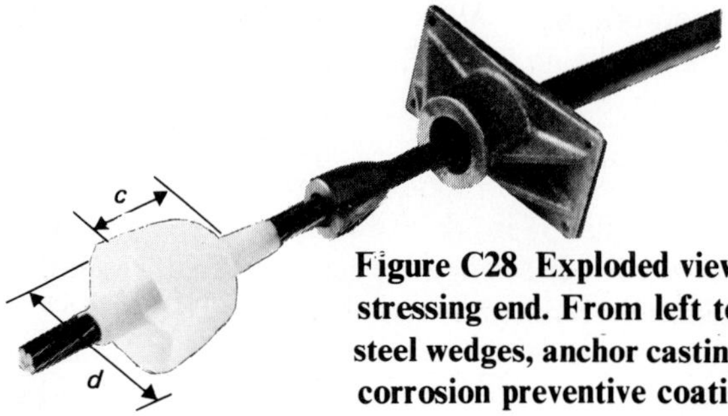

Figure C28 Exploded view of Cona Single-Strand anchor, stressing end. From left to right: Plastic pocket former, steel wedges, anchor casting, seven-wire strand with corrosion preventive coating, and plastic sheathing.

Table C20 Component dimensions

Strand	Dimensions			
	a	b	c	d
0.5 diameter standard	$4\frac{1}{2}$	$2\frac{1}{4}$	$1\frac{1}{2}$	$2\frac{1}{4}$
0.5 diameter oversized	$6\frac{1}{4}$	$2\frac{1}{4}$	$1\frac{1}{2}$	$2\frac{1}{4}$
0.6 diameter standard	$5\frac{3}{8}$	$2\frac{5}{8}$	$1\frac{1}{2}$	$2\frac{5}{8}$
0.6 diameter oversized	$7\frac{1}{8}$	$2\frac{5}{8}$	$1\frac{1}{2}$	$2\frac{5}{8}$

Dimensions in inches.

INRYCO CONA SINGLE-STRAND STRESSING EQUIPMENT

Open throat, twin ram Cona jack for stressing 0.5- or 0.6-in single strand is suitable for stressing at tendon ends or at intermediate tendon anchor.

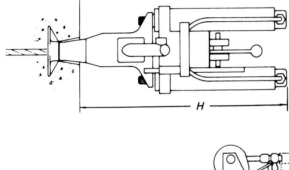

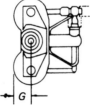

Figure C29.

Table C21

Jack force capacity, tons	17	26
Strand capacity, dia., in	0.5	0.6
Minimum clearance H, ft-in	2 ft 3 in	2 ft 3 in
Minimum clearance G, in	2	2
Jack stroke, in	6	$8\frac{1}{4}$
Jack weight, lb	77	87

For tendon elongation greater than the stroke of the jack, the wedges are temporarily seated after partial stressing and the jack is recycled.

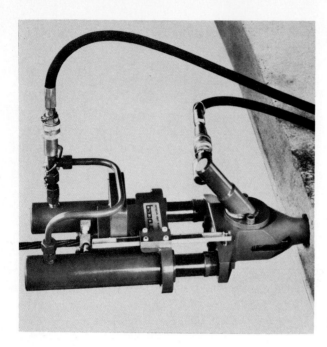

Figure C30.

INRYCO CONA MULTISTRAND POSTTENSIONING

(Courtesy Inryco, Inc., Posttensioning Division)

The Cona multistrand anchor system is most applicable to the posttensioning of heavy beams or structures requiring large-capacity tendons—cases in which a structure's irregularity dictates the use of tendons of varying size and length, or where accurate determination of length is difficult.

The system is made up of multiple units of seven-wire strands, each of which is 0.5 inches in diameter. Steel wedges anchor each strand separately and all strands of a given tendon are stressed simultaneously.

An important feature of the Cona multistrand system is the capability to hydraulically power-seat all wedges of the stressing anchor (WG), to assume a uniform initial force and to control seating loss.

Cona multistrand tendons are usually grouted, and are especially suited to the pass-through method of tendon installation. In this procedure, only empty rigid ducts are placed in the forms prior to concrete pouring. After pouring, site fabricated tendons are pulled through the ducts and are stressed and grouted in one continuous operation. The tendons can also be shop-fabricated, if desired.

Anchorages using 0.5-in-dia. strands are available in 7, 12, 19, 31 and 55 strand units. Consult your Incryo representative for specific recommendations.

Table C22 Cona Multistrand anchorage data, Type WG (stressing)

Anchor designation		7/0.5 WG	12/0.5 WG	19/0.5 WG	31/0.5 WG	55/0.5 WG
No. of 0.5-in strands (max.)		7	12	19	31	55
Bearing plate size, in		$8\frac{1}{2} \times 8\frac{1}{2}$	11×11	14×14	$17\frac{1}{2} \times 17\frac{1}{2}$	24×24
Trumpet o.d., in		$3\frac{1}{4}$	$4\frac{1}{2}$	$5\frac{3}{4}$	$7\frac{1}{2}$	10
Trumpet length, in		8	16	20	26	36
Conduit o.d., in	Pull-through	$2\frac{3}{8}$	$2\frac{3}{4}$	$3\frac{3}{8}$	$4\frac{3}{8}$	$5\frac{3}{4}$
	Assembled	2	$2\frac{3}{8}$	3	$3\frac{3}{4}$	$5\frac{1}{4}$

Type WG used as fixed anchorage for pull-through tendon installations. Wedges are pressed into anchorhead by special retaining plate.

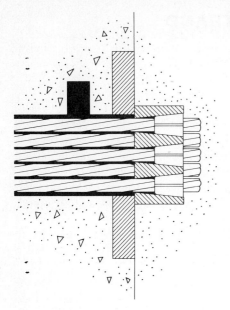

Figure C31.

Figure C32.

INRYCO CONA MULTISTRAND STRESSING EQUIPMENT

Center-hole hydraulic stressing jack is used to stress tendon at end projecting through bearing plate.

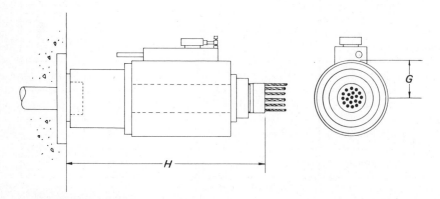

Figure C33.

Table C23

Jack force capacity, tons	115	225	340	550	1100
Strand capacity 0.5 in, number	7	12	19	31	55
Minimum clearance G, in	6	$7\frac{1}{4}$	$8\frac{1}{2}$	$11\frac{1}{2}$	14
Minimum clearance H, ft-in	3 ft	3 ft 4 in	4 ft 2 in	4 ft 4 in	4 ft 5 in
Jack stroke, in	12	12	16	16	12
Jack weight, lb	490	970	1700	3300	5500

Note: To engage and disengage jacking assembly, a space requirement of $2H$ is needed in a direct line with and perpendicular to the bearing plate. For tendon elongation greater than the stroke of the jack, the wedges are temporarily seated after partial stressing, and the jack is recycled.

Figure C34.

VSL MULTISTRAND SYSTEM ANCHORAGE COMPONENTS
(*Courtesy VSL Corporation*)

1. Introduction

Anchorage selection

The choice of anchorages is governed by their intended function, the requirements of the structure, and by the type and number of strands per tendon unit.

Like the tendon units, the anchorages are designated by the symbols 5-1 to 5-55 for 0.5-in-diameter strand, and 6-1 to 6-55 for 0.6-in-diameter strand. The number that follows the hyphen indicates the number of holes passing through the anchorage. It should be noted, however, that the number of strands in an anchorage unit does not have to correspond to the number of holes; in this way it is possible to provide tendons with initial forces up to 3000 kips by means of the same type of anchorage.

The range of standardized VSL anchorages and couplers is as follows:

Diameter of strand	Maximum number of strands per unit
0.5 in	1, 3, 4, 7, 12, 19, 22, 31, 55
0.6 in	1, 2, 3, 4, 7, 12, 19, 22, 31, 55

Stressing anchorages

The VSL stressing anchorages serve for the stressing and anchoring of the tendons. They are based upon the principle of individual gripping of each strand by means of wedges. The prestressing force is transmitted to the concrete through a rigid bearing plate. Mild reinforcement serves to contain the highly stressed concrete in the zone immediately behind the bearing plate. The bursting forces in the anchorage zone must be resisted by additional mild steel reinforcement designed as a function of the dimensions and arrangement of the bearing plate, and the dimensions of the end block. (Refer to Section 3.)

Stressing anchorages may also be used as fixed-end anchorages. When they are used in this way, the anchor head should remain accessible during stressing.

Fixed-end anchorages

There are various types of fixed-end anchorages; the choice will depend upon the characteristics of the structure, the position of the anchorage, and the profile and method of installation and stressing.

Couplers

Couplers serve either for connecting a new tendon to a stressed tendon or for extending an unstressed tendon.

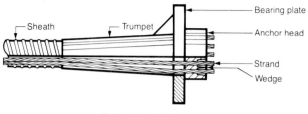

Typical Type E Anchorage

Figure C35 Typical type E anchorage.

Center stressing anchorages

Center stressing anchorages are employed where the ends of a tendon cannot be equipped with normal stressing anchorages. Center stressing anchorages float in a blockout and move along the axis of the tendon during stressing; stressing is carried out with the aid of a special curved chair, against which the jack is placed. Center stressing anchorages are especially suitable for:

1. Hoop posttensioning of pressure tunnels and shafts; the anchorages enable internal buttresses to be eliminated
2. The posttensioning of circular structures such as silos, tanks, etc., where they enable external or internal buttresses to be eliminated
3. The transverse posttensioning of bridge slabs, or where for aesthetic or construction reasons exterior anchorage blockouts are not desirable
4. The posttensioning of ribs of arches or frames, where access to the ends is difficult or impossible

2. VSL E Stressing Anchorage

This anchorage is composed of an anchor head, wedges, and a bearing plate with trumpet. The bearing plate is positioned in the structure concurrently with the formwork, whereas the anchor head is not placed until the time of stressing the tendons.

Figure C36.

Table C24 Data for E stressing anchorage

Anchorage type	VSL	5-3	5-7	5-12	5-19	5-31	5-55
Ultimate capacity, kips		124	289	496	785	1280	2272
A		5.50	8.25	10.75	13.50	17.50	24.00
B		5.50	8.25	10.75	13.50	17.50	24.00
C		0.75	1.25	1.50	1.75	2.25	3.00
D, dia.		1.88	2.88	4.13	5.25	6.75	9.00
E, dia.		3.38	4.50	6.00	7.00	9.00	12.50
F		2.40	2.40	2.40	2.95	3.95	6.00
G		3.40	3.40	3.40	4.00	5.00	7.00
H, dia.		2.12	2.12	2.81	3.56	4.50	6.00
J		4.00	8.00	12.00	23.00	34.00	42.00

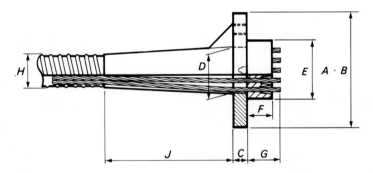

Figure C37.

Anchorage type	VSL	6-3	6-7	6-12	6-19	6-31
Ultimate capacity, kips		176	410	703	1114	1817
A		6.25	9.25	12.25	15.75	20.00
B		6.25	9.25	12.25	15.75	20.00
C		0.75	1.38	1.75	2.25	3.00
D, dia.		2.25	3.30	4.63	6.00	7.50
E, dia.		3.38	5.25	7.00	8.75	10.75
F		2.40	2.40	3.15	3.95	6.00
G		3.40	3.40	4.15	5.00	7.00
H, dia.		2.12	2.44	3.56	4.31	5.50
J		6.50	12.00	18.00	28.00	36.00

Dimensions in inches.

3. VSL EC Stressing Anchorage

This anchorage is distinguished from type E by the way in which it transmits the load to the concrete; this is done through an integral bearing plate incorporating the trumpet and an intermediate flange. This arrangement permits a reduction in the bearing area of the anchorage. It is therefore used where the space available at the end of the element is restricted.

Figure C38.

Table C25 Data for EC stressing anchorage

Anchorage type	VSL	5-3	5-7	5-12	5-19	5-31
Ultimate capacity, kips		124	289	496	785	1280
A		5.32	6.89	8.86	11.00	13.98
B		5.32	6.89	8.86	11.00	13.98
C		1.96	2.17	2.17	2.17	2.56
D, dia.		3.94	5.12	6.69	8.27	10.43
E, dia.		3.38	4.50	6.00	7.00	9.00
F		2.40	2.40	2.40	2.95	3.95
G		3.40	3.40	3.40	4.00	5.00
H, dia.		1.94	3.15	3.78	4.53	5.67
J		4.92	5.32	7.09	10.24	13.39

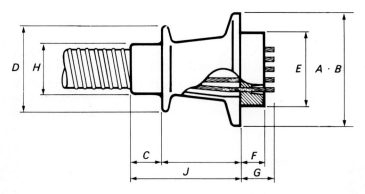

Figure C39.

Anchorage type	VSL	6-7	6-12	6-19	6-31
Ultimate capacity, kips		410	703	1114	1817
A		8.00	10.43	13.00	16.54
B		8.00	10.43	13.00	16.54
C		2.17	2.36	2.56	2.75
D, dia.		6.10	7.87	9.84	12.60
E, dia.		5.25	7.00	8.75	10.70
F		2.40	3.15	3.95	6.00
G		3.40	4.15	5.00	7.00
H, dia.		3.58	4.53	4.72	7.07
J		6.30	9.95	11.42	13.00

Dimensions in inches.

4. VSL H Fixed-End Anchorage

The posttensioning force is transmitted to the concrete by bond over the exposed length of the strands and by the bulb-shaped form at their ends. The bulbs may be arranged in either a square or a rectangular pattern. A rebar grid at the end of the

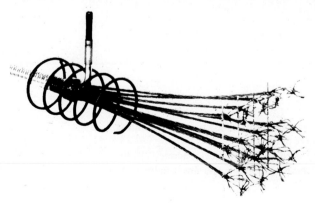

Figure C40.

anchorage keeps the strands in position. Mild reinforcement and a clamping ring prevent bursting of the concrete at the point where the strands start to diverge.

This dead-end anchorage is especially suitable where the tendons are placed before concreting.

The bulbs at the end of the strands are easily formed either in the plant or at the job site.

Table C26 Dimensions of H fixed-end anchorage

Anchorage type	VSL	5-4	5-7	5-12	5-19	5-31
A†		6.00	6.75	12.25	12.25	18.50
B†		6.75	7.50	10.63	15.50	17.00
A‡		2.75	2.75	7.50	7.50	8.63
B‡		12.25	14.50	13.75	18.50	26.50
C		36.00	48.00	48.00	48.00	48.00

Anchorage type	VSL	6-4	6-7	6-12	6-19	6-31
A†		7.50	8.25	15.30	15.50	22.50
B†		8.25	9.00	13.00	18.50	20.00
A‡		3.50	3.50	9.00	9.00	10.25
B‡		15.50	17.75	17.00	22.50	32.00
C		38.00	48.00	48.00	52.00	67.00

Dimensions in inches.
† Denotes square pattern.
‡ Denotes rectangular pattern.

Notes: 1. Dimensions and arrangements shown are the most common but can be altered where desired.

2. Spirals are generally used in lightweight concrete (110 lb/cu ft) and for tendons of 12 strands and larger, and are usually composed of 5 turns of $\frac{1}{2}$-inch bar.

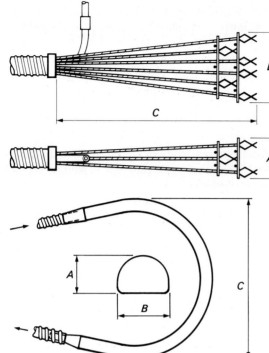

Figure C41 H fixed-end anchorage.

Figure C42 L fixed-end anchorage.

5. VSL L Fixed-End Anchorage

Table C27 Dimensions of L fixed-end anchorage

Anchorage type	5-3	5-7	5-12	5-19	5-31	5-55
A	2.50	2.50	2.50	3.50	4.50	7.00
B	3.00	3.00	4.00	6.25	9.50	15.00
C	48.00	48.00	48.00	50.00	51.00	56.00

Anchorage type	6-4	6-7	6-12	6-19	6-31
A	2.50	2.50	3.50	4.50	7.00
B	3.00	4.00	6.25	9.50	15.00
C	48.00	48.00	50.00	51.00	56.00

Dimensions in inches.

With this anchorage the profile of the cable describes an arc of 180°, enabling the cable to be brought back to the vicinity of its starting point. The anchorage consists of a rigid U-shaped tube of semicircular cross section. Normal reinforcement should be provided to prevent cracking of the concrete behind the anchorage. This anchorage is especially useful for vertical tendons where it is advantageous to place the tendon after concreting is completed.

6. VSL SO Stressing Anchorage

This anchorage is used for four-strand grouted tendons placed in flat sheathing. The strands are stressed individually by a monostrand ram and locked off in the anchor head which bears on the embedded plastic form.

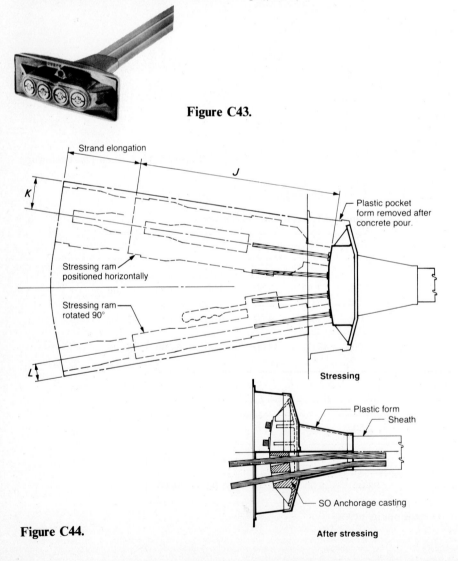

Figure C43.

Figure C44.

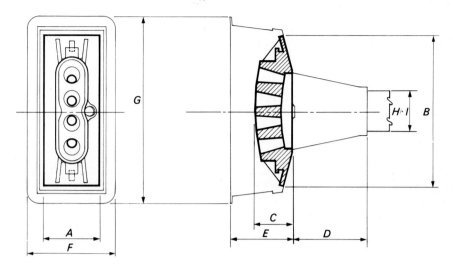

Figure C45.

Table C28

Anchorage type	VSL	SO5-4	SO6-4
A		3.50	3.50
B		11.00	11.00
C		2.87	2.87
D		6.25	6.25
E		5.00	5.00
F		5.62	5.62
G		13.00	13.00
H		3.00	3.00
I		1.00	1.00
J		24.00	24.00
K		4.00	4.00
L		2.50	2.50

Dimensions in inches.

ELASTOMERIC BEARING PADS AND FLAT JACKS

FREYSSI ELASTOMERIC BEARINGS
(*Courtesy Freyssinet International, Inc.*)

In 1954 Eugene Freyssinet first patented his invention of rubber layers, sandwiched between steel plates, for use as bridge bearings. Seventeen years of experience combined with research and testing have made Freyssinet a world leader in elastomeric bearings.

Freyssi elastomeric bearings offer an easy-to-install, maintenance-free cushion for the support of bridge girders, building members, and machinery. Freyssi elastomeric bearings will absorb a combination of vertical loads, horizontal loads, horizontal movements, and rotations. This is accomplished by the elastomer's flexure, rather than sliding.

Freyssi elastomeric bearings are composed of several neoprene or rubber layers, usually 0.250 to 0.750 in thick, bonded to metal plates. By using this laminated design, together with the correct number of layers and proper dimensions, Freyssi bearings will absorb all loads without exceeding the allowable bulge-out shear stresses.

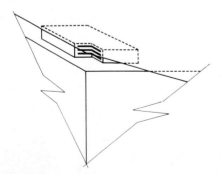

Figure D1.

In the case of a simply supported beam, two identical elastomeric bearing pads can be used, one under each end. The advantage of this arrangement is that the horizontal load on the beam, due to vehicle braking for instance, is then equally divided between the two supports.

Normally, it is not necessary to fix the pads to the adjacent parts of the structure because the maximum horizontal force is small in comparison to the minimum vertical reaction. Friction prevents any sliding between the bearing pads and the parts of the structure with which they are in contact.

Where horizontal forces are unusually large in comparison to the minimum vertical reaction, it is possible to provide a fixing arrangement.

Freyssi elastomeric bearings are ideal for concrete or steel bridge beams, building beams, walls, gantry cranes, and heavy machinery subjected to vibration.

1. Notations

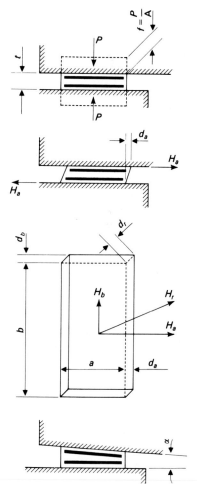

a	plan dimension parallel to girder, in
b	plan dimension perpendicular to girder, in
A	area $= a \times b$, in^2
t_e	thickness of one elastomer layer, in
$\sum t_e$	total thickness of elastomer—all layers, in
t	total bearing thickness, in
Δt	total vertical shortening, in
S	shape factor

$$S = \frac{a \times b}{2(a + b) \times t_e}$$

$P\,P_{max}\,P_{min}$	vertical loads, lb
$f\,f_{max}\,f_{min}$	average vertical stresses, psi
$d_a\,d_b\,d_r$	horizontal movements, in
$H_a\,H_b\,H_r$	horizontal forces, lb
α	total rotation around axis perpendicular to the girder, rad
α_e	rotation per layer, rad
G	elastomer shear modulus for short-term loading and deformation, psi
G'	elastomer shear modulus for permanent loading and deformations, psi
$v_P\,v_H\,v_\alpha$	shear stress, psi

Figure D2.

2. Design

The stiffness of an elastomer depends on the modulus G and the temperature. However the flexibility of a bearing is due mostly to a shape change, which makes the design calculation complicated. For detail computation procedure refer to "Design of Elastomer Bearings" by Charles Rejcha, published in PCI Journal of October, 1964. A simplified computation approach is shown below:

Vertical loads

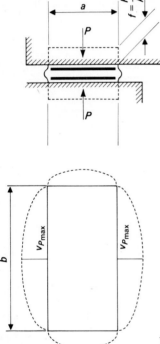

Figure D3.

The bulge out results in shear stresses, with a maximum value

$$v_{P_{max}} \simeq \frac{1.5 \times f}{S}$$

A typical interior elastomer layer, thickness t_e is to be considered. The top and bottom layers are usually thinner and do not need to be verified.

The ultimate shear value: $\qquad v_{P_{ult}} \simeq 500$ psi

The approximate vertical shortening under $f = 1000$ psi and $S = 6$ is:

$$\Delta t \simeq 0.006 \sum t_e$$

Horizontal loads and movements

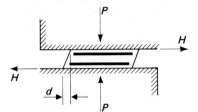

Figure D4.

The relation between the horizontal force and movement is as follows:

$$H = \frac{d \times A \times G}{\Sigma t_e}$$

Shear stress:

$$v_H = \frac{H}{A}$$

Sliding:

$$\frac{H_{max}}{P_{min}} > 0.2 \text{ or } 0.1$$

(see recommendations).

The ultimate shear value:

$$v_{H_{ult}} \simeq 200 \text{ psi}$$

Rotation

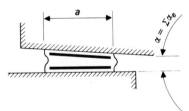

Figure D5.

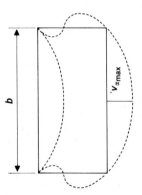

The rotation will cause a bulge-out on one side and bulge-in on the other side. This will generate a shear stress similar to the vertical load effect, with maximum value:

$$v_{\alpha_{max}} = \frac{G}{2} \times \frac{a^2}{t_e^2} \times \alpha_e$$

Steel plates

The plates are to be designed to balance all elastomer shear stresses.

3. Recommendations

Shear modulus for temperature above $-30°F$:

$$G_{50} \text{ durometer} = 155 \text{ psi}$$
$$G_{60} \text{ durometer} = 190 \text{ psi}$$
$$G_{70} \text{ durometer} = 225 \text{ psi}$$
$$G' \simeq 0.5 \text{ G}$$

Dimensional restrictions

$$S \geq 4.0$$

$$a \geq 4 \sum t_e \qquad\qquad b \geq 4 \sum t_e$$

(may be reduced in special cases)

$$a \geq 10 d_{a \text{ max}} \qquad\qquad b \geq 10 d_{b \text{ max}}$$
$$d_a / \sum t_e \leq 0.5 \qquad\qquad d_b / \sum t_e \leq 0.5$$
$$d_r / \sum t_e \leq 0.7$$

d_r due to braking and wind ≤ 0.2 in

$\alpha_{\text{min}} = 0.01$ rad (due to nonparallel bearing areas and girder deflection using modulus G').

Stress restriction

$$f_{\text{max}} \leq 170 \times S \leq 1500 \text{ psi}$$
$$v_{p\text{max}} + v\alpha \leq 300 \text{ psi}$$
$$v_{p\text{min}} \geq v\alpha \qquad\qquad v_{Hr} \leq 100 \text{ psi}$$

Sliding restriction

$$\text{Concrete girders:} \quad \frac{H_{r\text{max}}}{P_{\text{min}}} \leq 0.2$$

$$\text{Steel girders:} \quad \frac{H_{r\text{max}}}{P_{\text{min}}} \leq 0.1$$

The above might be reduced if a fixing arrangement is provided.

4. Tests

Elastomeric bearings may be tested on a simple testing frame as shown schematically in Fig. D6.

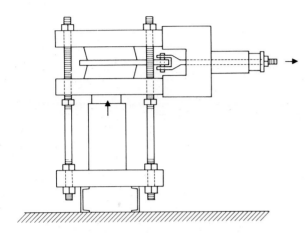

Figure D6.

5. Design tips

The bearing must be in "sandwich" contact throughout the entire area. Set the bearing level where possible.

Reduce the bearing area to a minimum. Use maximum allowable stress f in order to decrease the horizontal force due to horizontal movements.

Reduce the dimension a to a minimum in order to facilitate the rotation.

Use 50 or 60 durometer hardness. For material specifications, refer to current AASHTO standards.

In computing the horizontal movements and forces, take into account the support flexibility (tall piers for example).

For projects designed under current AASHTO standard specifications, reduction of the average vertical stress may be necessary.

6. Load Tables

Freyssinet Company Inc., can provide almost any bearing size as required by the designer for specific projects. Standard Freyssi bearings are shown in Table D1.

Table D1

Total thickness, in	Plan dimensions, in a ↓ b→	6	12	18	24	30	36
1.5 d_a max = ±1.77 in	6	31.5 4.38 0.080	72 8.75 0.052	108 13.13 0.045	144 17.51 0.042	180 21.88 0.040	216 26.26 0.038
	8	47.4 5.84 0.058	96 11.67 0.033	144 17.51 0.027	192 23.34 0.025	240 29.18 0.023	288 35.01 0.023
	10	59 7.29 0.040	120 14.59 0.025	180 21.88 0.020	240 29.18 0.017	300 36.47 0.016	360 43.76 0.015
	12	72 8.75 0.031	144 17.51 0.020	216 26.26 0.015	288 35.01 0.013	360 43.76 0.012	432 52.52 0.011
2 d_a max = ±1.31 in	8		96 8.75 0.047	144 13.13 0.038	192 17.51 0.036	240 21.88 0.033	288 26.26 0.033
	10		120 10.94 0.037	180 16.41 0.026	240 21.88 0.024	300 27.35 0.023	360 32.82 0.021
	12		144 13.13 0.028	216 19.69 0.021	288 26.26 0.018	360 32.82 0.017	432 39.39 0.016
	14		168 15.32 0.024	252 22.98 0.017	336 30.64 0.014	420 38.29 0.013	504 45.95 0.012
3 d_a max = ±0.85 in	12		144 8.50 0.077	216 12.75 0.057	288 17.00 0.050	360 21.26 0.047	432 25.51 0.044
	14		168 9.92 0.066	252 14.88 0.046	336 19.84 0.039	420 24.80 0.037	504 29.76 0.034
	16		192 11.34 0.056	288 17.00 0.041	384† 22.67† 0.032†	480 28.34 0.028	576 34.01 0.028
	18		216 12.75 0.047	324 19.13 0.034	432 25.51 0.026	540 31.89 0.024	648 38.26 0.022

Total thickness, in	Plan dimensions, in $a \downarrow \quad b \rightarrow$	6	12	18	24	30	36
d_a max $= \pm 0.64$ in 4	16			288 12.57 0.073	384 16.77 0.056	480 20.96 0.050	576 25.15 0.050
	18			324 14.15 0.060	432 18.86 0.047	540 23.58 0.042	648 28.29 0.039
	20			360 15.72 0.058	480 20.96 0.047	600 26.20 0.036	720 31.44 0.032
	22			396 17.29 0.050	528 23.05 0.040	660 28.82 0.032	792 34.58 0.028

† Example: 384   Maximum vertical load (kips)

22.67   Shear rating—horizontal force required to produce a horizontal movement of one inch using $G = 155$ psi (kips/inch)

0.032   Total vertical shortening under an average vertical stress of 1000 psi and $G = 155$ psi (in)

Design criteria: Hardness 50 + 5

Shear moduli $G = 155$ psi $G' = 77.5$ psi

Maximum rotation around axis parallel to b and $G' = 77.5$ psi:

$\alpha = 0.01$ rad

Maximum combined horizontal shear stress: 300 psi

Maximum average vertical stress: 1000 psi

Notes: 1. For long-term forces, reduce the shear rating by 50 percent and increase the total vertical shortening by 100 percent.

2. Check if $\dfrac{H_{max}}{P_{min}} < 0.2$ or 0.1 (see recommendations).

3. If the allowable horizontal movement for the bearing being considered exceeds its d_a maximum value shown on the table, a thicker bearing must be selected.

4. Check if the average vertical stresses under DL and DL + LL are in accordance with current AASHTO specifications.

FLAT JACKS

(*Courtesy Freyssinet International, Inc.*)

Invented by Freyssinet in 1938, flat jacks have been used since that time, on a world-wide basis, to solve an astonishing variety of structural and civil engineering problems:

Control of thrust forces
Prestressing between abutments
Adjustment of support reactions
Structural preloading
Structural lifting
Underpinning
Measurement of forces
Thrust maintenance

Essentially simple and compact, a Freyssinet flat jack is a thin pressure capsule capable of exerting, hydraulically, extremely large forces. Although the movement of a single jack is relatively small, larger movements can be economically obtained by "banking" jacks in series.

The normal range of Freyssinet flat jacks is of circular form and the characteristics of the regular sizes are given in Table D2. Nonstandard shapes, generally rectangular and oblong, may be custom-built for special applications.

Table D2

Diameter	mm in	70 2.75	120 4.72	150 5.90	220 8.66	250 9.84	270 10.6	300 11.8
Approximate maximum force†	kN kips	29.4 6.6	98 22	167 37.5	392 88.2	490 110	588 132	784 176
Maximum opening	mm in	15 6.8	25 1	25 1	25 1	25 1	25 1	25 1
Diameter	mm in	350 13.8	420 16.5	480 18.9	600 23.6	750 29.5	870 34.2	920 36.2
Approximate maximum force†	kN kips	1078 242	1666 375	2352 529	3626 816	5880 1323	8134 1830	8918 2006
Maximum opening	mm in	25 1	25 1	25 1	35 $1\frac{3}{8}$	35 $1\frac{3}{8}$	35 $1\frac{3}{8}$	35 $1\frac{3}{8}$

† With a pressure of 150 bars (2100 psi).

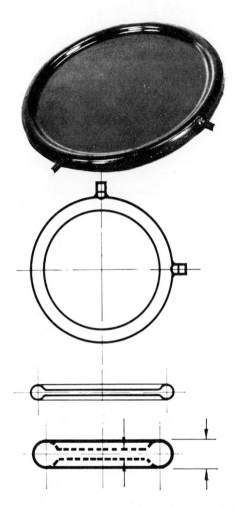

Figure D7.

Flat jacks can be operated with oil or water for temporary use. The hydraulic pressure can be raised in normal service to 150 kg/cm^2 (14.7 N/mm^2: 2.100 psi).

Where flat jacks, after utilization, are to remain permanently in the structure they are filled with a hard-setting matter such as current grout or epoxy resin.

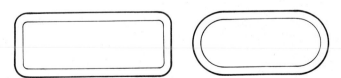

Figure D8.

EXCERPTS FROM
THE AASHTO STANDARD SPECIFICATIONS
FOR HIGHWAY BRIDGES
(*Ref. 14.19*)

Section 6—PRESTRESSED CONCRETE DESIGN

1.6.6—ALLOWABLE STRESSES

The design of precast prestressed members ordinarily shall be based on f'_c=5000 psi (34.4737 MPa). An increase to 6000 psi (41.3685 MPa) is permissible where, in the Engineer's judgment, it is reasonable to expect that this strength will be obtained consistently. Still higher concrete strengths may be considered on an individual area basis. In such cases, the Engineer shall satisfy himself completely that the controls over materials and fabrication procedures will provide the required strengths. The provisions of this Section are equally applicable to prestressed concrete structures and components designed with lower concrete strengths.

† *Courtesy American Association of State Highway and Transportation Officials, Washington D.C.*

(A) Prestressing steel

Temporary stress before loss due to creep and shrinkage . . . . $0.70f'_s$
Stress at service load* after losses . . . . $0.80f*_y$

(Overstressing to $0.80f'_s$ for short periods of time may be permitted provided the stress, after transfer to concrete in pretensioning or seating of anchorage in post-tensioning, does not exceed $0.70f'_s$).

(B) Concrete

(1) Temporary Stresses Before Losses Due to Creep and Shrinkage:

Compression
 Pretensioned members . . . . $0.60f'_{ci}$
 Post-tensioned members . . . . $0.55f'_{ci}$
Tension
 Precompressed tensile zone . . . . . No temporary allowable stresses are specified. See Article 1.6.6(B)(2) for allowable stresses after losses.
Other Areas
 In tension areas with no bonded reinforcement . . . 200 psi or $3\sqrt{f'_{ci}}$ or $(1.379$ MPa or $.249\sqrt{f'_{ci}})$ Where the calculated tensile stress exceeds this value, bonded reinforcement shall be provided to resist the total tension force in the concrete computed on the assumption of an uncracked section. The maximum tensile stress shall not exceed . . . . $7.5\sqrt{f'_{ci}}$ or $(.623\sqrt{f'_{ci}})$.

(2) Stress at Service Load After Losses have Occurred:

Compression . . . . $0.40f'_c$
Tension in the precompressed tensile zone
(a) For members with bonded reinforcement . . . . $6\sqrt{f'_c}$ or $(.498\sqrt{f'_c})$
 For severe corrosive exposure conditions, such as coastal areas . . . . $3\sqrt{f'_c}$ or $(.249\sqrt{f'_c})$
(b) For members without bonded reinforcement 0

Tension in other areas is limited by the allowable temporary stresses specified in Article 1.6.6(B)(1).

(3) Cracking Stress**

Modulus of rupture from tests or if not available:
 For normal weight concrete . . . . $7.5\sqrt{f'_c}$ or $(.623\sqrt{f'_c})$
 For sand-lightweight concrete . . . . $6.3\sqrt{f'_c}$ or $(.523\sqrt{f'_c})$
 For all other lightweight concrete . . . . $5.5\sqrt{f'_c}$ or $(.457\sqrt{f'_c})$

---

*Service load consists of all loads contained in Article 1.2.1 but does not include overload provisions.
**Refer to Article 1.6.10.

(4) Anchorage Bearing Stress:

Post-tensioned anchorage at service load . . . . 3000 psi (20.6842 MPa) (but not to exceed $0.9f'_{ci}$)

1.6.7—LOSS OF PRESTRESS

(A) Friction Losses

Friction losses in post-tensioned steel shall be based on experimentally determined wobble and curvature coefficients, and shall be verified during stressing operations. The values of coefficients assumed for design, and the acceptable ranges of jacking forces and steel elongations shall be shown on the plans. These friction losses shall be calculated as follows:

$$T_o = T_x \ e^{(KL + \mu\alpha)}$$

When $(KL + \mu\alpha)$ is not greater than 0.3, the following equation may be used:

$$T_o = T_x (1 + KL + \mu\alpha)$$

The following values for K and μ may be used when experimental data for the materials used are not available:

Type of Steel	Type of Duct	K/ft.	μ	(K/m)
Wire or ungalvanized strand	Bright Metal Sheathing	0.0020	0.30	0.0066
	Galvanized Metal Sheathing	0.0015	0.25	0.0049
	Greased or asphalt-coated and wrapped	0.0020	0.30	0.0066
	Galvanized rigid	0.0002	0.25	0.0007
High-strength bars	Bright Metal Sheathing	0.0003	0.20	0.0010
	Galvanized Metal Sheathing	0.0002	0.15	0.0007

Friction losses occur prior to anchoring but should be estimated for design and checked during stressing operations. Rigid ducts shall have sufficient strength to maintain their correct alignment without visible wobble during placement of concrete. Rigid ducts may be fabricated with either welded or interlocked seams. Galvanizing of the welded seam will not be required.

(B) Prestress Losses

(1) Loss of prestress due to all causes, excluding friction, may be deter-

mined by the following method.* The method is based on normal weight concrete and one of the following types of prestressing steel: 250 or 270 ksi, (1724 or 1862 MPa), seven-wire, stress-relieved strand: 240 ksi (1655 MPa) stress-relieved wires; or 145 to 160 ksi (1000 to 1103 MPa) smooth or deformed bars. For data regarding the properties and effects of lighweight aggregate concrete and low-relaxation tendons, refer to documented tests or see authorized suppliers.

TOTAL LOSS

$$\Delta f_s = SH + ES + CR_c + CR_s$$

where
$\Delta f_{s,}$ = total loss excluding friction in psi. (MPa)
SH = loss due to concrete shrinkage in psi. (MPa)
ES = loss due to elastic shortening in psi. (MPa)
CR_c = loss due to creep of concrete in psi. (MPa)
CR_s = loss due to relaxation of prestressing steel in psi. (MPa)

(a) SHRINKAGE
Pretensioned Members
SH = 17,000—150 RH
  or (117.21—1.034 RH)

Post-tensioned Members
SH = 0.80 (17,000—150 RH)
  or 0.80 (117.21—1.034 RH)

where RH = mean annual ambient relative humidity in percent (See Figure 1.6.7)

(b) ELASTIC SHORTENING
Pretensioned Members

$$ES = \frac{E_s}{E_{ci}} f_{cir}$$

Post-tensioned Members**

$$ES = 0.5 \frac{E_s}{E_{ci}} f_{cir}$$

---

*Should more exact prestress losses be desired, data representing the materials to be used, the methods of curing, the ambient service condition and any pertinent structural details should be determined for use in accordance with a method of calculating prestress losses that is supported by appropriate research data.

**Certain tensioning procedures may alter the elastic shortening losses.

where

E_s = modulus of elasticity of prestressing steel strand which can be assumed to be 28 x 10^6 psi (.193 x 10^6 MPa).

E_{ci} = modulus of elasticity of concrete in psi (MPa) at transfer of stress which can be calculated from:

$$E_{ci} = 33w^{3/2}\sqrt{f'_{ci}} \text{ or } (.0428\ w^{3/2}\sqrt{f'_{ci}})$$

where w is in lb/ft^3 (kg/m^3) and f'_{ci} is in psi (MPa)

f_{cir} = concrete stress at the center of gravity of the prestressing steel due to prestressing force and dead load of beam immediately after transfer, f_{cir} shall be computed at the section or sections of maximum moment. (At this stage, the initial stress in the tendon has been reduced by elastic shortening of the concrete and tendon relaxation during placing and curing the concrete for pretensioned members, or by elastic shortening of the concrete and tendon friction for post-tensioned members. The reductions to initial tendon stress due to these factors can be estimated, or the reduced tendon stress can be taken as 0.63 f'_s for typical pretensioned members.)

(c) CREEP OF CONCRETE
Pretensioned and Post-tensioned Members.

$$CR_c = 12f_{cir} - 7f_{cds}$$

where

f_{cds} = concrete stress at the center of gravity of the prestressing steel due to all dead loads except the dead load present at the time the prestressing force is applied.

(d) RELAXATION OF PRESTRESSING STEEL*
Pretensioned Members
250 to 270 ksi Strand (1724 to 1862 MPa)

$$CR_s = 20,000 - 0.4ES - 0.2\ (SH + CR_c)$$

or $(137.9 - 0.4\ ES - 0.2\ (SH + CR_c))$

Post-tensioned Members
250 to 270 ksi Strand (1724 to 1862 MPa)

$$CR_s = 20,000 - 0.3\ FR - 0.4\ ES - 0.2\ (SH + CR_c)$$

or $(137.9 - 0.3\ FR - 0.4\ ES - 0.2\ (SH + CR_c))$

---

*The relaxation losses are based on an initial stress of 0.70f$'_s$.

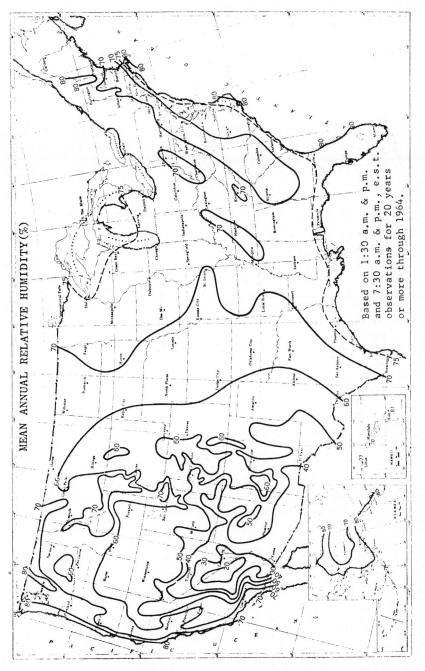

MEAN ANNUAL RELATIVE HUMIDITY (%)

Based on 1:30 a.m. & p.m. and 7:30 a.m. & p.m., e.s.t. observations for 20 years or more through 1964.

Figure 1.6.7

240 ksi Wire (1655 MPa)

$$CR_s = 18,000 - 0.3 \text{ FR} - 0.4 \text{ ES} - 0.2 (\text{SH} + CR_c)$$

or $(124.10 - 0.3 \text{ FR} - 0.4 \text{ ES} - 0.2 (\text{SH} + CR_c)$

145 to 160 ksi Bars (1000 to 1103 MPa)

$$CR_s = 3000 \,(20.68 \text{ MPa})$$

where

FR   = friction loss stress reduction in psi (MPa) below the level of 0.70 f'_s at the point under consideration, computed according to Article 1.6.7(A).

ES, SH, and CR_c = appropriate values as determined for either pre-tensioned or post-tensioned members.

(2) In lieu of the preceding method, the following estimates of total losses may be used for prestressed members or structures of usual design. These loss values are based on use of normal weight concrete, normal prestress levels, and average exposure conditions. For exceptionally long spans, or for unusual designs, the method in Article 1.6.7(B)(1) or a more exact method shall be used.

TYPE OF PRESTRESSING STEEL	TOTAL LOSS	
	f'_c = 4,000 psi (27.58 MPa)	f'_c = 5,000 psi (34.47 MPa)
PRETENSIONING Strand		45,000 psi (310.26 MPa)
POST-TENSIONING* Wire or Strand	32,000 psi (220.63 MPa)	33,000 psi (227.53 MPa)
Bars	22,000 psi (151.68 MPa)	23,000 psi (158.58 MPa)

*Losses due to friction are excluded. Friction losses should be computed according to Article 1.6.7(A).

ANSWERS TO SELECTED PROBLEMS

1.1 Examples include: Umbrella, tent, musical drum, trampoline, tempered glass, carriage wheels.

2.3 $(C_{CU})_{member} = K_{CH} K_{CA} K_{CS}(C_{CU})_{material}$

$$= 3 \times 1.25 t_A^{-0.118}(1.27 - 0.0067H)$$

t_A \ H	40%	60%	80%
7	2.99	2.59	2.19
28	2.54	2.20	1.86
90	2.21	1.91	1.62

2.5	Normal weight			Lightweight		
f'_c, psi	5000	6000	7000	5000	6000	7000
f'_{tc}, psi	-212	-232	-251	-141	-155	-167
f_r, psi	-530	-581	-627	-398	-436	-471
E_c, 10^6 psi	4.287	4.696	5.072	2.511	2.750	2.971

4.1a. $\sigma = \pm 1203$ psi. Failure will occur due to excess tension.
   b. $F = 259.85$ kips.
   c. $F = 129.9$ kips.
   d. $F = 86.62$ kips.

4.2 $W_L = 420$ plf.

4.3 $F_i \simeq 260$ kips; $k'_t = -1.08$ in; $k'_b = 2.99$ in.

4.4 $F_i \simeq 187.5$ kips; $e_o = 1.35$ in.

4.5 For a live load of 55 psf choose $h = 18$ in. For a live load of 75 psf, either $h = 24$ in or $h = 32$ in is acceptable. (Note for $h = 32$ in, the beam is lighter.)

4.6a. Minimum required $F = 51.9$ kips; actual $F \simeq 69.3$ kips (3 strands) at $e_o = 10.64$ in.
   c. $F \simeq 123.6$ kips (controlled by stress condition II).
   d.   Maximum live load $= 193$ psf.
   e. $k'_t = -6.54$ in; $k'_b = 8.67$ in.

   f.

x	0	4	8	12	16	20 ft
e_{ou}	-6.54	-1.58	2.29	5.05	6.71	7.26 in
e_{ol}	8.67	10.64	10.64	10.64	10.64	10.64 in

5.1 $\sigma_{ti} = 635.5$ psi; $\sigma_{ci} = 796.4$ psi; $\sigma_{cs} = 1216$ psi; $\sigma_{ts} = -70.6$ psi; $M_u = 1161.67$ kips-ft; $\phi M_n \simeq 1448$ kips-ft.

5.2a. $q_{min} = 0.05014$ (Eq. (5.54)); $\rho_{min} \simeq 0.00093$; $(\phi M_n)_{min} = 715.3$ kips-ft. (*Note:* $\rho_{min} = 0.00096$ from Eq. (5.4)).
   b. $q_{max} = 0.1856$; T-section behavior; $(\phi M_n)_{max} = 2390$ kips-ft.

5.3 Prestressed reinforcement only: $M_{cr} = 392$ kips-ft; $\phi M_n \simeq 559.5$ kips-ft. Prestressed and nonprestressed reinforcement: $\phi M_n \simeq 658.7$ kips-ft.

5.4a. $M_u = 122.45$ kips-ft; $M_{cr} = 104.77$ kips-ft; $\phi M_n = 152.4$ kips-ft; $\phi M_n > 1.2M_{cr}$.
   b. $f_{ps} = 211$ ksi; $\phi M_n = 121.56$ kips-ft $< 1.2M_{cr} = 125.7$ kips-ft (add some nonprestressed reinforcement).

5.6 $F = 504.9$ kips (22 strands); $e_o = 10.67$ in; $M_u = 1143$ kips-ft; $\phi M_n = 1224$ kips-ft; $M_{cr} = 799.6$ kips-ft; $\phi M_n > 1.2M_{cr}$.

5.7 $M_n = 501.7$ kips-ft; compressive steel yields.

5.8 $A_{ps} = 0.986$ in^2; $d_p = 21$ in.

6.1 For $\sigma_1 = 0$, $\sigma_y = 228.6$ psi; for $\sigma_1 = -100$ psi, $\sigma_y = 100$ psi.

6.2 Minimum shear reinforcement is adequate for the beam. It can be waived if the beam is part of a joist slab.

6.3 Minimum shear reinforcement is sufficient throughout.

6.4 $F = 367.2$ kips (16 strands); choose profile with two draping points at about 25 and 50 ft from support A; $e_o = 5.98$ in at A; $e_o = 17.98$ in at B; $e_o = 5.02$ in at C; $e_o = 0$ at D.

7.5a. 14 strands with $F_i = 374.85$ kips, $e_o = 17.01$ in at midspan, and $e_o = 10.73$ in at support. Single draping point at midspan.
   b. $\phi M_n = 12,751$ kips-in $> M_u = 11,205$ kips-in.
   c.   Shear reinforcement: #3 U stirrups at $s = 22$ in.
   d. $\Delta_{add} \simeq -1.58$ in.
   e.   Using incremental step method: $\Delta_{add} \simeq -1.95$ in.

8.1

t, days	1	3	7	30	60	365 days	5 yrs	40 yrs
$\sum \Delta f_{pT}(t_i, t_j)$ (ksi)	14.87	17.82	20.62	26.7	29.97	37.1	40.41	42.61

8.2 t, days	1	3	7	30	60	365	1214 days
$\sum \Delta f_{pT}(t_i, t_j)$ (ksi)	11.81	19.05	21.37	23.71	25.76	30.93	34.08

8.3 $F = 234.32$ kips.

8.5a. $(f_{pi})_A = 136.37$ ksi.
   b. $(f_{pi})_B = 77.92$ ksi.
   c.   Recommend prestressing from both ends.

9.1 Unshored: $F = 947.38$ kips; $N = 43$ strands. Shored: $F = 830$ kips; $N = 38$ strands. In both cases: select $e_o = 20.16$ in at midspan; $e_o = 3.87$ in at supports; ultimate strength is adequate; minimum shear reinforcement is needed; total long-term deflection is a camber.

9.2 $F = 183.6$ kips (8 strands); $e_o = 9.25$ in (also acceptable: 9 strands at $e_o = 9$ in).

10.1 The ZLC line coincides with the neutral axis of the member.

10.2 Fig. P10.2a. The ZLC line is a straight line with eccentricity $-0.15h$ at the left support and $0.30h$ at the right support.
   Fig. P10.2b. The ZLC line is a segmented line with eccentricity $-0.3375h$ at the left support, $0.28125h$ at midspan and 0 at the right support.
   Fig. P10.2c. The ZLC line is a parabola with eccentricity $-0.3689h$ at the left support, $0.1656h$ at midspan and 0 at the right support.
   Fig. P10.2d. The profile is concordant.

10.4 $M_{2C} = M_{2D} = 48$ kips-ft; $M_{2B} = 19.2$ kips-ft; $(M_{max})_C = 255$ kips-ft; $(M_{min})_C = 128$ kips-ft; $(M_{max})_B = 160$ kips-ft; $(M_{min})_B = 32$ kips-ft.

11.8 Typical exterior span: 1 strand (0.6-in diameter) at 18-in spacing and 1 strand (0.5-in diameter) at 36-in spacing. Typical interior span: 1 strand (0.6-in diameter) at 18-in spacing.

12.1 $A_g \simeq 271$ in^2; $A_{ps} \simeq 2.67$ in^2; a rectangular section 15×18 in will do.

12.3 From the feasible domain the least-weight section corresponds to $A_g \simeq 102$ in^2 and $A_{ps} = 1.80$ in^2.

13.1 $\phi P_{n,o} = 620.9$ kips; $\phi P_{n,ot} = 486.4$ kips, $\phi M_{n,ot} = 696.3$ kips-in, $e_{ot} = 143$ in, $\Phi_{ot} = 21.4 \times 10^{-5}$; $\phi P_{n,b} = 29.46$ kips, $\phi M_{n,b} = 724.6$ kips-in, $e_b = 24.6$ in, $\Phi_b = 113.2 \times 10^{-5}$; $\phi M_{n,f} = 814.43$ kips-in, $\Phi_f = 140.18 \times 10^{-5}$.

13.2 Assume $\phi = 0.7$ for square spirals; $\phi P_{n,o} = 574.2$ kips; $\phi P_{n,ot} = 433.54$ kips, $\phi M_{n,ot} = 739.5$ kips-in; $\phi P_{n,b} = 146.79$ kips, $\phi M_{n,b} = 1112.5$ kips-in; $\phi M_{n,f} = 693.23$ kips-in.

13.4 $\phi P_{n,o} = 1058.65$ kips; $\phi P_{n,ot} = 722.27$ kips, $\phi M_{n,ot} = 2129.46$ kips-in, $e_{ot} = 2.95$ in; $\phi P_{n,b} = 71.27$ kips, $\phi M_{n,b} = 2413.3$ kips-in, $e_b = 33.87$ in; $\phi M_{nf} = 2776.56$ kips-in. If slenderness is considered, the column becomes unsafe; $\delta = 1.35$. *Solution:* Increase steel content or cross section of column.

14.2 $M_{L+I} = 931.32$ kips-ft.

14.3 Maximum positive moment due to one lane loading $= 1321.24$ kips-ft. Maximum positive moment due to one truck loading $\simeq 1530.31$ kips-ft.

INDEX

CHARLESTON   X   X
1 2 3 4 5 6 7 8 9 0

CX
189   11
+101
+ TAX 4.51

CTO
189

R
105
32
6 630
370
5920

6900 IIT
2400 UICC

3250
1600

1650
2100